STUDENT'S SOLUTIONS MANUAL

CARRIE GREEN

BEGINNING ALGEBRA WITH APPLICATIONS AND VISUALIZATION

SECOND EDITION

Gary K. Rockswold
Minnesota State University, Mankato

Terry A. Krieger
Rochester Community and Technical College

Boston San Francisco New York
London Toronto Sydney Tokyo Singapore Madrid
Mexico City Munich Paris Cape Town Hong Kong Montreal

Reproduced by Pearson Addison-Wesley from electronic files supplied by the author.

Publishing as Pearson Addison-Wesley, 75 Arlington Street, Boston, MA 02116.

ISBN-13: 978-0-321-52336-9
ISBN-10: 0-321-52336-5

1 2 3 4 5 6 BB 11 10 09 08

Contents

Chapter 1 Introduction to Algebra 1

Chapter 2 Linear Equations and Inequalities 39

Chapter 3 Graphing Equations 77

Chapter 4 Systems of Linear Equations in Two Variables 140

Chapter 5 Polynomials and Exponents 178

Chapter 6 Factoring Polynomials and Solving Equations 214

Chapter 7 Rational Expressions 249

Chapter 8 Radical Expressions 303

Chapter 9 Quadratic Equations 338

Chapter 1: Introduction to Algebra

1.1: Numbers, Variables, and Expressions

Concepts

1. counting
2. zero
3. 1
4. composite
5. prime
6. composite
7. factors
8. formula
9. variable
10. equals sign
11. sum
12. product
13. quotient
14. difference

Prime Numbers and Composite Numbers

15. The number 4 is a composite number because it has factors other than itself and 1; $4 = 2\times 2$.

17. The number 1 is neither a prime nor a composite number.

19. The number 29 is a prime number because its only factors are itself and 1.

21. The number 92 is a composite number because it has factors other than itself and 1; $92 = 2\times 2\times 23$.

23. The number 225 is a composite number because it has factors other than itself and 1; $225 = 3\times 3\times 5\times 5$.

25. The number 149 is a prime number because its only factors are itself and 1.

27. $6 = 2\times 3$

29. $12 = 2\times 2\times 3$

31. $32 = 2\times 2\times 2\times 2\times 2$

33. $39 = 3\times 13$

35. $294 = 2\times 3\times 7\times 7$

37. $300 = 2\times 2\times 3\times 5\times 5$

39. Yes, the population of a country could be described by the whole numbers because we cannot have a fraction of a person.

41. No, a student's grade point average could not be described by the whole numbers because a grade point average usually contains a decimal point.

43. Yes, the number of bytes stored on a computer's hard drive could be described by the whole numbers because computer bytes do not contain a fraction or a decimal point.

45. Yes, the number of students in a class could be described by the whole numbers because we cannot have a fraction of a person.

Algebraic Expressions, Formulas, and Equations

47. The value of the expression $2x$, when $x=5$, is $2x=2(5)=10$.

49. The value of the expression $8-x$, when $x=1$, is $8-x=8-1=7$.

51. The value of the expression $\frac{x}{8}$, when $x=32$, is $\frac{x}{8}=\frac{32}{8}=4$.

53. The value of the expression $3(x+1)$, when $x=5$, is $3(x+1)=3(5+1)=(3)(6)=18$.

55. The value of the expression $\frac{x}{2}+1$, when $x=6$, is $\frac{x}{2}+1=\frac{6}{2}+1=3+1=4$.

57. When $x=5$ and $y=4$, $x+y=5+4=9$.

59. When $x=8$ and $y=4$, $4\cdot\frac{x}{y}=4\cdot\frac{8}{4}=4\cdot 2=8$.

61. When $x=5$ and $y=7$, $y(x-3)=7(5-3)=(7)(2)=14$.

63. When $x=0$, $y=x+1=0+1=1$.

65. When $x=7$, $y=4x=4\cdot 7=28$.

67. When $z=12$, $F=z-5=12-5=7$.

69. When $z=6$, $F=\frac{30}{z}=\frac{30}{6}=5$.

71. When $x=3$ and $z=15$, $y=x+z=3+15=18$.

73. When $x=9$ and $z=3$, $y=\frac{x}{z}=\frac{9}{3}=3$.

Translating Words to Symbols

75. If s is the cost of a soda, then three times this cost is $3s$.

77. If x is the number, $x+5$ is five more than the number.

79. If n is the number, $n+5$ is the sum of the number and 5.

81. If n is the number, $3n$ is triple the number.

83. If p is the population of a town, $p-200$ is two hundred less than the population.

85. If z is the number, $\frac{z}{6}$ is the number divided by 6.

87. If s is the speed and t is the time, then st is the product of the speed and the time.

89. If x is one number and y is another number, $\frac{x+7}{y}$ is a number plus 7, all divided by another number.

Applications

91. See Figure 91. $F = 3y$ because there are three feet in one yard.

Yards (y)	1	2	3	4	5	6	7
Feet (F)	3	6	9	12	15	18	21

Figure 91

93. $P = 100D$ because there are one hundred pennies in one dollar.

95. $M = 3x$; M is miles and x is minutes. If $x = 36$, then $M = 3x = 3(36) = 108$ mi.

97. $B = 6D$; for each drog there are 6 blims so there are 6 times as many blims as there are drogs.

99. Each CD costs $12. Thus, $C = 12x$ where C is cost and x is number of CDs.

101. Since the area of a rectangle equals its length times its width, $22 \text{ ft} \times 9 \text{ ft} = 198$ square feet.

1.2: Fractions

Concepts

1. The person ate $\frac{3}{4}$ of the pie. $\frac{1}{4}$ of the pie remains.

2. The numerator is 11, and the denominator is 21.

3. The variable b cannot equal 0.

4. $\frac{a}{a}$ with $a \neq 0$ equals 1.

5. numerators

6. denominators

7. $\frac{2}{8} = \frac{1 \cdot 2}{4 \cdot 2} = \frac{1}{4}$

8. $\frac{ac}{bc} = \frac{a}{b}$

9. multiply

10. The fractional part of half of a half is $\frac{1}{4}$.

11. The reciprocal of a with $a \neq 0$ is $\frac{1}{a}$.

12. $\frac{1}{5}$

13. $\frac{a}{b} \cdot \frac{c}{d} = \frac{ac}{bd}$

14. $\frac{a}{b} \div \frac{c}{d} = \frac{a}{b} \cdot \frac{d}{c} = \frac{ad}{bc}$.

15. $\frac{a}{b} + \frac{c}{b} = \frac{a+c}{b}$.

16. $\frac{a}{b}-\frac{c}{b}=\frac{a-c}{b}$.

17. The greatest common factor of 4 and 6 is 2 because 2 is the largest number that divides evenly into both 4 and 6.

18. The least common denominator for the fractions $\frac{1}{4}$ and $\frac{5}{6}$ is 12 because 12 is the smallest number that both 4 and 6 divide into evenly.

19. The least common denominator for $\frac{2}{9}$ and $7=\frac{7}{1}$ is 9.

20. $\frac{6}{6}$; $6\times 4=24$

Lowest Terms

21. The largest number that divides evenly into 4 and 12 is 4, so the GCF is 4.

23. The largest number that divides evenly into 50 and 75 is 25, so the GCF is 25.

25. The largest number that divides evenly into 100, 60 and 70 is 10, so the GCF is 10.

27. $\frac{3\cdot 4}{5\cdot 4}=\frac{3}{5}\cdot\frac{4}{4}=\frac{3}{5}\cdot 1=\frac{3}{5}$

29. $\frac{3\cdot 8}{8\cdot 5}=\frac{3\cdot 8}{5\cdot 8}=\frac{3}{5}\cdot\frac{8}{8}=\frac{3}{5}\cdot 1=\frac{3}{5}$

31. $\frac{4}{8}=\frac{1\cdot 4}{2\cdot 4}=\frac{1}{2}\cdot\frac{4}{4}=\frac{1}{2}\cdot 1=\frac{1}{2}$

33. $\frac{5}{15}=\frac{1\cdot 5}{3\cdot 5}=\frac{1}{3}\cdot\frac{5}{5}=\frac{1}{3}\cdot 1=\frac{1}{3}$

35. $\frac{10}{25}=\frac{2\cdot 5}{5\cdot 5}=\frac{2}{5}\cdot\frac{5}{5}=\frac{2}{5}\cdot 1=\frac{2}{5}$

37. $\frac{12}{36}=\frac{1\cdot 12}{3\cdot 12}=\frac{1}{3}\cdot\frac{12}{12}=\frac{1}{3}\cdot 1=\frac{1}{3}$

39. $\frac{12}{30}=\frac{2\cdot 6}{5\cdot 6}=\frac{2}{5}\cdot\frac{6}{6}=\frac{2}{5}\cdot 1=\frac{2}{5}$

41. $\frac{19}{76}=\frac{1\cdot 19}{4\cdot 19}=\frac{1}{4}\cdot\frac{19}{19}=\frac{1}{4}\cdot 1=\frac{1}{4}$

Multiplication and Division of Fractions

43. $\frac{1}{2}\cdot\frac{1}{3}=\frac{1\cdot 1}{2\cdot 3}=\frac{1}{6}$

45. $\frac{3}{4}\cdot\frac{1}{5}=\frac{3\cdot 1}{4\cdot 5}=\frac{3}{20}$

47. $\frac{5}{3}\cdot\frac{3}{5}=\frac{5\cdot 3}{3\cdot 5}=\frac{15}{15}=1$

49. $\frac{5}{6}\cdot\frac{18}{25}=\frac{5\cdot 18}{6\cdot 25}=\frac{90}{150}=\frac{3\cdot 30}{5\cdot 30}=\frac{3}{5}\cdot\frac{30}{30}=\frac{3}{5}\cdot 1=\frac{3}{5}$

51. $\frac{4}{1}\cdot\frac{3}{5}=\frac{4\cdot 3}{1\cdot 5}=\frac{12}{5}$

53. $\frac{2}{1}\cdot\frac{3}{8}=\frac{2\cdot 3}{1\cdot 8}=\frac{6}{8}=\frac{3\cdot 2}{4\cdot 2}=\frac{3}{4}\cdot\frac{2}{2}=\frac{3}{4}\cdot 1=\frac{3}{4}$

55. $\frac{x}{y}\cdot\frac{y}{x}=\frac{xy}{yx}=\frac{xy}{xy}=\frac{x}{x}\cdot\frac{y}{y}=\frac{x}{x}\cdot 1=\frac{x}{x}=1$

57. $\frac{a}{b}\cdot\frac{3}{2}=\frac{a\cdot 3}{b\cdot 2}=\frac{3a}{2b}$

59. $\frac{1}{4}\cdot\frac{3}{4}=\frac{1\cdot 3}{4\cdot 4}=\frac{3}{16}$

61. $\frac{2}{3}\cdot\frac{6}{1}=\frac{2\cdot 6}{3\cdot 1}=\frac{12}{3}=\frac{4\cdot 3}{1\cdot 3}=\frac{4}{1}\cdot\frac{3}{3}=\frac{4}{1}\cdot 1=4$

63. $\frac{1}{2}\cdot\frac{2}{3}=\frac{1\cdot 2}{2\cdot 3}=\frac{2}{6}=\frac{1\cdot 2}{3\cdot 2}=\frac{1}{3}\cdot\frac{2}{2}=\frac{1}{3}\cdot 1=\frac{1}{3}$

65. (a) $\frac{1}{5}$

(b) $\frac{1}{7}$

(c) $\frac{7}{4}$

(d) $\frac{8}{9}$

67. (a) $\frac{2}{1}=2$

(b) $\frac{9}{1}=9$

(c) $\frac{101}{12}$

(d) $\frac{17}{31}$

69. $\frac{1}{2}\div\frac{1}{3}=\frac{1}{2}\cdot\frac{3}{1}=\frac{1\cdot 3}{2\cdot 1}=\frac{3}{2}$

71. $\frac{3}{4}\div\frac{1}{2}=\frac{3}{4}\cdot\frac{2}{1}=\frac{3\cdot 2}{4\cdot 1}=\frac{6}{4}=\frac{3\cdot 2}{2\cdot 2}=\frac{3}{2}\cdot\frac{2}{2}=\frac{3}{2}\cdot 1=\frac{3}{2}$

73. $\frac{4}{3}\div\frac{1}{6}=\frac{4}{3}\cdot\frac{6}{1}=\frac{4\cdot 6}{3\cdot 1}=\frac{24}{3}=\frac{8\cdot 3}{1\cdot 3}=\frac{8}{1}\cdot\frac{3}{3}=\frac{8}{1}\cdot 1=8$

75. $\frac{32}{27}\div\frac{8}{9}=\frac{32}{27}\cdot\frac{9}{8}=\frac{32\cdot 9}{27\cdot 8}=\frac{288}{216}=\frac{4\cdot 72}{3\cdot 72}=\frac{4}{3}\cdot\frac{72}{72}=\frac{4}{3}\cdot 1=\frac{4}{3}$

77. $\frac{9}{1}\div\frac{8}{7}=\frac{9}{1}\cdot\frac{7}{8}=\frac{9\cdot 7}{1\cdot 8}=\frac{63}{8}$

79. $\frac{10}{1}\div\frac{5}{6}=\frac{10}{1}\cdot\frac{6}{5}=\frac{10\cdot 6}{1\cdot 5}=\frac{60}{5}=\frac{12\cdot 5}{1\cdot 5}=\frac{12}{1}\cdot\frac{5}{5}=\frac{12}{1}\cdot 1=12$

81. $\frac{9}{10}\div\frac{3}{1}=\frac{9}{10}\cdot\frac{1}{3}=\frac{9\cdot 1}{10\cdot 3}=\frac{9}{30}=\frac{3\cdot 3}{10\cdot 3}=\frac{3}{10}\cdot\frac{3}{3}=\frac{3}{10}\cdot 1=\frac{3}{10}$

83. $\frac{a}{b}\div\frac{2}{b}=\frac{a}{b}\cdot\frac{b}{2}=\frac{ab}{2b}=\frac{a}{2}\cdot\frac{b}{b}=\frac{a}{2}\cdot 1=\frac{a}{2}$

85. $\frac{x}{y}\div\frac{x}{y}=\frac{x}{y}\cdot\frac{y}{x}=\frac{xy}{xy}=1$

Addition and Subtraction of Fractions

87. (a) $\frac{2}{3}+\frac{1}{3}=\frac{2+1}{3}=\frac{3}{3}=1$

(b) $\frac{2}{3}-\frac{1}{3}=\frac{2-1}{3}=\frac{1}{3}$

89. (a) $\frac{3}{2}+\frac{1}{2}=\frac{3+1}{2}=\frac{4}{2}=2$

(b) $\frac{3}{2}-\frac{1}{2}=\frac{3-1}{2}=\frac{2}{2}=1$

91. (a) $\frac{5}{33}+\frac{2}{33}=\frac{5+2}{33}=\frac{7}{33}$

(b) $\frac{5}{33}-\frac{2}{33}=\frac{5-2}{33}=\frac{3}{33}=\frac{1\cdot 3}{11\cdot 3}=\frac{1}{11}\cdot\frac{3}{3}=\frac{1}{11}\cdot 1=\frac{1}{11}$

93. Prime factorizations are 5 and $10=5\times 2$. The LCD is $5\times 2=10$.

95. Prime factorizations are $9=3\times 3$ and $15=3\times 5$. The LCD is $3\times 3\times 5=45$.

97. Prime factorizations are 5 and $15=5\times 3$. The LCD is $5\times 3=15$.

99. Prime factorizations are $6=2\times 3$ and $8=2\times 2\times 2$. The LCD is $2\times 2\times 2\times 3=24$.

101. Prime factorizations are 2, 3 and $4=2\times 2$. The LCD is $2\times 2\times 3=12$.

103. Prime factorizations are $4=2\times 2$, $8=2\times 2\times 2$ and $12=2\times 2\times 3$. The LCD is $2\times 2\times 2\times 3=24$.

105. The LCD is 6. $\frac{1}{2}=\frac{1}{2}\cdot\frac{3}{3}=\frac{3}{6}$; $\frac{2}{3}=\frac{2}{3}\cdot\frac{2}{2}=\frac{4}{6}$

107. The LCD is 36. $\frac{7}{9}=\frac{7}{9}\cdot\frac{4}{4}=\frac{28}{36}$; $\frac{5}{12}=\frac{5}{12}\cdot\frac{3}{3}=\frac{15}{36}$

109. The LCD is 48. $\frac{1}{16}=\frac{1}{16}\cdot\frac{3}{3}=\frac{3}{48}$; $\frac{7}{12}=\frac{7}{12}\cdot\frac{4}{4}=\frac{28}{48}$

111. The LCD is 12. $\frac{1}{3}=\frac{1}{3}\cdot\frac{4}{4}=\frac{4}{12}$; $\frac{3}{4}=\frac{3}{4}\cdot\frac{3}{3}=\frac{9}{12}$; $\frac{5}{6}=\frac{5}{6}\cdot\frac{2}{2}=\frac{10}{12}$

113. $\frac{1}{2}+\frac{1}{3}=\frac{1}{2}\cdot\frac{3}{3}+\frac{1}{3}\cdot\frac{2}{2}=\frac{1\cdot 3}{2\cdot 3}+\frac{1\cdot 2}{3\cdot 2}=\frac{3}{6}+\frac{2}{6}=\frac{3+2}{6}=\frac{5}{6}$

115. $\frac{5}{8}+\frac{3}{16}=\frac{5}{8}\cdot\frac{2}{2}+\frac{3}{16}=\frac{5\cdot 2}{8\cdot 2}+\frac{3}{16}=\frac{10}{16}+\frac{3}{16}=\frac{10+3}{16}=\frac{13}{16}$

117. $\frac{1}{2}-\frac{1}{4}=\frac{1}{2}\cdot\frac{2}{2}-\frac{1}{4}=\frac{1\cdot 2}{2\cdot 2}-\frac{1}{4}=\frac{2}{4}-\frac{1}{4}=\frac{2-1}{4}=\frac{1}{4}$

119. $\frac{25}{24}-\frac{7}{8}=\frac{25}{24}-\frac{7}{8}\cdot\frac{3}{3}=\frac{25}{24}-\frac{7\cdot 3}{8\cdot 3}=\frac{25}{24}-\frac{21}{24}=\frac{4}{24}=\frac{1\cdot 4}{6\cdot 4}=\frac{1}{6}\cdot\frac{4}{4}=\frac{1}{6}\cdot 1=\frac{1}{6}$

121. $\frac{11}{14}+\frac{2}{35}=\frac{11}{14}\cdot\frac{5}{5}+\frac{2}{35}\cdot\frac{2}{2}=\frac{11\cdot 5}{14\cdot 5}+\frac{2\cdot 2}{35\cdot 2}=\frac{55}{70}+\frac{4}{70}=\frac{55+4}{70}=\frac{59}{70}$

123. $\frac{5}{12}-\frac{1}{18}=\frac{5}{12}\cdot\frac{3}{3}-\frac{1}{18}\cdot\frac{2}{2}=\frac{5\cdot 3}{12\cdot 3}-\frac{1\cdot 2}{18\cdot 2}=\frac{15}{36}-\frac{2}{36}=\frac{15-2}{36}=\frac{13}{36}$

125. $\frac{3}{100}+\frac{1}{300}-\frac{1}{200}=\frac{3}{100}\cdot\frac{6}{6}+\frac{1}{300}\cdot\frac{2}{2}-\frac{1}{200}\cdot\frac{3}{3}=\frac{3\cdot 6}{100\cdot 6}+\frac{1\cdot 2}{300\cdot 2}-\frac{1\cdot 3}{200\cdot 3}=$

$\frac{18}{600}+\frac{2}{600}-\frac{3}{600}=\frac{18+2-3}{600}=\frac{17}{600}$

127. $\frac{7}{8}-\frac{1}{6}+\frac{5}{12}=\frac{7}{8}\cdot\frac{3}{3}-\frac{1}{6}\cdot\frac{4}{4}+\frac{5}{12}\cdot\frac{2}{2}=\frac{7\cdot 3}{8\cdot 3}-\frac{1\cdot 4}{6\cdot 4}+\frac{5\cdot 2}{12\cdot 2}=$

$\frac{21}{24}-\frac{4}{24}+\frac{10}{24}=\frac{21-4+10}{24}=\frac{27}{24}=\frac{9\cdot 3}{8\cdot 3}=\frac{9}{8}\cdot\frac{3}{3}=\frac{9}{8}\cdot 1=\frac{9}{8}$

Applications

129. Find the product of $2\frac{1}{2}$ and $1\frac{9}{10}$. First convert $2\frac{1}{2}$ to $\frac{5}{2}$ and $1\frac{9}{10}$ to $\frac{19}{10}$. $\frac{5}{2}\cdot\frac{19}{10}=\frac{95}{20}=\frac{19\cdot 5}{4\cdot 5}=\frac{19}{4}=4\frac{3}{4}$ ft.

131. Find the sum of the motor vehicle deaths and the firearms deaths as a fraction of all accidental deaths.

$\frac{31}{42}+\frac{31}{1260}=\frac{31}{42}\cdot\frac{30}{30}+\frac{31}{1260}$. It follows that $\frac{31\cdot 30}{42\cdot 30}+\frac{31}{1260}=\frac{930}{1260}+\frac{31}{1260}=\frac{930+31}{1260}=\frac{961}{1260}$.

133. Find the value of one-half of $64\frac{5}{8}$. First convert $64\frac{5}{8}$ to $\frac{517}{8}$. $\frac{517}{8}\div 2=\frac{517}{8}\cdot\frac{1}{2}=\frac{517}{16}=32\frac{5}{16}$ in.

135. Convert the base of the triangle from $1\frac{2}{3}$ to $\frac{5}{3}$.

$\frac{1}{2}\cdot\frac{5}{3}\cdot\frac{3}{4}=\frac{1\cdot5\cdot3}{2\cdot3\cdot4}=\frac{15}{24}=\frac{5\cdot3}{8\cdot3}=\frac{5}{8}\cdot\frac{3}{3}=\frac{5}{8}\cdot1=\frac{5}{8}$ square yards.

137. Add the distance from Smalltown to Middletown and Middletown to Bigtown.

$3\frac{1}{2}+4\frac{3}{4}=\frac{7}{2}+\frac{19}{4}=\frac{7}{2}\cdot\frac{2}{2}+\frac{19}{4}=\frac{7\cdot2}{2\cdot2}+\frac{19}{4}=\frac{14}{4}+\frac{19}{4}=\frac{14+19}{4}=\frac{33}{4}=8\frac{1}{4}$ miles.

139. Multiply the fraction of the adult population who smoked in 2005 by the fraction of the people aged 25 to 44 who smoked in 2005 to find the fraction of the entire adult population who were aged 25 to 44 and who smoked in 2005.

$\frac{1}{5}\cdot\frac{6}{25}=\frac{1\cdot6}{5\cdot25}=\frac{6}{125}$.

Checking Basic Concepts for Sections 1.1 & 1.2

1. (a) Prime
   (b) Composite; $28=2\times2\times7$
   (c) Neither
   (d) Composite; $180=2\times2\times3\times3\times5$

2. $\frac{10}{3+2}=\frac{10}{5}=2$

3. $y=6\cdot5=30$

4. $x+5$

5. $I=12F$, because there are 12 inches in 1 foot.

6. (a) The largest number that divides evenly into 3 and 18 is 3, so the GCF is 3.
   (b) The largest number that divides evenly into 40 and 72 is 8, so the GCF is 8.

7. (a) $\frac{25}{35}=\frac{5\cdot5}{7\cdot5}=\frac{5}{7}\cdot\frac{5}{5}=\frac{5}{7}\cdot1=\frac{5}{7}$

   (b) $\frac{26}{39}=\frac{2\cdot13}{3\cdot13}=\frac{2}{3}\cdot\frac{13}{13}=\frac{2}{3}\cdot1=\frac{2}{3}$

8. $\frac{3}{4}$

9. (a) $\frac{2}{3}\cdot\frac{3}{4}=\frac{2\cdot3}{3\cdot4}=\frac{6}{12}=\frac{1\cdot6}{2\cdot6}=\frac{1}{2}\cdot\frac{6}{6}=\frac{1}{2}\cdot1=\frac{1}{2}$

   (b) $\frac{5}{6}\div\frac{10}{3}=\frac{5}{6}\cdot\frac{3}{10}=\frac{5\cdot3}{6\cdot10}=\frac{15}{60}=\frac{1\cdot15}{4\cdot15}=\frac{1}{4}\cdot\frac{15}{15}=\frac{1}{4}\cdot1=\frac{1}{4}$

   (c) $\frac{3}{10}+\frac{1}{10}=\frac{3+1}{10}=\frac{4}{10}=\frac{2\cdot2}{5\cdot2}=\frac{2}{5}\cdot\frac{2}{2}=\frac{2}{5}\cdot1=\frac{2}{5}$

(d) $\frac{3}{4}-\frac{1}{6}=\frac{3}{4}\cdot\frac{3}{3}-\frac{1}{6}\cdot\frac{2}{2}=\frac{3\cdot3}{4\cdot3}-\frac{1\cdot2}{6\cdot2}=\frac{9}{12}-\frac{2}{12}=\frac{9-2}{12}=\frac{7}{12}$

10. Multiply $1\frac{2}{3}$ by 2. $\frac{5}{3}\cdot\frac{2}{1}=\frac{5\cdot2}{3\cdot1}=\frac{10}{3}=3\frac{1}{3}$ cups.

1.3: Exponents and Order of Operations

Concepts

1. add
2. multiply
3. six
4. 2
5. a^6
6. base; exponent
7. 6^2
8. 8^3
9. 4^5
10. x^3
11. 17; multiplication; addition
12. 2; exponents; subtraction
13. 4; left; right
14. 2; left; right
15. No; $2^3=8$, but $3^2=9$.
16. $5\cdot5$

Natural Number Exponents

17. 2^5

19. 3^4

21. $\left(\frac{1}{2}\right)^4$

23. a^5

25. $(x+3)^2$

27. (a) $2^4=2\cdot2\cdot2\cdot2=16$

 (b) $4^2=4\cdot4=16$

29. (a) $6^1=6$

 (b) $1^6=1$

31. (a) $2^5 = 2 \cdot 2 \cdot 2 \cdot 2 \cdot 2 = 32$

(b) $10^3 = 10 \cdot 10 \cdot 10 = 1000$

33. (a) $\left(\frac{2}{3}\right)^2 = \frac{2}{3} \cdot \frac{2}{3} = \frac{2 \cdot 2}{3 \cdot 3} = \frac{4}{9}$

(b) $\left(\frac{1}{2}\right)^5 = \frac{1}{2} \cdot \frac{1}{2} \cdot \frac{1}{2} \cdot \frac{1}{2} \cdot \frac{1}{2} = \frac{1 \cdot 1 \cdot 1 \cdot 1 \cdot 1}{2 \cdot 2 \cdot 2 \cdot 2 \cdot 2} = \frac{1}{32}$

35. (a) $\left(\frac{2}{5}\right)^3 = \frac{2}{5} \cdot \frac{2}{5} \cdot \frac{2}{5} = \frac{2 \cdot 2 \cdot 2}{5 \cdot 5 \cdot 5} = \frac{8}{125}$

(b) $\left(\frac{9}{7}\right)^2 = \frac{9}{7} \cdot \frac{9}{7} = \frac{9 \cdot 9}{7 \cdot 7} = \frac{81}{49}$

37. $8 = 2 \cdot 2 \cdot 2 = 2^3$

39. $25 = 5 \cdot 5 = 5^2$

41. $49 = 7 \cdot 7 = 7^2$

43. $1000 = 10 \cdot 10 \cdot 10 = 10^3$

45. $\frac{1}{16} = \frac{1}{2} \cdot \frac{1}{2} \cdot \frac{1}{2} \cdot \frac{1}{2} = \left(\frac{1}{2}\right)^4$

47. $\frac{32}{243} = \frac{2}{3} \cdot \frac{2}{3} \cdot \frac{2}{3} \cdot \frac{2}{3} \cdot \frac{2}{3} = \left(\frac{2}{3}\right)^5$

Order of Operations

49. Perform multiplication before addition: $5 + 4 \cdot 6 = 5 + 24 = 29$

51. Perform division before addition: $6 \div 3 + 2 = 2 + 2 = 4$

53. Perform division before subtraction: $100 - \frac{50}{5} = 100 - 10 = 90$

55. $10 - 6 - 1 = 3$

57. $20 \div 5 \div 2 = 4 \div 2 = 2$

59. $3 + 2^4 = 3 + 2 \cdot 2 \cdot 2 \cdot 2 = 3 + 16 = 19$

61. $4 \cdot 2^3 = 4 \cdot 2 \cdot 2 \cdot 2 = 4 \cdot 8 = 32$

63. $(3+2)^3 = 5^3 = 5 \cdot 5 \cdot 5 = 125$

65. $\frac{4+8}{1+3} = \frac{12}{4} = \frac{3 \cdot 4}{1 \cdot 4} = \frac{3}{1} \cdot \frac{4}{4} = \frac{3}{1} \cdot 1 = 3$

67. $\frac{2^3}{4-2} = \frac{2 \cdot 2 \cdot 2}{2} = \frac{8}{2} = 4$

69. $10^2-(30-2\cdot5)=10^2-(30-10)=10^2-20=10\cdot10-20=100-20=80$

71. $\left(\frac{1}{2}\right)^4+\frac{5+4}{3}=\frac{1}{2}\cdot\frac{1}{2}\cdot\frac{1}{2}\cdot\frac{1}{2}+\frac{9}{3}=\frac{1}{16}+3=\frac{1}{16}+\frac{3}{1}\cdot\frac{16}{16}=\frac{1}{16}+\frac{48}{16}=\frac{1+48}{16}=\frac{49}{16}$

Translating Words to Symbols

73. $2^3-8=2\cdot2\cdot2-8=8-8=0$

75. $30-4\cdot3=30-12=18$

77. $\frac{4^2}{2^3}=\frac{4\cdot4}{2\cdot2\cdot2}=\frac{16}{8}=\frac{2\cdot8}{1\cdot8}=\frac{2}{1}\cdot\frac{8}{8}=\frac{2}{1}\cdot1=2$

79. $\frac{40}{10}+2=\frac{4\cdot10}{1\cdot10}+2=\frac{4}{1}\cdot\frac{10}{10}+2=\frac{4}{1}\cdot1+2=4+2=6$

81. $100(2+3)=100\cdot5=500$

Applications

83. $512\text{ MB}=512\cdot2^{20}\text{ bytes}=536{,}870{,}912\text{ bytes}$

85. (a) Because $2^7=128$, $k=7$.

(b) $\frac{128}{100}=\frac{32\cdot4}{25\cdot4}=\frac{32}{25}$. There were 32 males for every 25 females.

87. (a) $72\div9=8$ years

(b) The investment doubles every $72\div12=6$ years, so in 18 years the amount will double 3 times. The investment will increase by $2^3=8$ times \$10,000 and will equal \$80,000.

1.4: Real Numbers and the Number Line

Concepts

1. $-b$
2. 7
3. b
4. -9
5. natural
6. rational
7. 4
8. principal
9. real
10. irrational
11. rational
12. irrational
13. $0.\overline{27}$

14. 1; 4

15. $\sqrt{2}$ is one example; *Answers may vary.*

16. not equal

17. approximately equal

18. 3

19. 0

20. left

21. origin

22. b

23. $>$

24. $>$

25. $=$

26. $=$

Signed Numbers

27. (a) The opposite of 9 is -9.

(b) The opposite of -9 is $-(-9)=9$.

29. (a) The opposite of $\frac{2}{3}$ is $-\frac{2}{3}$.

(b) The opposite of $-\frac{2}{3}$ is $-\left(-\frac{2}{3}\right)=\frac{2}{3}$.

31. (a) $-(-8)=8$, so the opposite of 8 is -8.

(b) $-(-(-8))=-8$, so the opposite of -8 is $-(-8)=8$.

33. (a) The opposite of a is $-a$.

(b) The opposite of $-a$ is $-(-a)=a$.

35. The additive inverse of t is $-t. -t=6$

37. The additive inverse of $-b$ is $-(-b)=b. b=\frac{1}{2}$

Numbers and the Number Line

39. $\frac{1}{4}=0.25$

41. $\frac{7}{8}=0.875$

43. $\frac{3}{2}=1.5$

45. $\frac{1}{20}=0.05$

47. $\frac{2}{3} = 0.\overline{6}$

49. $\frac{7}{9} = 0.\overline{7}$

51. 8 is a natural, whole and rational number, and is an integer.

53. $\frac{16}{4} = 4$ is a natural, whole and rational number, and is an integer.

55. 0 is a whole and rational number, and is an integer.

57. $-4.5 = -4\frac{1}{2} = -\frac{9}{2}$ is a rational number.

59. $\frac{3}{7}$ is a rational number.

61. $\sqrt{11}$ is an irrational number.

63. $\frac{8}{4} = 2$ is a natural and rational number, and is an integer.

65. $\sqrt{49} = 7$ is a natural and rational number, and is an integer.

67. $1.\overline{8}$ is a rational number because its decimal repeats the same number.

69. (b) (a) (c)
−5 −4 −3 −2 −1 0 1 2 3 4 5

71. (b) (a) (c)
−5 −4 −3 −2 −1 0 1 2 3 4 5

73. (b) (c)(a)
−5 −4 −3 −2 −1 0 1 2 3 4 5

75. (b) (a) (c)
−50 −40 −30 −20 −10 0 10 20 30 40 50

77. (c) (b) (a)
−5 −4 −3 −2 −1 0 1 2 3 4 5

Absolute Values

79. $|5.23| = 5.23$

81. $|-7| = 7$

83. $|2-6| = |-4| = 4$

85. $|\pi - 3| = \pi - 3$

87. $|b|$, if b is negative, $= -b$

89. $5 < 7$

91. $-5 > -7$

93. $-\frac{1}{3} > -\frac{2}{3}$

95. $-1.9 < -1.3$

97. $|-8| = 8$ and $8 > 3$ so $|-8| > 3$.

99. $|-2| = 2$ and $|-7| = 7$ and $2 < 7$ so $|-2| < |-7|$.

101. $-9, -2^3, -3, 0, 1$

103. $-2, -\frac{3}{2}, \frac{1}{3}, \sqrt{5}, \pi$

105. $-4^2, -\frac{17}{28}, -\frac{4}{7}, \sqrt{2}, \sqrt{7}$

Applications

107. (a) The percentage in 2000 was 16.5%.

(b) *Answers may vary.*

(c) The average percentage was $\frac{17.2+17.4+17.4+16.5}{4} = \frac{68.5}{4} = 17.125\%$.

Checking Basic Concepts for Sections 1.3 & 1.4

1. (a) $5 \cdot 5 \cdot 5 \cdot 5 = 5^4$

(b) $7 \cdot 7 \cdot 7 \cdot 7 \cdot 7 = 7^5$

2. (a) $2^3 = 2 \cdot 2 \cdot 2 = 8$

(b) $10^4 = 10 \cdot 10 \cdot 10 \cdot 10 = 10{,}000$

(c) $\left(\frac{2}{3}\right)^3 = \frac{2}{3} \cdot \frac{2}{3} \cdot \frac{2}{3} = \frac{2 \cdot 2 \cdot 2}{3 \cdot 3 \cdot 3} = \frac{8}{27}$

(d) $-3^4 = -(3 \cdot 3 \cdot 3 \cdot 3) = -81$

3. (a) $64 = 4 \cdot 4 \cdot 4 = 4^3$

(b) $64 = 2 \cdot 2 \cdot 2 \cdot 2 \cdot 2 \cdot 2 = 2^6$

4. (a) $6 + 5 \cdot 4 = 6 + 20 = 26$

(b) $6 + 6 \div 2 = 6 + 3 = 9$

(c) $5 - 2 - 1 = 2$

(d) $\frac{6-3}{2+4} = \frac{3}{6} = \frac{1 \cdot 3}{2 \cdot 3} = \frac{1}{2} \cdot \frac{3}{3} = \frac{1}{2} \cdot 1 = \frac{1}{2}$

(e) $12 \div (6 \div 2) = 12 \div 3 = 4$

(f) $2^3 - 2\left(2 + \frac{4}{2}\right) = 8 - 2(2+2) = 8 - 2(4) = 8 - 8 = 0$

5. $5^3 \div 3$, or $\frac{5^3}{3}$

6. (a) The opposite of -17 is $-(-17)=17$.

(b) The opposite of a is $-a$.

7. (a) $\frac{3}{20}=\frac{3\cdot 5}{20\cdot 5}=\frac{15}{100}=0.15$

(b) $\frac{5}{8}=\frac{5\cdot 125}{8\cdot 125}=\frac{625}{1000}=0.625$

8. (a) $\frac{10}{2}=5$ is a natural and rational number, and is an integer.

(b) -5 is a rational number and is an integer.

(c) $\sqrt{5}$ is an irrational number.

(d) $-\frac{5}{6}$ is a rational number.

9. (b) (e) (a)(d) (c)

–5 –4 –3 –2 –1 0 1 2 3 4 5

10. (a) $|-12|=-(-12)=12$

(b) $|-a|$, if $a>0$, $=-(-a)=a$

11. (a) $4<9$

(b) $-1.3<-0.5$

(c) $|-3|=3$ and $|-5|=5$ and $3<5$ so $|-3|<|-5|$.

12. $-7, -1.6, 0, \frac{1}{3}, \sqrt{3}, 3^2$

1.5: Addition and Subtraction of Real Numbers

Concepts

1. two
2. addends
3. sum
4. zero
5. positive
6. negative
7. absolute value
8. difference
9. addition
10. opposite; $(-b)$
11. addition
12. subtraction

Addition and Subtraction of Real Numbers

13. The opposite of 25 is -25. $25+(-25)=0$

15. The opposite of $-\sqrt{21}$ is $\sqrt{21}$. $-\sqrt{21}+\sqrt{21}=0$

17. The opposite of 5.63 is -5.63. $5.63+(-5.63)=0$

19. 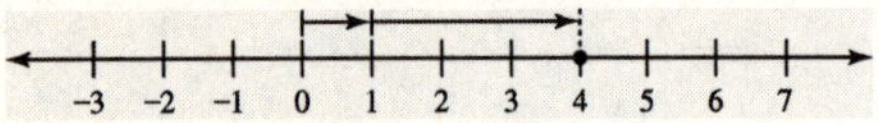 ; $1+3=4$

21. 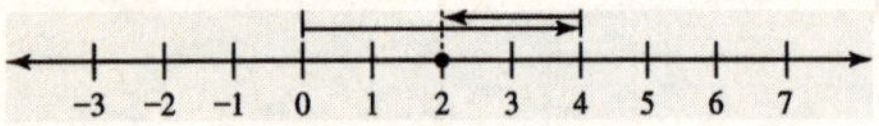 ; $4+(-2)=2$

23. 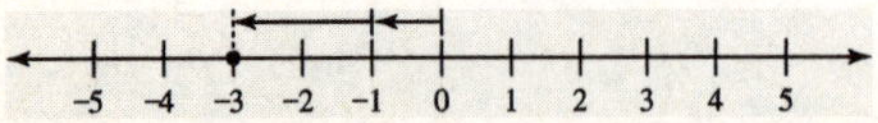 ; $-1+(-2)=-3$

25. 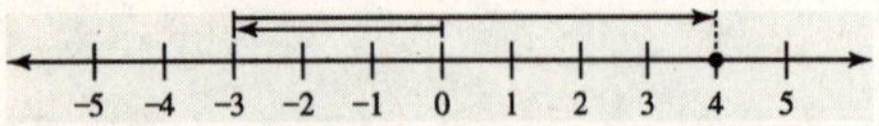 ; $-3+7=4$

27. 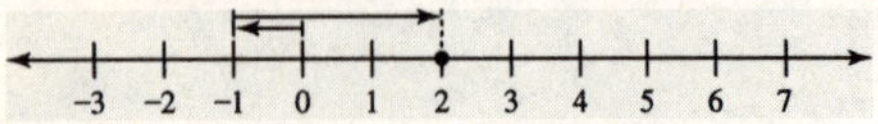 ; $-1+3=2$

29. 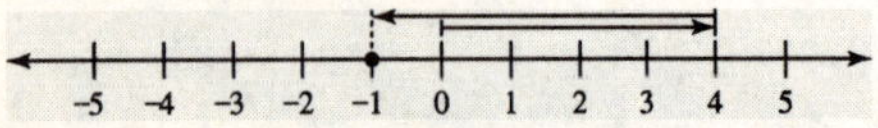 ; $4+(-5)=-1$

31. 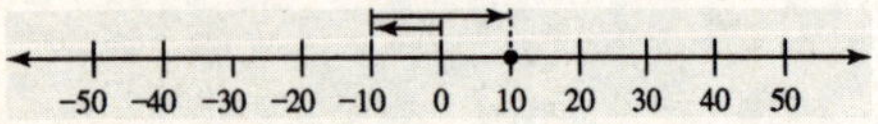 ; $-10+20=10$

33. 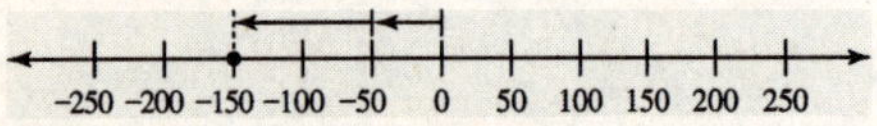 ; $-50+(-100)=-150$

35. $5+(-4)=5-|-4|=5-4=1$

37. $-1+(-6)=-1-|-6|=-1-6=-7$

39. $\frac{3}{4}+\left(-\frac{1}{2}\right)=\frac{3}{4}-\left|-\frac{1}{2}\right|=\frac{3}{4}-\frac{1}{2}=\frac{3}{4}-\frac{1}{2}\cdot\frac{2}{2}=\frac{3}{4}-\frac{2}{4}=\frac{1}{4}$

41. $-\frac{6}{7}+\frac{3}{14}=\frac{3}{14}-\left|-\frac{6}{7}\right|=\frac{3}{14}-\frac{6}{7}=\frac{3}{14}-\frac{6}{7}\cdot\frac{2}{2}=\frac{3}{14}-\frac{12}{14}=-\frac{9}{14}$

43. $-\frac{1}{2}+\left(-\frac{3}{4}\right)=-\frac{1}{2}-\left|-\frac{3}{4}\right|=-\frac{1}{2}-\frac{3}{4}=-\frac{1}{2}\cdot\frac{2}{2}-\frac{3}{4}=-\frac{2}{4}-\frac{3}{4}=-\frac{5}{4}$

45. $0.6+(-1.7)=0.6-|-1.7|=0.6-1.7=-1.1$

47. $-52+86=86-|-52|=86-52=34$

49. $8+(-7)+(-2)=8-|-2|-|-7|=8-2-7=-1$

51. $\frac{1}{2}+\frac{3}{4}+\left(-\frac{1}{2}\right)+\left(-\frac{3}{4}\right)=\frac{1}{2}+\frac{3}{4}-\left|-\frac{1}{2}\right|-\left|-\frac{3}{4}\right|=\frac{1}{2}+\frac{3}{4}-\frac{1}{2}-\frac{3}{4}=\frac{1}{2}-\frac{1}{2}+\frac{3}{4}-\frac{3}{4}=0+0=0$

53. $5-8=5+(-8)=-3$

55. $-2-(-9)=-2+9=7$

57. $\frac{1}{3}-\left(-\frac{2}{3}\right)=\frac{1}{3}+\frac{2}{3}=\frac{3}{3}=1$

59. $\frac{6}{7}-\frac{13}{14}=\frac{12}{14}-\frac{13}{14}=\frac{12}{14}+\left(-\frac{13}{14}\right)=-\frac{1}{14}$

61. $-\frac{1}{10}-\left(-\frac{3}{5}\right)=-\frac{1}{10}-\left(-\frac{6}{10}\right)=-\frac{1}{10}+\frac{6}{10}=\frac{5}{10}=\frac{1}{2}$

63. $0.8-(-2.1)=0.8+2.1=2.9$

65. $-73-91=-73+(-91)=-164$

67. $-7-(-6)-10=-7+6-10=-11$

69. $10-19=10+(-19)=-9$

71. $19-(-22)+1=19+22+1=42$

73. $-3+4-6=-3+4+(-6)=4-|-3|+(-6)=4-3+(-6)=-5$

75. $100-200+100-(-50)=100+100-200+50=200-200+50=50$

77. $1.5-2.3+9.6=1.5+(-2.3)+9.6=8.8$

79. $-\frac{1}{2}+\frac{1}{4}-\left(-\frac{3}{4}\right)=-\frac{2}{4}+\frac{1}{4}+\frac{3}{4}=\frac{2}{4}=\frac{1}{2}$

81. $|4-9|-|1-7|=|-5|-|-6|=5-6=-1$

83. $2+(-5)=-3$

85. $-5+7=2$

87. $2^3=8$; the opposite of 8 is -8.

89. $-6-7=-6+(-7)=-13$

91. $6+(-10)-5=6+(-10)+(-5)=-9$

Applications

93. Take the initial balance and then add to it the deposits and subtract from it the withdrawals.

$358-45+37+120-240=358+37+120+(-45)+(-240)=\$230$

95. Add the yardage gained or lost in each play.

$9+(-2)+(-1)+14+5=25$ yd.

97. The word height is in reference to sea level, the height of Mount Everest is 29,029 feet and the height of the Mariana Trench is $(-35,839)$.

To find the difference take $29,029-(-35,839)=29,029+35,839=64,868$ feet.

1.6: Multiplication and Division of Real Numbers

Concepts

1. factors
2. product
3. negative
4. positive
5. quotient
6. dividend; divisor
7. $\frac{1}{a}$
8. reciprocal
9. $-\frac{4}{3}$
10. reciprocal or multiplicative inverse
11. $-a$
12. $(-5)^2 = (-5)(-5) = 25$
13. $-5^2 = -5 \cdot 5 = -25$
14. $\frac{1}{b}$
15. positive
16. negative
17. subtraction
18. 5; 8

Multiplication and Division of Real Numbers

19. $-3 \cdot 4 = -12$

21. $6 \cdot (-3) = -18$

23. $0 \cdot (-2.13) = 0$

25. $-6 \cdot (-10) = 60$

27. $-\frac{1}{2} \cdot \left(-\frac{2}{4}\right) = \frac{2}{8} = \frac{1}{4}$

29. $-\frac{3}{7} \cdot \frac{7}{3} = -\frac{21}{21} = -1$

31. $-10 \cdot (-20) = 200$

33. $-50 \cdot 100 = -5000$

35. $-2\cdot 3\cdot(-4)\cdot 5=-6\cdot(-20)=120$

37. $-6\cdot\frac{1}{6}\cdot\frac{7}{9}\cdot\left(-\frac{9}{7}\right)\cdot\left(-\frac{3}{2}\right)=-\frac{6}{6}\cdot\left(-\frac{63}{63}\right)\cdot\left(-\frac{3}{2}\right)=-1\cdot(-1)\cdot\left(-\frac{3}{2}\right)=1\cdot\left(-\frac{3}{2}\right)=-\frac{3}{2}$

39. $(-1)\cdot(-1)\cdot(-1)\cdot(-1)=1\cdot 1=1$

41. Negative, because there is an odd number of negative factors

43. $(-1)^3=(-1)(-1)(-1)=-1$

45. $-2^4=-2\cdot 2\cdot 2\cdot 2=-16$

47. $-(-2)^3=-(-2)(-2)(-2)=-(-8)=8$

49. $5\cdot(-2)^3=5\cdot[(-2)(-2)(-2)]=5\cdot(-8)=-40$

51. $-10\div 5=-\frac{10}{1}\cdot\frac{1}{5}=-\frac{10}{5}=-2$

53. $-20\div(-2)=-\frac{20}{1}\cdot\left(-\frac{1}{2}\right)=\frac{20}{2}=10$

55. $-\frac{12}{3}=-4$

57. $\frac{39}{-13}=-3$

59. $-16\div\frac{1}{2}=-\frac{16}{1}\cdot\frac{2}{1}=-\frac{32}{1}=-32$

61. $0\div 3=0$

63. $\frac{0}{-2}=0$

65. $\frac{1}{2}\div(-11)=\frac{1}{2}\cdot-\frac{1}{11}=-\frac{1}{22}$

67. $-\frac{4}{5}\div(-3)=-\frac{4}{5}\cdot\left(-\frac{1}{3}\right)=\frac{4}{15}$

69. $\frac{5}{6}\div\left(-\frac{8}{9}\right)=\frac{5}{6}\cdot\left(-\frac{9}{8}\right)=-\frac{45}{48}=-\frac{15}{16}$

71. $-\frac{1}{2}\div 0=$ undefined, because division by 0 is not allowed.

73. $-0.5\div\frac{1}{2}=-\frac{5}{10}\div\frac{1}{2}=-\frac{5}{10}\cdot\frac{2}{1}=-\frac{10}{10}=-1$

75. $-\frac{2}{3}\div 0.5=-\frac{2}{3}\div\frac{5}{10}=-\frac{2}{3}\cdot\frac{10}{5}=-\frac{20}{15}=-\frac{4}{3}$

Converting Between Fractions and Decimals

77. $\frac{1}{2} = 0.5$

79. $\frac{3}{16} = 0.1875$

81. Because $1 \div 2 = 0.5$, $3\frac{1}{2} = 3.5$

83. Because $2 \div 3 = 0.\overline{6}$, $5\frac{2}{3} = 5.\overline{6}$

85. Because $7 \div 16 = 0.4375$, $1\frac{7}{16} = 1.4375$

87. $\frac{7}{8} = 7 \div 8 = 0.875$

89. $0.25 = \frac{25}{100} = \frac{1 \cdot 25}{4 \cdot 25} = \frac{1}{4}$

91. $0.16 = \frac{16}{100} = \frac{4 \cdot 4}{25 \cdot 4} = \frac{4}{25}$

93. $0.625 = \frac{625}{1000} = \frac{5 \cdot 125}{8 \cdot 125} = \frac{5}{8}$

95. $0.6875 = \frac{6875}{10{,}000} = \frac{11 \cdot 625}{16 \cdot 625} = \frac{11}{16}$

97. $\left(\frac{1}{3} + \frac{5}{6}\right) \div \frac{1}{2} = 2.\overline{3}$ or $\frac{7}{3}$

99. $\frac{4}{5} \div \frac{2}{3} \cdot \frac{7}{4} = 2.1$ or $\frac{21}{10}$

101. $\frac{15}{2} - 4 \cdot \frac{7}{3} = -1.8\overline{3}$ or $-\frac{11}{6}$

103. $\frac{17}{40} + 3 \div 8 = 0.8$ or $\frac{4}{5}$

Applications

105. Multiply the real numbers 202 and $\frac{13}{20}$ to obtain $202 \cdot \frac{13}{20} = 131.3$. Total admissions for *The Ten Commandments* were about 131 million.

107. $\frac{39}{250} = 39 \div 250 = 0.156$

Checking Basic Concepts for Sections 1.5 & 1.6

1. (a) $-4+4=0$

   (b) $-10+(-12)+3=-22+3=-19$

2. (a) $\frac{2}{3}-\left(-\frac{2}{9}\right)=\frac{2}{3}+\frac{2}{9}=\frac{6}{9}+\frac{2}{9}=\frac{8}{9}$

   (b) $-1.2-5.1+3.1=-1.2+(-5.1)+3.1=-6.3+3.1=-3.2$

3. (a) $-1+5=4$

   (b) $4-(-3)=4+3=7$

4. $98-(-46)=98+46=144°\text{F}$ is the difference between these two temperatures.

5. (a) $-5\cdot(-7)=35$

   (b) $-\frac{1}{2}\cdot\frac{2}{3}\cdot\left(-\frac{4}{5}\right)=-\frac{2}{6}\cdot\left(-\frac{4}{5}\right)=\frac{8}{30}=\frac{4}{15}$

6. (a) $-3^2=-3\cdot3=-9$

   (b) $4\cdot(-2)^3=4\cdot[(-2)(-2)(-2)]=4\cdot(-8)=-32$

   (c) $(-5)^2=(-5)(-5)=25$

7. (a) $-5\div\frac{2}{3}=-\frac{5}{1}\cdot\frac{3}{2}=-\frac{15}{2}$

   (b) $-\frac{5}{8}\div\left(-\frac{4}{3}\right)=-\frac{5}{8}\cdot\left(-\frac{3}{4}\right)=\frac{15}{32}$

8. The reciprocal of $-\frac{7}{6}$ is $-\frac{6}{7}$.

9. (a) $\frac{-10}{2}=\frac{-5\cdot2}{1\cdot2}=\frac{-5}{1}=-5$

   (b) $\frac{10}{-2}=\frac{5\cdot2}{-1\cdot2}=\frac{5}{-1}=-5$

   (c) $-\frac{10}{2}=-\frac{5\cdot2}{1\cdot2}=-\frac{5}{1}=-5$

   (d) $\frac{-10}{-2}=\frac{-5\cdot2}{-1\cdot2}=\frac{-5}{-1}=5$

10. (a) $\frac{3}{5}=0.6$

    (b) Because $\frac{7}{8}=0.875$, $3\frac{7}{8}=3.875$

1.7: Properties of Real Numbers

Concepts

1. commutative; addition
2. commutative; multiplication
3. associative; addition
4. associative; multiplication
5. subtraction; division
6. subtraction; division
7. distributive
8. distributive
9. b
10. a
11. identity; addition
12. identity; multiplication
13. $-a$
14. $\frac{1}{a}$

Properties of Real Numbers

15. $-6+10=10+(-6)$

17. $-5\cdot 6=6\cdot(-5)$

19. $a+10=10+a$

21. $b\cdot 7=7\cdot b$

23. $(1+2)+3=1+(2+3)$

25. $2\cdot(3\cdot 4)=(2\cdot 3)\cdot 4$

27. $(a+5)+c=a+(5+c)$

29. $(x\cdot 3)\cdot 4=x\cdot(3\cdot 4)$

31. $$\begin{aligned} a+b+c &= (a+b)+c \\ &= c+(a+b) \\ &= c+(b+a) \\ &= c+b+a \end{aligned}$$

33. $4(3+2)=(4\cdot 3)+(4\cdot 2)=12+8=20$

35. $a(b-8)=ab-8a$

37. $-1(t+z)=-1t-1z=-t-z$

39. $-(5-a)=-(1)(5)-1(-a)=-5+a$

41. $(a+5)3=3a+(3)(5)=3a+15$

43. $(6-z)(-3)=-18+3z=3z-18$

45. $a\cdot(b+c+d)=a\cdot((b+c)+d)$

$$=a\cdot(b+c)+ad$$

$$=ab+ac+ad$$

47. $6x+5x=(6+5)x=11x$

49. $-4b+3b=(-4+3)b=(-1)b=-b$

51. $3a-a=(3-1)a=2a$

53. $13w-27w=(13-27)w=(13+(-27))w=-14w$

55. Commutative (multiplication)

57. Associative (addition)

59. Distributive

61. Distributive, Commutative (multiplication)

63. Distributive

65. Associative (multiplication)

67. Distributive

Identity and Inverse Properties

69. Identity (addition)

71. Identity (multiplication)

73. Identity (multiplication)

75. Inverse (multiplication)

77. Inverse (addition)

Mental Calculations

79. $(4+2)+(9+8)+(1+6)=30$

81. $(45+43)+(5+7)=100$

83. $129+50-1=179-1=178$

85. $379+100-2=479-2=477$

87. $178-100+1=78+1=79$

89. $6\cdot10+6\cdot5=60+30=90$

91. $8\cdot100+8\cdot2=800+16=816$

93. $\left(\frac{1}{2}\cdot\frac{1}{2}\cdot\frac{1}{2}\right)\cdot2\cdot2\cdot2=\left(\frac{1}{8}\right)\cdot8=1$

95. $\left(\frac{7}{6}\cdot\frac{1}{2}\right)\cdot\left(\frac{1}{2}\cdot\frac{1}{2}\right)\cdot\frac{8}{7}=\left(\frac{7}{12}\cdot\frac{1}{4}\right)\cdot\frac{8}{7}=\frac{7}{48}\cdot\frac{8}{7}=\frac{56}{336}=\frac{1}{6}$

Multiplying and Dividing by Powers of 10 Mentally

97. (a) $10\times 41=410$

(b) $10\times 997=9970$

(c) $-630\times 10=-6300$

(d) $-14{,}000\times 10=-140{,}000$

99. (a) $1000\times 19=19{,}000$

(b) $100\times(-451)=-45{,}100$

(c) $10{,}000\times 6=60{,}000$

(d) $-79\times 100{,}000=-7{,}900{,}000$

101. (a) $12.56\div 10=1.256$

(b) $9.6\div 10=.96$

(c) $0.987\div 10=0.0987$

(d) $-0.056\div 10=-0.0056$

(e) $1200\div 10=120$

(f) $4578\div 10=457.8$

Applications

103. Because $100+75=75+100$, this shows the commutative property of addition.

105. Because 1 gallon is $10\div 10$ gallons, divide 198 by 10 to get 19.8 miles.

107. (a) Because multiplying by 10 is easy, multiply $13\cdot(5\cdot 2)=13\cdot 10=130$ ft^3.

(b) The associative property of multiplication.

1.8: Simplifying and Writing Algebraic Expressions

Concepts

1. term
2. is not; is
3. coefficient
4. −1
5. 6
6. factors; terms
7. like
8. like, unlike
9. like
10. distributive

Like Terms

11. The expression 91 is a term; its coefficient is 91.

13. The expression $-6b$ is a term; its coefficient is -6.

15. The expression $x+10$ is not a term because it is the sum of two terms.

17. The expression x^2 is a term; its coefficient is 1.

19. The expression $4x-5$ is not a term because it is the difference of two terms.

21. The expression $-9xyz$ is a term; its coefficient is -9.

23. The terms 6 and -8 are like because neither term contains a variable.

25. The terms $5x$ and $-22x$ are like because each term contains the same variable raised to the same power.

27. The terms 14 and $14a$ are unlike because one contains a variable while the other does not.

29. The terms $18x$ and $18y$ are unlike because the first term contains a different variable than the second term.

31. The terms $x^2, -15x^2$ and $6x^2$ are like because each term contains the same variable raised to the same power.

33. The terms $3x^2y$ and $5xy^2$ are unlike because the variables are the same but they are raised to different powers.

35. The terms xy, xz and $2xy$ are unlike because each term does not contain the same variables.

37. The terms $4ab$ and $-3ba$ are like because each term contains the same variables raised to the same powers.

39. $3x+5x=(3+5)x=8x$

41. $19y-5y=(19-5)y=14y$

43. $28a+13a=(28+13)a=41a$

45. $11z-11z=(11-11)z=0z=0$

47. It is not possible to combine terms because the first term contains a different variable than the second term.

49. It is not possible to combine terms because the first term does not contain a variable and the second term does.

51. $5x^2-2x^2=(5-2)x^2=3x^2$

53. $8xy-10xy+xy=(8-10+1)xy=-1xy=-xy$

55. $12ab-7ab-5ab=(12-7-5)ab=0ab=0$

Simplifying and Writing Expressions

57. $5+x-3+2x=5-3+x+2x=5-3+(1+2)x=2+3x=3x+2$

59. $-\frac{3}{4}+z-3z+\frac{5}{4}=z-3z+\frac{5}{4}-\frac{3}{4}=(1-3)z+\frac{5}{4}-\frac{3}{4}=-2z+\frac{5}{4}-\frac{3}{4}=-2z+\frac{1}{2}$

61. $4y-y+8y=(4-1+8)y=11y$

63. $-3+6z+2-2z=6z-2z-3+2=(6-2)z-3+2=4z-1$

65. $-2(3z-6y)-z=-6z+12y-z=-6z-z+12y=(-6-1)z+12y=-7z+12y=12y-7z$

67. $2-\frac{3}{4}(4x+8)=2-\frac{3\cdot 4}{4}x-\frac{3\cdot 8}{4}=2-\frac{12}{4}x-\frac{24}{4}=2-3x-6=2-6-3x=$

$-4-3x=-3x-4$

69. $-x-(5x+1)=-x-5x-1=(-1-5)x-1=-6x-1$

71. $1-\frac{1}{3}(x+1)=1-\frac{1}{3}x-\frac{1}{3}=-\frac{1}{3}x+1-\frac{1}{3}=-\frac{1}{3}x+\frac{3}{3}-\frac{1}{3}=-\frac{1}{3}x+\frac{2}{3}$

73. $\frac{3}{5}(x+y)-\frac{1}{5}(x-1)=\frac{3}{5}x+\frac{3}{5}y-\frac{1}{5}x+\frac{1}{5}=\frac{3}{5}x-\frac{1}{5}x+\frac{3}{5}y+\frac{1}{5}=\left(\frac{3}{5}-\frac{1}{5}\right)x+\frac{3}{5}y+\frac{1}{5}=$

$\frac{2}{5}x+\frac{3}{5}y+\frac{1}{5}$

75. $0.2x^2+0.3x^2-0.1x^2=(0.2+0.3-0.1)x^2=0.4\,x^2$

77. $2x^2-3x+5x^2-4x=2x^2+5x^2-3x-4x=(2+5)x^2-(3+4)x=7x^2-7x$

79. $a+3b-a-b=a-a+3b-b=(1-1)a+(3-1)b$

$=0a+2b=2b$

81. $8x^3+7-x^3-5=8x^3-x^3+7-5=(8-1)x^3+7-5$

$=7x^3+2$

83. $\frac{8x}{8}=\frac{8}{8}\cdot\frac{x}{1}=1\cdot\frac{x}{1}=1\cdot x=x$

85. $\frac{-3y}{-y}=\frac{-3}{1}\cdot\frac{y}{-y}=-3\cdot(-1)=3$

87. $\frac{-108z}{-108}=\frac{-108}{-108}\cdot\frac{z}{1}=1\cdot\frac{z}{1}=1\cdot z=z$

89. $\frac{9x-6}{3}=\frac{9x}{3}-\frac{6}{3}=\frac{9}{3}\cdot\frac{x}{1}-\frac{6}{3}=3x-2$

91. $\frac{14z+21}{7}=\frac{14z}{7}+\frac{21}{7}=\frac{14}{7}\cdot\frac{z}{1}+\frac{21}{7}=2z+3$

93. $5x+6x=(5+6)x=11x$

95. $x^2+2x^2=(1+2)x^2=3x^2$

97. $6x-4x=(6-4)x=2x$

Applications

99. (a) Let w be the constant width of the street in feet. The area of each street section equals its length times its width.

The total area of the street is $400w+350w+220w+600w=(400+350+220+600)w=1570w$.

(b) If the width = 42 feet, $w=42$. Then, $1570w=1570\cdot 42=65{,}940\text{ ft}^2$.

101. (a) Let x be the number of minutes. Then, $20x+30x=50x=$ the number of cubic feet of snow removed in x minutes.

(b) Let $x=48$. Then, the total number of cubic feet of snow removed in 48 minutes is

$50x=50\cdot 48=2400\text{ ft}^3$.

(c) First, calculate how many cubic feet of snow the driveway contains. Volume equals length times width times height; thus, the driveway contains $30\cdot 20\cdot 2=1200\text{ ft}^3$ of snow. If 50 cubic feet of snow are removed in 1 minute, then $\frac{1200}{50}$ cubic feet of snow are removed in 24 minutes.

Checking Basic Concepts for Sections 1.7 & 1.8

1. (a) $y\cdot 18=18y$

(b) $10+x=x+10$

2. $5\cdot(y\cdot 4)=5\cdot(4y)=20y$

3. (a) $10-(5+x)=10-5-x=5-x$

(b) $5(x-7)=5x-35$

4. Because $5x+3x=(5+3)x=8x$, this equation illustrates the distributive property.

5. $-4xy+4xy=(-4+4)xy=0xy=0$

6. (a) $32+17+8+3=60$

(b) $\left(\frac{5}{6}\cdot\frac{6}{5}\right)\cdot\left(\frac{7}{8}\cdot\frac{8}{1}\right)=1\cdot 7=7$

(c) $567-200+1=367+1=368$

7. (a) The terms $-3xy$ and $-3yz$ are unlike because the first term contains a different variable than the second term.

(b) The terms $4x^2$ and $-2x^2$ are like because each term contains the same variable raised to the same power.

8. (a) $5z+9z=(5+9)z=14z$

(b) $5y-4-8y+7=5y-8y-4+7=(5-8)y-4+7=-3y+3$

9. (a) $2y-(5y+3)=2y-5y-3-(2-5)y-3=-3y-3$

(b) $-4(x+3y)+2(2x-y)=-4x-12y+4x-2y=-4x+4x-12y-2y=$

$(-4+4)x-(12+2)y=0x-14y=-14y$

(c) $\frac{20x}{20}=\frac{20}{20}\cdot x=1\cdot x=x$

(d) $\frac{35x^2}{x^2}=35\cdot\frac{x^2}{x^2}=35\cdot 1=35$

10. $3x+5x=(3+5)x=8x$

Chapter 1 Review Exercises

Section 1.1

1. The number 29 is prime because its only factors are itself and 1.
2. The number 27 is a composite number because it has factors other than itself and 1; $27=3\times 3\times 3$.
3. The number 108 is a composite number because it has factors other than itself and 1; $108=2\times 2\times 3\times 3\times 3$.
4. The number 91 is a composite number because it has factors other than itself and 1; $91=7\times 13$.
5. The number 0 is neither a prime nor a composite number.
6. The number 1 is neither a prime nor a composite number.
7. $2x-5$, when $x=4$, is $2\cdot 4-5=8-5=3$
8. $7-\frac{10}{x}$, when $x=5$, is $7-\frac{10}{5}=7-\frac{2\cdot 5}{1\cdot 5}=7-\frac{2}{1}\cdot\frac{5}{5}=7-2\cdot 1=7-2=5$
9. $9x-2y$, when $x=2$ and $y=3$, is $9\cdot 2-2\cdot 3=18-6=12$
10. $\frac{2x}{x-y}$, when $x=6$ and $y=4$, is $\frac{2\cdot 6}{6-4}=\frac{12}{2}=\frac{6\cdot 2}{1\cdot 2}=\frac{6}{1}\cdot\frac{2}{2}=6\cdot 1=6$
11. $y=x-5$, when $x=12$, is $y=12-5\Rightarrow y=7$
12. $y=xz+1$, when $x=2$ and $z=3$, is $y=2\cdot 3+1\Rightarrow y=6+1\Rightarrow y=7$
13. $y=4(x-z)$, when $x=7$ and $z=5$, is $y=4(7-5)\Rightarrow y=4(2)\Rightarrow y=8$
14. $y=\frac{x+z}{4}$, when $x=14$ and $z=10$, is $y=\frac{14+10}{4}\Rightarrow y=\frac{24}{4}\Rightarrow y=6$
15. Let c be the cost of the CD. Then, five times the cost of the CD is $5\cdot c=5c$.
16. Let x be the number. Then, five less than the number is $x-5$.
17. Three squared increased by five is 3^2+5.
18. Two cubed divided by the quantity three plus one is $2^3\div(3+1)$.
19. Let x be the number. Then, the product of three and the number is $3x$.
20. Let x be the number. Then, the difference between the number and four is $x-4$.

Section 1.2

21. (a) $\frac{5\cdot 7}{8\cdot 7}=\frac{5}{8}\cdot\frac{7}{7}=\frac{5}{8}\cdot 1=\frac{5}{8}$

(b) $\frac{3a}{4a}=\frac{3}{4}\cdot\frac{a}{a}=\frac{3}{4}\cdot 1=\frac{3}{4}$

22. (a) $\frac{9}{12}=\frac{3\cdot 3}{4\cdot 3}=\frac{3}{4}\cdot\frac{3}{3}=\frac{3}{4}\cdot 1=\frac{3}{4}$

(b) $\frac{36}{60}=\frac{3\cdot 12}{5\cdot 12}=\frac{3}{5}\cdot\frac{12}{12}=\frac{3}{5}\cdot 1=\frac{3}{5}$

23. $\frac{3}{4}\cdot\frac{5}{6}=\frac{3\cdot 5}{4\cdot 6}=\frac{15}{24}=\frac{5\cdot 3}{8\cdot 3}=\frac{5}{8}\cdot\frac{3}{3}=\frac{5}{8}\cdot 1=\frac{5}{8}$

24. $\frac{1}{2}\cdot\frac{4}{9}=\frac{1\cdot 4}{2\cdot 9}=\frac{4}{18}=\frac{2\cdot 2}{9\cdot 2}=\frac{2}{9}\cdot\frac{2}{2}=\frac{2}{9}\cdot 1=\frac{2}{9}$

25. $\frac{2}{3}\cdot\frac{5}{11}\cdot\frac{9}{10}=\frac{2\cdot 5\cdot 9}{3\cdot 11\cdot 10}=\frac{90}{330}=\frac{3\cdot 30}{11\cdot 30}=\frac{3}{11}\cdot\frac{30}{30}=\frac{3}{11}\cdot 1=\frac{3}{11}$

26. $\frac{12}{11}\cdot\frac{22}{23}\cdot\frac{1}{2}=\frac{12\cdot 22\cdot 1}{11\cdot 23\cdot 2}=\frac{264}{506}=\frac{12\cdot 22}{23\cdot 22}=\frac{12}{23}\cdot\frac{22}{22}=\frac{12}{23}\cdot 1=\frac{12}{23}$

27. $\frac{x}{3}\cdot\frac{6}{x}=\frac{x\cdot 6}{3\cdot x}=\frac{6}{3}\cdot\frac{x}{x}=2\cdot 1=2$

28. $\frac{2}{3}\cdot\frac{9x}{4y}=\frac{2\cdot 9x}{3\cdot 4y}=\frac{18x}{12y}=\frac{3x\cdot 6}{2y\cdot 6}=\frac{3x}{2y}\cdot 1=\frac{3x}{2y}$

29. One-fifth of three-sevenths equals $\frac{1}{5}\cdot\frac{3}{7}=\frac{1\cdot 3}{5\cdot 7}=\frac{3}{35}$.

30. (a) The reciprocal of 8 equals the reciprocal of $\frac{8}{1}=\frac{1}{8}$.

(b) The reciprocal of 1 equals the reciprocal of $\frac{1}{1}=\frac{1}{1}$.

(c) The reciprocal of $\frac{5}{19}=\frac{19}{5}$.

(d) The reciprocal of $\frac{3}{2}=\frac{2}{3}$.

31. $\frac{3}{2}\div\frac{1}{6}=\frac{3}{2}\cdot\frac{6}{1}=\frac{3\cdot 6}{2\cdot 1}=\frac{18}{2}=\frac{9\cdot 2}{1\cdot 2}=\frac{9}{1}\cdot\frac{2}{2}=9\cdot 1=9$

32. $\frac{9}{10}\div\frac{7}{5}=\frac{9}{10}\cdot\frac{5}{7}=\frac{9\cdot 5}{10\cdot 7}=\frac{45}{70}=\frac{9\cdot 5}{14\cdot 5}=\frac{9}{14}\cdot\frac{5}{5}=\frac{9}{14}\cdot 1=\frac{9}{14}$

33. $8\div\frac{2}{3}=\frac{8}{1}\cdot\frac{3}{2}=\frac{8\cdot 3}{1\cdot 2}=\frac{24}{2}=\frac{12\cdot 2}{1\cdot 2}=\frac{12}{1}\cdot\frac{2}{2}=12\cdot 1=12$

34. $\frac{3}{4}\div 6=\frac{3}{4}\cdot\frac{1}{6}=\frac{3\cdot 1}{4\cdot 6}=\frac{3}{24}=\frac{1\cdot 3}{8\cdot 3}=\frac{1}{8}\cdot\frac{3}{3}=\frac{1}{8}\cdot 1=\frac{1}{8}$

35. The least common denominator for the fractions $\frac{1}{8}$ and $\frac{5}{12}$ is 24, because 24 is the smallest number that both 8 and 12 divide into evenly.

36. The least common denominator for the fractions $\frac{3}{14}$ and $\frac{1}{21}$ is 42, because 42 is the smallest number that both 14 and 21 divide into evenly.

37. $\frac{2}{15}+\frac{3}{15}=\frac{2+3}{15}=\frac{5}{15}=\frac{1\cdot 5}{3\cdot 5}=\frac{1}{3}$

38. $\frac{5}{4}-\frac{3}{4}=\frac{5-3}{4}=\frac{2}{4}=\frac{1\cdot 2}{2\cdot 2}=\frac{1}{2}$

39. $\frac{11}{12}-\frac{1}{8}=\frac{11\cdot 2}{12\cdot 2}-\frac{1\cdot 3}{8\cdot 3}=\frac{22}{24}-\frac{3}{24}=\frac{22-3}{24}=\frac{19}{24}$

40. $\frac{6}{11}-\frac{3}{22}=\frac{6\cdot 2}{11\cdot 2}-\frac{3}{22}=\frac{12}{22}-\frac{3}{22}=\frac{12-3}{22}=\frac{9}{22}$

41. $\frac{2}{3}-\frac{1}{2}+\frac{1}{4}=\frac{2\cdot 4}{3\cdot 4}-\frac{1\cdot 6}{2\cdot 6}+\frac{1\cdot 3}{4\cdot 3}=\frac{8}{12}-\frac{6}{12}+\frac{3}{12}=\frac{8-6+3}{12}=\frac{5}{12}$

42. $\frac{1}{6}+\frac{2}{3}-\frac{1}{9}=\frac{1\cdot 3}{6\cdot 3}+\frac{2\cdot 6}{3\cdot 6}-\frac{1\cdot 2}{9\cdot 2}=\frac{3}{18}+\frac{12}{18}-\frac{2}{18}=\frac{3+12-2}{18}=\frac{13}{18}$

Section 1.3

43. $5\cdot 5\cdot 5\cdot 5\cdot 5\cdot 5=5^6$

44. $\frac{7}{6}\cdot\frac{7}{6}\cdot\frac{7}{6}=\left(\frac{7}{6}\right)^3$

45. $3\cdot 3\cdot 3\cdot 3=3^4$

46. $x\cdot x\cdot x\cdot x\cdot x=x^5$

47. $(x+1)\cdot(x+1)=(x+1)^2$

48. $(a-5)\cdot(a-5)\cdot(a-5)=(a-5)^3$

49. (a) $4^3=4\cdot 4\cdot 4=64$

(b) $7^2=7\cdot 7=49$

(c) $8^1=8$

50. $2^n=32$; $2\cdot 2\cdot 2\cdot 2\cdot 2=32$, thus $2^5=32$ and $n=5$

51. $7+3\cdot 6=7+18=25$

52. $15-5-3=(15-5)-3=10-3=7$

53. $24\div 4\div 2=(24\div 4)\div 2=6\div 2=3$

54. $30-15\div 3=30-(15\div 3)=30-5=25$

55. $18\div 6-2=(18\div 6)-2=3-2=1$

56. $\frac{18}{4+5}=\frac{18}{9}=\frac{2\cdot 9}{1\cdot 9}=\frac{2}{1}\cdot\frac{9}{9}=2\cdot 1=2$

57. $9-3^2=9-9=0$

58. $2^3-8=8-8=0$

59. $2^4-8+\frac{4}{2}=(2\cdot 2\cdot 2\cdot 2)-8+2=16-8+2=(16-8)+2=8+2=10$

60. $3^2-4(5-3)=3^2-4(2)=9-8=1$

61. $7-\frac{4+6}{2+3}=7-\frac{10}{5}=7-2=5$

62. $3^3-2^3=(3\cdot 3\cdot 3)-(2\cdot 2\cdot 2)=27-8=19$

Section 1.4

63. (a) The opposite of -8 is $-(-8)=8$.

(b) The opposite of $-(-(-3))$ is $-(-(-(-3)))=3$.

64. (a) The opposite of $-\left(\frac{-3}{7}\right)$ is $-\left(-\left(\frac{-3}{7}\right)\right)=-\frac{3}{7}$.

(b) The opposite of $\frac{-2}{-5}$ is $-\left(\frac{-2}{-5}\right)=-\frac{2}{5}$.

65. (a) $\frac{4}{5}=\frac{4}{5}\cdot\frac{2}{2}=\frac{8}{10}=0.8$

(b) $\frac{3}{20}=\frac{3}{20}\cdot\frac{5}{5}=\frac{15}{100}=0.15$

66. (a) $\frac{5}{9}=0.\overline{5}$

(b) $\frac{7}{11}=0.\overline{63}$

67. The number 0 is a whole number and a rational number, and is an integer.

68. The number $-\frac{5}{6}$ is a rational number.

69. The number -7 is a rational number and is an integer.

70. The number $\sqrt{17}$ is an irrational number.

71. The number π is an irrational number.

72. The number 3.4 is a rational number.

73. 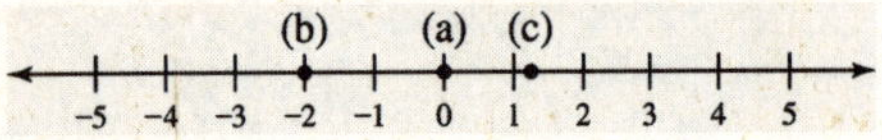

74. (a) $|-5| = 5$

(b) $|-\pi| = \pi$

(c) $|\sqrt{2}-1| = \sqrt{2}-1$

75. (a) $-5 < 4$

(b) $-\frac{1}{2} > -\frac{5}{2}$

(c) $|-9| = 9$ and $-3 < 9$ so $-3 < |-9|$.

(d) $|-8| = 8$ and $|-1| = 1$ and $8 > 1$ so $|-8| > |-1|$.

76. $-3, -\frac{2}{3}, \sqrt{3}, \pi - 1, 3$

Sections 1.5 and 1.6

77. 

$-5+9=4$

78. 

$4+(-7)=-3$

79. 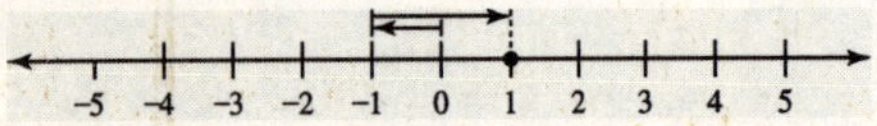

$-1+2=1$

80. 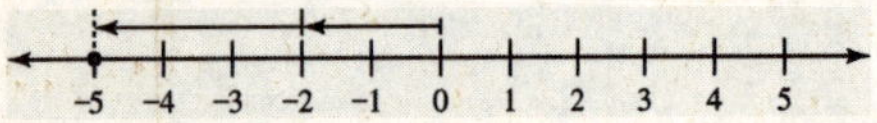

$-2+(-3)=-5$

81. $5+(-4)=1$

82. $-9-(-7)=-9+7=-2$

83. $11\cdot(-4)=-44$

84. $-8\cdot(-5)=40$

85. $11\div(-4)=-\frac{11}{4}$

86. $-4\div\frac{4}{7}=-\frac{4}{1}\cdot\frac{7}{4}=-\frac{4\cdot 7}{4}=-\frac{28}{4}=-\frac{7\cdot 4}{1\cdot 4}=-\frac{7}{1}\cdot\frac{4}{4}=-7\cdot 1=-7$

87. $-\frac{5}{9}-\left(-\frac{1}{3}\right)=-\frac{5}{9}+\frac{1}{3}=-\frac{5}{9}+\frac{1\cdot 3}{3\cdot 3}=-\frac{5}{9}+\frac{3}{9}=-\frac{2}{9}$

88. $-\frac{1}{2}+\left(-\frac{3}{4}\right)=-\frac{1\cdot 2}{2\cdot 2}+\left(-\frac{3}{4}\right)=-\frac{2}{4}+\left(-\frac{3}{4}\right)=-\frac{5}{4}$

89. $-\frac{1}{3}\cdot\left(-\frac{6}{7}\right)=\frac{1\cdot 6}{3\cdot 7}=\frac{6}{21}=\frac{2\cdot 3}{7\cdot 3}=\frac{2}{7}\cdot\frac{3}{3}=\frac{2}{7}\cdot 1=\frac{2}{7}$

90. $\frac{\frac{4}{5}}{-7}=\frac{4}{5}\div(-7)=\frac{4}{5}\cdot\left(-\frac{1}{7}\right)=-\frac{4}{35}$

91. $-\frac{3}{2}\div\left(-\frac{3}{8}\right)=-\frac{3}{2}\cdot\left(-\frac{8}{3}\right)=\frac{24}{6}=4$

92. $\frac{3}{8}\div(-0.5)=\frac{3}{8}\div\left(-\frac{1}{2}\right)=\frac{3}{8}\cdot\left(-\frac{2}{1}\right)=-\frac{6}{8}=-\frac{3\cdot 2}{4\cdot 2}=-\frac{3}{4}\cdot\frac{2}{2}=-\frac{3}{4}\cdot 1=-\frac{3}{4}$

93. $3+(-5)=-2$

94. $2-(-4)=2+4=6$

95. $\frac{7}{9}=0.\overline{7}$

96. $2\frac{1}{5}=2.2$

97. $0.6=\frac{6}{10}=\frac{3\cdot 2}{5\cdot 2}=\frac{3}{5}\cdot\frac{2}{2}=\frac{3}{5}\cdot 1=\frac{3}{5}$

98. $0.375=\frac{375}{1000}=\frac{3\cdot 125}{8\cdot 125}=\frac{3}{8}\cdot\frac{125}{125}=\frac{3}{8}\cdot 1=\frac{3}{8}$

Section 1.7

99. Commutative (multiplication)

100. Associative (addition)

101. Distributive

102. Commutative (addition)

103. Identity (multiplication)

104. Associative (multiplication)

105. Distributive

106. Identity (addition)

107. Inverse (addition)

108. Inverse (multiplication)

109. $7+9+12+8+1+3=(7+3)+(9+1)+(12+8)=10+10+20=40$

110. $500-199=500-200+1=300+1=301$

111. $25 \cdot 99 = 25(100-1) = 25(100) - 25(1) = 2500 - 25 = 2475$

112. $4581 + 2000 - 1 = 6581 - 1 = 6580$

113. $54.98 \times 10 = 549.8$ because we move the decimal point to the right one place.

114. $4356 \div 100 = 43.56$ because we move the decimal point to the left two places.

Section 1.8

115. $55x$ is a term; its coefficient is 55.

116. $-xy$ is a term; its coefficient is -1.

117. $9xy + 2z$ is not a term because it is the sum of two terms.

118. $x-7$ is not a term because it is the difference of two terms.

119. $-10x + 4x = (-10+4)x = -6x$

120. $19z - 4z = (19-4)z = 15z$

121. $3x^2 + x^2 = (3+1)x^2 = 4x^2$

122. $7 + 2x - 6 + x = 7 - 6 + 2x + x = 7 - 6 + (2+1)x = 1 + 3x = 3x + 1$

123. $-\frac{1}{2} + \frac{3}{2}z - z + \frac{5}{2} = -\frac{1}{2} + \frac{5}{2} + \frac{3}{2}z - z = \frac{-1+5}{2} + \left(\frac{3}{2} - 1\right)z = \frac{4}{2} + \left(\frac{3}{2} - \frac{2}{2}\right)z = 2 + \frac{1}{2}z = \frac{1}{2}z + 2$

124. $5(x-3) - (4x+3) = 5x - 15 - 4x - 3 = 5x - 4x - 15 - 3 = (5-4)x - 15 - 3 = x - 18$

125. $4x^2 - 3 + 5x^2 - 3 = 4x^2 + 5x^2 - 3 - 3$

$= (4+5)x^2 - (3+3) = 9x^2 - 6$

126. $3x^2 + 4x^2 - 7x^2 = (3+4-7)x^2 = 0x^2 = 0$

127. $\frac{35a}{7a} = \frac{35}{7} \cdot \frac{a}{a} = \frac{35}{7} \cdot 1 = 5 \cdot 1 = 5$

128. $\frac{0.5c}{0.5} = \frac{0.5}{0.5} \cdot c = 1 \cdot c = c$

129. $\frac{15y+10}{5} = \frac{15y}{5} + \frac{10}{5} = \frac{15}{5} \cdot \frac{y}{1} + \frac{10}{5} = 3y + 2$

130. $\frac{24x-60}{12} = \frac{24x}{12} - \frac{60}{12} = \frac{24}{12} \cdot \frac{x}{1} - \frac{60}{12} = 2x - 5$

Applications

131. (a) Let x be the number of minutes. The first person then paints $3 \cdot x$ or $3x$, square feet in x minutes. The second person paints $4 \cdot x$ or $4x$, square feet in x minutes. Working together the two people paint $3x + 4x = (3+4)x = 7x$ square feet in x minutes.

(b) First convert hours to minutes; 1 hour = 60 minutes. Replace x with 60 to get $7x = 7 \cdot 60 = 420\ \text{ft}^2$.

(c) First find the number of square feet the wall contains. The area of the wall is width times height. The area $= 8 \cdot 21 = 168$ ft^2. Because the two people paint 7 square feet in 1 minute, $168 \div 7 = 24$ minutes.

132. The area of a triangle is $\frac{1}{2} \cdot$ base $\cdot$ height. The base of the triangle is 8 and the height is 4. Thus,

$\frac{1}{2} \cdot 8 \cdot 4 = 16$ ft^2.

133. See Figure 133. Since there are 8 pints in 1 gallon, $P = 8G$, where P = pints and G = gallons.

Gallons (G)	1	2	3	4	5	6
Pints (P)	8	16	24	32	40	48

Figure 133

134. Because each CD costs 25 cents, $C = 0.25x$, where C represents cost and x represents the number of CD's purchased.

135. Find the fraction of the population that will be over the age of 65 but under the age of 85. Thus,

$\frac{1}{5} - \frac{1}{20} = \frac{1 \cdot 4}{5 \cdot 4} - \frac{1}{20} = \frac{4}{20} - \frac{1}{20} = \frac{3}{20}$.

136. The investment doubles every $72 \div 9 = 8$ years, so in 24 years the investment will double 3 times. The investment will increase by $2^3 = 8$ times \$25,000 and will equal \$200,000.

137. Find the length of each piece when the board is cut into five equal lengths. Because the board measures

$5\frac{3}{4} = \frac{23}{4}$, $\frac{23}{4} \div 5 = \frac{23}{4} \cdot \frac{1}{5} = \frac{23}{20} = 1\frac{3}{20}$ feet per piece.

138. Add the individual lengths to find the total distance. Thus, $3\frac{1}{8} + 4\frac{3}{8} + 6\frac{1}{4} + 1\frac{5}{8} = \frac{25}{8} + \frac{35}{8} + \frac{25}{4} + \frac{13}{8} =$

$\frac{25}{8} + \frac{35}{8} + \frac{50}{8} + \frac{13}{8} = \frac{25 + 35 + 50 + 13}{8} = \frac{123}{8} = 15\frac{3}{8}$ miles.

139. Subtract the withdrawals and add the deposits to the initial balance.

Thus, $1652 - 78 - 91 + 256 - 638 = \$1101$.

140. The difference between the two temperatures is $108 - (-16) = 108 + 16 = 124$°F.

141. Multiply the real numbers 202 and $\frac{16}{25}$ to obtain $202 \cdot \frac{16}{25} = 129.28$. Total admissions for *Titanic* were about 129 million people.

Chapter 1 Test

1. (a) The number 29 is a prime number because its only factors are itself and 1.

(b) The number 56 is a composite number because it has factors other than itself and 1; $56 = 2 \times 2 \times 2 \times 7$.

2. $\frac{5x}{2x - 1}$, when $x = -3$, is $\frac{5 \cdot (-3)}{2 \cdot (-3) - 1} = \frac{-15}{-6 - 1} = \frac{-15}{-7} = \frac{15}{7}$

3. $4^2 - 3 = 16 - 3 = 13$

4. $\dfrac{24}{32}=\dfrac{3\cdot 8}{4\cdot 8}=\dfrac{3}{4}$

5. (a) $\dfrac{5}{8}+\dfrac{1}{8}=\dfrac{5+1}{8}=\dfrac{6}{8}=\dfrac{3\cdot 2}{4\cdot 2}=\dfrac{3}{4}$

(b) $\dfrac{5}{9}-\dfrac{3}{15}=\dfrac{5\cdot 5}{9\cdot 5}-\dfrac{3\cdot 3}{15\cdot 3}=\dfrac{25}{45}-\dfrac{9}{45}=\dfrac{25-9}{45}=\dfrac{16}{45}$

(c) $\dfrac{3}{5}\cdot\dfrac{10}{21}=\dfrac{3\cdot 10}{5\cdot 21}=\dfrac{30}{105}=\dfrac{2\cdot 15}{7\cdot 15}=\dfrac{2}{7}$

(d) $6\div\dfrac{8}{5}=\dfrac{6}{1}\cdot\dfrac{5}{8}=\dfrac{6\cdot 5}{1\cdot 8}=\dfrac{30}{8}=\dfrac{15\cdot 2}{4\cdot 2}=\dfrac{15}{4}$

(e) $\dfrac{5}{12}+\dfrac{4}{9}=\dfrac{5\cdot 3}{12\cdot 3}+\dfrac{4\cdot 4}{9\cdot 4}=\dfrac{15}{36}+\dfrac{16}{36}=\dfrac{15+16}{36}=\dfrac{31}{36}$

(f) $\dfrac{10}{13}\div 5=\dfrac{10}{13}\cdot\dfrac{1}{5}=\dfrac{10\cdot 1}{13\cdot 5}=\dfrac{10}{65}=\dfrac{2\cdot 5}{13\cdot 5}=\dfrac{2}{13}$

6. $y\cdot y\cdot y\cdot y=y^4$

7. (a) $6+10\div 5=6+2=8$

(b) $4^3-(3-5\cdot 2)=64-(3-10)=64-(-7)=64+7=71$

(c) $-6^2-6+\dfrac{4}{2}=-36-6+2=-40$

(d) $11-\dfrac{1+3}{6-4}=11-\dfrac{4}{2}=11-2=9$

8. (a) -1 is an integer and a rational number.

(b) $\sqrt{5}$ is an irrational number.

9.

(a) (b) (c)

−5 −4 −3 −2 −1 0 1 2 3 4 5

10. (a) $2<|-5|$ because $|-5|=5$ and $2<5$.

(b) $|-1|>|0|$ because $|-1|=1, |0|=0,$ and $1>0$.

11. (a) $-5\div\dfrac{5}{6}=\dfrac{-5}{1}\cdot\dfrac{6}{5}=\dfrac{-5\cdot 6}{1\cdot 5}=\dfrac{-6}{1}=-6$

(b) $-7\cdot(-3)=21$

12. (a) Distributive

(b) Associative (multiplication)

(c) Commutative (addition)

13. $17\cdot 102=17(100+2)=17(100)+17(2)$

$=1700+34=1734$

14. (a) $5-5z+7+z=5+7-5z+z=12+(-5+1)z=12+(-4)z=12-4z$

(b) $12x-(6-3x)=12x-6+3x=12x+3x-6=(12+3)x-6=15x-6$

(c) $5-4(x+6)+\frac{15x}{3}=5-4x-24+\frac{15}{3}x=5-4x-24+5x=5x-4x+5-24=$

$(5-4)x-19=1x-19=x-19$

15. (a) Let x be the number of hours. Then, the first person can mow $\frac{4}{3}\cdot x$ acres in x hours, and the second person can mow $\frac{1}{4}\cdot x$ acres in x hours. Thus, $\frac{4}{3}x+\frac{1}{4}x=\frac{16}{12}x+\frac{3}{12}x=\left(\frac{16}{12}+\frac{3}{12}\right)=\frac{19}{12}x.$

(b) Let $x=8$. Then, $\frac{19}{12}\cdot 8=\frac{19\cdot 8}{12}=\frac{152}{12}=\frac{38\cdot 4}{3\cdot 4}=\frac{38}{3}=12\frac{2}{3}$ acres.

16. Find $7\frac{4}{5}\div 3$ to find the length of three equal parts.

Because $7\frac{4}{5}=\frac{39}{5}, \frac{39}{5}\div 3=\frac{39}{5}\cdot\frac{1}{3}=\frac{39}{15}=\frac{13\cdot 3}{5\cdot 3}=\frac{13}{5}=2\frac{3}{5}$ feet.

17. (a) Because $39\div 3=13$, each ticket costs $13.

Then, let $C=$ cost and x be the number of tickets to obtain $C=13x$.

(b) Let $x=17$. Thus, $13\cdot 17=\$221$.

18. Subtract the withdrawals from and add the deposits to the initial balance.

Thus, $892-57+150-345=\$640$.

Chapter 1 Extended and Discovery Exercises

1. $2+2-2-2=4-4=0$; $3\times 3+3\div 3=9+1=10$; $4\div 4+4-4=1+0=1$;

$6\times 6+6-6=36+0=36$; $7\times 7+7+7=49+14=63$ *Answers may vary.*

2.

16	2	3	13
5	11	10	8
9	7	6	12
4	14	15	1

Critical Thinking Solutions for Chapter 1

Section 1.1

- Because you can not find a fraction of a population, this is an example of natural or whole number use. *Answers may vary*
- Because a prime number has as its only factors itself and 1, the tree diagram will consist of a single dot.

Section 1.2

- Fractions are needed when baking a cake from scratch, because the measurements of cake ingredients are in fractions. *Answers may vary*

Section 1.3

- A gigabyte is $2^{30} = 1,073,741,824$ bytes, which is more than 1 billion bytes.

Section 1.4

- $-\pi + \pi = 0$

Section 1.5

- Change the problem to an addition problem. Then, if you are adding a negative number, you move left on the number line, and if adding a positive number you move right on the number line. *Answers may vary.*

Section 1.7

- Subtract 200 from 5283 to obtain 5083 and then add 2. The result is 5085.

Section 1.8

- Draw rectangles x by y and $2x$ by y. Their combined areas equal $3xy$.

Chapter 2: Linear Equations and Inequalities

2.1: Introduction to Equations

Concepts

1. true
2. false
3. solution
4. solution set
5. solutions
6. Equivalent
7. $b+c$
8. subtraction
9. bc
10. division
11. equivalent
12. given

The Addition Property of Equality

13. 22
15. 3
17. $x+5=0 \Rightarrow x+5-5=0-5 \Rightarrow x+0=-5 \Rightarrow x=-5$; To check your answer, substitute -5 for x in the original equation, $-5+5=0$. This statement is true. Thus, the solution $x=-5$ is correct.
19. $x-7=1 \Rightarrow x-7+7=1+7 \Rightarrow x+0=8 \Rightarrow x=8$
21. $a+41=7 \Rightarrow a+41-41=7-41 \Rightarrow a+0=7+(-41) \Rightarrow a=-34$
23. $a-12=-3 \Rightarrow a-12+12=-3+12 \Rightarrow a+0=9 \Rightarrow a=9$
25. $9=y-8 \Rightarrow 9+8=y-8+8 \Rightarrow 17=y+0 \Rightarrow 17=y \Rightarrow y=17$
27. $\frac{1}{2}=z-\frac{3}{2} \Rightarrow \frac{1}{2}+\frac{3}{2}=z-\frac{3}{2}+\frac{3}{2} \Rightarrow \frac{4}{2}=z+0 \Rightarrow 2=z \Rightarrow z=2$
29. $t-0.8=4.2 \Rightarrow t-0.8+0.8=4.2+0.8 \Rightarrow t+0=5 \Rightarrow t=5$
31. $25+x=10 \Rightarrow 25-25+x=10-25 \Rightarrow 0+x=-15 \Rightarrow x=-15$
33. $1989=26+y \Rightarrow 1989+(-26)=26+(-26)+y \Rightarrow 1963=0+y \Rightarrow 1963=y \Rightarrow y=1963$
35. a

The Multiplication Property of Equality

37. $\frac{1}{5}$
39. 6

41. $5x = 15 \Rightarrow \frac{5x}{5} = \frac{15}{5} \Rightarrow x = 3$

43. $-7x = 0 \Rightarrow \frac{-7x}{-7} = \frac{0}{-7} \Rightarrow x = 0$

45. $-5a = -35 \Rightarrow \frac{-5a}{-5} = \frac{-35}{-5} \Rightarrow a = 7$

47. $3a = -18 \Rightarrow \frac{3a}{3} = \frac{-18}{3} \Rightarrow a = -6$

49. $\frac{1}{2}x = \frac{3}{2} \Rightarrow \frac{2}{1} \cdot \frac{1}{2}x = \frac{3}{2} \cdot \frac{2}{1} \Rightarrow x = 3$

51. $\frac{1}{2} = \frac{2}{5}z \Rightarrow \frac{5}{2} \cdot \frac{1}{2} = \frac{5}{2} \cdot \frac{2}{5}z \Rightarrow \frac{5}{4} = z \Rightarrow z = \frac{5}{4}$

53. $25 = 5z \Rightarrow \frac{25}{5} = \frac{5z}{5} \Rightarrow 5 = z \Rightarrow z = 5$

55. $0.5t = 3.5 \Rightarrow \frac{0.5t}{0.5} = \frac{3.5}{0.5} \Rightarrow t = 7$

57. $\frac{3}{8} = \frac{1}{4}y \Rightarrow \frac{3}{8} \cdot \frac{4}{1} = \frac{4}{1} \cdot \frac{1}{4}y \Rightarrow \frac{3}{2} = y \Rightarrow y = \frac{3}{2}$

59. *a*

Applications

61. (a) See Figure 61.

(b) Let R represent total rainfall and let x represent the number of hours past noon. Start with 3 inches of rain and then add $\frac{1}{2}$, or 0.5, inches per hour after noon, $3 + 0.5x = R$, or equivalently $R = 0.5x + 3$.

(c) At 3 pm, $x = 3$. Substituting x with 3 in the formula, $R = 0.5 \cdot 3 + 3 \Rightarrow R = 4.5$ inches. This answer agrees with the table from part (a).

(d) At 2:15 pm, $x = 2.25$. Substituting x with 2.25 in the formula, $R = 0.5 \cdot 2.25 + 3 \Rightarrow R = 4.125$ inches.

Hours (x)	0	1	2	3	4	5	6
Rainfall (R)	3	3.5	4	4.5	5	5.5	6

Figure 61

63. (a) Let L be the length of the football fields and x be the number of fields. Because each field x contains 300 feet, $L = 300x$.

(b) Substitute L with 870. Then $870 = 300x$.

(c) $870 = 300x \Rightarrow \frac{870}{300} = \frac{300x}{300} \Rightarrow x = \frac{870}{300} \Rightarrow x = 2.9$

65. Let x represent the total number of cubic miles that have melted and let y represent the number of years. Because the glaciers are melting at a rate of 24 cubic miles per year, $x = 24y$. Substitute x with 420.

Then, $420 = 24y \Rightarrow \frac{420}{24} = \frac{24y}{24} \Rightarrow y = \frac{420}{24} \Rightarrow y = 17.5$. Thus, it takes 17.5 years for 420 cubic miles of the glacier to melt.

67. Let x represent the cost of the car to obtain the equation $0.07x = 1750$. Then the solution is $\frac{0.07x}{0.07} = \frac{1750}{0.07} \Rightarrow x = \frac{1750}{0.07} \Rightarrow x = 25{,}000$. Thus, the cost of the car is \$25,000.

2.2: Linear Equations

Concepts

1. $ax + b = 0$
2. 3; 2
3. Exactly one
4. is not
5. addition, multiplication
6. distributive
7. LCD
8. 100
9. None
10. Infinitely many
11. Infinitely many
12. None

Identifying Linear Equations

13. $3x - 7 = 0$ is a linear equation. $a = 3$ and $b = -7$.

15. $\frac{1}{2}x = 0$ is a linear equation. $a = \frac{1}{2}$ and $b = 0$.

17. $4x^2 - 6 = 11$ is not a linear equation because it cannot be written in the form $ax + b = 0$. It has a non-zero term containing x^2.

19. $\frac{6}{x} - 4 = 2$ is not a linear equation because it cannot be written in the form $ax + b = 0$. It has the variable x in the denominator.

21. $1.1x = 0.9$ is a linear equation. $1.1x = 0.9 \Rightarrow 1.1x - 0.9 = 0.9 - 0.9 \Rightarrow 1.1x - 0.9 = 0$. $a = 1.1$ and $b = -0.9$.

23. $2(x - 3) = 0$ is a linear equation. Use the distributive property to obtain $2x - 6 = 0$. $a = 2$ and $b = -6$.

25. $|3x| + 2 = 1$ is not a linear equation because it cannot be written in the form $ax + b = 0$. It has a non-zero term containing $|x|$.

Solving Linear Equations

27. For $x=0$, substitute 0 for x and solve: $-3(0)+7=0+7=7$.

For $x=1$, substitute 1 for x and solve: $-3(1)+7=-3+7=4$.

For $x=2$, substitute 2 for x and solve: $-3(2)+7=-6+7=1$.

For $x=3$, substitute 3 for x and solve: $-3(3)+7=-9+7=-2$.

For $x=4$, substitute 4 for x and solve: $-3(4)+7=-12+7=-5$.

See Figure 27. From the table, we see that the equation $-3x+7=1$ is true when $x=2$. Therefore, the solution to the equation $-3x+7=1$ is $x=2$.

x	0	1	2	3	4
$-3x+7$	7	4	1	−2	−5

Figure 27

x	−2	−1	0	1	2
$4-2x$	8	6	4	2	0

Figure 29

29. For $x=-2$, substitute -2 for x and solve: $4-2(-2)=4+4=8$.

For $x=-1$, substitute -1 for x and solve: $4-2(-1)=4+2=6$.

For $x=0$, substitute 0 for x and solve: $4-2(0)=4-0=4$.

For $x=1$, substitute 1 for x and solve: $4-2(1)=4-2=2$.

For $x=2$, substitute 2 for x and solve: $4-2(2)=4-4=0$.

See Figure 29. From the table, we see that the equation $4-2x=6$ is true when $x=-1$. Therefore, the solution to the equation $4-2x=6$ is $x=-1$.

31. $11x=3 \Rightarrow \frac{11x}{11}=\frac{3}{11} \Rightarrow x=\frac{3}{11}$

33. $x-18=5 \Rightarrow x-18+18=5+18 \Rightarrow x=23$

35. $2x-1=13 \Rightarrow 2x-1+1=13+1 \Rightarrow 2x=14 \Rightarrow \frac{2x}{2}=\frac{14}{2} \Rightarrow x=7$

37. $5x+5=-6 \Rightarrow 5x+5-5=-6-5 \Rightarrow 5x=-11 \Rightarrow \frac{5x}{5}=\frac{-11}{5} \Rightarrow x=-\frac{11}{5}$

39. $3z+2=z-5 \Rightarrow 3z+2-2=z-5-2 \Rightarrow 3z=z-7 \Rightarrow 3z-z=z-z-7 \Rightarrow 2z=-7 \Rightarrow \frac{2z}{2}=\frac{-7}{2} \Rightarrow z=-\frac{7}{2}$

41. $12y-6=33-y \Rightarrow 12y-6+6=33+6-y \Rightarrow 12y=39-y \Rightarrow 12y+y=39-y+y \Rightarrow$

$13y=39 \Rightarrow \frac{13y}{13}=\frac{39}{13} \Rightarrow y=3$

43. $4(x-1)=5 \Rightarrow 4x-4=5 \Rightarrow 4x-4+4=5+4 \Rightarrow 4x=9 \Rightarrow \frac{4x}{4}=\frac{9}{4} \Rightarrow x=\frac{9}{4}$

45. $1-(3x+1)=5-x \Rightarrow 1-3x-1=5-x \Rightarrow -3x=5-x \Rightarrow -3x+x=5-x+x \Rightarrow$

$-2x=5 \Rightarrow \frac{-2x}{-2}=\frac{5}{-2} \Rightarrow x=-\frac{5}{2}$

47. $(5t-6)+2(t+1)=0 \Rightarrow 5t-6+2t+2=0 \Rightarrow 7t-4=0 \Rightarrow 7t-4+4=0+4 \Rightarrow 7t=4 \Rightarrow \frac{7t}{7}=\frac{4}{7} \Rightarrow t=\frac{4}{7}$

49. $3(4z-1)-2(z+2)=2(z+1) \Rightarrow 12z-3-2z-4=2z+2 \Rightarrow 10z-7=2z+2 \Rightarrow$

$10z-7-2=2z+2-2 \Rightarrow 10z-9=2z \Rightarrow 10z-9+9=2z+9 \Rightarrow 10z=2z+9 \Rightarrow$

$10z-2z=2z-2z+9 \Rightarrow 8z=9 \Rightarrow \frac{8z}{8}=\frac{9}{8} \Rightarrow z=\frac{9}{8}$

51. $7.3x-1.7=5.6 \Rightarrow 7.3x-1.7+1.7=5.6+1.7 \Rightarrow 7.3x=7.3 \Rightarrow \frac{7.3x}{7.3}=\frac{7.3}{7.3} \Rightarrow x=1$

53. $-9.5x-0.05=10.5x+1.05 \Rightarrow -9.5x-10.5x-0.05=10.5x-10.5x+1.05 \Rightarrow$

$-20x-0.05=1.05 \Rightarrow -20x-0.05+0.05=1.05+0.05 \Rightarrow -20x=1.1 \Rightarrow \frac{-20x}{-20}=\frac{1.1}{-20} \Rightarrow x=-0.055$

55. $\frac{1}{2}x-\frac{3}{2}=\frac{5}{2} \Rightarrow \frac{1}{2}x-\frac{3}{2}+\frac{3}{2}=\frac{5}{2}+\frac{3}{2} \Rightarrow \frac{1}{2}x=4 \Rightarrow 2\cdot\frac{1}{2}x=4\cdot 2 \Rightarrow x=8$

57. $-\frac{3}{8}x+\frac{1}{4}=\frac{1}{2}x+\frac{1}{8} \Rightarrow -\frac{3}{8}x-\frac{1}{2}x+\frac{1}{4}=\frac{1}{2}x-\frac{1}{2}x+\frac{1}{8} \Rightarrow -\frac{7}{8}x+\frac{1}{4}=\frac{1}{8} \Rightarrow$

$-\frac{7}{8}x+\frac{1}{4}-\frac{1}{4}=\frac{1}{8}-\frac{1}{4} \Rightarrow -\frac{7}{8}x=-\frac{1}{8} \Rightarrow \left(-\frac{8}{7}\right)\left(-\frac{7}{8}\right)x=\left(-\frac{1}{8}\right)\left(-\frac{8}{7}\right) \Rightarrow x=\frac{8}{56} \Rightarrow x=\frac{1}{7}$

59. $4y-2(y+1)=0 \Rightarrow 4y-2y-2=0 \Rightarrow 2y-2=0 \Rightarrow 2y-2+2=0+2 \Rightarrow 2y=2 \Rightarrow \frac{2y}{2}=\frac{2}{2} \Rightarrow y=1$

61. $ax+b=0 \Rightarrow ax+b-b=0-b$

$\Rightarrow ax=-b$

$\Rightarrow \frac{ax}{a}=\frac{-b}{a}$

$\Rightarrow x=-\frac{b}{a}$

63. $5x=5x+1 \Rightarrow 5x-5x=5x-5x+1 \Rightarrow 0=1$

Because the equation $0=1$ is always false, there are no solutions.

65. $8x=0 \Rightarrow \frac{8x}{8}=\frac{0}{8} \Rightarrow x=0$ Thus, there is only one solution.

67. $5(2x+7)-(10x+5)=30 \Rightarrow 10x+35-10x-5=30 \Rightarrow 30=30$

Since the equation $30=30$ is always true, there are infinitely many solutions.

69. $4x=5(x+3)-x \Rightarrow 4x=5x+15-x \Rightarrow 4x=4x+15 \Rightarrow 4x-4x=4x-4x+15 \Rightarrow 0=15$

Because the equation $0=15$ is always false, there are no solutions.

71. $2x-(x+5)=x-5 \Rightarrow 2x-x-5=x-5 \Rightarrow x-5=x-5 \Rightarrow x-x-5=x-x-5 \Rightarrow -5=-5$ Since the equation $-5=-5$ is always true, there are infinitely many solutions.

Applications

73. (a) See Figure 73.

(b) Let D represent the distance from home and x represent the number of hours. Then $D=4+8x$.

(c) Substitute 3 for x. Then, $D=4+8(3)=28$ miles. This agrees with the value found in the table.

(d) Using the formula $D=4+8x$, substitute 22 for D. Then, $22=4+8x$. Then, solving for x:

$22-4=4-4+8x \Rightarrow 18=8x \Rightarrow \frac{18}{8}=\frac{8x}{8} \Rightarrow \frac{9}{4}=x \Rightarrow x=2.25$ hours. Thus, the bicyclist is 22 miles from home after 2 hours and 15 minutes.

Hours (x)	0	1	2	3	4
Distance (D)	4	12	20	28	36

Figure 73

75. Using the formula, substitute 490 for I and solve for x. $490=116x-231{,}627$

$\Rightarrow 490+231{,}627=116x-231{,}627+231{,}627$

$\Rightarrow 232{,}117=116x \Rightarrow \frac{232{,}117}{116}=\frac{116x}{116} \Rightarrow x \approx 2001$

77. Using the formula, substitute 1.5 for N and solve for x.

$1.5=0.03x-58.62 \Rightarrow 1.5+58.62=0.03x-58.62+58.62$

$\Rightarrow 60.12=0.03x \Rightarrow \frac{60.12}{0.03}=\frac{0.03x}{0.03} \Rightarrow x=2004$

Checking Basic Concepts for Sections 2.1 & 2.2

1. (a) $4x^3-2=0$ is not linear because it cannot be written in the form $ax+b=0$. It has a non-zero term containing x^3.

(b) $2(x+1)=4$ is a linear equation. Use the distributive property to obtain $2x+2=4$, then

$2x+2-4=4-4$

$\Rightarrow 2x-2=0$. $a=2$ and $b=-2$.

2. For $x=3$, substitute 3 for x and solve: $4(3)-3=12-3=9$

For $x=3.5$, substitute 3.5 for x and solve: $4(3.5)-3=14-3=11$

For $x=4$, substitute 4 for x and solve: $4(4)-3=16-3=13$

For $x=4.5$, substitute 4.5 for x and solve: $4(4.5)-3=18-3=15$

For $x=5$, substitute 5 for x and solve: $4(5)-3=20-3=17$

See Figure 2. To solve $4x-3=13$, the table tells us that when $x=4$, $4x-3=13$.

x	3	3.5	4	4.5	5
$4x-3$	9	11	13	15	17

Figure 2

3. (a) $x-12=6 \Rightarrow x-12+12=6+12 \Rightarrow x=18$ To check the answer, substitute 18 for x in the original equation $x-12=6$. $18-12=6 \Rightarrow 6=6$. Since this is true $x=18$ is correct.

(b) $\frac{3}{4}z=\frac{1}{8} \Rightarrow \frac{4}{3}\cdot\frac{3}{4}z=\frac{1}{8}\cdot\frac{4}{3} \Rightarrow z=\frac{4}{24} \Rightarrow z=\frac{1}{6}$

(c) $0.6t+0.4=2 \Rightarrow 0.6t+0.4-0.4=2-0.4 \Rightarrow 0.6t=1.6 \Rightarrow \frac{0.6t}{0.6}=\frac{1.6}{0.6} \Rightarrow t=2.\overline{6}$

(d) $5-2(x-2)=3(4-x) \Rightarrow 5-2x+4=12-3x \Rightarrow 9-2x=12-3x \Rightarrow$

$9-2x+3x=12-3x+3x \Rightarrow 9+x=12 \Rightarrow 9-9+x=12-9 \Rightarrow x=3$

4. (a) $x-5=6x \Rightarrow x-x-5=6x-x \Rightarrow -5=5x \Rightarrow \frac{-5}{5}=\frac{5x}{5} \Rightarrow -1=x$. Thus, the equation has 1 solution.

(b) $-2(x-5)=10-2x \Rightarrow -2x+10=10-2x \Rightarrow -2x+2x+10=10-2x+2x \Rightarrow 10=10$

Since $10=10$ is always true, the equation has infinitely many solutions.

(c) $-(x-1)=-x-1 \Rightarrow -x+1=-x-1 \Rightarrow -x+x+1=-x+x-1 \Rightarrow 1=-1$

Since this is never true, the equation has no solutions.

5. (a) Let D represent distance from home and x represent hours driven. Note that the driver is initially 300 miles from home and that each hour driven the driver gets closer to home by 75 miles. Thus, the formula is $D=300-75x$.

(b) Since the distance from home, when the drive is home, is 0, use the formula and set D equal to 0. Thus, $0=300-75x$.

(c) $0=300-75x \Rightarrow 0+75x=300-75x+75x \Rightarrow 75x=300 \Rightarrow \frac{75x}{75}=\frac{300}{75} \Rightarrow x=4$ hours.

2.3: Introduction to Problem Solving

Concepts

1. Check your solution.
2. $n+1$ and $n+2$
3. $\frac{x}{100}$
4. 0.01
5. left
6. right
7. $\frac{P_2-P_1}{P_1}\times 100$

8. increase; decrease

9. *rt*

10. distance; time

Number Problems

11. Let x represent the number. $2+x=12 \Rightarrow 2-2+x=12-2 \Rightarrow x=10$

13. $\frac{x}{5}=x-24 \Rightarrow \frac{x}{5}\cdot 5=5(x-24) \Rightarrow x=5x-120 \Rightarrow x-5x=5x-5x-120 \Rightarrow -4x=-120 \Rightarrow$

$\frac{-4x}{-4}=\frac{-120}{-4} \Rightarrow x=30$

15. $\frac{x+5}{2}=7 \Rightarrow \frac{x+5}{2}\cdot 2=7\cdot 2 \Rightarrow x+5=14 \Rightarrow x+5-5=14-5 \Rightarrow x=9$

17. $\frac{x}{2}=17 \Rightarrow \frac{x}{2}\cdot 2=17\cdot 2 \Rightarrow x=34$

19. Let the smallest natural number be represented by x. $x+(x+1)+(x+2)=96 \Rightarrow 3x+3=96 \Rightarrow$

$3x+3-3=96-3 \Rightarrow 3x=93 \Rightarrow \frac{3x}{3}=\frac{93}{3} \Rightarrow x=31$ Thus, the numbers are 31, 32 and 33.

21. $3x=102 \Rightarrow \frac{3x}{3}=\frac{102}{3} \Rightarrow x=34$

23. $5x=2x+24 \Rightarrow 5x-2x=2x-2x+24 \Rightarrow 3x=24 \Rightarrow \frac{3x}{3}=\frac{24}{3} \Rightarrow x=8$

25. $\frac{6x}{7}=18 \Rightarrow \frac{6x}{7}\cdot 7=18\cdot 7 \Rightarrow 6x=126 \Rightarrow \frac{6x}{6}=\frac{126}{6} \Rightarrow x=21$

27. $4(x+5)=64 \Rightarrow 4x+20=64 \Rightarrow 4x+20-20=64-20 \Rightarrow 4x=44 \Rightarrow \frac{4x}{4}=\frac{44}{4} \Rightarrow x=11$

29. $x+10=2x+3 \Rightarrow x-x+10=2x-x+3 \Rightarrow 10=x+3$

$\Rightarrow 10-3=x+3-3 \Rightarrow 7=x \Rightarrow x=7$

31. $24=2x-2 \Rightarrow 24+2=2x-2+2 \Rightarrow 26=2x$

$\Rightarrow \frac{26}{2}=\frac{2x}{2} \Rightarrow 13=x \Rightarrow x=13$

33. $311=3x+23 \Rightarrow 311-23=3x+23-23 \Rightarrow 288=3x$

$\Rightarrow \frac{288}{3}=\frac{3x}{3} \Rightarrow 96=x \Rightarrow x=96$ billion kilowatt-hours

Percent Problems

35. $37\%=\frac{37}{100}$

$37\%=37\times 0.01=0.37$

37. $148\% = \frac{148}{100} = \frac{37 \cdot 4}{25 \cdot 4} = \frac{37}{25}$

$148\% = 148 \times 0.01 = 1.48$

39. $6.9\% = \frac{6.9}{100} = \frac{6.9}{100} \cdot \frac{10}{10} = \frac{69}{1000}$

$6.9\% = 6.9 \times 0.01 = 0.069$

41. $0.05\% = \frac{0.05}{100} = \frac{0.05}{100} \cdot \frac{100}{100} = \frac{5}{10{,}000} = \frac{1 \cdot 5}{2000 \cdot 5} = \frac{1}{2000}$

$0.05\% = 0.05 \times 0.01 = 0.0005$

43. $0.45 = 0.45 \times 100\% = 45\%$

45. $1.8 = 1.8 \times 100\% = 180\%$

47. $0.006 = 0.006 \times 100\% = 0.6\%$

49. $\frac{2}{5} = 0.4 = 0.4 \times 100\% = 40\%$

51. $\frac{3}{4} = 0.75 = 0.75 \times 100\% = 75\%$

53. $\frac{5}{6} = 0.83\overline{3} = 0.83\overline{3} \times 100\% = 83.\overline{3}\%$

55. Let P_2 represent voters in 2004 and let P_1 represent voters in 1980. Then, the percent change in the number of voters is $\frac{P_2 - P_1}{P_1} = \frac{122.3 - 86.5}{86.5} = \frac{35.8}{86.5} \approx 0.414$ or about 41.4%.

57. Calculate the value of 4% of 950, then add the value to 950. Thus, 4% of $950 = .04(950) = 38$. Then, $950 + 38 = \$988$ per credit.

59. Let x represent the number of returns in 2003. Then, $x + 0.0046x = 131 \Rightarrow 1.0046x = 131 \Rightarrow x \approx 130.4$ Thus, about 130.4 million returns were processed in 2003.

61. Let x represent the number of AIDS deaths in 1995. Then, $0.345x = 18{,}017 \Rightarrow \frac{0.345x}{0.345} = \frac{18{,}017}{0.345} \Rightarrow$ $x \approx 52{,}223$. Thus, there were about 52,223 AIDS deaths in 1995.

63. Let x represent the percent change. Then, $\frac{1.50 - 1.20}{1.20} = x \Rightarrow \frac{1.50 - 1.20}{1.20} \cdot 1.20 = x \cdot 1.20 \Rightarrow$

$1.50 - 1.20 = 1.20x \Rightarrow 0.3 = 1.20x \Rightarrow \frac{0.3}{1.20} = \frac{1.20x}{1.20} \Rightarrow 0.25 = x \Rightarrow x = 0.25$ Thus, the percent change from \$1.20 to \$1.50 is 25%.

To calculate the percent change from \$1.50 to \$1.20, let P_1 represent \$1.50 and let P_2 represent \$1.20 and let x represent percent change. Then, $\frac{P_2 - P_1}{P_1} = x \Rightarrow \frac{1.20 - 1.50}{1.50} = x \Rightarrow$

$\frac{1.20-1.50}{1.50} \cdot 1.50 = x \cdot 1.50 \Rightarrow 1.20 - 1.50 = 1.5x \Rightarrow -0.30 = 1.5x \Rightarrow \frac{-0.30}{1.5} = \frac{1.5x}{1.5} \Rightarrow$

$-0.2 = x \Rightarrow x = -0.2$. Thus, the percent change from \$1.50 to \$1.20 is −20%.

Distance Problems

65. $d = rt \Rightarrow d = 4 \cdot 2 \Rightarrow d = 8$ miles

67. $d = rt \Rightarrow 1000 = r \cdot 50 \Rightarrow \frac{1000}{50} = \frac{r \cdot 50}{50} \Rightarrow 20 = r \Rightarrow r = 20$ feet/second

69. $d = rt \Rightarrow 200 = 40t \Rightarrow \frac{200}{40} = \frac{40t}{40} \Rightarrow 5 = t \Rightarrow t = 5$ hours

71. Given that the distance traveled (d) is 255 and that the time spent traveling (t) is 4.25 hours, calculate the speed of the car (r). Then, $d = rt \Rightarrow 255 = r \cdot 4.25 \Rightarrow \frac{255}{4.25} = \frac{r \cdot 4.25}{4.25} \Rightarrow 60 = r \Rightarrow r = 60$ miles/hour.

73. Let the slower runner be standing still. Then, the faster runner will be traveling at $0 + 2$ mph. Then, this problem is equivalent to solving how long it takes the faster runner to travel $\frac{3}{4}$ of a mile. Using the $d = rt$ formula: $\frac{3}{4} = 2t \Rightarrow \frac{3}{4} \cdot \frac{1}{2} = \frac{2}{1} \cdot \frac{1}{2} t \Rightarrow \frac{3}{8} = t \Rightarrow t = \frac{3}{8}$. So, in $\frac{3}{8}$ hour the faster runner will be $\frac{3}{4}$ mile ahead of the slower runner.

75. Let t represent the amount of time spent running 5 mph. Since the total time spent running was 1.1 hours, let $1.1 - t$ represent the amount of time running at 8 mph. Using the $d = rt$ formula, the distance run will equal the sum of $5t$ and $8(1.1 - t)$. Thus, $7 = 5t + 8(1.1 - t) \Rightarrow 7 = 5t + 8.8 - 8t \Rightarrow 7 = 8.8 - 3t \Rightarrow$

$3t + 7 = 8.8 - 3t + 3t \Rightarrow 3t + 7 = 8.8 \Rightarrow 3t + 7 - 7 = 8.8 - 7 \Rightarrow 3t = 1.8 \Rightarrow t = \frac{1.8}{3} = 0.6$. Therefore, the athlete ran at 5 mph for 0.6 hour and ran at 8 mph for $(1.1 - 0.6) = 0.5$ hour.

77. Since the plane is already 300 miles west of Chicago, it will have to fly $2175 - 300 = 1875$ miles to be 2175 miles west of Chicago. The plane is traveling at 500 mph. Then,

$1875 = 500t \Rightarrow \frac{1875}{500} = \frac{500t}{500} \Rightarrow 3.75 = t \Rightarrow t = 3.75$. Therefore, it will take the plane 3.75 hours to be 2175 miles west of Chicago.

Other Types of Problems

79. Let x represent the amount of water that should be added. Note that there is no salt in pure water and that we will add the amount of pure water to the 3% salt solution to obtain a 1.2% solution. Therefore, set up the equation so that the amount of salt on both sides of the equation is equal. Thus,

$$x(0.00)+20(0.03)=(x+20)(0.012)\Rightarrow 0.00x+0.6=0.012x+0.24\Rightarrow$$

$$0.6-0.24=0.012x+0.24-0.24\Rightarrow 0.36=0.012x\Rightarrow \frac{0.36}{0.012}=\frac{0.012x}{0.012}\Rightarrow 30=x\Rightarrow x=30.$$

Therefore, 30 ounces of water should be added.

81. Let x represent the amount of the loan at 6% and let $x+1000$ represent the amount of the loan at 5%. The total interest for one year is \$215 and this is the sum of the interest paid on the two loans. Therefore:

$$0.06x+0.05(x+1000)=215\Rightarrow 0.06x+0.05x+50=215\Rightarrow 0.11x+50-50=215-50\Rightarrow$$

$0.11x=165\Rightarrow \frac{0.11x}{0.11}=\frac{165}{0.11}\Rightarrow x=1500.$ Therefore, the amount of the loan at 6% interest is \$1500 and the amount of the loan at 5% interest is $1500+1000=\$2500$.

83. Let x represent the amount of 70% antifreeze. Then, the 45% antifreeze mixture is the sum of the 70% mixture and the 30% mixture. Therefore: $0.7x+10(0.3)=(x+10)(0.45)\Rightarrow 0.7x+3=0.45x+4.5\Rightarrow$

$$0.7x+3-3=0.45x+4.5-3\Rightarrow 0.7x=0.45x+1.5\Rightarrow 0.7x-0.45x=0.45x-0.45x+1.5\Rightarrow$$

$0.25x=1.5\Rightarrow \frac{0.25x}{0.25}=\frac{1.5}{0.25}\Rightarrow x=6.$ Therefore, 6 gallons of 70% antifreeze should be mixed with 10 gallons of 30% antifreeze to obtain the 45% mixture.

85. Let x represent the amount of 2.5% cream. Then, the 1% cream is the sum of the 2.5% cream and the base cream (0%). Therefore $0.025x+15(0)=(x+15)(0.01)$

$$\Rightarrow 0.025x=0.01x+0.15\Rightarrow 0.025x-0.01x=0.01x-0.01x+0.15$$

$$\Rightarrow 0.015x=0.15\Rightarrow \frac{0.015x}{0.015}=\frac{0.15}{0.015}\Rightarrow x=10.$$

Therefore, 10 grams of 2.5% hydrocortisone cream should be mixed with the base to obtain the 1% cream.

2.4: Formulas

Concepts

1. formula
2. LW
3. $\frac{D}{G}$
4. $\frac{1}{2}bh$

5. $\frac{1}{360}$

6. 360

7. 180

8. LWH

9. $2LW + 2WH + 2LH$

10. $2\pi r$

11. πr^2

12. $\pi r^2 h$

13. $\frac{x}{5}$

14. $\frac{1}{2}(a+b)h$

Formulas from Geometry

15. $A = LW$. Thus, $A = 6 \cdot 3 = 18 \text{ ft}^2$.

17. $A = \frac{1}{2}bh$. Thus, $A = \frac{1}{2} \cdot 6 \cdot 3 = 9 \text{ in}^2$.

19. $A = \pi r^2$. Thus, $A = \pi 4^2 \approx 50.3 \text{ ft}^2$.

21. $A = \frac{1}{2}(a+b)h$. Thus, $A = \frac{1}{2}(5+6)2 = \frac{1}{2}(11)2 = 11 \text{ ft}^2$.

23. $A = LW$. Thus, $A = 13 \cdot 7 = 91 \text{ in}^2$.

25. $A = \frac{1}{2}bh$. Thus, $A = \frac{1}{2} \cdot 12 \cdot 6 = 36 \text{ in}^2$.

27. $C = 2\pi r$. Because the circle has a diameter of 8 inches, the radius is $= \frac{8}{2} = 4$ inches. Thus,

$C = 2\pi r = 2\pi 4 = 8\pi \approx 25.1$ inches.

29. The total area of the lot is the sum of the area of the square and the area of the triangle. The area of the square is $LW = 52 \cdot 52 = 2704 \text{ ft}^2$. The area of the triangle is $\frac{1}{2}bh = \frac{1}{2} \cdot 73 \cdot 52 = 1898 \text{ ft}^2$. Thus, the area of the lot is $1898 + 2704 = 4602 \text{ ft}^2$.

31. The sum of the angles of a triangle is 180°. Let the unknown angle be represented by x. Then,

$x + 75 + 40 = 180 \Rightarrow x + 75 + 40 - 75 - 40 = 180 - 75 - 40 \Rightarrow x = 65$. Thus, the third angle is 65°.

33. The sum of the angles of a triangle is 180°. Let the unknown angle be represented by x. Then,

$x + 23 + 76 = 180 \Rightarrow x + 23 + 76 - 23 - 76 = 180 - 23 - 76 \Rightarrow x = 81$. Thus, the third angle is 81°.

35. Because the sum of the angles of a triangle is 180, $x+2x+3x=180 \Rightarrow 6x=180 \Rightarrow \frac{6x}{6}=\frac{180}{6} \Rightarrow$ $x=30$. Thus, the value of x is 30°.

37. Let x represent the largest angle. Then, $x+\frac{1}{3}x+\frac{1}{3}x=180 \Rightarrow \frac{5}{3}x=180 \Rightarrow \frac{3}{5}\cdot\frac{5}{3}x=\frac{180}{1}\cdot\frac{3}{5} \Rightarrow$ $x=\frac{540}{5} \Rightarrow x=108$. Thus, the largest angle has a measure of 108° and the two small angles each have measure $\frac{1}{3}\cdot\frac{108}{1}=\frac{108}{3}=36°$.

39. $C=2\pi r$. Since the diameter of the circle is 12 inches, the radius is $\frac{12}{2}=6$ inches. Then, $C=2\pi 6=12\pi \approx 37.7$ inches. Then, $A=\pi r^2=\pi 6^2=36\pi \approx 113.1\ \text{in}^2$.

41. $C=2\pi r$. Then, set C equal to 2π and solve for r. $2\pi=2\pi r \Rightarrow \frac{2\pi}{2\pi}=\frac{2\pi r}{2\pi} \Rightarrow 1=r \Rightarrow r=1$ inch. Then, $A=\pi r^2$. Because $r=1$, substitute 1 for r and solve for A. $A=\pi(1)^2=\pi$. Thus, the area is equal to π, which is approximately equal to $3.14\ \text{in}^2$.

43. $V=LWH$. Thus, $V=22\cdot 12\cdot 10=2640\ \text{in}^3$. Surface area equals $2LW+2LH+2WH$. Thus, $S=2\cdot 22\cdot 12+2\cdot 22\cdot 10+2\cdot 12\cdot 10=528+440+240=1208\ \text{in}^2$.

45. Convert yards to feet. Then, $\frac{2}{3}$ yard $=\frac{2}{3}\cdot 3=2$ feet. Then, $V=LWH=2\cdot\frac{2}{3}\cdot\frac{3}{2}=2\ \text{ft}^3$. Surface area equals $2LW+2LH+2WH$. Thus, $S=2\cdot 2\cdot\frac{2}{3}+2\cdot 2\cdot\frac{3}{2}+2\cdot\frac{2}{3}\cdot\frac{3}{2}=\frac{8}{3}+6+2=8\frac{8}{3}=10\frac{2}{3}\ \text{ft}^2$.

47. $V=\pi r^2 h$. Thus, $V=\pi 2^2\cdot 5=\pi\cdot 4\cdot 5=20\pi\ \text{in}^3$.

49. Convert feet to inches to obtain $h=2$ feet $=2\cdot 12=24$ inches. Then, $V=\pi r^2 h$. Thus, $V=\pi 5^2\cdot 24=$ $\pi\cdot 25\cdot 24=600\pi\ \text{in}^3$.

51. The volume formula for a cylindrical container is given by $V=\pi r^2 h$. Because the diameter of the barrel is $1\frac{3}{4}=\frac{7}{4}$ feet, the radius is $\left(\frac{1}{2}\right)\left(\frac{7}{4}\right)=\frac{7}{8}$ feet. Thus, $V=\pi r^2 h=\pi\left(\frac{7}{8}\right)^2(3)=\pi\left(\frac{49}{64}\right)(3)=\pi\left(\frac{147}{64}\right)=$ $\frac{147}{64}\pi \approx 7.2\ \text{ft}^3$.

Solving for a Variable

53. The formula is given as $A=LW$. To solve for W, proceed as follows: $A=LW \Rightarrow \frac{A}{L}=\frac{LW}{L} \Rightarrow W=\frac{A}{L}$.

55. The formula is given as $V=\pi r^2 h$. To solve for h, proceed as follows: $V=\pi r^2 h \Rightarrow \frac{V}{\pi r^2}=\frac{\pi r^2 h}{\pi r^2} \Rightarrow h=\frac{V}{\pi r^2}$.

57. The formula is given as $\frac{1}{2}(a+b)h = A$. To solve for a, proceed as follows: $A = \frac{1}{2}(a+b)h \Rightarrow$

$2A = \frac{1}{2}(2)(a+b)h \Rightarrow 2A = (a+b)h \Rightarrow \frac{2A}{h} = \frac{(a+b)h}{h} \Rightarrow \frac{2A}{h} = a+b \Rightarrow \frac{2A}{h} - b = a+b-b \Rightarrow a = \frac{2A}{h} - b.$

59. The formula is given as $V = LWH$. To solve for W, proceed as follows: $V = LWH \Rightarrow \frac{V}{LH} = \frac{LWH}{LH} \Rightarrow$

$W = \frac{V}{LH}.$

61. $s = \frac{a+b+c}{2} \Rightarrow 2s = \frac{a+b+c}{2} \cdot 2 \Rightarrow 2s = a+b+c \Rightarrow 2s-a-c = a+b+c-a-c \Rightarrow$

$2s-a-c = b \Rightarrow b = 2s-a-c$

63. $\frac{a}{b} - \frac{c}{b} = 1 \Rightarrow b\left(\frac{a}{b} - \frac{c}{b}\right) = 1(b) \Rightarrow a-c = b \Rightarrow b = a-c$

65. $ab = cd + ad \Rightarrow ab - ad = cd + ad - ad \Rightarrow ab - ad = cd \Rightarrow a(b-d) = cd \Rightarrow \frac{a(b-d)}{(b-d)} = \frac{cd}{(b-d)} \Rightarrow a = \frac{cd}{b-d}$

67. Because the perimeter equals the lengths of the four sides, $P = 2W + 2L$. Thus, $P = 2W + 2L \Rightarrow$

$40 = 2(5) + 2L \Rightarrow 40 = 10 + 2L \Rightarrow 40 - 10 = 10 + 2L - 10 \Rightarrow 2L = 30 \Rightarrow \frac{2L}{2} = \frac{30}{2} \Rightarrow L = 15$ Thus, the length of the rectangle is 15 inches.

Other Formulas and Applications

69. The formula for GPA is given by $\frac{4a+3b+2c+d}{a+b+c+d+f}$.

$\frac{4(30)+3(45)+2(12)+1(4)}{30+45+12+4+4} = \frac{120+135+24+4}{95} = \frac{283}{95} \approx 2.98.$ Thus, the GPA is 2.98.

71. The formula for GPA is given by $\frac{4a+3b+2c+d}{a+b+c+d+f}$.

$\frac{4(0)+3(60)+2(80)+1(10)}{0+60+80+10+6} = \frac{0+180+160+10}{156} = \frac{350}{156} \approx 2.24.$ Thus, the GPA is 2.24.

73. To convert Celsius to Fahrenheit temperature, the formula given is $\frac{9}{5}C + 32 = F$.

$\frac{9}{5}(25) + 32 = F \Rightarrow \frac{225}{5} + \frac{160}{5} = F \Rightarrow \frac{385}{5} = F \Rightarrow F = 77°\text{F}.$

75. To convert Celsius to Fahrenheit temperature, the formula given is $\frac{9}{5}C + 32 = F$.

$\frac{9}{5}(-40) + 32 = F \Rightarrow \frac{-360}{5} + \frac{160}{5} = F \Rightarrow \frac{-200}{5} = F \Rightarrow F = -40°\text{F}.$

77. To convert Fahrenheit to Celsius temperature, the formula given is $C = \frac{5}{9}(F - 32)$.

$C = \frac{5}{9}(23 - 32) \Rightarrow C = \frac{5}{9}(-9) \Rightarrow C = \frac{-45}{9} \Rightarrow C = -5°\text{C}$.

79. To convert Fahrenheit to Celsius temperature, the formula given is $C = \frac{5}{9}(F - 32)$.

$C = \frac{5}{9}(-4 - 32) \Rightarrow C = \frac{5}{9}(-36) \Rightarrow C = \frac{-180}{9} \Rightarrow C = -20°\text{C}$.

81. The formula for calculating gas mileage is $M = \frac{D}{G}$, where D is the distance and G is the gasoline used.

Therefore, $M = \frac{88,043 - 87,625}{38} \Rightarrow M = \frac{418}{38} \Rightarrow M = 11$ mpg.

83. The formula given for calculating the delay between seeing lightning and hearing the thunder is $D = \frac{x}{5}$ where

D represents the distance from the lightning and x represents the delay. Therefore, $D = \frac{x}{5} \Rightarrow D = \frac{12}{5} \Rightarrow$

$D = 2\frac{2}{5} = 2.4$ miles.

Checking Basic Concepts for Sections 2.3 & 2.4

1. (a) $3x = 36 \Rightarrow \frac{3x}{3} = \frac{36}{3} \Rightarrow x = 12$

   (b) $35 - x = 43 \Rightarrow 35 - 35 - x = 43 - 35 \Rightarrow -x = 8 \Rightarrow -1(x) = 8(-1) \Rightarrow x = -8$

2. $x + (x+1) + (x+2) = -93 \Rightarrow 3x + 3 = -93 \Rightarrow 3x + 3 - 3 = -93 - 3 \Rightarrow 3x = -96 \Rightarrow$

   $\frac{3x}{3} = \frac{-96}{3} \Rightarrow x = -32$. The three consecutive integers are $-32, -31, -30$.

3. $9.5\% = 0.095$

4. $\frac{5}{4} = 1\frac{1}{4} = 1.25 = 125\%$

5. Convert 25% to the decimal 0.25 and let x represent the unknown rate. Therefore, $x - 0.25x = 2652$. Thus,

   $x - 0.25x = 2652 \Rightarrow 0.75x = 2652 \Rightarrow \frac{0.75x}{0.75} = \frac{2652}{0.75} \Rightarrow x = 3536$. Thus, the rate in 2000 was about 3536.

6. Use the formula $D = rt$, where D is distance, r is the speed and t is the time. Thus, $390 = 60t \Rightarrow$

   $\frac{390}{60} = \frac{60t}{60} \Rightarrow 6.5 = t$. Thus, the travel time is 6.5 hours.

7. Let x represent the amount of the loan at 7% and $x+2000$ represent the amount of the loan at 6%. Thus, $0.07x+0.06(x+2000)=510 \Rightarrow 0.07x+0.06x+120=510 \Rightarrow 0.13x+120-120=510-120 \Rightarrow$

$0.13x=390 \Rightarrow \frac{0.13x}{0.13}=\frac{390}{0.13} \Rightarrow x=3000.$ Thus, the loan at 7% was \$3000 and the loan at 6% was \$5000.

8. The formula for gas mileage is $M=\frac{D}{G}$, where D is the distance and G is the gasoline used. Therefore,

$28=\frac{504}{G} \Rightarrow 28G=\left(\frac{504}{G}\right)G \Rightarrow 28G=504 \Rightarrow \frac{28G}{28}=\frac{504}{28} \Rightarrow G=18$ gal.

9. The area of a triangle is given by $A=\frac{1}{2}bh.$ Thus, $A=\frac{1}{2}bh \Rightarrow 36=\frac{1}{2}(6)h \Rightarrow 36=3h \Rightarrow \frac{36}{3}=\frac{3h}{3} \Rightarrow$

$12=h.$ Thus, the height of the triangle is 12 inches.

10. The area of a circle is given by $A=\pi r^2$. Thus, $A=\pi r^2 \Rightarrow A=\pi(3)^2 \Rightarrow A=9\pi \approx 28.3\text{ ft}^2$.

The circumference of a circle is given by $C=2\pi r$. Thus, $C=2\pi r \Rightarrow C=2\pi 3 \Rightarrow C=6\pi \approx 18.8$ feet.

11. $x+2x+3x=180 \Rightarrow 6x=180 \Rightarrow \frac{6x}{6}=\frac{180}{6} \Rightarrow x=30.$ Thus, the value of x is 30°.

12. $A=\pi r^2+\pi rl \Rightarrow A-\pi r^2=\pi r^2-\pi r^2+\pi rl \Rightarrow A-\pi r^2=\pi rl \Rightarrow \frac{A-\pi r^2}{\pi r}=\frac{\pi rl}{\pi r} \Rightarrow l=\frac{A-\pi r^2}{\pi r}$

2.5: Linear Inequalities

Concepts

1. equals sign; $<$; $\le$; $>$; $\ge$
2. greater than; less than
3. solution
4. equivalent
5. one
6. infinitely many
7. number line
8. is not; $3(5)<10$ is not true.
9. $>$
10. $<$
11. $>$
12. They are not equivalent inequalities because, $-4x<8 \Rightarrow \frac{-4x}{-4}>\frac{8}{-4} \Rightarrow x>-2$, which is not equivalent to $x<-2$.

Solutions and Number Line Graphs

13. 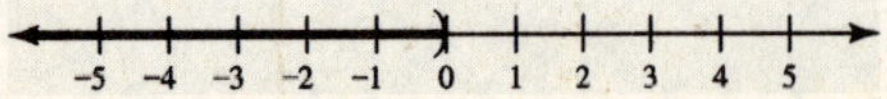

15. 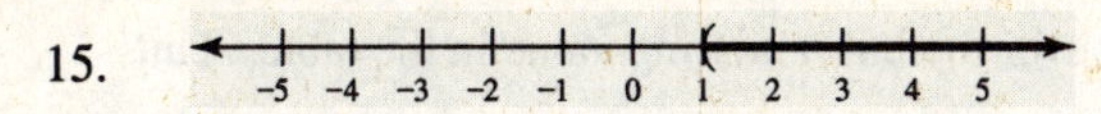

17.

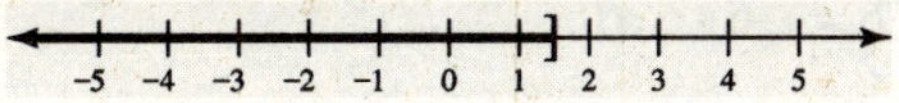

19.

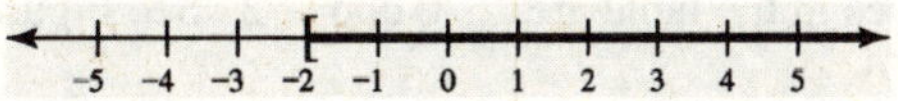

21. $x<0$

23. $x\leq 3$

25. $x\geq 10$

27. $[6,\infty)$

29. $(-2,\infty)$

31. $(-\infty,7]$

33. Substitute x with 4 in order to test the inequality. Thus, $x+5>5\Rightarrow 4+5>5\Rightarrow 9>5$. Because this inequality statement is true, $x=4$ is a solution.

35. First, isolate x on one side of the statement. Thus, $5x\geq 25\Rightarrow \frac{5x}{5}\geq \frac{25}{5}\Rightarrow x\geq 5$. Then substitute x with 5 in order to test the inequality. Thus, $x\geq 5\Rightarrow 5\geq 5$. Because this inequality statement is true, $x=5$ is a solution.

37. First, isolate y on one side of the statement. Thus, $4y-3\leq 5\Rightarrow 4y-3+3\leq 5+3\Rightarrow 4y\leq 8\Rightarrow$ $\frac{4y}{4}\leq \frac{8}{4}\Rightarrow y\leq 2$. Then substitute y with -3 in order to test the inequality. Thus, $y\leq 2\Rightarrow -3\leq 2$. Because this inequality statement is true, $y=-3$ is a solution.

39. First, isolate z on one side of the statement. Thus, $5(z+1)<3z-7\Rightarrow 5z+5<3z-7\Rightarrow$ $5z-3z+5<3z-3z-7\Rightarrow 2z+5<-7\Rightarrow 2z+5-5<-7-5\Rightarrow 2z<-12\Rightarrow \frac{2z}{2}<\frac{-12}{2}\Rightarrow\ z<-6$. Then substitute z with -7 in order to test the inequality. Thus, $z<-6\Rightarrow -7<-6$. Because this inequality statement is true, $z=-7$ is a solution.

41. First, isolate t on one side of the statement. Thus, $\frac{3}{2}t-\frac{1}{2}\geq 1-t\Rightarrow \frac{3}{2}t+t-\frac{1}{2}\geq 1-t+t\Rightarrow$ $\frac{5}{2}t-\frac{1}{2}\geq 1\Rightarrow \frac{5}{2}t-\frac{1}{2}+\frac{1}{2}\geq 1+\frac{1}{2}\Rightarrow \frac{5}{2}t\geq \frac{3}{2}\Rightarrow \frac{2}{5}\cdot\frac{5}{2}t\geq \frac{3}{2}\cdot\frac{2}{5}\Rightarrow t\geq \frac{3}{5}$. Then substitute t with -2 in order to test the inequality. Thus, $t\geq \frac{3}{5}\Rightarrow -2\geq \frac{3}{5}$. Because this inequality statement is not true, $t=-2$ is not a solution.

Tables and Linear Inequalities

43. $x>-2$

45. $x<1$

47. To complete the table, insert the x value into $-2x+6$ whenever there is a missing value in the table. Thus,

$-2x+6 \Rightarrow -2(2)+6 \Rightarrow -4+6=2;$ $\quad -2x+6 \Rightarrow -2(3)+6 \Rightarrow -6+6=0;$

$-2x+6 \Rightarrow -2(4)+6 \Rightarrow -8+6=-2.$ Thus, the missing values in the table are 2, 0 and -2. See Figure 47.

From the table, we see that $-2x+6 \le 0$ whenever $x \ge 3$. Thus, the solution to the inequality is $x \ge 3$.

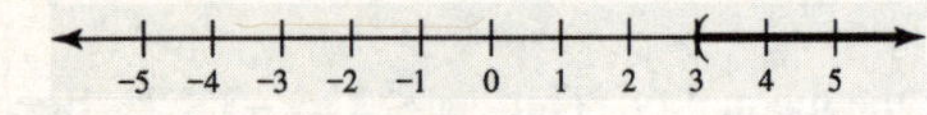

x	1	2	3	4	5
$-2x+6$	4	2	0	-2	-4

Figure 47

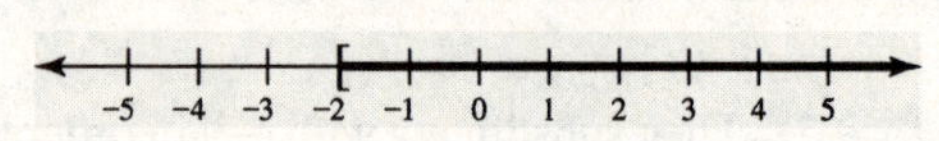

x	-3	-2	-1	0	1
$5-x$	8	7	6	5	4
$x+7$	4	5	6	7	8

Figure 49

49. To complete the table, insert the x value into $5-x$ and $x+7$ whenever there is a missing value in the table.

$5-x \Rightarrow 5-(-2)=7;\ 5-x \Rightarrow 5-(-1)=6;\ 5-x \Rightarrow 5-(0)=5.$

Thus, the missing values in the table that correspond to $5-x$ are 7, 6 and 5.

$x+7 \Rightarrow (-2)+7=5;\ x+7 \Rightarrow (-1)+7=6;\ x+7 \Rightarrow (0)+7=7.$

Thus, the missing values in the table that correspond to $x+7$ are 5, 6 and 7. See Figure 49.

From the table, we see that $5-x > x+7$ whenever $x<-1$. Thus, the solution to the inequality is $x<-1$.

Solving Linear Inequalities

51. $x-3>0 \Rightarrow x-3+3>0+3 \Rightarrow x>3.$ See Figure 51.

53. $3-y \le 5 \Rightarrow 3-3-y \le 5-3 \Rightarrow -y \le 2 \Rightarrow -1(-y) \ge 2(-1) \Rightarrow y \ge -2.$ See Figure 53.

Figure 51

Figure 53

55. $12<4+z \Rightarrow 12-4<4-4+z \Rightarrow 8<z \Rightarrow z>8.$ See Figure 55.

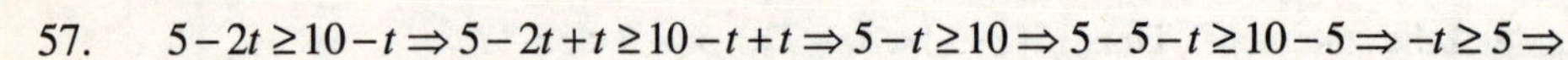

57. $5-2t \ge 10-t \Rightarrow 5-2t+t \ge 10-t+t \Rightarrow 5-t \ge 10 \Rightarrow 5-5-t \ge 10-5 \Rightarrow -t \ge 5 \Rightarrow$

$-1(-t) \le 5(-1) \Rightarrow t \le -5.$ See Figure 57.

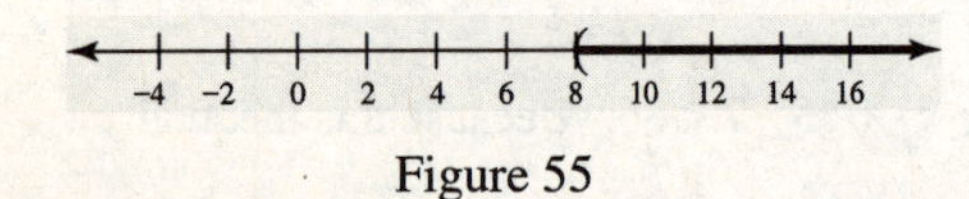

Figure 55

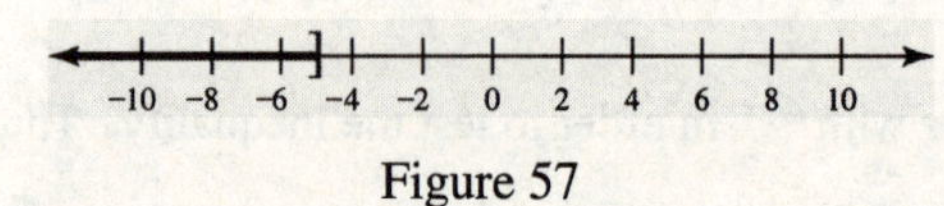

Figure 57

59. $2x<10 \Rightarrow \dfrac{2x}{2}<\dfrac{10}{2} \Rightarrow x<5.$ See Figure 59.

61. $-\dfrac{1}{2}t \ge 1 \Rightarrow \dfrac{-\frac{1}{2}t}{-\frac{1}{2}} \le \dfrac{1}{-\frac{1}{2}} \Rightarrow t \le -2.$ See Figure 61.

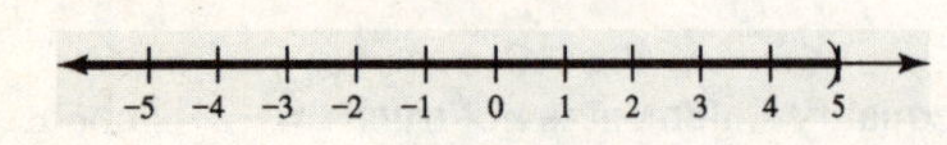

Figure 59

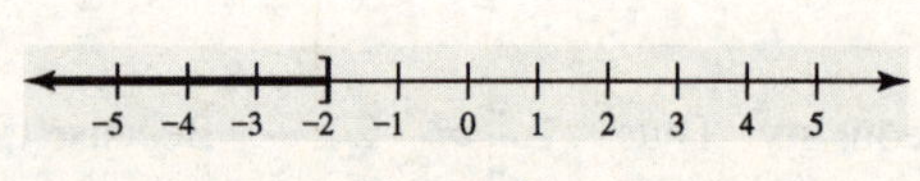

Figure 61

63. $\dfrac{3}{4}>-5y \Rightarrow -5y<\dfrac{3}{4} \Rightarrow \dfrac{-5y}{-5}>\dfrac{\frac{3}{4}}{-5} \Rightarrow y>\dfrac{3}{4}\cdot\left(-\dfrac{1}{5}\right) \Rightarrow y>-\dfrac{3}{20}.$ See Figure 63.

65. $-\dfrac{2}{3} \le \dfrac{1}{7}z \Rightarrow \dfrac{1}{7}z \ge -\dfrac{2}{3} \Rightarrow \dfrac{7}{1}\left(\dfrac{1}{7}z\right) \ge -\dfrac{2}{3}\left(\dfrac{7}{1}\right) \Rightarrow z \ge -\dfrac{14}{3}.$ See Figure 65.

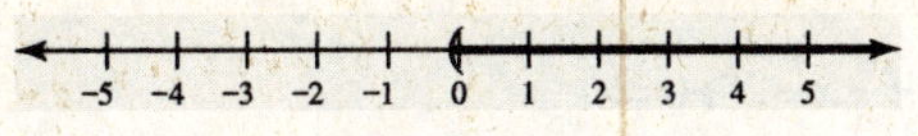

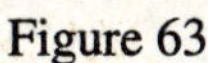

Figure 63

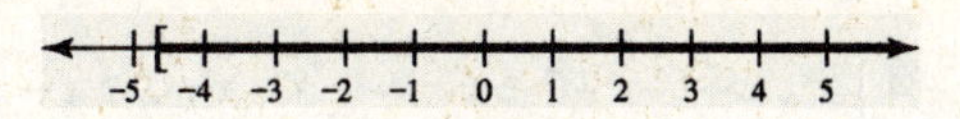

Figure 65

67. $x+6>7 \Rightarrow x+6-6>7-6 \Rightarrow x>1 \Rightarrow \{x|x>1\}$

69. $-3x \le 21 \Rightarrow \dfrac{-3x}{-3} \ge \dfrac{21}{-3} \Rightarrow x \ge -7 \Rightarrow \{x|x \ge -7\}$

71. $2x-3<9 \Rightarrow 2x-3+3<9+3 \Rightarrow 2x<12 \Rightarrow \dfrac{2x}{2}<\dfrac{12}{2}$

$\Rightarrow x<6 \Rightarrow \{x|x<6\}$

73. $3x+1<22 \Rightarrow 3x+1-1<22-1 \Rightarrow 3x<21 \Rightarrow \dfrac{3x}{3}<\dfrac{21}{3} \Rightarrow x<7$

75. $5-\dfrac{3}{4}x \ge 6 \Rightarrow 5-5-\dfrac{3}{4}x \ge 6-5 \Rightarrow -\dfrac{3}{4}x \ge 1 \Rightarrow -\dfrac{4}{3}\left(-\dfrac{3}{4}x\right) \le 1\left(-\dfrac{4}{3}\right) \Rightarrow x \le -\dfrac{4}{3}$

77. $45>6-2x \Rightarrow 6-2x<45 \Rightarrow 6-6-2x<45-6 \Rightarrow -2x<39 \Rightarrow \dfrac{-2x}{-2}>\dfrac{39}{-2} \Rightarrow x>-\dfrac{39}{2}$

79. $5x-2 \le 3x+1 \Rightarrow 5x-3x-2 \le 3x-3x+1 \Rightarrow 2x-2 \le 1 \Rightarrow 2x-2+2 \le 1+2 \Rightarrow 2x \le 3 \Rightarrow \dfrac{2x}{2} \le \dfrac{3}{2} \Rightarrow x \le \dfrac{3}{2}$

81. $-x+24<x+23 \Rightarrow -x-x+24<x-x+23 \Rightarrow -2x+24<23 \Rightarrow -2x+24-24<23-24 \Rightarrow$

$-2x<-1 \Rightarrow \dfrac{-2x}{-2}>\dfrac{-1}{-2} \Rightarrow x>\dfrac{1}{2}$

83. $-(x+1) \ge 3(x-2) \Rightarrow -x-1 \ge 3x-6 \Rightarrow -x-3x-1 \ge 3x-3x-6 \Rightarrow -4x-1 \ge -6 \Rightarrow$

$-4x-1+1 \ge -6+1 \Rightarrow -4x \ge -5 \Rightarrow \dfrac{-4x}{-4} \le \dfrac{-5}{-4} \Rightarrow x \le \dfrac{5}{4}$

85. $3(2x+1)>-(5-3x) \Rightarrow 6x+3>-5+3x \Rightarrow 6x-3x+3>-5+3x-3x \Rightarrow 3x+3>-5 \Rightarrow$

$3x+3-3>-5-3 \Rightarrow 3x>-8 \Rightarrow \dfrac{3x}{3}>\dfrac{-8}{3} \Rightarrow x>-\dfrac{8}{3}$

87. $-(7x+5)+1 \ge 3x-1 \Rightarrow -7x-5+1 \ge 3x-1 \Rightarrow -7x-4 \ge 3x-1 \Rightarrow$

$-7x-3x-4 \ge 3x-3x-1 \Rightarrow -10x-4 \ge -1 \Rightarrow -10x-4+4 \ge -1+4 \Rightarrow -10x \ge 3 \Rightarrow \dfrac{-10x}{-10} \le \dfrac{3}{-10} \Rightarrow x \le -\dfrac{3}{10}$

89. $1.6x+0.4 \le 0.4x \Rightarrow 1.6x-0.4x+0.4 \le 0.4x-0.4x \Rightarrow 1.2x+0.4 \le 0 \Rightarrow$

$1.2x+0.4-0.4 \le 0-0.4 \Rightarrow 1.2x \le -0.4 \Rightarrow \dfrac{1.2x}{1.2} \le \dfrac{-0.4}{1.2} \Rightarrow x \le -\dfrac{1}{3}$

91. $0.8x-0.5<x+1-0.5x \Rightarrow 0.8x-0.5<0.5x+1 \Rightarrow 0.8x-0.5x-0.5<0.5x-0.5x+1 \Rightarrow$

$0.3x-0.5<1 \Rightarrow 0.3x-0.5+0.5<1+0.5 \Rightarrow 0.3x<1.5 \Rightarrow \dfrac{0.3x}{0.3}<\dfrac{1.5}{0.3} \Rightarrow x<5$

93. $-\frac{1}{2}\left(\frac{2}{3}x+4\right)\geq x \Rightarrow -\frac{1}{3}x-2\geq x \Rightarrow -\frac{1}{3}x-x-2\geq x-x \Rightarrow -\frac{4}{3}x-2\geq 0 \Rightarrow$

$-\frac{4}{3}x-2+2\geq 0+2 \Rightarrow -\frac{4}{3}x\geq 2 \Rightarrow -\frac{3}{4}\left(-\frac{4}{3}x\right)\leq 2\left(-\frac{3}{4}\right) \Rightarrow x\leq -\frac{3}{2}$

95. $\frac{3}{7}x+\frac{2}{7}>-\frac{1}{7}x-\frac{5}{14} \Rightarrow \frac{3}{7}x+\frac{1}{7}x+\frac{2}{7}>-\frac{1}{7}x+\frac{1}{7}x-\frac{5}{14} \Rightarrow \frac{4}{7}x+\frac{2}{7}>-\frac{5}{14} \Rightarrow$

$\frac{4}{7}x+\frac{2}{7}-\frac{2}{7}>-\frac{5}{14}-\frac{2}{7} \Rightarrow \frac{4}{7}x>-\frac{5}{14}-\frac{4}{14} \Rightarrow \frac{4}{7}x>-\frac{9}{14} \Rightarrow \frac{7}{4}\left(\frac{4}{7}x\right)>-\frac{9}{14}\left(\frac{7}{4}\right) \Rightarrow x>-\frac{63}{56} \Rightarrow x>-\frac{9}{8}$

97. $\frac{x}{3}+\frac{5x}{6}\leq\frac{2}{3} \Rightarrow \frac{2x}{6}+\frac{5x}{6}\leq\frac{2}{3} \Rightarrow \frac{7x}{6}\leq\frac{2}{3} \Rightarrow 6\left(\frac{7x}{6}\right)\leq\frac{2}{3}(6) \Rightarrow 7x\leq 4 \Rightarrow \frac{7x}{7}\leq\frac{4}{7} \Rightarrow x\leq\frac{4}{7}$

99. $\frac{6x}{7}<\frac{1}{3}x+1 \Rightarrow \frac{6x}{7}-\frac{1}{3}x<\frac{1}{3}x-\frac{1}{3}x+1 \Rightarrow \frac{6x}{7}-\frac{x}{3}<1 \Rightarrow \frac{18x}{21}-\frac{7x}{21}<1 \Rightarrow \frac{11x}{21}<1 \Rightarrow$

$21\left(\frac{11x}{21}\right)<1(21) \Rightarrow 11x<21 \Rightarrow \frac{11x}{11}<\frac{21}{11} \Rightarrow x<\frac{21}{11}$

Translating Inequalities

101. $x>60$

103. $x\geq 21$

105. $x>40{,}000$

107. $x\leq 70$

Applications

109. $2(x+5)+2x<50 \Rightarrow 2x+10+2x<50 \Rightarrow 4x+10<50 \Rightarrow 4x+10-10<50-10 \Rightarrow$

$4x<40 \Rightarrow \frac{4x}{4}<\frac{40}{4} \Rightarrow x<10$ feet.

111. The area of a triangle is $\frac{1}{2}bh$. Substitute 12 for h and solve. $\frac{1}{2}bh<120 \Rightarrow \frac{1}{2}b(12)<120 \Rightarrow$

$6b<120 \Rightarrow \frac{6b}{6}<\frac{120}{6} \Rightarrow b<20$. Thus, the base of the triangle must be less than 20 inches.

113. Let x represent the unknown test score. Then, $\frac{74+x}{2}\geq 80 \Rightarrow 2\left(\frac{74+x}{2}\right)\geq 80(2) \Rightarrow 74+x\geq 160 \Rightarrow$

$74-74+x\geq 160-74 \Rightarrow x\geq 86$. Thus, the student needs a score of 86 or more to maintain an average of at least 80.

115. Let x represent the number of hours parked after the first half hour. We see that there is a $2.00 cost for the first half hour and $1.25 cost for each hour after that. Therefore,

$2+1.25x \le 8 \Rightarrow 2-2+1.25x \le 8-2 \Rightarrow 1.25x \le 6 \Rightarrow \dfrac{1.25x}{1.25} \le \dfrac{6}{1.25} \Rightarrow x \le 4.8.$ This result would indicate that the student can park for as long as 4.8 hours beyond the first half hour for $8.00. However, because a partial hour of parking is charged as a full hour, the longest amount of time that the student could park for $8.00 is 4.5 hours.

117. Let x represent the number of days. Then, $25x+0.20(90)x \le 200 \Rightarrow 25x+18x \le 200 \Rightarrow 43x \le 200 \Rightarrow$

$\dfrac{43x}{43} \le \dfrac{200}{43} \Rightarrow x \le 4.65.$ Because the car can not be rented for a partial day, the person can rent the car for 4 days.

119. (a) $C = 1.5x + 2000$

(b) $R = 12x$

(c) $P = 12x - (1.5x + 2000) \Rightarrow P = 10.5x - 2000$

(d) To yield a positive profit, revenue must be greater than cost. Then, $12x > 1.5x + 2000 \Rightarrow$

$12x - 1.5x > 1.5x - 1.5x + 2000 \Rightarrow 10.5x > 2000 \Rightarrow \dfrac{10.5x}{10.5} > \dfrac{2000}{10.5} \Rightarrow x > 190.476.$ Thus, 191 or more compact discs must be sold to yield a profit.

121. (a) Set the distances equal and then solve for x. Then, $\dfrac{1}{6}x = \dfrac{1}{8}x + 2 \Rightarrow 4x = 3x + 48 \Rightarrow x = 48.$

Thus, at 48 minutes the athletes are the same distance from the parking lot.

(b) $\dfrac{1}{6}x > \dfrac{1}{8}x + 2 \Rightarrow 4x > 3x + 48 \Rightarrow x > 48.$ Thus, after more than 48 minutes, the first athlete is farther from the parking lot than the second athlete.

123. Because $T = 90 - 19x,$ set an inequality statement with T equal to 4.5 and solve. Then,

$90 - 19x < 4.5 \Rightarrow 90 - 90 - 19x < 4.5 - 90 \Rightarrow -19x < -85.5 \Rightarrow \dfrac{-19x}{-19} > \dfrac{-85.5}{-19} \Rightarrow x > 4.5.$

Thus, at altitudes more than 4.5 miles, the air temperature is less than 4.5°F.

125. $0.11(x-1980)+14.1 \ge 16.3 \Rightarrow 0.11x - 217.8 + 14.1 \ge 16.3 \Rightarrow 0.11x - 203.7 \ge 16.3 \Rightarrow$

$0.11x - 203.7 + 203.7 \ge 16.3 + 203.7 \Rightarrow 0.11x \ge 220 \Rightarrow \dfrac{0.11x}{0.11} \ge \dfrac{220}{0.11} \Rightarrow x \ge 2000.$ Thus, in year 2000 and later, a 65-year-old man could expect to live an additional 16.3 years or more.

Checking Basic Concepts for Section 2.5

1. −5 −4 −3 −2 −1 0 1 2 3 4 5

2. $x<1$

3. $4x-5\le -15 \Rightarrow 4x-5+5\le -15+5 \Rightarrow 4x\le -10 \Rightarrow \frac{4x}{4}\le \frac{-10}{4} \Rightarrow x\le -\frac{5}{2}.$

   Since $-3<-\frac{5}{2}$, -3 is a solution to the inequality.

4. When $x=-2$, then $5-2(-2)=5+4=9$; When $x=-1$, then $5-2(-1)=5+2=7$;

   When $x=0$, then $5-2(0)=5-0=5$; When $x=1$, then $5-2(1)=5-2=3$. Therefore, the numbers that complete the table are 9, 7, 5 and 3. See Figure 4.

x	−2	−1	0	1	2
$5-2x$	9	7	5	3	1

Figure 4

From the table, we see that $5-2x\le 7$ whenever $x\ge -1$. Thus, the solution to the inequality is $x\ge -1$.

5. (a) $x+5>8 \Rightarrow x+5-5>8-5 \Rightarrow x>3$

   (b) $-\frac{5}{7}x\le 25 \Rightarrow -\frac{7}{5}\left(-\frac{5}{7}x\right)\ge 25\left(-\frac{7}{5}\right) \Rightarrow x\ge -\frac{175}{5} \Rightarrow x\ge -35$

   (c) $3x\ge -2(1-2x)+3 \Rightarrow 3x\ge -2+4x+3 \Rightarrow 3x\ge 1+4x \Rightarrow 3x-4x\ge 1+4x-4x \Rightarrow$

   $-x\ge 1 \Rightarrow -1(-x)\le -1(1) \Rightarrow x\le -1$

6. $x\le 12$

7. Let l represent length and w represent width. Then, $l=2w+5$. Therefore, $2(2w+5)+2w>88 \Rightarrow$

   $4w+10+2w>88 \Rightarrow 6w+10>88 \Rightarrow 6w+10-10>88-10 \Rightarrow 6w>78 \Rightarrow \frac{6w}{6}>\frac{78}{6} \Rightarrow w>13$. Thus, the possible widths must be more than 13 inches.

Chapter 2 Review Exercises

Section 2.1

1. $x+9=3 \Rightarrow x+9-9=3-9 \Rightarrow x=-6$

2. $x-4=-2 \Rightarrow x-4+4=-2+4 \Rightarrow x=2$

3. $x-\frac{3}{4}=\frac{3}{2} \Rightarrow x-\frac{3}{4}+\frac{3}{4}=\frac{3}{2}+\frac{3}{4} \Rightarrow x=\frac{6}{4}+\frac{3}{4} \Rightarrow x=\frac{9}{4}$

4. $x+0.5=0 \Rightarrow x+0.5-0.5=0-0.5 \Rightarrow x=-0.5 \Rightarrow x=-\frac{1}{2}$

5. $4x=12 \Rightarrow \frac{4x}{4}=\frac{12}{4} \Rightarrow x=3$

6. $3x = -7 \Rightarrow \frac{3x}{3} = \frac{-7}{3} \Rightarrow x = -\frac{7}{3}$

7. $-0.5x = 1.25 \Rightarrow \frac{-0.5x}{-0.5} = \frac{1.25}{-0.5} \Rightarrow x = -2.5$

8. $-\frac{1}{3}x = \frac{7}{6} \Rightarrow -\frac{3}{1}\left(-\frac{1}{3}x\right) = \frac{7}{6}\left(-\frac{3}{1}\right) \Rightarrow x = -\frac{21}{6} \Rightarrow x = -\frac{7}{2}$

Section 2.2

9. The equation $5x - 3 = 0$ is linear; $a = 5,\ b = -3$

10. The equation $-4x + 3 = 2$ is linear.

$-4x + 3 = 2 \Rightarrow -4x + 3 - 2 = 2 - 2 \Rightarrow -4x + 1 = 0;\quad a = -4,\ b = 1$

11. The equation $\frac{1}{x} + 3 = 0$ is not a linear equation because it cannot be written in the form $ax + b = 0$.

12. The equation $\frac{3}{8}x^2 - x = \frac{1}{4}$ is not a linear equation because it cannot be written in the form $ax + b = 0$.

13. $4x - 5 = 3 \Rightarrow 4x - 5 + 5 = 3 + 5 \Rightarrow 4x = 8 \Rightarrow \frac{4x}{4} = \frac{8}{4} \Rightarrow x = 2$. To check the solution, substitute 2 for x in the original equation: $4x - 5 = 3 \Rightarrow 4(2) - 5 = 3 \Rightarrow 8 - 5 = 3$. Because this statement is true, the solution checks.

14. $7 - \frac{1}{2}x = -4 \Rightarrow 7 - 7 - \frac{1}{2}x = -4 - 7 \Rightarrow -\frac{1}{2}x = -11 \Rightarrow -\frac{2}{1}\left(-\frac{1}{2}x\right) = -\frac{11}{1}\left(-\frac{2}{1}\right) \Rightarrow x = 22$. To check the solution, substitute 22 for x in the original equation: $7 - \frac{1}{2}x = -4 \Rightarrow 7 - \frac{1}{2}(22) = -4 \Rightarrow 7 - 11 = -4$. Because this statement is true, the solution checks.

15. $5(x-3) = 12 \Rightarrow 5x - 15 = 12 \Rightarrow 5x - 15 + 15 = 12 + 15 \Rightarrow 5x = 27 \Rightarrow \frac{5x}{5} = \frac{27}{5} \Rightarrow x = \frac{27}{5}$. To check the solution, substitute $\frac{27}{5}$ for x in the original equation: $5\left(\frac{27}{5} - 3\right) = 12 \Rightarrow 5\left(\frac{27}{5} - \frac{15}{5}\right) = 12 \Rightarrow$

$5\left(\frac{12}{5}\right) = 12 \Rightarrow \frac{60}{5} = 12$. Because this statement is true, the solution checks.

16. $3 + x = 2x - 4 \Rightarrow 3 + x - x = 2x - x - 4 \Rightarrow 3 = x - 4$

$\Rightarrow 3 + 4 = x - 4 + 4 \Rightarrow 7 = x \Rightarrow x = 7$.

To check the solution, substitute 7 for x in the original equation: $3 + x = 2x - 4 \Rightarrow 3 + 7 = 2(7) - 4$

$\Rightarrow 10 = 14 - 4 \Rightarrow 10 = 10$. Because this statement is true, the solution checks.

17. $2(x-1) = 4(x+3) \Rightarrow 2x - 2 = 4x + 12$

$\Rightarrow 2x - 2x - 2 = 4x - 2x + 12 \Rightarrow -2 = 2x + 12$

$\Rightarrow -2 - 12 = 2x + 12 - 12 \Rightarrow -14 = 2x \Rightarrow \frac{-14}{2} = \frac{2x}{2}$

$\Rightarrow -7 = x \Rightarrow x = -7.$

To check the solution, substitute -7 for x in the original equation:

$2(x-1) = 4(x+3) \Rightarrow 2(-7-1) = 4(-7+3)$

$\Rightarrow 2(-8) = 4(-4) \Rightarrow -16 = -16.$ Because this statement is true, the solution checks.

18. $1-(x-3) = 6+2x \Rightarrow 1-x+3 = 6+2x \Rightarrow 4-x = 6+2x \Rightarrow 4-x+x = 6+2x+x \Rightarrow$

$4 = 6+3x \Rightarrow 4-6 = 6-6+3x \Rightarrow -2 = 3x \Rightarrow \frac{-2}{3} = \frac{3x}{3} \Rightarrow -\frac{2}{3} = x \Rightarrow x = -\frac{2}{3}.$ To check the solution,

substitute $-\frac{2}{3}$ for x in the original equation: $1-\left(-\frac{2}{3}-3\right) = 6+2\left(-\frac{2}{3}\right) \Rightarrow$

$1+\frac{2}{3}+3 = 6-\frac{4}{3} \Rightarrow 4\frac{2}{3} = 4\frac{2}{3}.$ Because this statement is true, the solution checks.

19. $3.4x-4 = 5-0.6x \Rightarrow 3.4x-4+4 = 5+4-0.6x \Rightarrow 3.4x = 9-0.6x \Rightarrow$

$3.4x+0.6x = 9-0.6x+0.6x \Rightarrow 4x = 9 \Rightarrow \frac{4x}{4} = \frac{9}{4} \Rightarrow x = \frac{9}{4}.$ To check the solution, substitute $\frac{9}{4}$ for x in the original equation: $3.4\left(\frac{9}{4}\right)-4 = 5-0.6\left(\frac{9}{4}\right) \Rightarrow 3.4(2.25)-4 = 5-0.6(2.25) \Rightarrow$

$7.65-4 = 5-1.35 \Rightarrow 3.65 = 3.65.$ Because this statement is true, the solution checks.

20. $-\frac{1}{3}(3-6x) = -(x+2)+1 \Rightarrow -1+2x = -x-2+1 \Rightarrow 2x-1 = -x-1 \Rightarrow$

$2x-1+1 = -x-1+1 \Rightarrow 2x = -x \Rightarrow 2x+x = -x+x \Rightarrow 3x = 0 \Rightarrow \frac{3x}{3} = \frac{0}{3} \Rightarrow x = 0.$ To check the solution,

substitute 0 for x in the original equation: $-\frac{1}{3}(3-6(0)) = -(0+2)+1 \Rightarrow$

$-\frac{1}{3}(3) = -2+1 \Rightarrow -1 = -1.$ Because this statement is true, the solution checks.

21. $\frac{2}{3}x-\frac{1}{6} = \frac{5}{12} \Rightarrow \frac{2}{3}x-\frac{1}{6}+\frac{1}{6} = \frac{5}{12}+\frac{1}{6} \Rightarrow \frac{2}{3}x = \frac{5}{12}+\frac{2}{12} \Rightarrow \frac{2}{3}x = \frac{7}{12} \Rightarrow \frac{3}{2}\left(\frac{2}{3}x\right) = \frac{7}{12}\left(\frac{3}{2}\right) \Rightarrow$

$x = \frac{21}{24} \Rightarrow x = \frac{7}{8}.$ To check the solution, substitute $\frac{7}{8}$ for x in the original equation: $\frac{2}{3}\left(\frac{7}{8}\right)-\frac{1}{6} = \frac{5}{12} \Rightarrow$

$\frac{14}{24}-\frac{4}{24} = \frac{10}{24} \Rightarrow \frac{10}{24} = \frac{10}{24}.$ Because this statement is true, the solution checks.

22. $2y-3(2-y) = 5+y \Rightarrow 2y-6+3y = 5+y \Rightarrow 5y-6 = 5+y \Rightarrow 5y-y-6 = 5+y-y \Rightarrow$

$4y-6 = 5 \Rightarrow 4y-6+6 = 5+6 \Rightarrow 4y = 11 \Rightarrow \frac{4y}{4} = \frac{11}{4} \Rightarrow y = \frac{11}{4}.$ To check the solution, substitute $\frac{11}{4}$ for y

in the original equation: $2\left(\frac{11}{4}\right)-3\left(2-\frac{11}{4}\right) = 5+\frac{11}{4} \Rightarrow \frac{22}{4}-6+\frac{33}{4} = \frac{20}{4}+\frac{11}{4} \Rightarrow$

$\frac{22}{4}-\frac{24}{4}+\frac{33}{4}=\frac{20}{4}+\frac{11}{4}\Rightarrow\frac{31}{4}=\frac{31}{4}$. Because this statement is true, the solution checks.

23. First, solve for x: $4(3x-2)=2(6x+5)\Rightarrow 12x-8=12x+10\Rightarrow$

$12x-12x-8=12x-12x+10\Rightarrow -8=10$. Because this statement is not true, the equation has no solutions.

24. First, solve for x: $5(3x-1)=15x-5\Rightarrow 15x-5=15x-5$. Because this statement is true for any value of x, the equation has infinitely many solutions.

25. First, solve for x: $8x=5x+3x\Rightarrow 8x=8x$. Because this statement is true for any value of x, the equation has infinitely many solutions.

26. First solve for x: $9x-2=8x-2\Rightarrow 9x-8x-2=8x-8x-2\Rightarrow x-2=-2\Rightarrow$

$x-2+2=-2+2\Rightarrow x=0$. Thus, there is one solution to the equation.

27. When $x=1.0$, then $-2(1.0)+3=-2+3=1$; When $x=1.5$, then $-2(1.5)+3=-3+3=0$;

When $x=2.0$, then $-2(2.0)+3=-4+3=-1$; When $x=2.5$, then $-2(2.5)+3=-5+3=-2$;

Thus, the missing values in the table are 1, 0, -1 and -2. See Figure 27. From the table we see that when $x=1.5$, the value of $-2x+3$ is 0.

28. When $x=-2$, then $-(-2+1)+3=2-1+3=4$; When $x=-1$, then $-(-1+1)+3=1-1+3=3$;

When $x=0$, then $-(0+1)+3=0-1+3=2$; When $x=1$, then $-(1+1)+3=-2+3=1$;

Thus, the missing values in the table are 4, 3, 2 and 1. See Figure 28. From the table we see that when $x=0$, the value of $-(x+1)+3$ is 2.

x	0.5	1.0	1.5	2.0	2.5
$-2x+3$	2	1	0	−1	−2

Figure 27

x	−2	−1	0	1	2
$-(x+1)+3$	4	3	2	1	0

Figure 28

Section 2.3

29. $6x=72\Rightarrow\frac{6x}{6}=\frac{72}{6}\Rightarrow x=12$

30. $x+18=-23\Rightarrow x+18-18=-23-18\Rightarrow x=-41$

31. $2x-5=x+4\Rightarrow 2x-5+5=x+4+5\Rightarrow 2x=x+9\Rightarrow 2x-x=x-x+9\Rightarrow x=9$

32. $x+4=3x\Rightarrow x-3x+4=3x-3x\Rightarrow -2x+4=0\Rightarrow -2x+4-4=0-4\Rightarrow -2x=-4\Rightarrow \frac{-2x}{-2}=\frac{-4}{-2}\Rightarrow x=2$

33. $x+(x+1)+(x+2)+(x+3)=70\Rightarrow 4x+6=70\Rightarrow 4x+6-6=70-6\Rightarrow 4x=64\Rightarrow \frac{4x}{4}=\frac{64}{4}\Rightarrow x=16$. The numbers are 16, 17, 18 and 19.

34. $x+(x+1)+(x+2)=-153\Rightarrow 3x+3=-153\Rightarrow 3x+3-3=-153-3\Rightarrow 3x=-156\Rightarrow$

$\frac{3x}{3}=\frac{-156}{3}\Rightarrow x=-52$. The numbers are -52, -51 and -50.

35. $85\% = \frac{85}{100} = \frac{17}{20}$; $85\% = 0.85$

36. $5.6\% = \frac{56}{1000} = \frac{7}{125}$; $5.6\% = 0.056$

37. $0.03\% = \frac{0.03}{100} = \frac{3}{10,000}$; $0.03\% = 0.0003$

38. $342\% = \frac{342}{100} = \frac{171}{50}$; $342\% = 3.42$

39. $0.89 = 89\%$

40. $0.005 = 0.5\%$

41. $2.3 = 230\%$

42. $1 = 100\%$

43. $d = rt \Rightarrow d = 8(3) \Rightarrow d = 24$ miles.

44. $d = rt \Rightarrow d = 70(55) \Rightarrow d = 3850$ feet.

45. $d = rt \Rightarrow 500 = r(20) \Rightarrow \frac{500}{20} = r\frac{20}{20} \Rightarrow \frac{500}{20} = r \Rightarrow r = 25$ yd/sec.

46. $d = rt \Rightarrow 125 = 15t \Rightarrow \frac{125}{15} = \frac{15t}{15} \Rightarrow \frac{125}{15} = t \Rightarrow t = \frac{25}{3}$ hours.

Section 2.4

47. The area of a triangle is given as $\frac{1}{2}(\text{base})(\text{height})$. Thus, $A = \frac{1}{2}(b)(h) = \frac{1}{2}(5)(3) = 7.5 \text{ m}^2$.

48. The area of a circle is given as πr^2 where r represents radius. Thus, $A = \pi r^2 = \pi(6^2) = 36\pi \approx 113.1 \text{ ft}^2$.

49. The area of a rectangle is given as length (l) times width (w). Thus, $A = lw = (36)(24) = 864 \text{ in}^2$ or 6 ft^2.

50. $A = \frac{1}{2}bh$, where A represents area, b represents the length of the base and h represents the height. Thus,

$A = \frac{1}{2}bh = \frac{1}{2}(13)(7) = \frac{1}{2}(91) = 45\frac{1}{2} = 45.5 \text{ in}^2$.

51. The circumference of a circle is given as $2\pi r$, where r represents radius. Thus,

$r = \frac{1}{2}(\text{diameter}) = \frac{1}{2}(18) = 9$.

$C = 2\pi r = 2\pi(9) = 18\pi \approx 56.5$ feet.

52. $A = \pi r^2$, where A represents area and r represents radius. Thus, $A = \pi r^2 = \pi(5^2) = 25\pi \approx 78.5 \text{ in}^2$.

53. The angles in a triangle must add up to 180°. Let x represent the unknown angle. Thus,

$90 + 40 + x = 180 \Rightarrow 130 + x = 180 \Rightarrow 130 - 130 + x = 180 - 130 \Rightarrow x = 50°$.

54. The angles in a triangle must add up to 180°. Thus, $x+3x+4x=180 \Rightarrow 8x=180 \Rightarrow \frac{8x}{8}=\frac{180}{8} \Rightarrow$

$x=22.5°$.

55. $V=\pi r^2h=\pi(5^2)(25)=\pi(25)(25)=625\pi\approx 1963.5 \text{ in}^3$.

56. First, convert height (h) and base (b) to inches. $h=5$ feet $=5(12)=60$ inches and

$b=3$ feet $=3(12)=36$ inches. Then, $A=\frac{1}{2}(a+b)h=\frac{1}{2}(36+18)60=\frac{1}{2}(54)60=(27)60=1620 \text{ in}^2$.

Or, convert the base in inches to feet. $b=18$ inches $=\frac{18}{12}=1.5$ feet. Then,

$A=\frac{1}{2}(a+b)h=\frac{1}{2}(3+1.5)5=\frac{1}{2}(4.5)5=(2.25)5=11.25 \text{ ft}^2$.

57. The total area of the figure is the sum of the area of the triangle and the area of the rectangle. The area of the triangle is $\frac{1}{2}bh=\frac{1}{2}\cdot 8\cdot 6=24 \text{ in}^2$.

The area of the rectangle is $LW=25\cdot 6=150 \text{ in}^2$.

Thus, the total area is $24+150=174 \text{ in}^2$.

58. The total area of the figure is the sum of the area of the rectangle and the area of a circle. The area of the rectangle is $LW=12\cdot 4=48 \text{ ft}^2$.

The area of the circle is $\pi r^2=\pi\left(\frac{4}{2}\right)^2=\pi(2)^2=4\pi\approx 12.6 \text{ ft}^2$. Thus, the total area is about

$48+12.6=60.6 \text{ ft}^2$.

59. $a=x+y \Rightarrow a-y=x+y-y \Rightarrow a-y=x \Rightarrow x=a-y$

60. $P=2x+2y \Rightarrow P-2y=2x+2y-2y \Rightarrow P-2y=2x \Rightarrow \frac{P-2y}{2}=\frac{2x}{2} \Rightarrow \frac{P-2y}{2}=x \Rightarrow x=\frac{P-2y}{2}$

61. $z=2xy \Rightarrow \frac{z}{2x}=\frac{2xy}{2x} \Rightarrow \frac{z}{2x}=y \Rightarrow y=\frac{z}{2x}$

62. $S=\frac{a+b+c}{3} \Rightarrow 3S=\frac{a+b+c}{3}\cdot 3 \Rightarrow 3S=a+b+c \Rightarrow 3S-a-c=a+b+c-a-c \Rightarrow$

$3S-a-c=b \Rightarrow b=3S-a-c$

63. $T=\frac{a}{3}+\frac{b}{4} \Rightarrow \frac{12T}{12}=\frac{4a}{12}+\frac{3b}{12} \Rightarrow 12\left(\frac{12T}{12}\right)=12\left(\frac{4a}{12}\right)+12\left(\frac{3b}{12}\right) \Rightarrow 12T=4a+3b \Rightarrow$

$12T-4a=4a-4a+3b \Rightarrow 12T-4a=3b \Rightarrow \frac{12T-4a}{3}=\frac{3b}{3} \Rightarrow \frac{12T-4a}{3}=b \Rightarrow b=\frac{12T-4a}{3}$

64. $cd=ab+bc \Rightarrow cd-bc=ab+bc-bc \Rightarrow cd-bc=ab \Rightarrow c(d-b)=ab \Rightarrow \frac{c(d-b)}{(d-b)}=\frac{ab}{(d-b)} \Rightarrow c=\frac{ab}{d-b}$

65. The formula for GPA is given by $\dfrac{4a+3b+2c+d}{a+b+c+d+f}$. $\dfrac{4(20)+3(25)+2(12)+4}{20+25+12+4+4}=\dfrac{80+75+24+4}{65}=\dfrac{183}{65}\approx 2.82$.

Thus, the GPA is 2.82.

66. The formula for GPA is given by $\dfrac{4a+3b+2c+d}{a+b+c+d+f}$.

$\dfrac{4(64)+3(32)+2(20)+10}{64+32+20+10+3}=\dfrac{256+96+40+10}{129}=\dfrac{402}{129}\approx 3.12$. Thus, the GPA is 3.12.

67. To convert Celsius to Fahrenheit temperature, the formula given is $\dfrac{9}{5}C+32=F$.

$\dfrac{9}{5}(15)+32=F\Rightarrow\dfrac{135}{5}+32=F\Rightarrow 27+32=F\Rightarrow F=59°\text{F}$.

68. To convert Fahrenheit to Celsius temperature, the formula given is $C=\dfrac{5}{9}(F-32)$.

$C=\dfrac{5}{9}(113-32)\Rightarrow C=\dfrac{5}{9}(81)\Rightarrow C=\dfrac{405}{9}\Rightarrow C=45°\text{C}$.

Section 2.5

69. 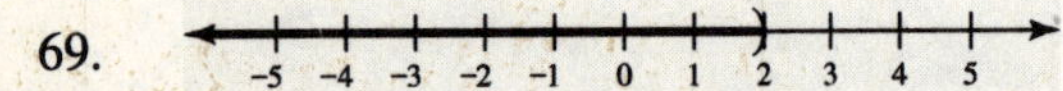

70. -5 -4 -3 -2 -1 0 1 2 3 4 5

71. -5 -4 -3 -2 -1 0 1 2 3 4 5

72. -5 -4 -3 -2 -1 0 1 2 3 4 5

73. $x<3$

74. $x\geq -1$

75. Substitute -3 for x and check for accuracy: $2x+1\leq 5\Rightarrow 2(-3)+1\leq 5\Rightarrow -6+1\leq 5\Rightarrow -5\leq 5$.

Because this statement is true, $x=-3$ is a solution to the inequality.

76. Substitute 4 for x and check for accuracy: $5-\dfrac{1}{2}(x)>-1\Rightarrow 5-\dfrac{1}{2}(4)>-1\Rightarrow 5-2>-1\Rightarrow 3>-1$.

Because this statement is true, $x=4$ is a solution to the inequality.

77. Substitute -2 for x and check for accuracy: $1-(x+3)\geq x\Rightarrow 1-(-2+3)\geq -2\Rightarrow 1-1\geq -2\Rightarrow$

$0\geq -2$. Because this statement is true, $x=-2$ is a solution to the inequality.

78. Substitute -1 for x and check for accuracy: $4(x+1)<-(5-x)\Rightarrow 4(-1+1)<-(5-(-1))\Rightarrow$

$4(0)<-(6)\Rightarrow 0<-6$. Because this statement is not true, $x=-1$ is not a solution to the inequality.

79. When $x=1$, then $5-x=5-1=4$; When $x=2$, then $5-x=5-2=3$;

When $x=3$, then $5-x=5-3=2$; When $x=4$, then $5-x=5-4=1$;

Thus, the missing values in the table are 4, 3, 2 and 1. See Figure 79. From the table we see that $5-x>3$, when $x<2$.

80. When $x=1.5$, then $2x-5=2(1.5)-5=3-5=-2$; When $x=2$, then $2x-5=2(2)-5=4-5=-1$;

When $x=2.5$, then $2x-5=2(2.5)-5=5-5=0$; When $x=3$, then $2x-5=2(3)-5=6-5=1$;

Thus, the missing values in the table are -2, -1, 0 and 1. See Figure 80. From the table we see that $2x-5\le 0$ when $x\le 2.5$.

x	0	1	2	3	4
$5-x$	5	4	3	2	1

Figure 79

x	1	1.5	2	2.5	3
$2x-5$	−3	−2	−1	0	1

Figure 80

81. $x-3>0\Rightarrow x-3+3>0+3\Rightarrow x>3$

82. $-2x\le 10\Rightarrow \dfrac{-2x}{-2}\ge\dfrac{10}{-2}\Rightarrow x\ge -5$

83. $5-2x\ge 7\Rightarrow 5-5-2x\ge 7-5\Rightarrow -2x\ge 2\Rightarrow \dfrac{-2x}{-2}\le\dfrac{2}{-2}\Rightarrow x\le -1$

84. $3(x-1)<20\Rightarrow 3x-3<20\Rightarrow 3x-3+3<20+3\Rightarrow 3x<23\Rightarrow \dfrac{3x}{3}<\dfrac{23}{3}\Rightarrow x<\dfrac{23}{3}$

85. $5x\le 3-(4x+2)\Rightarrow 5x\le 3-4x-2\Rightarrow 5x+4x\le 3-4x+4x-2\Rightarrow 9x\le 1\Rightarrow \dfrac{9x}{9}\le\dfrac{1}{9}\Rightarrow x\le\dfrac{1}{9}$

86. $3x-2(4-x)\ge x+1\Rightarrow 3x-8+2x\ge x+1\Rightarrow 5x-8+8\ge x+1+8\Rightarrow 5x\ge x+9\Rightarrow$

$5x-x\ge x-x+9\Rightarrow 4x\ge 9\Rightarrow \dfrac{4x}{4}\ge\dfrac{9}{4}\Rightarrow x\ge\dfrac{9}{4}$

87. $x<50$

88. $x\le 45{,}000$

89. $x\ge 16$

90. $x<1995$

Applications

91. (a) See Figure 91.

(b) $R=2+\dfrac{3}{4}x$

(c) At 5 PM, $x=5$; $R=2+\dfrac{3}{4}(5)=2+\dfrac{15}{4}=\dfrac{23}{4}=5\dfrac{3}{4}$ inches. This value does agree with the table.

(d) At 3:45 PM, $x=3.75$; $R=2+\dfrac{3}{4}\left(3\dfrac{3}{4}\right)=2+\dfrac{3}{4}\left(\dfrac{15}{4}\right)=2+\dfrac{45}{16}=\dfrac{32}{16}+\dfrac{45}{16}=\dfrac{77}{16}=4\dfrac{13}{16}$ inches.

Time	12:00	1:00	2:00	3:00	4:00	5:00
Rainfall (*R*)	2	2.75	3.5	4.25	5	5.75

Figure 91

92. Let x represent the cost of the laptop. $0.05x = 106.25 \Rightarrow \frac{0.05x}{0.05} = \frac{106.25}{0.05} \Rightarrow x = 2125$. Thus, the cost of the laptop is \$2125.

93. (a) See Figure 93.

(b) $D = 50 - 10x$

(c) $D = 50 - 10x = 50 - 10(3) = 50 - 30 = 20$ miles. This value does agree with the table.

(d) $D \geq 20$. Thus, $50 - 10x \geq 20 \Rightarrow 50 - 50 - 10x \geq 20 - 50 \Rightarrow -10x \geq -30 \Rightarrow \frac{-10x}{-10} \leq \frac{-30}{-10} \Rightarrow x \leq 3$. Thus, the bicyclist was at least 20 miles from home when he had traveled for 3 or fewer hours, or from noon to 3 PM.

Hours (x)	1	2	3	4	5
Distance (D)	40	30	20	10	0

Figure 93

94. $N = \frac{1}{15}x - 130.4$. Substitute 2.8 for N and solve for x: $2.8 = \frac{1}{15}x - 130.4 \Rightarrow$

$2.8 + 130.4 = \frac{1}{15}x - 130.4 + 130.4 \Rightarrow 133.2 = \frac{1}{15}x \Rightarrow 133.2\left(\frac{15}{1}\right) = \frac{15}{1}\left(\frac{1}{15}x\right) \Rightarrow 1998 = x \Rightarrow x = 1998$. Thus, in the year 1998, the number reached 2.8 million.

95. $0 = 3x - 12 \Rightarrow 0 + 12 = 3x - 12 + 12 \Rightarrow 12 = 3x \Rightarrow \frac{12}{3} = \frac{3x}{3} \Rightarrow 4 = x \Rightarrow x = 4$

96. First, subtract the smaller number from the larger to obtain the difference between them: $419{,}401 - 230{,}500 = 188{,}901$. Then, determine what percentage 188,901 is of 230,500. Do this by dividing the smaller number by the larger: $\frac{188{,}901}{230{,}500} \approx 0.82$. Thus, there was about an 82% change in master's degrees received between 1971 and 1997.

97. Use the distance $(d) = \text{rate } (r) \times \text{time } (t)$ formula. Determine how long it takes the faster car to be 2 miles ahead of the slower car, let $(r + 12)$ be the rate of the faster car and r be the rate of the slower car. Thus,

$d = rt \Rightarrow 2 = (r + 12 - r)t \Rightarrow 2 = 12t \Rightarrow \frac{2}{12} = \frac{12t}{12} \Rightarrow \frac{1}{6} = t \Rightarrow t = \frac{1}{6}$ hour, or 10 minutes.

98. Use the gas mileage formula $M = \frac{D}{G}$.

$18 = \frac{504}{G} \Rightarrow 18\,G = \left(\frac{504}{G}\right)G \Rightarrow 18\,G = 504 \Rightarrow \frac{18\,G}{18} = \frac{504}{18} \Rightarrow G = 28$ gal.

99. Let x represent the amount of water. The amount of salt on one side of the equation must equal the amount of salt on the other side. Thus, $100(0.03)+x(0.00)=(100+x)(0.02)\Rightarrow 3+0=2+0.02x\Rightarrow$ $3-2=2-2+0.02x\Rightarrow 1=0.02x\Rightarrow \frac{1}{0.02}=\frac{0.02x}{0.02}\Rightarrow 50=x\Rightarrow x=50.$ Thus, 50 ml of water must be added.

100. Let x represent the higher interest rate. Then, $800(x)+500(x-0.02)=55\Rightarrow 800x+500x-10=55\Rightarrow$ $1300x-10+10=55+10\Rightarrow 1300x=65\Rightarrow \frac{1300x}{1300}=\frac{65}{1300}\Rightarrow x=0.05.$ Thus, the interest rate on the \$800 loan is 5% and the interest rate on the \$500 loan is 3%.

101. Perimeter $(P)=2\times$width $(W)+2\times$length (L). Then, $W=L-10\Rightarrow 2(L-10)+2L=112\Rightarrow$ $2L-20+2L=112\Rightarrow 2L-20+20+2L=112+20\Rightarrow 4L=132\Rightarrow \frac{4L}{4}=\frac{132}{4}\Rightarrow L=33.$ Because the length is 33, the width is $(L-10)=23.$ Thus, the dimensions are 33 by 23 inches.

102. Use the formula $D=\frac{x}{5}$. $D=\frac{9}{5}\Rightarrow D=1\frac{4}{5}$ or 1.8 mi.

103. Area (A) of a triangle is $\frac{1}{2}\times$base $(b)\times$height (h). Thus, $A=\frac{1}{2}bh\Rightarrow \frac{1}{2}bh\le 100\Rightarrow \frac{1}{2}b(8)\le 100\Rightarrow$ $4b\le 100\Rightarrow \frac{4b}{4}\le\frac{100}{4}\Rightarrow b\le 25.$ Therefore, the base must be 25 inches or less.

104. Let x represent the unknown test score. Then, $\frac{75+91+x}{3}\ge 80\Rightarrow \frac{166+x}{3}\ge 80\Rightarrow$ $3\left(\frac{166+x}{3}\right)\ge 3(80)\Rightarrow 166+x\ge 240\Rightarrow 166-166+x\ge 240-166\Rightarrow x\ge 74.$ Thus, the student must score 74 or more.

105. Let x represent the unknown number of hours after the first hour. Then, $2.25+1.25x=9\Rightarrow$ $2.25-2.25+1.25x=9-2.25\Rightarrow 1.25x=6.75\Rightarrow \frac{1.25x}{1.25}=\frac{6.75}{1.25}\Rightarrow x=5.4.$ Because each partial hour is charged as a full hour, the person can park for $1+5=6$ hours.

106. (a) $C=150{,}000+85x$

(b) $R=225x$

(c) $P=225x-(150{,}000+85x)\Rightarrow P=140x-150{,}000$

(d) $140x-150{,}000<0\Rightarrow 140x-150{,}000+150{,}000<0+150{,}000\Rightarrow 140x<150{,}000\Rightarrow$ $\frac{140x}{140}<\frac{150{,}000}{140}\Rightarrow x<1071.43.$ Therefore, if 1071 or fewer DVD players are sold, there will be a loss.

Chapter 2 Test

1. $9=3-x \Rightarrow 9-3=3-3-x \Rightarrow 6=-x \Rightarrow 6(-1)=(-x)(-1) \Rightarrow -6=x \Rightarrow x=-6$

To check the solution: $9=3-(-6) \Rightarrow 9=9$. The solution checks.

2. $4x-3=7 \Rightarrow 4x-3+3=7+3 \Rightarrow 4x=10 \Rightarrow \frac{4x}{4}=\frac{10}{4} \Rightarrow x=\frac{5}{2}$

To check the solution: $4\left(\frac{5}{2}\right)-3=7 \Rightarrow \frac{20}{2}-\frac{6}{2}=7 \Rightarrow \frac{14}{2}=7 \Rightarrow 7=7$. The solution checks.

3. $4x-(2-x)=-3(2x+6) \Rightarrow 4x-2+x=-6x-18 \Rightarrow 5x-2=-6x-18 \Rightarrow$

$5x-2+2=-6x-18+2 \Rightarrow 5x=-6x-16 \Rightarrow 5x+6x=-6x+6x-16 \Rightarrow 11x=-16 \Rightarrow$

$\frac{11x}{11}=\frac{-16}{11} \Rightarrow x=-\frac{16}{11}$. To check the solution: $4\left(-\frac{16}{11}\right)-\left(2-\left(-\frac{16}{11}\right)\right)=-3\left(2\left(-\frac{16}{11}\right)+6\right) \Rightarrow$

$-\frac{64}{11}-2-\frac{16}{11}=-6\left(-\frac{16}{11}\right)-18 \Rightarrow -\frac{64}{11}-\frac{22}{11}-\frac{16}{11}=\frac{96}{11}-\frac{198}{11} \Rightarrow -\frac{102}{11}=-\frac{102}{11}$. The solution checks.

4. $\frac{1}{12}x-\frac{2}{3}=\frac{1}{2}\left(\frac{3}{4}-\frac{1}{3}x\right) \Rightarrow \frac{1}{12}x-\frac{2}{3}=\frac{3}{8}-\frac{1}{6}x \Rightarrow \frac{1}{12}x+\frac{1}{6}x-\frac{2}{3}=\frac{3}{8}-\frac{1}{6}x+\frac{1}{6}x \Rightarrow$

$\frac{3}{12}x-\frac{2}{3}+\frac{2}{3}=\frac{3}{8}+\frac{2}{3} \Rightarrow \frac{3}{12}x=\frac{9}{24}+\frac{16}{24} \Rightarrow \frac{3}{12}x=\frac{25}{24} \Rightarrow \frac{12}{3}\left(\frac{3}{12}x\right)=\frac{12}{3}\left(\frac{25}{24}\right) \Rightarrow x=\frac{300}{72}=\frac{25}{6}$.

To check the solution: $\frac{1}{12}\left(\frac{25}{6}\right)-\frac{2}{3}=\frac{1}{2}\left(\frac{3}{4}-\frac{1}{3}\left(\frac{25}{6}\right)\right) \Rightarrow \frac{25}{72}-\frac{48}{72}=\frac{1}{2}\left(\frac{27}{36}-\frac{50}{36}\right) \Rightarrow$

$-\frac{23}{72}=\frac{1}{2}\left(-\frac{23}{36}\right) \Rightarrow -\frac{23}{72}=-\frac{23}{72}$. The solution checks.

5. First, solve for x: $6(2x-1)=-4(3-3x) \Rightarrow 12x-6=-12+12x \Rightarrow$

$12x-12x-6=-12+12x-12x \Rightarrow -6=-12$. Because this statement is not true, there are no solutions.

6. When $x=1$, then $6-2x=6-2(1)=4$; When $x=2$, then $6-2x=6-2(2)=2$;

When $x=3$, then $6-2x=6-2(3)=0$; When $x=4$, then $6-2x=6-2(4)=-2$;

Thus, the missing values in the table are 4, 2, 0 and -2. See Figure 6. From the table we see that $6-2x=0$, when $x=3$.

x	0	1	2	3	4
$6-2x$	6	4	2	0	-2

Figure 6

7. $x+(-7)=6 \Rightarrow x-7=6 \Rightarrow x-7+7=6+7 \Rightarrow x=13$

8. $2x+6=x-7 \Rightarrow 2x-x+6=x-x-7 \Rightarrow x+6=-7 \Rightarrow x+6-6=-7-6 \Rightarrow x=-13$

9. $x+(x+1)+(x+2)=336 \Rightarrow 3x+3=336 \Rightarrow 3x+3-3=336-3 \Rightarrow 3x=333 \Rightarrow \frac{3x}{3}=\frac{333}{3} \Rightarrow x=111$. Thus, the three numbers are 111, 112 and 113.

10. $3.2\% = 0.032;\ 3.2\% = \frac{32}{1000}=\frac{16}{500}=\frac{8}{250}=\frac{4}{125}$

11. $0.345 = 34.5\%$

12. Let x represent the unknown number. To find 7.5% of \$500, multiply 500 by 0.075. Then, $500(0.075)=x \Rightarrow\ 37.5=x \Rightarrow x=37.5$. Thus, 7.5% of \$500 is \$37.50.

13. $\frac{5280}{5}=1056$ ft/sec.

14. Area $(A)=\frac{1}{2}\times$ base $(b)\times$ height (h). Thus, $A=\frac{1}{2}bh=\frac{1}{2}(5)(3)=7.5\text{ in}^2$.

15. Circumference of a circle is given as $C=2\pi r$. Then, $C=2\pi r=2\pi\left(\frac{30}{2}\right)=2\pi(15)=30\pi\approx 94.2$ inches.

Area of a circle is given as $A=\pi r^2$. Then, $A=\pi r^2=\pi(15)^2=225\pi\approx 706.9\text{ in}^2$.

16. The angles in a triangle must add up to 180°.

Then, $x+2x+3x=180 \Rightarrow 6x=180 \Rightarrow \frac{6x}{6}=\frac{180}{6} \Rightarrow x=30$. Thus, the angles are 30°, 60° and 90°.

17. $z=y-3xy \Rightarrow z-y=y-y-3xy \Rightarrow z-y=-3xy \Rightarrow \frac{z-y}{-3y}=\frac{-3xy}{-3y} \Rightarrow \frac{z-y}{-3y}=x \Rightarrow x=\frac{y-z}{3y}$

18. $R=\frac{x}{4}+\frac{y}{5} \Rightarrow 20R=5x+4y \Rightarrow 20R-4y=5x \Rightarrow x=\frac{20R-4y}{5}$

19. $-3x+9\ge x-15 \Rightarrow -3x+9-9\ge x-15-9 \Rightarrow -3x\ge x-24 \Rightarrow -3x-x\ge x-x-24$

$\Rightarrow -4x\ge -24 \Rightarrow \frac{-4x}{-4}\le\frac{-24}{-4} \Rightarrow x\le 6$

20. $3(6-5x)<20-x \Rightarrow 18-15x<20-x \Rightarrow 18-18-15x<20-18-x \Rightarrow -15x<2-x \Rightarrow$

$-15x+x<2-x+x \Rightarrow -14x<2 \Rightarrow \frac{-14x}{-14}>\frac{2}{-14} \Rightarrow x>-\frac{1}{7}$

21. (a) $S=5+2x$, where x represents hours past noon.

(b) $x=8$. Thus, $S=5+2(8)=21$ inches.

(c) $x=6.25$. Thus, $S=5+2(6.25)=17.5$ inches.

22. The amount of acid on the left side of the equation must equal the amount of acid on the right side of the equation. Let x represent the unknown amount of water. Then, $1000(0.45)+x(0)=(1000+x)(0.15) \Rightarrow$

$450=150+0.15x \Rightarrow 450-150=150-150+0.15x \Rightarrow 300=0.15x \Rightarrow \frac{300}{0.15}=\frac{0.15x}{0.15} \Rightarrow$

$2000 = x \Rightarrow x = 2000$. Thus, 2000 ml of water must be added.

23. Subtract the lesser amount from the larger amount and then calculate the percentage difference as compared to the smaller amount. Then, $32-8=24$; $\frac{24}{8}=3=300\%$. Therefore, there was a 300% increase in premiums from 1998 to 2003.

Chapter 2 Extended and Discovery Exercises

1. For the first hour, the distance traveled was $d=rt$ such that $d=(50)(1)=50$ miles. For the second hour, the distance traveled was $d=rt$ such that $d=(70)(1)=70$ miles. Thus, for the two hours $r=\frac{d}{t}$ such that $r=\frac{70+50}{1+1}=\frac{120}{2}=60$. Thus, the average speed of the car was 60 mph.

2. Uphill, $t=\frac{d}{r}$ such that $t=\frac{1}{5}=\frac{1}{5}$ of an hour. Downhill, $t=\frac{d}{r}$ such that $t=\frac{1}{10}=\frac{1}{10}$ of an hour. Thus, the average speed $r=\frac{d}{t}$ is $r=\frac{1+1}{\frac{1}{5}+\frac{1}{10}}=\frac{2}{\frac{3}{10}}=\frac{20}{3}=6.\overline{6}$ mph. *Answers may vary.*

3. For the first two miles, $t=\frac{d}{r}$ such that $t=\frac{2}{8}=\frac{1}{4}$ of an hour. For the third mile, $t=\frac{d}{t}$ such that $t=\frac{1}{10}=\frac{1}{10}$ of an hour. Thus, the average speed of the athlete is

$$r=\frac{d}{t}=\frac{3}{\frac{1}{4}+\frac{1}{10}}=\frac{3}{\frac{5}{20}+\frac{2}{20}}=\frac{3}{\frac{7}{20}}=\frac{60}{7}\approx 8.6 \text{ mph}.$$

4. Choose a distance of 400 miles as the distance between the two cities (the distance is arbitrary because any distance gives the same average speed). Then, the pilot flew at 200 mph for 1 hour and at 100 mph for 2 hours.

Then, $r=\frac{d}{t}=\frac{400}{1+2}=\frac{400}{3}=133.\overline{3}$. Thus, the average speed is $133.\overline{3}$ mph.

5. The lighter coin can be found in two weighings as follows: Place two coins on each pan of the balance and set three coins off to the side. Case 1: The pans balance and the lighter coin is one of the three coins that were set off to the side. Case 2: The pans do not balance and the lighter coin is one of the two coins on the higher pan. To find the lighter coin in Case 1, work only with the three remaining coins. Place one coin on each side of the balance and set one coin off to the side. If the pans do not balance, the lighter coin is the one on the higher pan. To find the lighter coin in Case 2, work with only the two coins from the higher pan. Place one coin on each side of the balance. The lighter coin is on the higher pan.

6. (a) Surface area $(A)=4\pi r^2=4\pi(3960)^2\approx 197{,}060{,}797 \text{ mi}^2$.

   (b) $0.71(197{,}060{,}797)\approx 139{,}913{,}166 \text{ mi}^2$.

(c) $\frac{680,000}{139,913,166} \approx 0.00486$ miles. To convert 0.00486 miles to feet: $0.00486(5280) \approx 25.7$ feet.

(d) They would be flooded.

(e) Divide the volume of the Antarctic ice cap by the surface area of the oceans: $\frac{6,300,000}{139,913,166} \approx 0.045$ miles.

To convert 0.045 miles to feet: $0.045(5280) \approx 237.7$ feet.

Chapters 1–2 Cumulative Review Exercises

1. The number 45 is a composite number because it has factors other than itself and 1; $45 = 3 \times 3 \times 5$

2. The number 37 is a prime number because its only factors are itself and 1.

3. $\frac{4}{3} \cdot \frac{3}{8} = \frac{4 \cdot 3}{3 \cdot 8} = \frac{12}{24} = \frac{1 \cdot 12}{2 \cdot 12} = \frac{1}{2} \cdot \frac{12}{12} = \frac{1}{2} \cdot 1 = \frac{1}{2}$

4. $\frac{2}{3} \div 6 = \frac{2}{3} \div \frac{6}{1} = \frac{2}{3} \cdot \frac{1}{6} = \frac{2 \cdot 1}{3 \cdot 6} = \frac{2}{18} = \frac{1 \cdot 2}{9 \cdot 2} = \frac{1}{9} \cdot \frac{2}{2} = \frac{1}{9} \cdot 1 = \frac{1}{9}$

5. $\frac{11}{12} - \frac{3}{8} = \frac{11}{12} \cdot \frac{2}{2} - \frac{3}{8} \cdot \frac{3}{3} = \frac{11 \cdot 2}{12 \cdot 2} - \frac{3 \cdot 3}{8 \cdot 3} = \frac{22}{24} - \frac{9}{24} = \frac{22-9}{24} = \frac{13}{24}$

6. $\frac{2}{3} + \frac{1}{5} = \frac{2}{3} \cdot \frac{5}{5} + \frac{1}{5} \cdot \frac{3}{3} = \frac{2 \cdot 5}{3 \cdot 5} + \frac{1 \cdot 3}{5 \cdot 3} = \frac{10}{15} + \frac{3}{15} = \frac{10+3}{15} = \frac{13}{15}$

7. -1 is a rational number, and is an integer.

8. $\sqrt{3}$ is an irrational number.

9. $15 - 4 \cdot 3 = 15 - 12 = 3$

10. $30 \div 6 \cdot 2 = 5 \cdot 2 = 10$

11. $23 - 4^2 \div 2 = 23 - 16 \div 2 = 23 - 8 = 15$

12. $11 - \frac{3+1}{6-4} = 11 - \frac{4}{2} = 11 - 2 = 9$

13. $-14 - (-7) = -14 + 7 = -7$

14. $-\frac{2}{3} \cdot \left(-\frac{9}{14}\right) = \frac{2 \cdot 9}{3 \cdot 14} = \frac{18}{42} = \frac{3 \cdot 6}{7 \cdot 6} = \frac{3}{7} \cdot \frac{6}{6} = \frac{3}{7} \cdot 1 = \frac{3}{7}$

15. $5x^3 - x^3 = (5-1)x^3 = 4x^3$

16. $4 + 2x - 1 + 3x = 4 - 1 + 2x + 3x = 4 - 1 + (2+3)x = 3 + 5x$

17. $x - 3 = 11 \Rightarrow x - 3 + 3 = 11 + 3 \Rightarrow x = 14$

18. $4x - 6 = -22 = 4x - 6 + 6 = -22 + 6 \Rightarrow 4x = -16 \Rightarrow \frac{4x}{4} = \frac{-16}{4} \Rightarrow x = -4$

19. $5(6y+2) = 25 \Rightarrow 5(6y) + 5(2) = 25 \Rightarrow 30y + 10 = 25 \Rightarrow 30y + 10 - 10 = 25 - 10 \Rightarrow 30y = 15 \Rightarrow$

$\frac{30y}{30} = \frac{15}{30} \Rightarrow y = \frac{1}{2}$

20. $11-(y+2)=3y+5 \Rightarrow 11-y-2=3y+5 \Rightarrow 11-2-y=3y+5$

$\Rightarrow 9-y=3y+5 \Rightarrow 9-y+y=3y+y+5 \Rightarrow 9=4y+5 \Rightarrow 9-5=4y+5-5 \Rightarrow 4=4y \Rightarrow \frac{4}{4}=\frac{4y}{4} \Rightarrow 1=y$

$\Rightarrow y=1$

21. First, solve for x: $6x+2=2(3x+1) \Rightarrow 6x+2=6x+2 \Rightarrow 6x-6x+2=6x-6x+2 \Rightarrow 2=2.$ Because this statement is true for any value of x, the equation has infinitely many solutions.

22. First, solve for x: $2(3x-4)=6(x-1) \Rightarrow 6x-8=6x-6 \Rightarrow 6x-6x-8=6x-6x-6 \Rightarrow -8=-6.$ Because this statement is not true, the equation has no solutions.

23. Let x represent the first integer. Then $x+1$ and $x+2$ represent the other two integers. The sum of the integers is $x+x+1+x+2,$ so $x+x+1+x+2=90 \Rightarrow 3x+3=90 \Rightarrow 3x+3-3=90-3 \Rightarrow 3x=87$

$\Rightarrow \frac{3x}{3}=\frac{87}{3} \Rightarrow x=29,\ x+1=30,\ x+2=31.$ The integers are 29, 30, and 31.

24. $4.7\% = 4.7 \times 0.01 = 0.047$

25. $0.17 = 0.17 \times 100\% = 17\%$

26. Given that the distance traveled (d) is 325 miles and that the time spent traveling (t) is 5 hours, calculate the speed of the car (r). Then, $d=rt \Rightarrow 325 = r\cdot 5 \Rightarrow \frac{325}{5}=\frac{r\cdot 5}{5} \Rightarrow 65=r \Rightarrow r=65$ miles/hour.

27. The sum of the measures of the angles of a triangle is 180°. Then

$2x+3x+4x=180 \Rightarrow (2+3+4)x=180 \Rightarrow 9x=180 \Rightarrow \frac{9x}{9}=\frac{180}{9} \Rightarrow x=20$

28. Since the diameter of the circle is 10 inches, the radius is $\frac{10}{2}=5$ inches. Then,

$A=\pi r^2=\pi 5^2=25\pi \approx 78.5 \text{ in}^2.$

29. $a=3xy-4 \Rightarrow a+4=3xy-4+4 \Rightarrow a+4=3xy \Rightarrow \frac{a+4}{3y}=\frac{3xy}{3y} \Rightarrow \frac{a+4}{3y}=x$

30. $A=\frac{x+y+z}{3} \Rightarrow 3A=3\left(\frac{x+y+z}{3}\right) \Rightarrow 3A=x+y+z \Rightarrow 3A-y-z=x+y-y+z-z \Rightarrow 3A-y-z=x$

31. $7-3x>4 \Rightarrow 7-7-3x>4-7 \Rightarrow -3x>4+(-7) \Rightarrow -3x>-3 \Rightarrow \frac{-3x}{-3}<\frac{-3}{-3} \Rightarrow x<1$

−3 −2 −1 0 1 2 3

32. $6x \le 5-(x-9) \Rightarrow 6x \le 5-x+9 \Rightarrow 6x \le 14-x \Rightarrow 6x+x \le 14-x+x \Rightarrow 7x \le 14 \Rightarrow \frac{7x}{7} \le \frac{14}{7} \Rightarrow x \le 2$

−3 −2 −1 0 1 2 3

33. $I = 36Y$

34. Start with the initial balance, subtract the amounts withdrawn and add the amounts deposited:

$468 - 14 + 200 - 73 - 21 + 58 = 454 + 200 - 73 - 21 + 58 = 654 - 73 - 21$

$+ 58 = 581 - 21 + 58 = 560 + 58 = \$618$

35. Let x represent the amount of 4% acid. Then, the 6% acid solution is the sum of the 4% acid and the 10% acid. Therefore

$0.04x + 0.10(150) = 0.06\,(x + 150)$

$\Rightarrow 0.04x + 15 = 0.06x + 9$

$\Rightarrow 0.04x - 0.04x + 15 = 0.06x - 0.04x + 9$

$\Rightarrow 15 = 0.02x + 9 \Rightarrow 15 - 9 = 0.02x + 9 - 9 \Rightarrow 6 = 0.02x$

$$\Rightarrow \frac{6}{0.02} = \frac{0.02x}{0.02} \Rightarrow 300 = x$$

300 mL of 4% acid solution should be mixed with the 10% acid solution to dilute it to a 6% acid solution.

36. Let x represent the amount borrowed at 4%. Then the amount borrowed at 6% is $x + 250$. The total interest for one year will equal the sum of the interest for each loan. Therefore:

$0.04x + 0.06(x + 250) = 165$

$\Rightarrow 0.04x + 0.06x + 15 = 165$

$\Rightarrow 0.10x + 15 = 165 \Rightarrow 0.10x + 15 - 15 = 165 - 15$

$$\Rightarrow 0.10x = 150 \Rightarrow \frac{0.10x}{0.10} = \frac{150}{0.10} \Rightarrow x = 1500, \quad x + 250 = 1500 + 250 = 1750.$$

Therefore, the amount borrowed at 4% interest is \$1500 and the amount borrowed at 6% interest is \$1750.

Critical Thinking Solutions for Chapter 2

Section 2.1

- If an error is made, the resulting equations may not be equivalent to the given equation.

Section 2.2

- Solve for x: $bx - 2 = dx + 7 \Rightarrow bx - 2 + 2 = dx + 7 + 2 \Rightarrow bx = dx + 9 \Rightarrow$

  $bx - dx = dx - dx + 9 \Rightarrow bx - dx = 9 \Rightarrow x(b - d) = 9 \Rightarrow \dfrac{x(b-d)}{b-d} = \dfrac{9}{b-d} \Rightarrow x = \dfrac{9}{b-d}$. Thus, if

  $b = d$, then $x = \dfrac{9}{0}$. Because dividing by 0 is not allowed, there are no solutions. If $b \neq d$, then there is one solution.

Section 2.3

- Let x represent the lower salary. Then, $x + 2x$ equals the increased salary amount. Thus, because $x + 2x = 3x$, the lower salary increased by a factor of 3.

Section 2.4

- $C = 2\pi r = \pi 2r = \pi d;\quad A = \pi r^2 = \pi\left(\frac{1}{2}d\right)^2 = \frac{1}{4}\pi d^2$

- Yes. Multiply one expression by 1 in the form $\frac{-1}{-1}$ to transform it to the other.

Section 2.5

- $-5-3x > -2x+7 \Rightarrow -5-3x+3x > -2x+3x+7 \Rightarrow -5 > x+7 \Rightarrow -5-7 > x+7-7 \Rightarrow\ -12 > x$

Chapter 3: Graphing Equations

3.1: Introduction to Graphing

Concepts

1. xy-plane
2. origin
3. $(0, 0)$
4. 4
5. II
6. III
7. axes
8. x; y
9. scatterplot
10. line

Cartesian Coordinate Plane

11. $(-2, -2)$, $(-2, 2)$, $(0, 0)$, $(2, 2)$

13. $(-1, 0)$, $(0, -3)$, $(0, 2)$, $(2, 0)$

15. (a) Quadrant I

    (b) Quadrant III

17. (a) None, because the point is on the axis.

    (b) Quadrant I

19. (a) Quadrant II

    (b) Quadrant IV

21. I and III

23. See Figure 23.

25. See Figure 25.

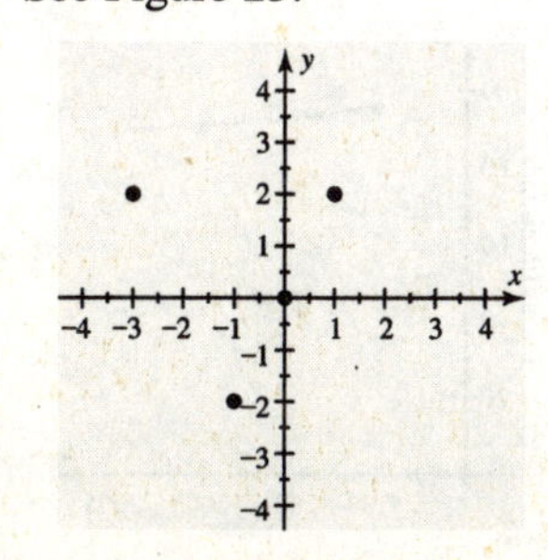

Figure 23

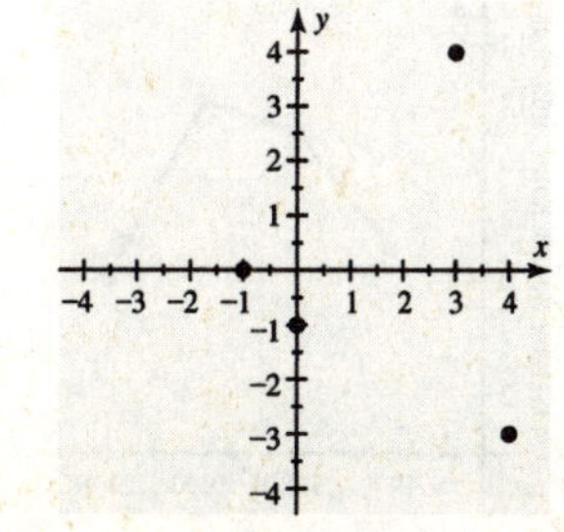

Figure 25

27. See Figure 27.

29. See Figure 29.

31. See Figure 31.

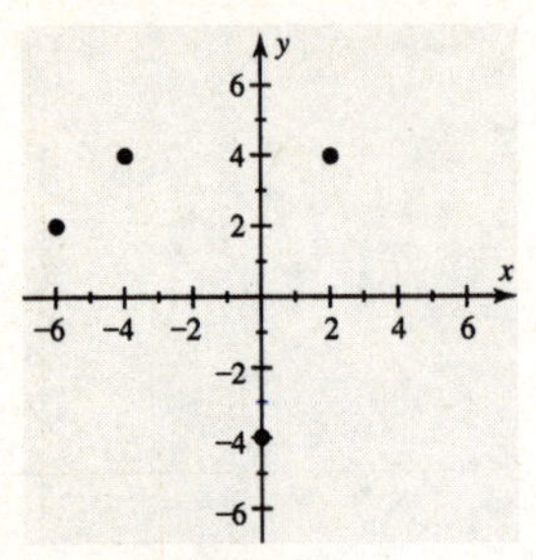

Figure 27

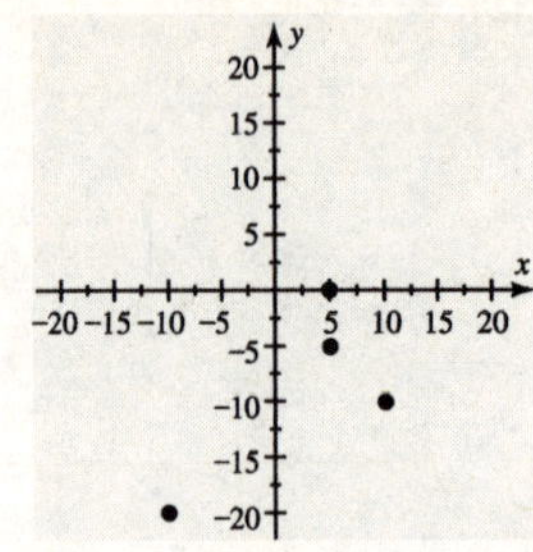

Figure 29

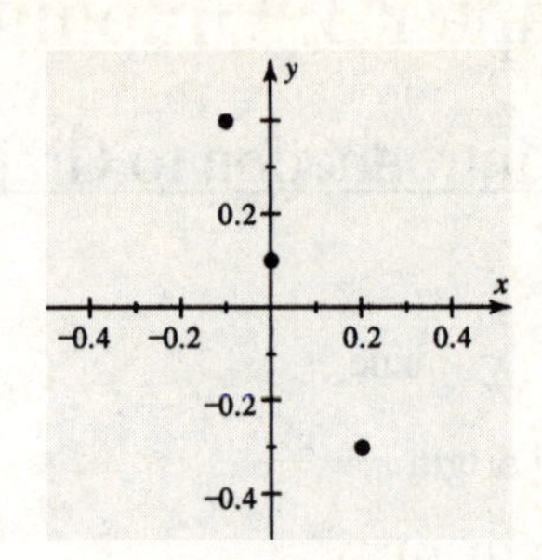

Figure 31

33. See Figure 33.

35. See Figure 35.

37. See Figure 37.

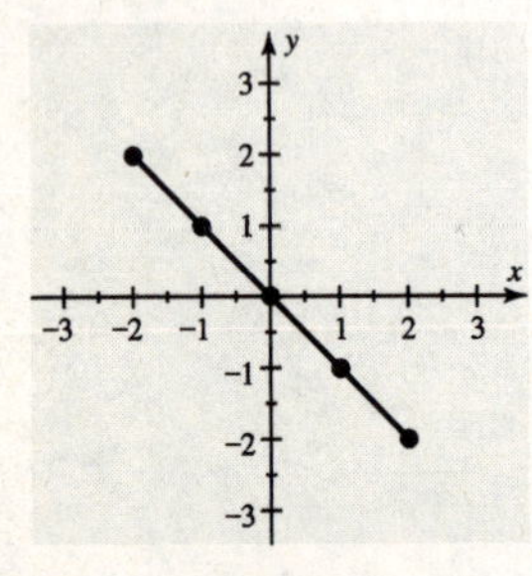

Figure 33

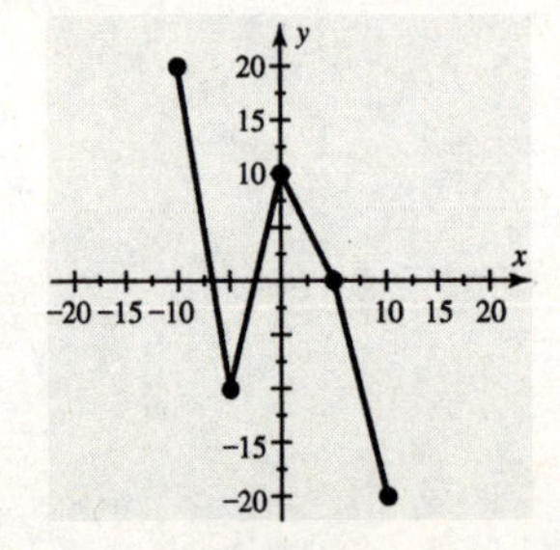

Figure 35

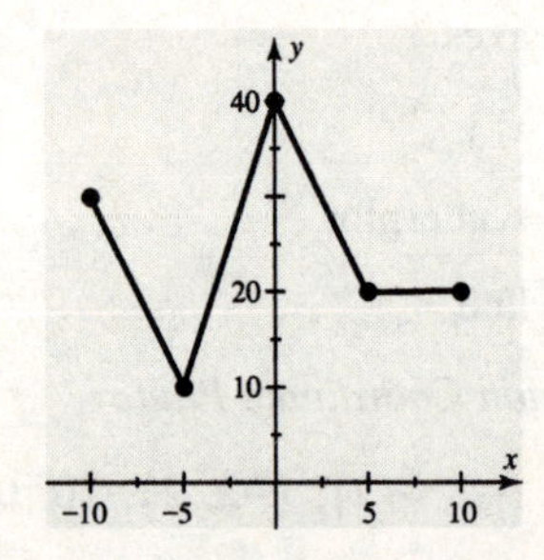

Figure 37

39. (1940, 182), (1960, 484), (1980, 632), (2000, 430); In 1940, there were 182 billion cigarettes consumed in the U.S. *Answers may vary slightly.*

Graphing Real Data

41. (a) See Figure 41.

    (b) Federal income tax receipts increased.

43. (a) See Figure 43.

    (b) The number of welfare beneficiaries increased and then decreased.

45. (a) See Figure 45.

    (b) The number of farms decreased.

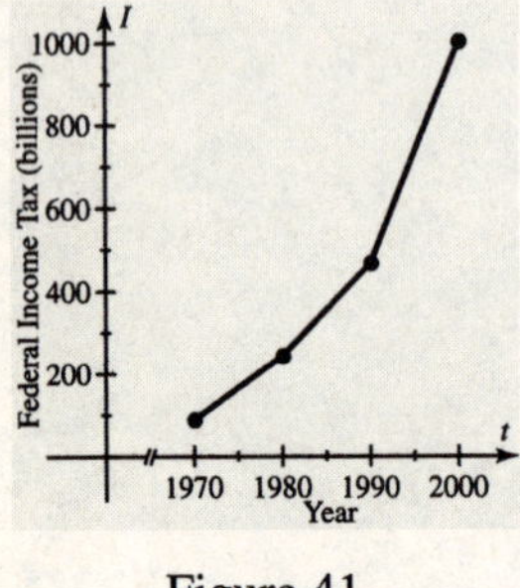

Figure 41

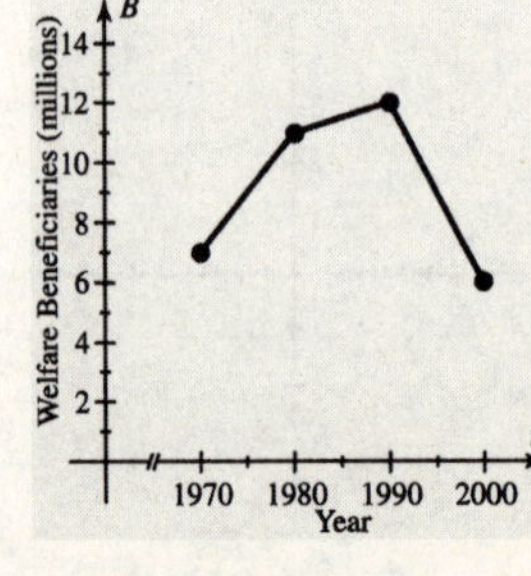

Figure 43

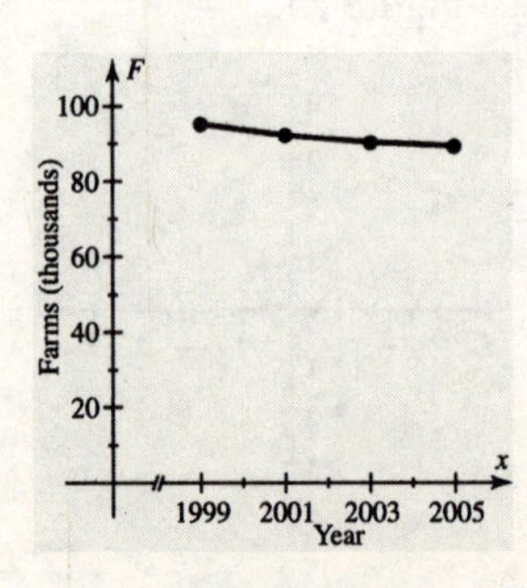

Figure 45

47. (a) The rate decreased.

    (b) From the graph, the infant mortality rate in 1990 was 9 per 1000 births.

(c) From the graph, the rate in 1960 was about 26 and the rate in 2000 was about 7. The percent change was

$\frac{7-26}{26}\times 100 \approx -73.1\%.$

3.2: Linear Equations in Two Variables

Concepts

1. two
2. ordered pair
3. solution
4. linear
5. is
6. solutions
7. graph
8. line

Solutions to Equations

9. Substitute 5 for x and 6 for y: $y = x+1 \Rightarrow 6 = 5+1 \Rightarrow 6 = 6$. This is a true statement, so the ordered pair (5, 6) is a solution.

11. Substitute 2 for x and 13 for y: $y = 4x+7 \Rightarrow 13 = 4(2)+7 \Rightarrow 13 = 8+7 \Rightarrow 13 = 15$. This is not a true statement, so the ordered pair (2, 13) is not a solution.

13. Substitute -2 for x and 3 for y: $4x - y = -13 \Rightarrow 4(-2)-3 = -13 \Rightarrow -8-3 = -13 \Rightarrow -11 = -13$. This is not a true statement, so the ordered pair $(-2, 3)$ is not a solution.

15. Substitute $\frac{1}{2}$ for x and 2 for y: $y - 6x = -1 \Rightarrow 2 - 6\left(\frac{1}{2}\right) = -1 \Rightarrow 2-3 = -1 \Rightarrow -1 = -1$. This is a true statement, so the ordered pair $\left(\frac{1}{2}, 2\right)$ is a solution.

17. Substitute 100 for x and 100 for y: $0.31x - 0.42y = -9 \Rightarrow 0.31(100) - 0.42(100) = -9 \Rightarrow$ $31 - 42 = -9 \Rightarrow -11 = -9$. This is not a true statement, so the ordered pair (100, 100) is not a solution.

19. When $x = -1$: $y = 4x \Rightarrow y = 4(-1) \Rightarrow y = -4$; when $x = 0$: $y = 4x \Rightarrow y = 4(0) \Rightarrow y = 0$; when $x = 1$: $y = 4x \Rightarrow y = 4(1) \Rightarrow y = 4$; when $x = 2$: $y = 4x \Rightarrow y = 4(2) \Rightarrow y = 8$.

    Thus, the missing values in the table are -4, 0, 4 and 8. See Figure 19.

x	−2	−1	0	1	2
y	−8	−4	0	4	8

Figure 19

x	−8	−4	0	4	8
y	−4	0	4	8	12

Figure 21

21. When $y = 0$: $y = x+4 \Rightarrow 0 = x+4 \Rightarrow 0-4 = x+4-4 \Rightarrow -4 = x \Rightarrow x = -4$;

    when $y = 4$: $y = x+4 \Rightarrow 4 = x+4 \Rightarrow 4-4 = x+4-4 \Rightarrow 0 = x \Rightarrow x = 0$;

when $y = 8$: $y = x+4 \Rightarrow 8 = x+4 \Rightarrow 8-4 = x+4-4 \Rightarrow 4 = x \Rightarrow x = 4$;

when $y = 12$: $y = x+4 \Rightarrow 12 = x+4 \Rightarrow 12-4 = x+4-4 \Rightarrow 8 = x \Rightarrow x = 8$.

Thus, the missing values in the table are –4, 0, 4 and 8. See Figure 21.

23. When $y = -2$: $3y+2x = 6 \Rightarrow 3(-2)+2x = 6 \Rightarrow -6+2x = 6 \Rightarrow -6+6+2x = 6+6 \Rightarrow$

$2x = 12 \Rightarrow \frac{2x}{2} = \frac{12}{2} \Rightarrow x = 6$; when $y = 0$: $3y+2x = 6 \Rightarrow 3(0)+2x = 6 \Rightarrow 2x = 6 \Rightarrow \frac{2x}{2} = \frac{6}{2} \Rightarrow$

$x = 3$; when $y = 2$: $3y+2x = 6 \Rightarrow 3(2)+2x = 6 \Rightarrow 6+2x = 6 \Rightarrow 6-6+2x = 6-6 \Rightarrow$

$2x = 0 \Rightarrow \frac{2x}{2} = \frac{0}{2} \Rightarrow x = 0$; when $y = 4$: $3y+2x = 6 \Rightarrow 3(4)+2x = 6 \Rightarrow 12+2x = 6 \Rightarrow$

$12-12+2x = 6-12 \Rightarrow 2x = -6 \Rightarrow \frac{2x}{2} = \frac{-6}{2} \Rightarrow x = -3$; when $y = 8$: $3y+2x = 6 \Rightarrow$

$3(8)+2x = 6 \Rightarrow 24+2x = 6 \Rightarrow 24-24+2x = 6-24 \Rightarrow 2x = -18 \Rightarrow \frac{2x}{2} = \frac{-18}{2} \Rightarrow x = -9$.

Thus, the missing values in the table are 6, 3, 0, −3 and −9. See Figure 23.

25. See Figure 25.

27. See Figure 27.

x	6	3	0	−3	−9
y	−2	0	2	4	8

Figure 23

x	−3	0	3	6
y	−9	0	9	18

Figure 25

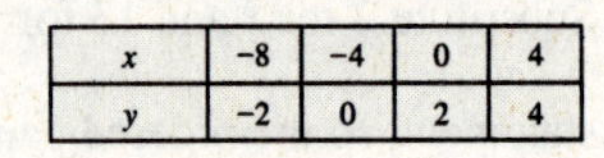

x	−8	−4	0	4
y	−2	0	2	4

Figure 27

29. See Figure 29.

31. See Figure 31.

x	8	6	4	2
y	−2	0	2	4

Figure 29

x	$-\frac{1}{2}$	$-\frac{1}{4}$	0	$\frac{1}{4}$
y	−2	−1	0	1

Figure 31

33. They must be multiples of 5.

35.

x	−1	0	1
y	2	0	−2

Table values may vary.

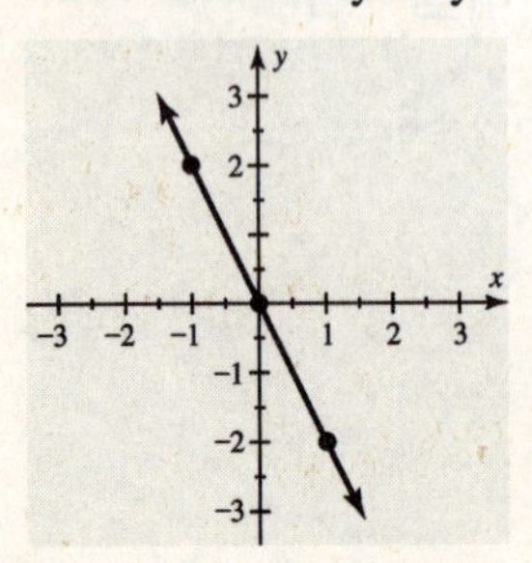

37.

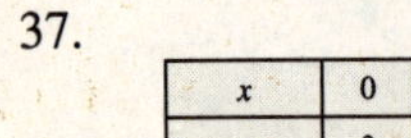

x	0	1	2
y	3	2	1

Table values may vary.

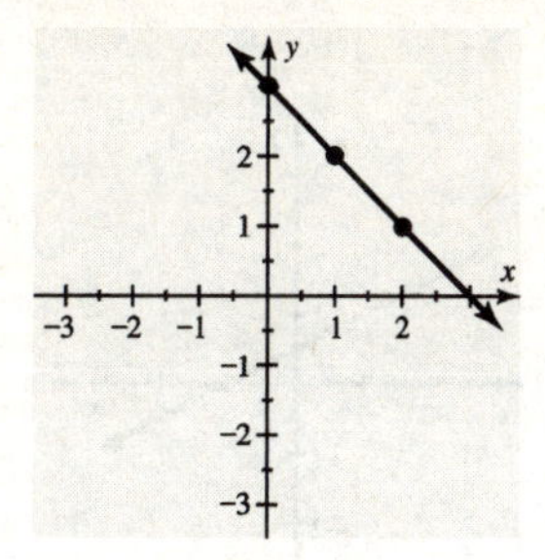

39.

x	−2	0	2
y	3	2	1

Table values may vary.

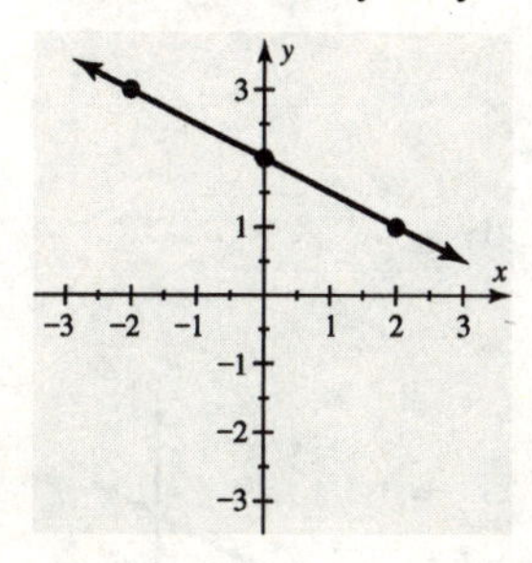

Graphing Equations

41. See Figure 41.

43. See Figure 43.

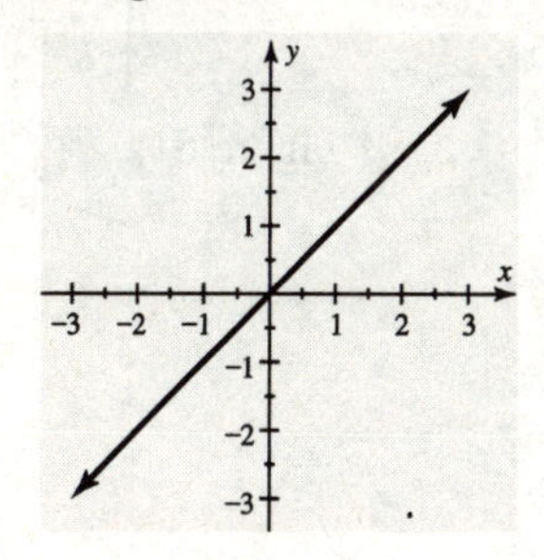

Figure 41

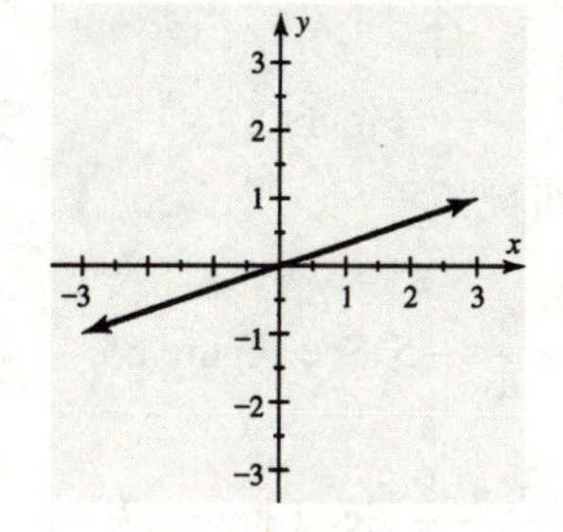

Figure 43

45. See Figure 45.

47. See Figure 47.

49. See Figure 49.

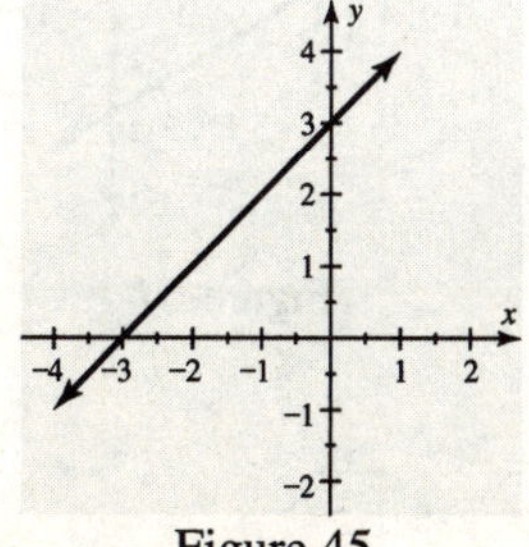

Figure 45

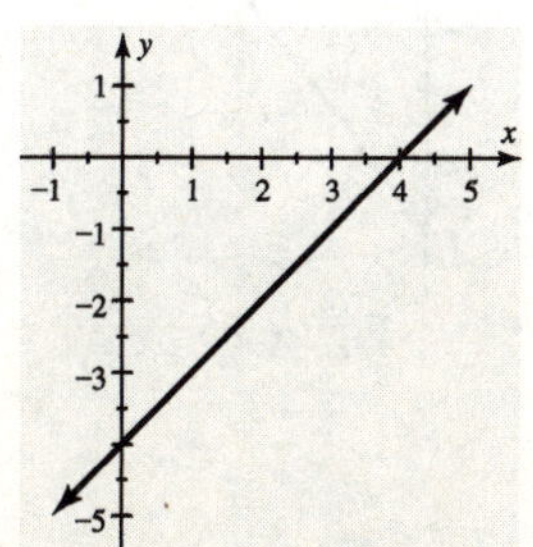

Figure 47

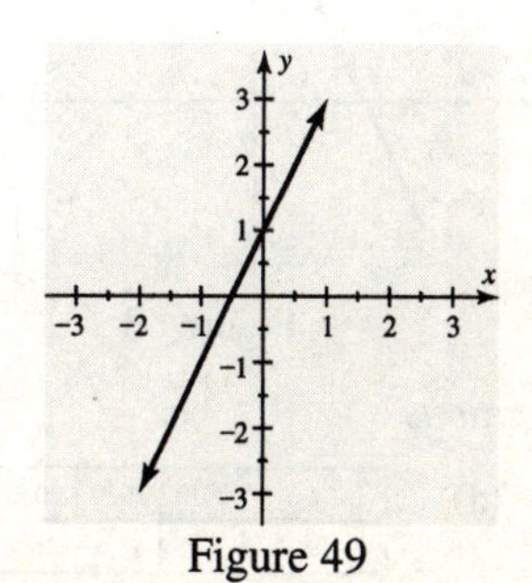

Figure 49

51. See Figure 51.

53. See Figure 53.

55. See Figure 55.

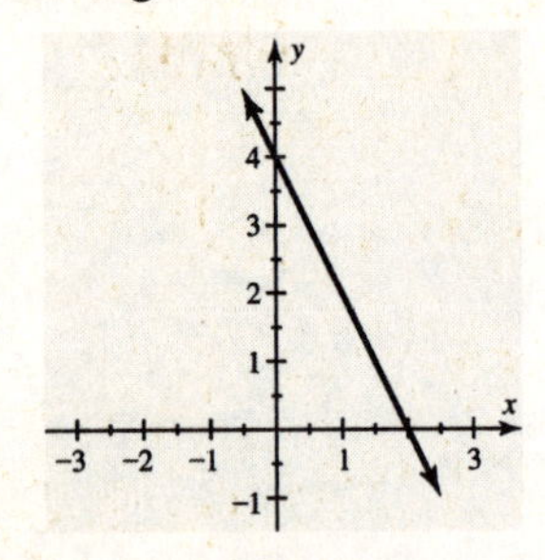

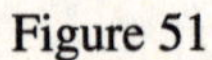
Figure 51

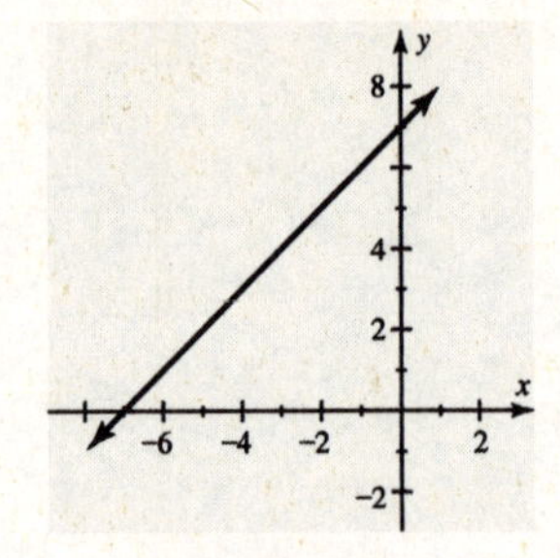

Figure 53

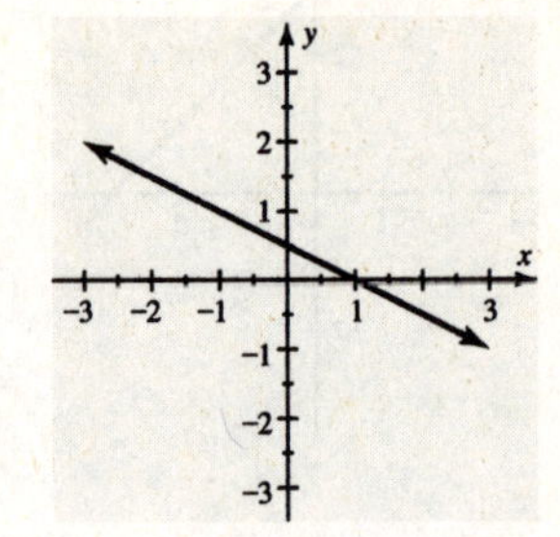

Figure 55

57. $2x+3y=6 \Rightarrow 3y=-2x+6 \Rightarrow y=-\dfrac{2}{3}x+2$ See Figure 57.

59. $x+4y=4 \Rightarrow 4y=-x+4 \Rightarrow y=-\dfrac{1}{4}x+1$ See Figure 59.

61. $-x+2y=8 \Rightarrow 2y=x+8 \Rightarrow y=\dfrac{1}{2}x+4$ See Figure 61.

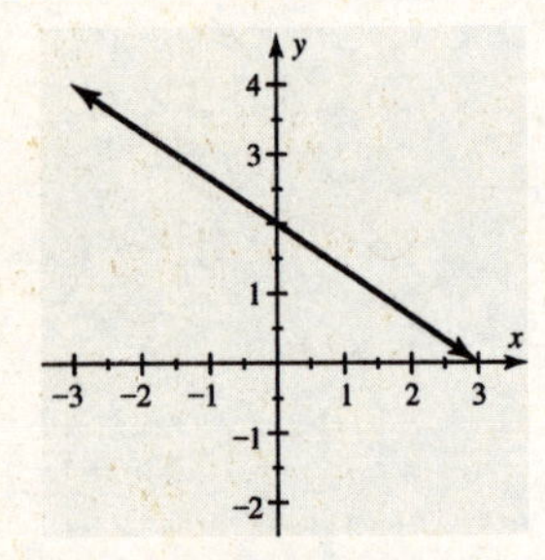

Figure 57

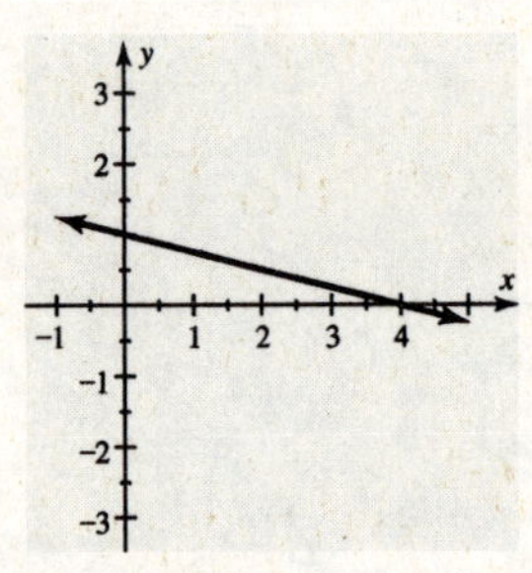

Figure 59

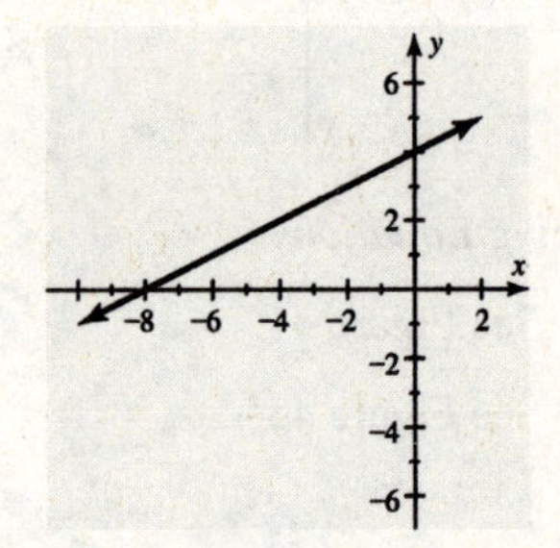

Figure 61

63. $y-2x=7 \Rightarrow y=2x+7$ See Figure 63.

65. $5x-4y=20 \Rightarrow -4y=-5x+20 \Rightarrow y=\dfrac{5}{4}x-5$ See Figure 65.

67. $3x+5y=-9 \Rightarrow 5y=-3x-9 \Rightarrow y=-\dfrac{3}{5}x-\dfrac{9}{5}$ See Figure 67.

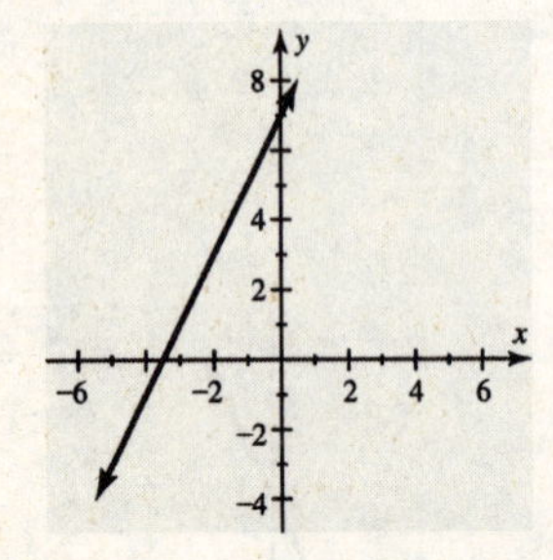

Figure 63

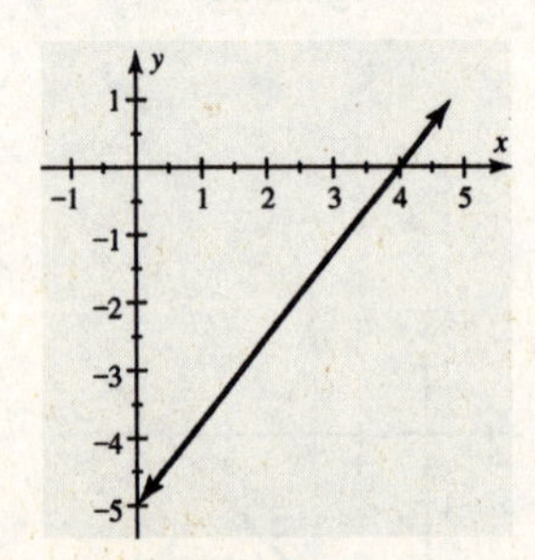

Figure 65

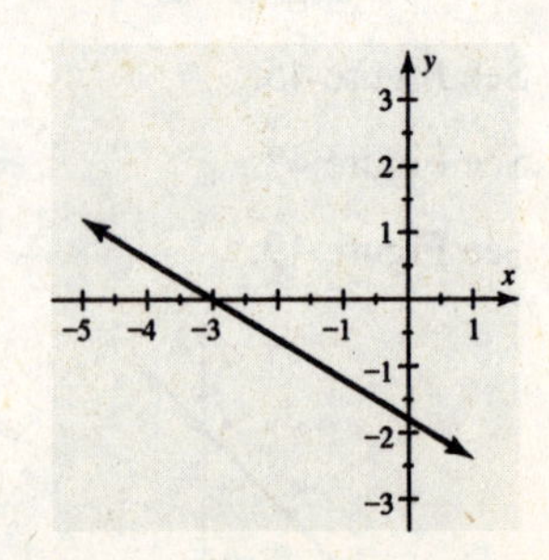

Figure 67

Applications

69. (a)

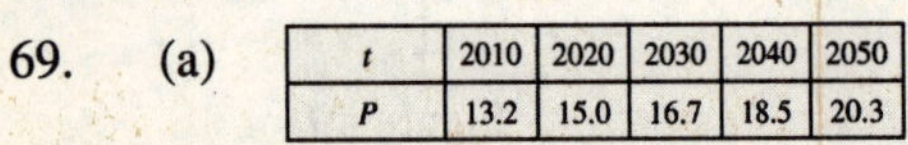

t	2010	2020	2030	2040	2050
P	13.2	15.0	16.7	18.5	20.3

(b) From the table, the percentage is 16.7% in 2030.

71. (a) See Figure 71.

(b) Set $A = 100$ and solve for t: $A = 2.7t \Rightarrow 100 = 2.7t \Rightarrow \dfrac{100}{2.7} = \dfrac{2.7t}{2.7} \Rightarrow t \approx 37$. Therefore, about 37 days.

73. (a) For year 1985, set $t =$ 1985 and solve for P: $P = 5.9t - 11{,}709 = 5.9(1985) - 11{,}709 =$

$11{,}711.5 - 11{,}709 = 2.5$. Therefore, in year 1985 it was 2.5%.

For year 2000, set $t =$ 2000 and solve for P: $P = 5.9t - 11{,}709 = 5.9(2000) - 11{,}709 =$

$11{,}800 - 11{,}709 = 91$. Therefore, in year 2000 it was 91%.

(b) See Figure 73.

(c) P was 79.2% in year 1998.

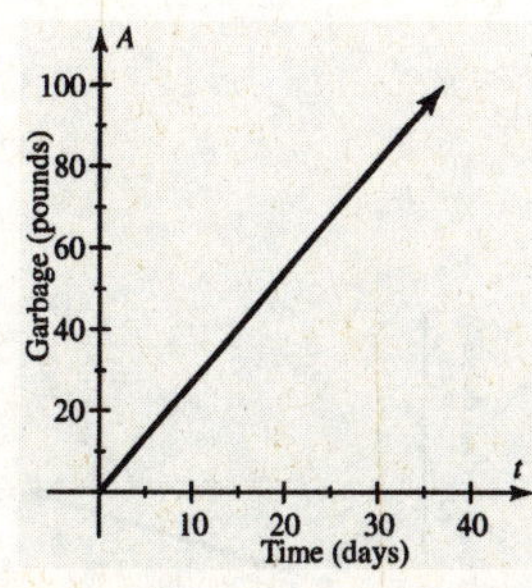

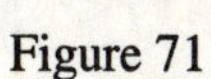
Figure 71

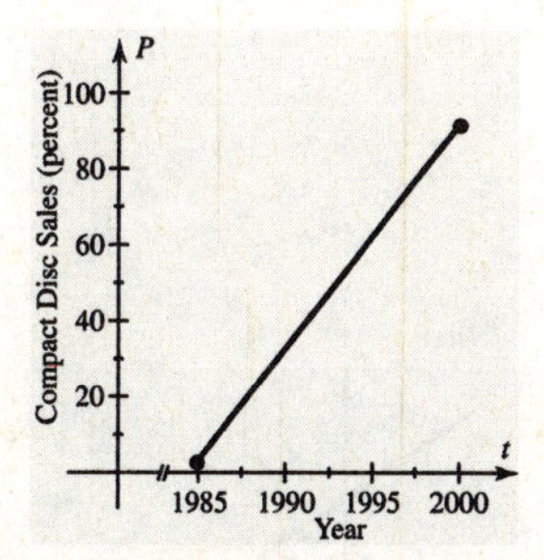

Figure 73

Checking Basic Concepts for Sections 3.1 & 3.2

1. $(-2, 2)$, II; $(-1, -2)$, III; $(1, 3)$, I; $(3, 0)$, none.

2. See Figure 2.

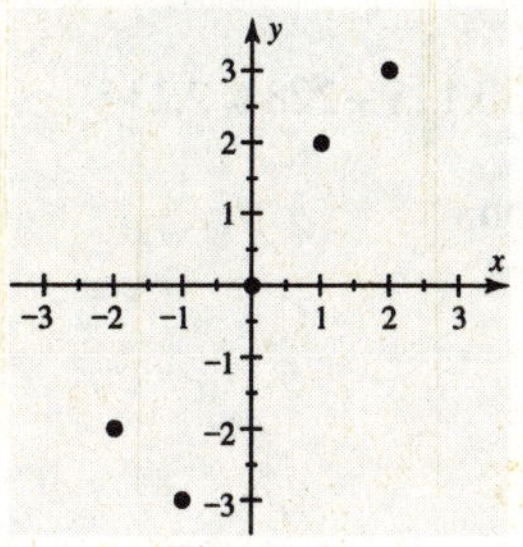

Figure 2

3. See Figure 3.

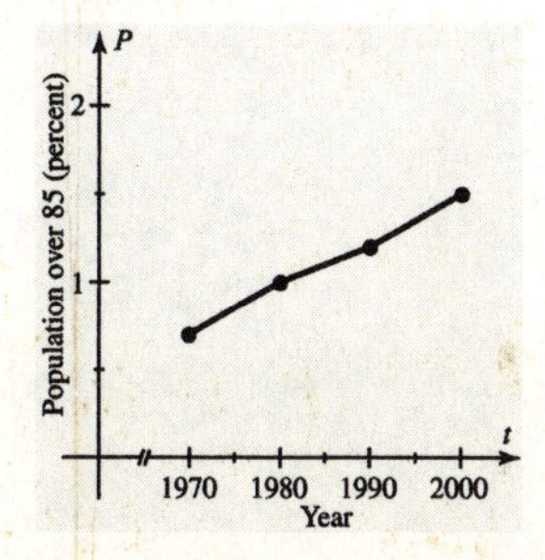

The percentage increased.

4. Substitute -2 for x and -3 for y: $-2x-y=7 \Rightarrow -2(-2)-(-3)=7 \Rightarrow 4+3=7 \Rightarrow 7=7$. This is a true statement, so the ordered pair $(-2, -3)$ is a solution.

5. When $x=-1$: $y=-2x+1=-2(-1)+1=2+1=3$;

   when $x=0$: $y=-2x+1=-2(0)+1=1$;

   when $x=1$: $y=-2x+1=-2(1)+1=-2+1=-1$;

   when $x=2$: $y=-2x+1=-2(2)+1=-4+1=-3$.

   Thus, the missing values in the table are $3,\ 1,\ -1,\ -3$. See Figure 5.

x	-2	-1	0	1	2
y	5	3	1	-1	-3

Figure 5

6. (a) See Figure 6a.

   (b) See Figure 6b.

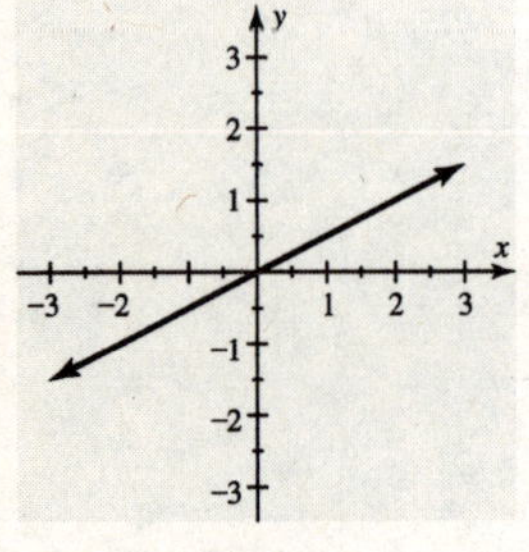

Figure 6a

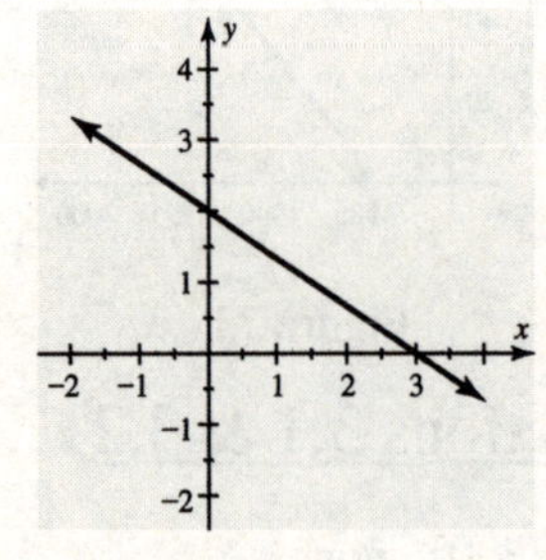

Figure 6b

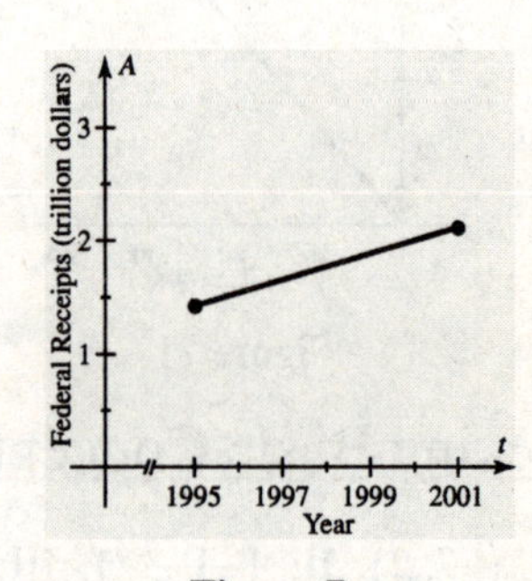

Figure 7

7. (a) Set $t=1995$ and solve for A: $A=0.115t-228=0.115(1995)-228=229.425-228=1.425$

   Set $t=2001$ and solve for A: $A=0.115t-228=0.115(2001)-228=230.115-228=2.115$

   In 1995, receipts were \$1.425 trillion; in 2001, receipts were \$2.115 trillion.

   (b) See Figure 7.

   (c) 1996

3.3: More Graphing of Lines

Concepts

1. two
2. one; one
3. x-intercept
4. 0
5. y-intercept
6. 0
7. 3
8. horizontal
9. $y=b$

10. 3

11. vertical

12. $x = k$

Finding Intercepts

13. Because the graph crosses the x-axis at $x = 3$, $x = 3$ is the x-intercept.

Because the graph crosses the y-axis at $y = -2$, $y = -2$ is the y-intercept.

15. Because the graph crosses the x-axis at $x = 0$, $x = 0$ is the x-intercept.

Because the graph crosses the y-axis at $y = 0$, $y = 0$ is the y-intercept.

17. Because the graph crosses the x-axis at $x = -2$ and at $x = 2$, $x = -2$ and $x = 2$ are the x-intercepts.

Because the graph crosses the y-axis at $y = 4$, $y = 4$ is the y-intercept.

19. Because the graph touches the x-axis at $x = 1$, $x = 1$ is the x-intercept.

Because the graph crosses the y-axis at $y = 1$, $y = 1$ is the y-intercept.

21. When $x = -2$: $y = x + 2 = -2 + 2 = 0$; when $x = -1$: $y = x + 2 = -1 + 2 = 1$;

when $x = 0$: $y = x + 2 = 0 + 2 = 2$; when $x = 1$: $y = x + 2 = 1 + 2 = 3$;

when $x = 2$: $y = x + 2 = 2 + 2 = 4$. Therefore, the missing values in the table are 0, 1, 2, 3 and 4.

See Figure 21. The x-intercept is at $x = -2$, because at $x = -2$ the value of y *is* 0.

The y-intercept is at $y = 2$, because at $y = 2$ the value of x is 0.

x	−2	−1	0	1	2
y	0	1	2	3	4

Figure 21

x	−4	−2	0	2	4
y	−6	−4	−2	0	2

Figure 23

23. When $x = -4$: $-x + y = -2 \Rightarrow -(-4) + y = -2 \Rightarrow 4 + y = -2 \Rightarrow 4 - 4 + y = -2 - 4 \Rightarrow y = -6$;

when $x = -2$: $-x + y = -2 \Rightarrow -(-2) + y = -2 \Rightarrow 2 + y = -2 \Rightarrow 2 - 2 + y = -2 - 2 \Rightarrow y = -4$;

when $x = 0$: $-x + y = -2 \Rightarrow -(0) + y = -2 \Rightarrow y = -2$;

when $x = 2$: $-x + y = -2 \Rightarrow -(2) + y = -2 \Rightarrow -2 + y = -2 \Rightarrow -2 + 2 + y = -2 + 2 \Rightarrow y = 0$;

when $x = 4$: $-x + y = -2 \Rightarrow -(4) + y = -2 \Rightarrow -4 + y = -2 \Rightarrow -4 + 4 + y = -2 + 4 \Rightarrow y = 2$.

Therefore, the missing values in the table are −6, −4, −2, 0 and 2. See Figure 23.

The x-intercept is at $x = 2$, because at $x = 2$ the value of y is 0.

The y-intercept is at $y = -2$, because at $y = -2$ the value of x is 0.

25. To find the x-intercept, set $y = 0$ and solve for x: $-2x + 3y = -6 \Rightarrow -2x + 3(0) = -6 \Rightarrow$

$-2x = -6 \Rightarrow \frac{-2x}{-2} = \frac{-6}{-2} \Rightarrow x = 3$. Therefore, the x-intercept is at $x = 3$.

To find the y-intercept, set $x=0$ and solve for y:

$-2x+3y=-6 \Rightarrow -2(0)+3y=-6 \Rightarrow 3y=-6 \Rightarrow \frac{3y}{3}=\frac{-6}{3} \Rightarrow y=-2$. Therefore, the y-intercept is at $y=-2$. See Figure 25.

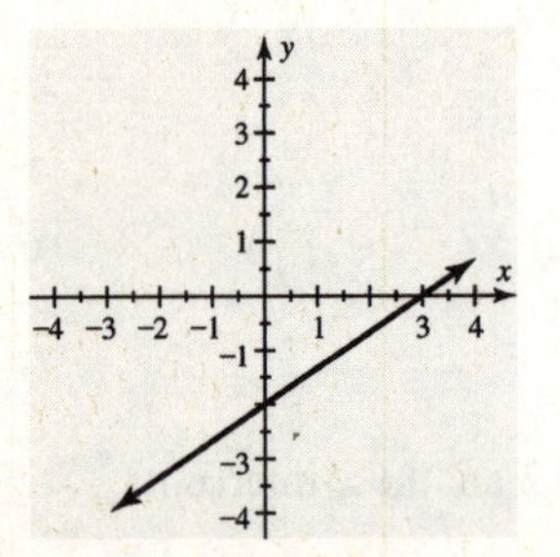

Figure 25

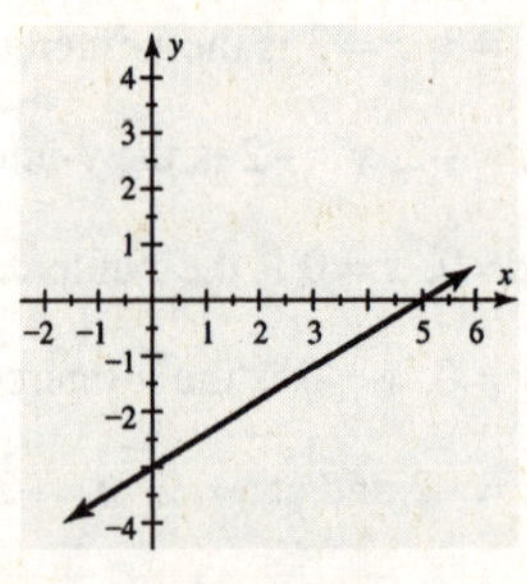

Figure 27

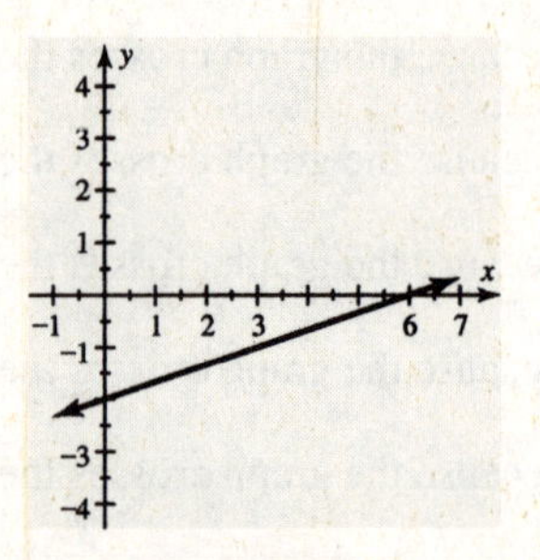

Figure 29

27. To find the x-intercept, set $y=0$ and solve for x: $3x-5y=15 \Rightarrow 3x-5(0)=15 \Rightarrow$

$3x=15 \Rightarrow \frac{3x}{3}=\frac{15}{3} \Rightarrow x=5$. Therefore, the x-intercept is at $x=5$.

To find the y-intercept, set $x=0$ and solve for y: $3x-5y=15 \Rightarrow 3(0)-5y=15 \Rightarrow$

$-5y=15 \Rightarrow \frac{-5y}{-5}=\frac{15}{-5} \Rightarrow y=-3$. Therefore, the y-intercept is at $y=-3$. See Figure 27.

29. To find the x-intercept, set $y=0$ and solve for x: $x-3y=6 \Rightarrow x-3(0)=6 \Rightarrow x=6$.

Therefore, the x-intercept is at $x=6$.

To find the y-intercept, set $x=0$ and solve for y: $x-3y=6 \Rightarrow 0-3y=6 \Rightarrow$

$-3y=6 \Rightarrow \frac{-3y}{-3}=\frac{6}{-3} \Rightarrow y=-2$. Therefore, the y-intercept is at $y=-2$. See Figure 29.

31. To find the x-intercept, set $y=0$ and solve for x: $6x-y=-6 \Rightarrow 6x-0=-6 \Rightarrow$

$6x=-6 \Rightarrow \frac{6x}{6}=\frac{-6}{6} \Rightarrow x=-1$. Therefore, the x-intercept is at $x=-1$.

To find the y-intercept, set $x=0$ and solve for y: $6x-y=-6 \Rightarrow 6(0)-y=-6 \Rightarrow -y=-6 \Rightarrow$

$-1(-y)=-1(-6) \Rightarrow y=6$. Therefore, the y-intercept is at $y=6$. See Figure 31.

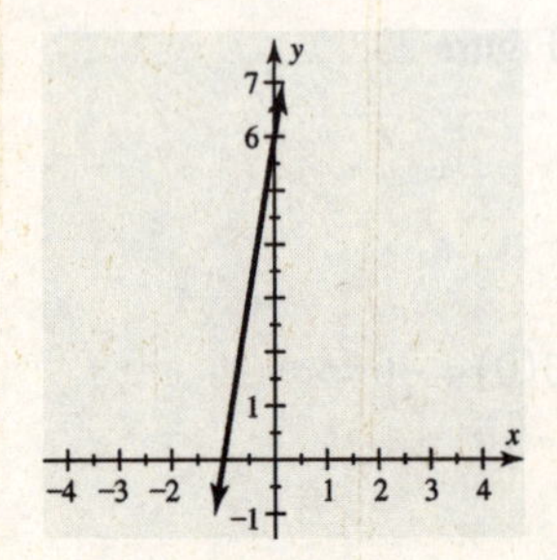

Figure 31

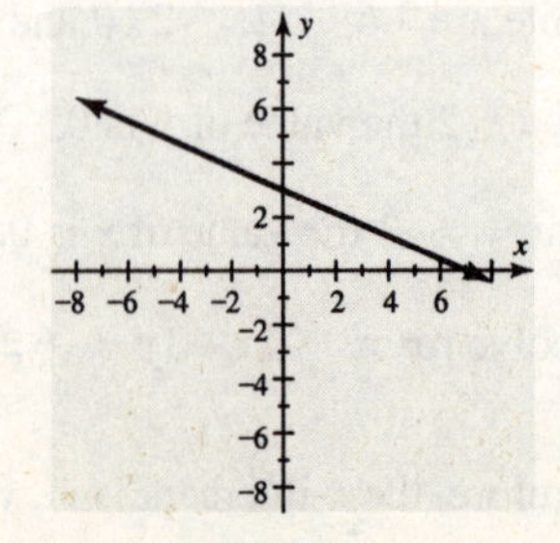

Figure 33

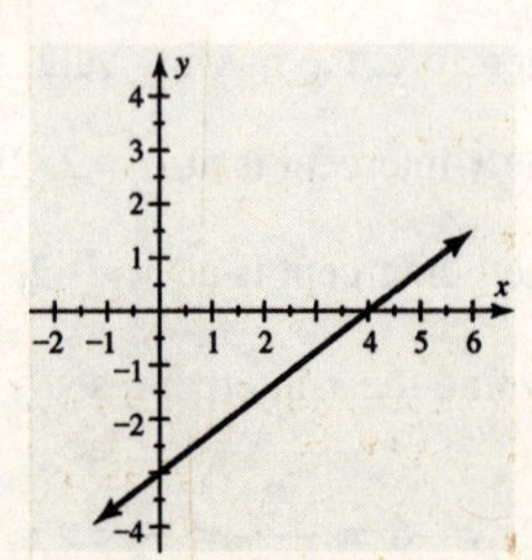

Figure 35

33. To find the x-intercept, set $y=0$ and solve for x: $3x+7y=21 \Rightarrow 3x+7(0)=21 \Rightarrow$

$3x=21 \Rightarrow \frac{3x}{3}=\frac{21}{3} \Rightarrow x=7$. Therefore, the x-intercept is at $x=7$.

To find the y-intercept, set $x=0$ and solve for y: $3x+7y=21 \Rightarrow 3(0)+7y=21 \Rightarrow$

$7y=21 \Rightarrow \frac{7y}{7}=\frac{21}{7} \Rightarrow y=3$. Therefore, the y-intercept is at $y=3$. See Figure 33.

35. To find the x-intercept, set $y=0$ and solve for x: $40y-30x=-120 \Rightarrow 40(0)-30x=-120 \Rightarrow$

$-30x=-120 \Rightarrow \frac{-30x}{-30}=\frac{-120}{-30} \Rightarrow x=4$. Therefore, the x-intercept is at $x=4$.

To find the y-intercept, set $x=0$ and solve for y: $40y-30x=-120 \Rightarrow 40y-30(0)=-120 \Rightarrow$

$40y=-120 \Rightarrow \frac{40y}{40}=\frac{-120}{40} \Rightarrow y=-3$. Therefore, the y-intercept is at $y=-3$. See Figure 35.

37. To find the x-intercept, set $y=0$ and solve for x: $\frac{1}{2}x-y=2 \Rightarrow \frac{1}{2}x-0=2 \Rightarrow$

$\frac{1}{2}x=2 \Rightarrow 2\left(\frac{1}{2}x\right)=2(2) \Rightarrow x=4$. Therefore, the x-intercept is at $x=4$.

To find the y-intercept, set $x=0$ and solve for y: $\frac{1}{2}x-y=2 \Rightarrow \frac{1}{2}(0)-y=2 \Rightarrow$

$-y=2 \Rightarrow -1(-y)=2(-1) \Rightarrow y=-2$. Therefore, the y-intercept is at $y=-2$. See Figure 37.

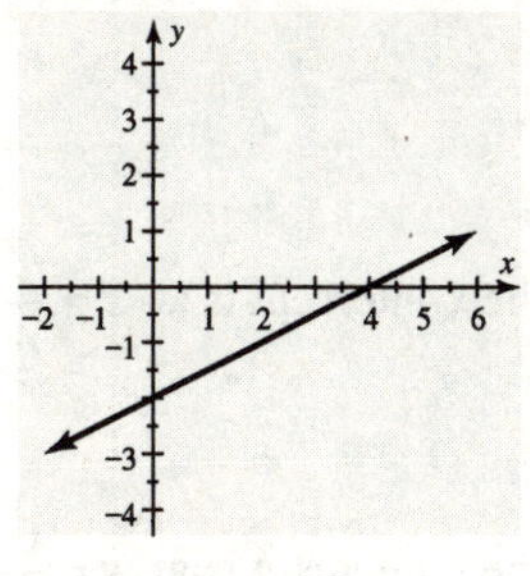

Figure 37

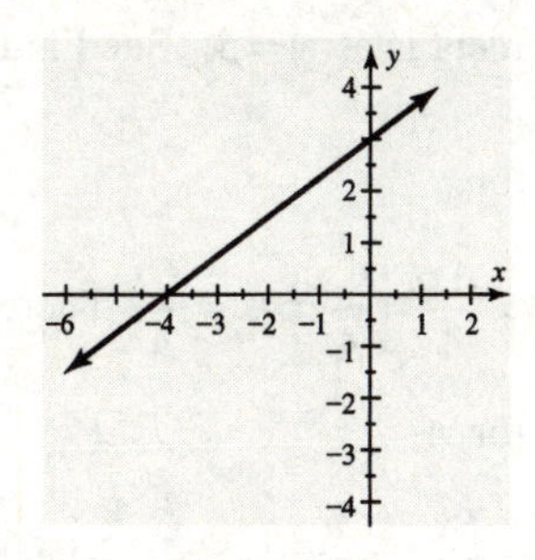

Figure 39

39. To find the x-intercept, set $y=0$ and solve for x: $-\frac{x}{4}+\frac{y}{3}=1 \Rightarrow -\frac{x}{4}+\frac{0}{3}=1 \Rightarrow$

$-\frac{x}{4}=1 \Rightarrow -4\left(-\frac{x}{4}\right)=1(-4) \Rightarrow x=-4$. Therefore, the x-intercept is at $x=-4$.

To find the y-intercept, set $x=0$ and solve for y: $-\frac{x}{4}+\frac{y}{3}=1 \Rightarrow -\frac{0}{4}+\frac{y}{3}=1 \Rightarrow$

$\frac{y}{3}=1 \Rightarrow 3\left(\frac{y}{3}\right)=3(1) \Rightarrow y=3$. Therefore, the y-intercept is at $y=3$. See Figure 39.

41. To find the x-intercept, set $y=0$ and solve for x: $\frac{x}{3}+\frac{y}{2}=1 \Rightarrow \frac{x}{3}+\frac{0}{2}=1 \Rightarrow$

$\frac{x}{3}=1 \Rightarrow 3\left(\frac{x}{3}\right)=1(3) \Rightarrow x=3$. Therefore, the x-intercept is at $x=3$.

To find the y-intercept, set $x=0$ and solve for y: $\frac{x}{3}+\frac{y}{2}=1 \Rightarrow \frac{0}{3}+\frac{y}{2}=1 \Rightarrow$

$\frac{y}{2}=1 \Rightarrow 2\left(\frac{y}{2}\right)=2(1) \Rightarrow y=2$. Therefore, the y-intercept is at $y=2$. See Figure 41.

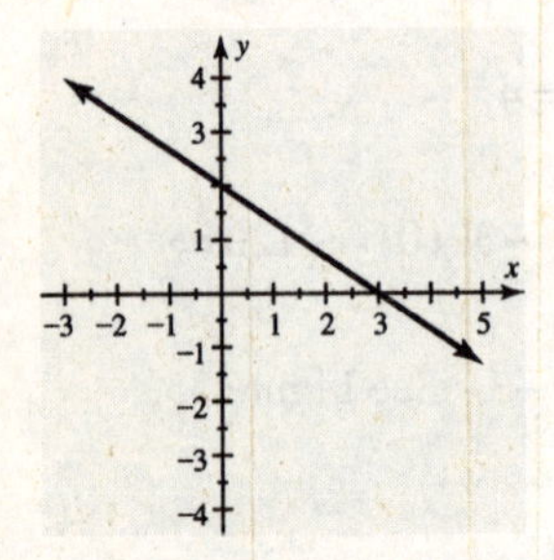

Figure 41

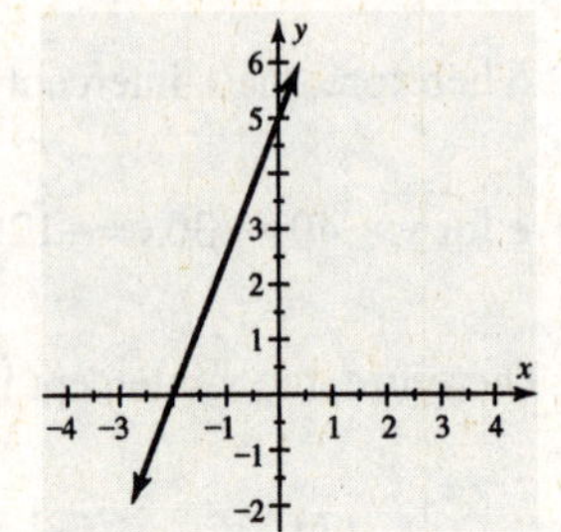

Figure 43

43. To find the x-intercept, set $y=0$ and solve for x: $0.6y-1.5x=3 \Rightarrow 0.6(0)-1.5x=3 \Rightarrow -1.5x=3 \Rightarrow$

$\frac{-1.5x}{-1.5}=\frac{3}{-1.5} \Rightarrow x=-2$. Therefore, the x-intercept is at $x=-2$.

To find the y-intercept, set $x=0$ and solve for y: $0.6y-1.5x=3 \Rightarrow 0.6y-1.5(0)=3 \Rightarrow 0.6y=3 \Rightarrow$

$\frac{0.6y}{0.6}=\frac{3}{0.6} \Rightarrow y=5$. Therefore, the y-intercept is at $y=5$. See Figure 43.

45. To find the x-intercept, set $y=0$ and solve for x:

$Ax+By=C \Rightarrow Ax+B(0)=C \Rightarrow Ax=C \Rightarrow \frac{Ax}{A}=\frac{C}{A} \Rightarrow x=\frac{C}{A}$. Therefore, the x-intercept is at $x=\frac{C}{A}$.

To find the y-intercept, set $x=0$ and solve for y:

$Ax+By=C \Rightarrow A(0)+By=C \Rightarrow By=C \Rightarrow \frac{By}{B}=\frac{C}{B} \Rightarrow y=\frac{C}{B}$. Therefore, the y-intercept is at $y=\frac{C}{B}$.

Horizontal and Vertical Lines

47. (a) See Figure 47a.

(b) See Figure 47b.

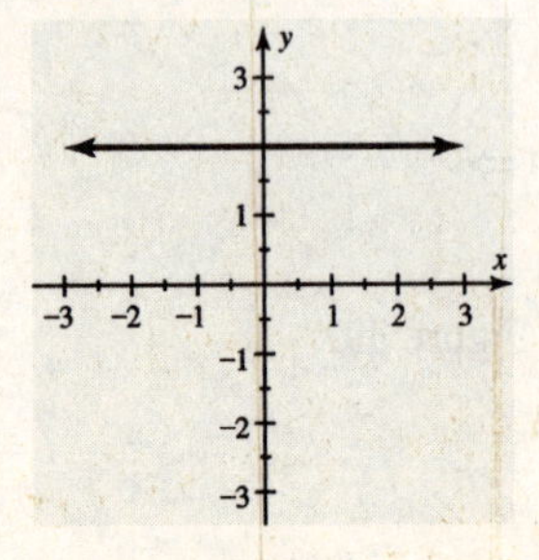

Figure 47a

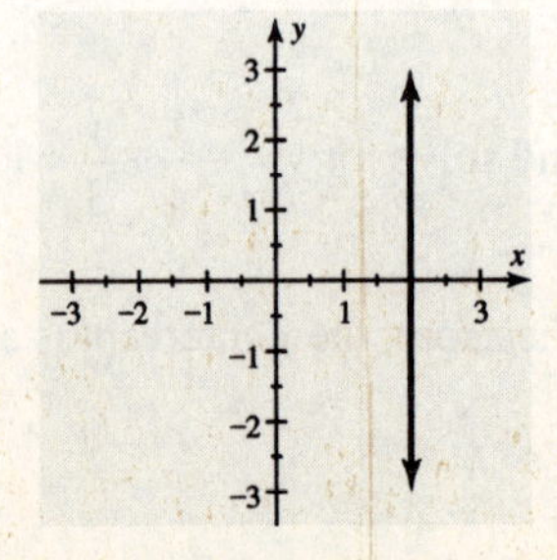

Figure 47b

49. (a) See Figure 49a.

(b) See Figure 49b.

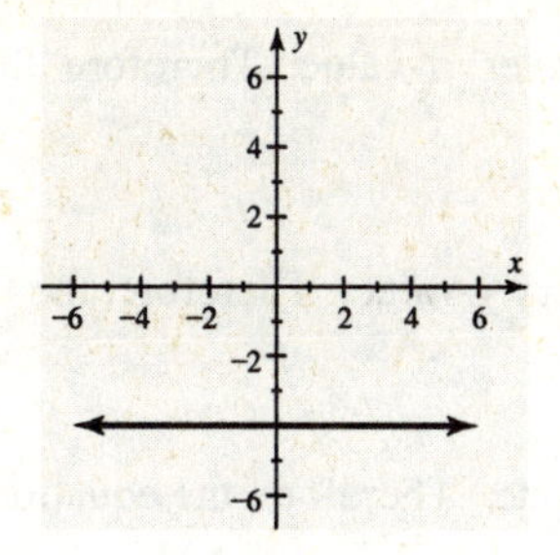

Figure 49a

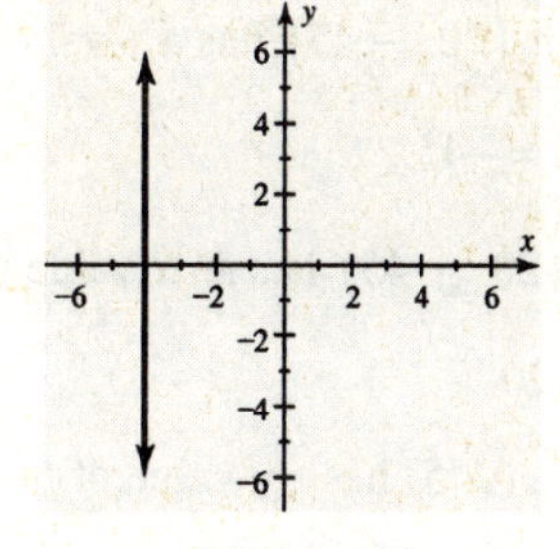

Figure 49b

51. (a) See Figure 51a.

(b) See Figure 51b.

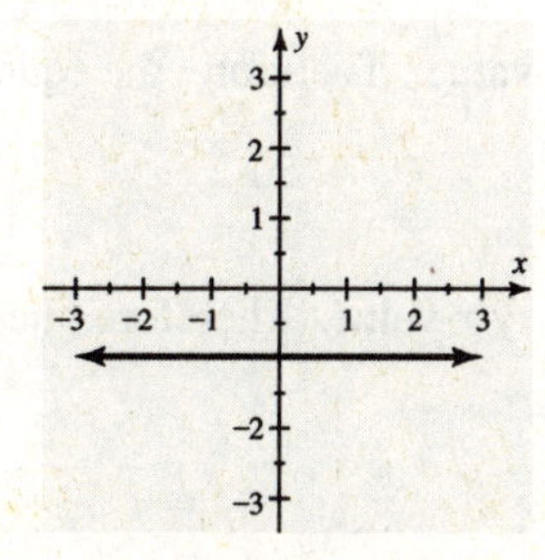

Figure 51a

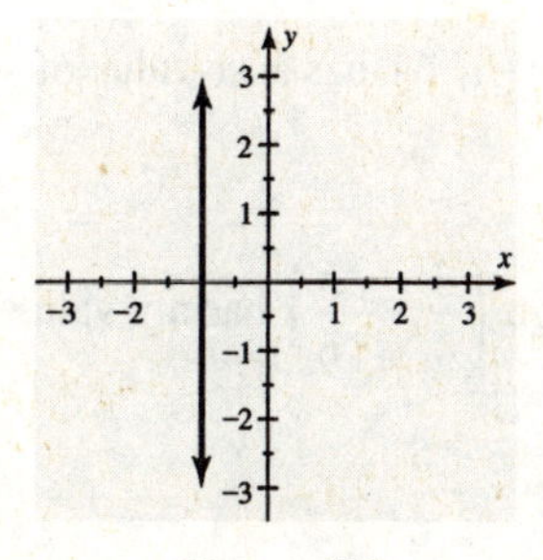

Figure 51b

53. (a) See Figure 53a.

(b) See Figure 53b.

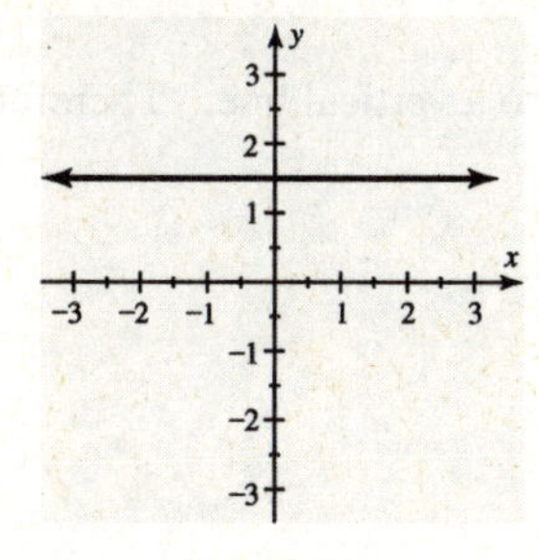

Figure 53a

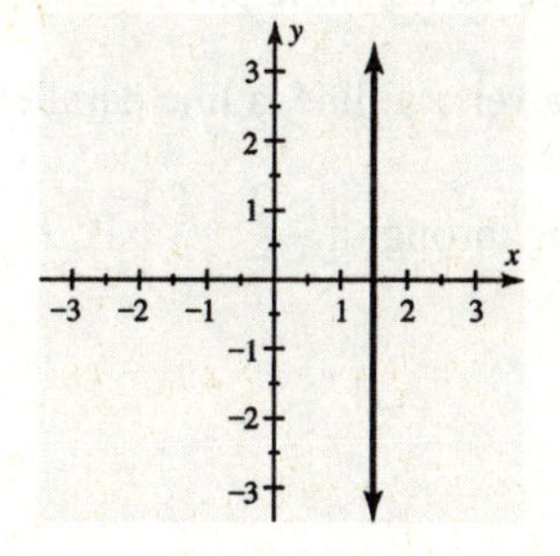

Figure 53b

55. $y = 4$

57. $x = -1$

59. $y = -6$

61. $x = 5$

63. Because $y = 1$ for every x-value, the graph is a horizontal line. The equation for the line is $y = 1$.

65. Because $x = -6$ for every y-value, the graph is a vertical line. The equation for the line is $x = -6$.

67. A horizontal line that passes through (1, 2) has a y-value of 2 for every x-value. Therefore, the equation for the horizontal line is $y = 2$.

A vertical line that passes through (1, 2) has an x-value of 1 for every y-value. Therefore, the equation for the vertical line is $x = 1$.

69. A horizontal line that passes through $(20, -45)$ has a y-value of -45 for every x-value. Therefore, the equation for the horizontal line is $y = -45$.

A vertical line that passes through $(20, -45)$ has an x-value of 20 for every y-value. Therefore, the equation for the vertical line is $x = 20$.

71. A horizontal line that passes through (0, 5) has a y-value of 5 for every x-value. Therefore, the equation for the horizontal line is $y = 5$.

A vertical line that passes through (0, 5) has an x-value of 0 for every y-value. Therefore, the equation for the vertical line is $x = 0$.

73. A vertical line that passes through $(-1, 6)$ has an x-value of -1 for every y-value. Therefore, the equation for the vertical line is $x = -1$.

75. A horizontal line that passes through $\left(\frac{3}{4}, -\frac{5}{6}\right)$ has a y-value of $-\frac{5}{6}$ for every x-value. Therefore, the equation for the horizontal line is $y = -\frac{5}{6}$.

77. Because $y = \frac{1}{2}$ is the equation of a horizontal line, a line perpendicular to $y = \frac{1}{2}$ is a vertical line. Therefore, the equation of the vertical line passing through $(4, -9)$ is $x = 4$.

79. Because $x = 4$ is the equation of a vertical line, a line parallel to $x = 4$ is also a vertical line. Therefore, the equation of the vertical line passing through $\left(-\frac{2}{3}, \frac{1}{2}\right)$ is $x = -\frac{2}{3}$.

81. $y = 0$

Applications

83. (a) y-intercept, 200; x-intercept, 4.

(b) The driver was initially 200 mi from home; the driver arrived home after 4 hr.

85. (a) To find the v-intercept, set $t = 0$ and solve for v: $v = 128 - 32t = 128 - 32(0) = 128$.

To find the t-intercept, set $v = 0$ and solve for t: $v = 128 - 32t \Rightarrow 0 = 128 - 32t \Rightarrow$

$0 + 32t = 128 - 32t + 32t \Rightarrow 32t = 128 \Rightarrow \frac{32t}{32} = \frac{128}{32} \Rightarrow t = 4$. See Figure 85.

(b) The initial velocity was 128 ft/sec; the velocity after 4 seconds was 0.

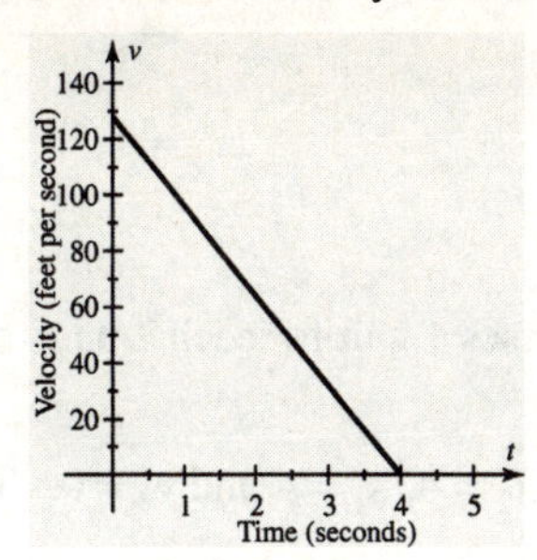

Figure 85

87. (a) y-intercept, 2000; x-intercept, 4.

(b) The pool initially contained 2000 gal; the pool was empty after 4 hr.

3.4: Slope and Rates of Change

Concepts

1. horizontal
2. vertical
3. rise; run
4. horizontal
5. vertical
6. $\dfrac{y_2 - y_1}{x_2 - x_1}$
7. positive
8. negative
9. Because the graph of the line rises as it goes from left to right, the slope is positive.
10. Because the graph of the line falls as it goes from left to right, the slope is negative.
11. Because the graph of the line neither falls nor rises as it goes from left to right, the slope is zero.
12. Because the graph of the line rises as it goes from left to right, the slope is positive.
13. Because the graph of the line falls as it goes from left to right, the slope is negative.
14. Because the graph is a vertical line, the run is 0. Thus, the slope is undefined.

Finding Slopes of Lines

15. Because the graph is a horizontal line, the slope is 0. Thus, the rise always equals 0.

17. A point on the graph is $(0,\ -2)$ and another point is $(2,\ 0)$. Let $x_1 = 0$, $x_2 = 2$, $y_1 = -2$ and $y_2 = 0$. Then:

$m\ (\text{slope}) = \dfrac{y_2 - y_1}{x_2 - x_1} = \dfrac{0-(-2)}{2-0} = \dfrac{2}{2} = 1.$ Thus, the slope is 1; the graph rises 1 unit for each unit of run.

19. A point on the graph is $(-1, -1)$ and another point is $(0, 1)$. Let $x_1 = -1$, $x_2 = 0$, $y_1 = -1$ and $y_2 = 1$. Then:

$m\ (\text{slope}) = \dfrac{y_2 - y_1}{x_2 - x_1} = \dfrac{1-(-1)}{0-(-1)} = \dfrac{2}{1} = 2$. Thus, the slope is 2; the graph rises 2 units for each unit of run.

21. Because the run always equals 0, the slope is undefined.

23. A point on the graph is $(0,\ -2)$ and another point is $(2,\ -1)$. Let $x_1 = 0$, $x_2 = 2$, $y_1 = -2$ and $y_2 = -1$. Then:

$m\ (\text{slope}) = \dfrac{y_2 - y_1}{x_2 - x_1} = \dfrac{-1-(-2)}{2-0} = \dfrac{1}{2}$. Thus, the slope is $\dfrac{1}{2}$; the graph rises 1 unit for each 2 units of run.

25. A point on the graph is $(0,\ 4)$ and another point is $(4,\ 0)$. Let $x_1 = 0$, $x_2 = 4$, $y_1 = 4$ and $y_2 = 0$. Then:

$m\ (\text{slope}) = \dfrac{y_2 - y_1}{x_2 - x_1} = \dfrac{0-4}{4-0} = \dfrac{-4}{4} = -1$. Thus, the slope is -1; the graph falls 1 unit for each unit of run.

27. $m = \dfrac{y_2 - y_1}{x_2 - x_1} = \dfrac{4-2}{2-1} = \dfrac{2}{1} = 2$. Thus, the slope of the line is 2. See Figure 27.

29. $m = \dfrac{y_2 - y_1}{x_2 - x_1} = \dfrac{4-1}{2-2} = \dfrac{3}{0}$. Because the denominator is 0, the slope is undefined. See Figure 29.

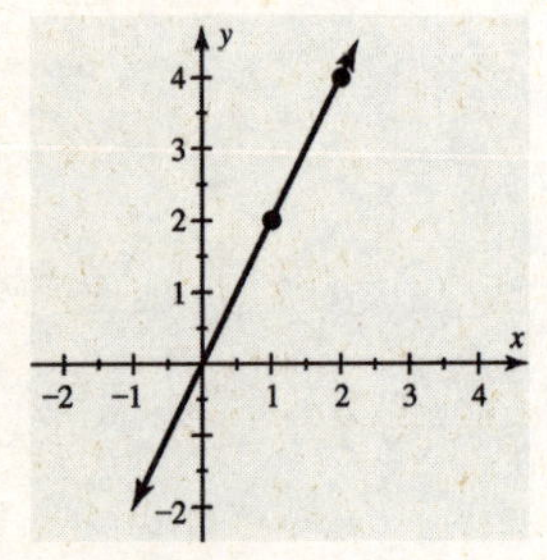

Figure 27

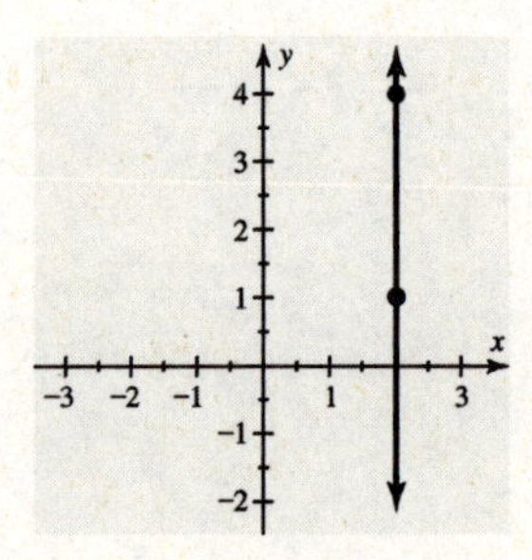

Figure 29

31. $m = \dfrac{y_2 - y_1}{x_2 - x_1} = \dfrac{5-3}{-2-1} = \dfrac{2}{-3} = -\dfrac{2}{3}$. Thus, the slope of the line is $-\dfrac{2}{3}$. See Figure 31.

33. $m = \dfrac{y_2 - y_1}{x_2 - x_1} = \dfrac{-1-(-1)}{-2-2} = \dfrac{-1+1}{-2+(-2)} = \dfrac{0}{-4} = 0$. Thus, the slope of the line is 0. See Figure 33.

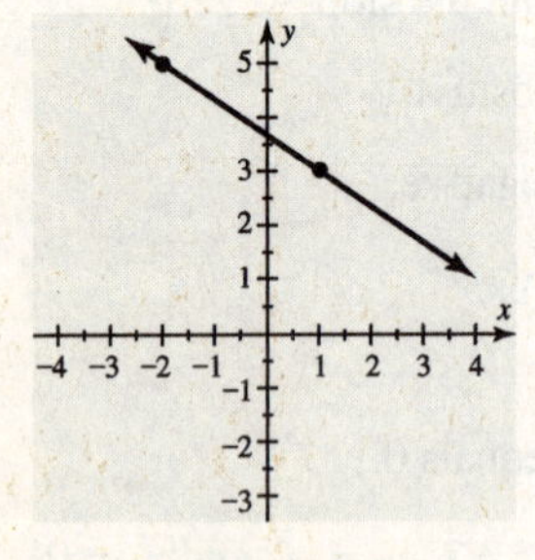

Figure 31

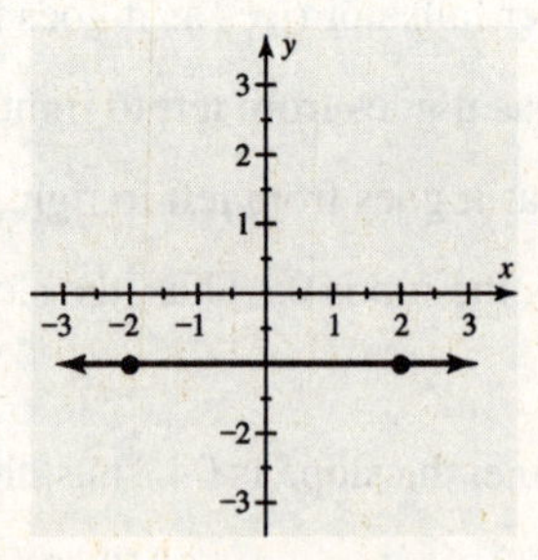

Figure 33

35. $m = \dfrac{y_2 - y_1}{x_2 - x_1} = \dfrac{-7-(-2)}{-3-4} = \dfrac{-5}{-7} = \dfrac{5}{7}$. Thus, the slope of the line is $\dfrac{5}{7}$.

37. $m = \dfrac{y_2 - y_1}{x_2 - x_1} = \dfrac{-2-4}{4-(-3)} = \dfrac{-6}{7} = -\dfrac{6}{7}$. Thus, the slope of the line is $-\dfrac{6}{7}$.

39. $m = \dfrac{y_2 - y_1}{x_2 - x_1} = \dfrac{5-5}{2-(-3)} = \dfrac{0}{5} = 0.$ Thus, the slope of the line is 0.

41. $m = \dfrac{y_2 - y_1}{x_2 - x_1} = \dfrac{-4-6}{-1-(-1)} = \dfrac{-10}{0}.$ Because the denominator is 0, the slope is undefined.

43. $m = \dfrac{y_2 - y_1}{x_2 - x_1} = \dfrac{18-5}{2000-1980} = \dfrac{13}{20}.$ Thus, the slope of the line is $\dfrac{13}{20}$.

45. $m = \dfrac{y_2 - y_1}{x_2 - x_1} = \dfrac{10.6-6.1}{2000-1950} = \dfrac{4.5}{50} = \dfrac{9}{100}.$ Thus, the slope of the line is $\dfrac{9}{100}$.

47. $m = \dfrac{y_2 - y_1}{x_2 - x_1} = \dfrac{\frac{3}{7}-\left(-\frac{2}{7}\right)}{-\frac{2}{3}-\frac{1}{3}} = \dfrac{\frac{5}{7}}{-\frac{3}{3}} = -\dfrac{5}{7}.$ Thus, the slope of the line is $-\dfrac{5}{7}$.

49. $m = \dfrac{y_2 - y_1}{x_2 - x_1} = \dfrac{64-(-34)}{14-12} = \dfrac{98}{2} = 49.$ Thus, the slope of the line is 49.

51. See Figure 51.

53. See Figure 53.

55. See Figure 55.

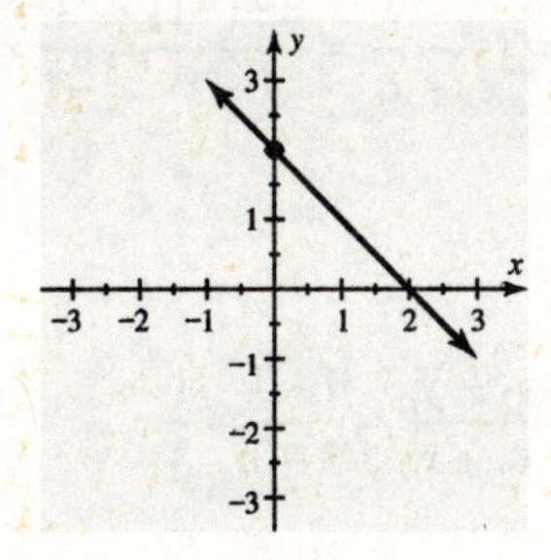

Figure 51

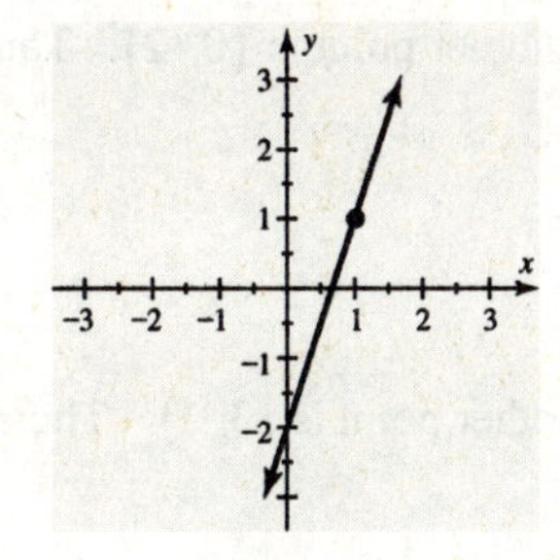

Figure 53

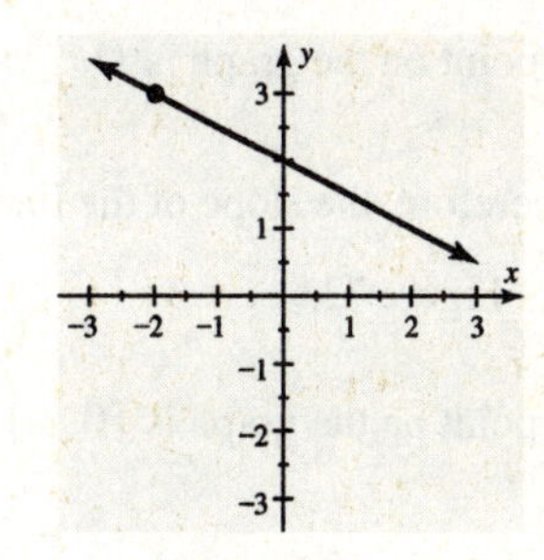

Figure 55

57. See Figure 57.

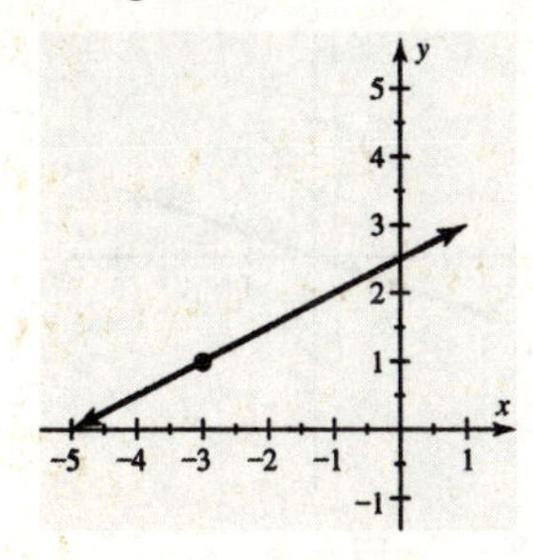

Figure 57

59. Because the y-value increases by 2 when the x-value increases by 1, the slope is 2. Because $y = 0$ when $x = 1$, the x-intercept is at $x = 1$. Because $x = 0$ when $y = -2$, the y-intercept is at $y = -2$.

61. Because the y-value decreases by 9 when the x-value increases by 1, the slope is -9. Because $y = 0$ when $x = -1$, the x-intercept is at $x = -1$. Because $x = 0$ when $y = -9$, the y-intercept is at $y = -9$.

63. Because the slope is 2, the y-value increases by 2 when the x-value increases by 1. Thus, when $x = 1$, $y = -4 + 2 = -2$; when $x = 2$, $y = -2 + 2 = 0$; when $x = 3$, $y = 0 + 2 = 2$. See Figure 63.

65. Because the slope is -3, the y-value decreases by 3 when the x-value increases by 1. Thus, when $x=2$, $y=4-3=1$; when $x=3$, $y=1-3=-2$; when $x=4$, $y=-2-3=-5$; See Figure 65.

67. Because the slope is $\frac{3}{2}$, the y-value increases by $\frac{3}{2}$ when the x-value increases by 1 and the y-value increases by $2\left(\frac{3}{2}\right)=3$ when the x-value increases by 2. Thus, when $x=-2$, $y=0+3=3$; when $x=0$, $y=3+3=6$; when $x=2$, $y=6+3=9$. See Figure 67.

x	0	1	2	3
y	−4	−2	0	2

Figure 63

x	1	2	3	4
y	4	1	−2	−5

Figure 65

x	−4	−2	0	2
y	0	3	6	9

Figure 67

69. (a) See Figure 69.

(b) A point on the graph is $(0,-1)$ and another point is $\left(\frac{1}{2},0\right)$. Thus: $m=\frac{y_2-y_1}{x_2-x_1}=\frac{0-(-1)}{\frac{1}{2}-0}=\frac{1}{\frac{1}{2}}=2.$

Therefore, the slope of the line is 2.

71. (a) See Figure 71.

(b) A point on the graph is $(1,\ -1)$ and another point is $(0,\ 2)$. Thus: $m=\frac{y_2-y_1}{x_2-x_1}=\frac{2-(-1)}{0-1}=\frac{3}{-1}=-3.$

Therefore, the slope of the line is -3.

73. (a) See Figure 73.

(b) A point on the graph is $(0,\ 0)$ and another point is $(3,\ 1)$. Thus: $m=\frac{y_2-y_1}{x_2-x_1}=\frac{1-0}{3-0}=\frac{1}{3}.$

Therefore, the slope of the line is $\frac{1}{3}$.

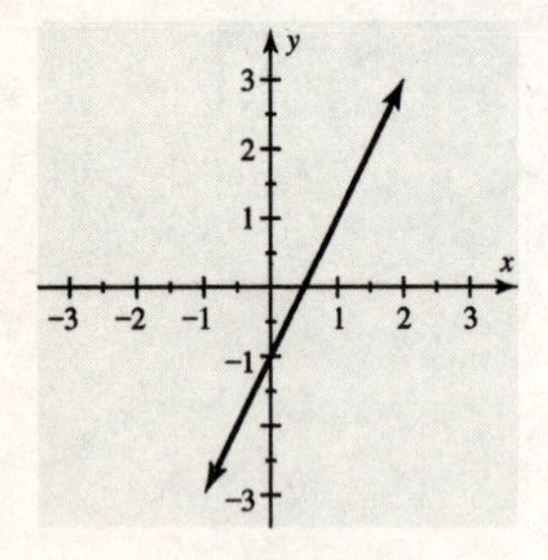

Figure 69

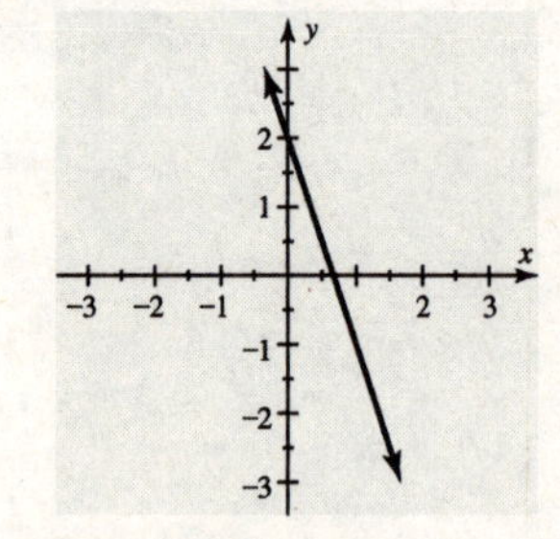

Figure 71

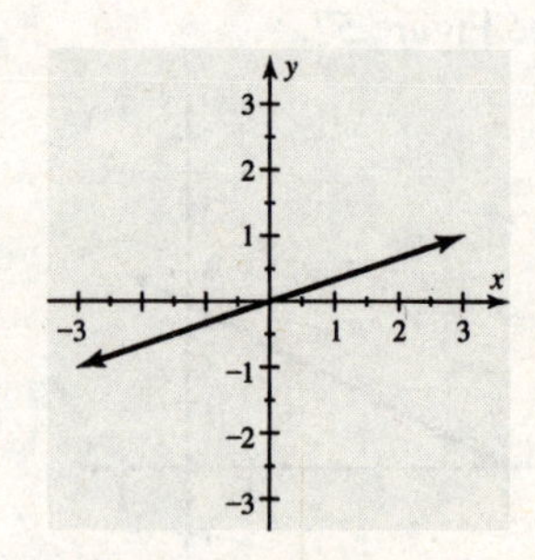

Figure 73

75. (a) See Figure 75.

(b) Because the line is horizontal, the rise is always 0 and therefore the slope is 0.

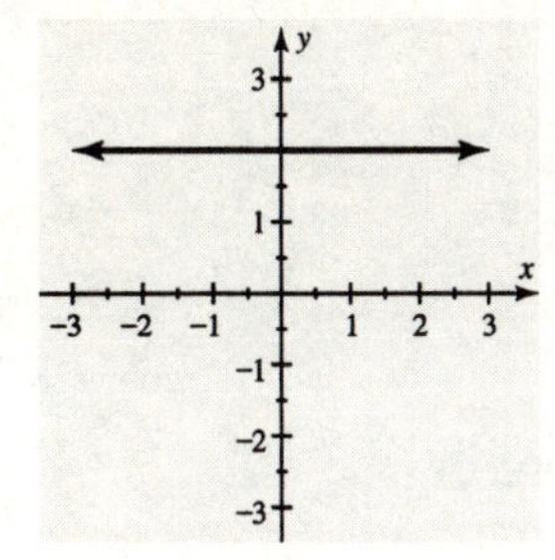

Figure 75

Slope as a Rate of Change

77. Because zero gumballs cost zero cents, the graph must pass through (0, 0). c

79. The average cost of a new car has increased over the past 30 years. b

81. (a) When $0 \le x \le 3$, the y-value increases by 1000 when x increases by 1. Therefore, $m_1 = 1000$.

When $3 \le x \le 5$, the y-value decreases by 1000 when x increases by 1. Therefore, $m_2 = -1000$.

(b) $m_1 = 1000$: Water is being added to the pool at a rate of 1000 gallons per hour.

$m_2 = -1000$: Water is being removed from the pool at a rate of 1000 gallons per hour.

(c) Initially the pool contained 2000 gallons of water. Over the first 3 hours, water was pumped into the pool at a rate of 1000 gallons per hour. For the next 2 hours, water was pumped out of the pool at a rate of 1000 gallons per hour.

83. (a) When $0 \le x \le 2$, the y-value increases by 50 when x increases by 1. Therefore, $m_1 = 50$.

When $2 \le x \le 3$, the y-value remains constant when x increases. Therefore, $m_2 = 0$.

When $3 \le x \le 4$, the y-value decreases by 50 when x increases by 1. Therefore, $m_3 = -50$.

(b) $m_1 = 50$: The car is moving away from home at a rate of 50 mph.

$m_2 = 0$: The car is not moving.

$m_3 = -50$: The car is moving toward home at a rate of 50 mph.

(c) Initially the car is at home. Over the first 2 hours, the car travels away from home at a rate of 50 mph. Then the car is parked for 1 hour. Finally, the car travels toward home at a rate of 50 mph.

85. m^3/min

87. See Figure 87.

89. See Figure 89.

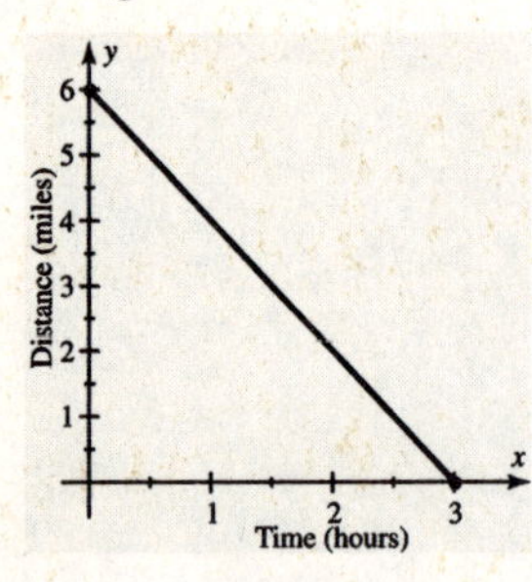

Figure 87

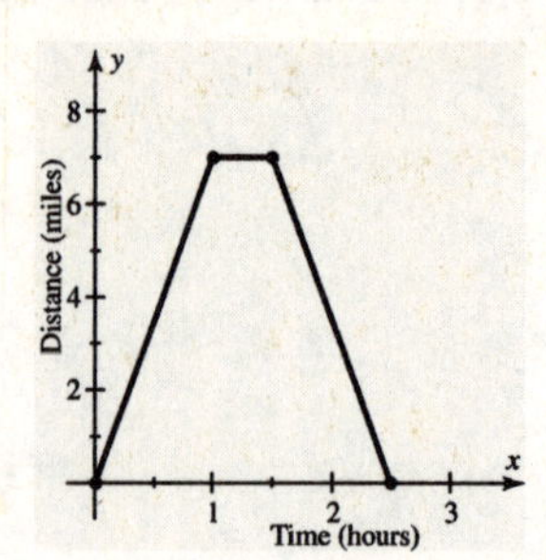

Figure 89

Applications

91. (a) $m = \dfrac{y_2 - y_1}{x_2 - x_1} = \dfrac{110 - 70}{2004 - 2000} = \dfrac{40}{4} = 10$. Thus, the slope of the line is 10.

(b) The birth rate increased on average by 10,000 children per year.

93. (a) A point on the line is $(40, 100)$ and another point is $(0, 0)$. Thus: $m = \dfrac{y_2 - y_1}{x_2 - x_1} = \dfrac{100 - 0}{40 - 0} = \dfrac{100}{40} = 2.5$.

Thus, the slope of the line is 2.5.

(b) The revenue is \$2.50 per screwdriver.

95. (a) Substitute 50 for G and solve for O: $O = \dfrac{1}{50}G = \dfrac{1}{50}(50) = 1$. Thus, 1 gallon of oil should be added to 50 gallons of gasoline. Then substitute 100 for G and solve for O: $O = \dfrac{1}{50}G = \dfrac{1}{50}(100) = 2$. Thus, 2 gallons of oil should be added to 100 gallons of gasoline.

(b) When G increases by 1, O increases by $\dfrac{1}{50}$. Thus, the slope of the graph is $\dfrac{1}{50}$.

(c) Oil should be added at a rate of 1 gallon of oil per 50 gallons of gasoline.

97. (a) See Figure 97.

(b) $m_1 = \dfrac{y_2 - y_1}{x_2 - x_1} = \dfrac{100 - 0}{5 - 0} = \dfrac{100}{5} = 20$; $m_2 = \dfrac{y_2 - y_1}{x_2 - x_1} = \dfrac{250 - 100}{10 - 5} = \dfrac{150}{5} = 30$;

$m_3 = \dfrac{y_2 - y_1}{x_2 - x_1} = \dfrac{450 - 250}{15 - 10} = \dfrac{200}{5} = 40$.

(c) $m_1 = 20$: Each mile between 0 and 5 miles is worth \$20/mile.

$m_2 = 30$: Each mile between 5 and 10 miles is worth \$30/mile.

$m_3 = 40$: Each mile between 10 and 15 miles is worth \$40/mile.

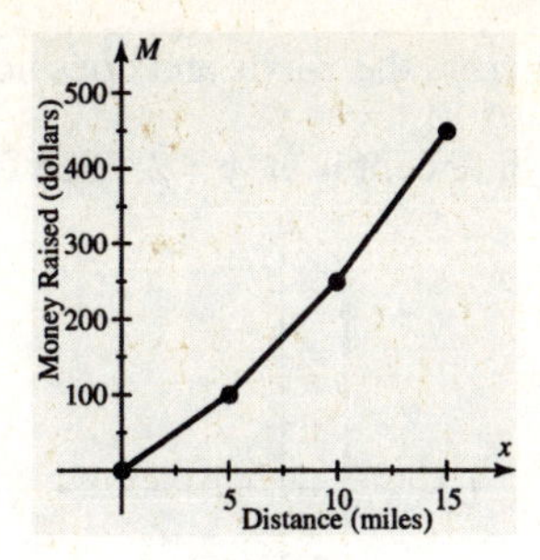

Figure 97

99. (a) $m=\dfrac{y_2-y_1}{x_2-x_1}=\dfrac{42{,}000-18{,}000}{2000-1980}=\dfrac{24{,}000}{20}=1200$. Thus, the slope of the line is 1200.

(b) Median family income increased on average by \$1200/year over this period of time.

(c) $42{,}000+5(1200)=\$48{,}000$

101. (a) Because the slope of the line is negative, the ant is moving toward the stone.

(b) When the ant starts moving, $x=0$, so $-2(0)+10=0+10=10$. The ant is 10 feet from the stone.

(c) The slope is -2, so the ant is moving 2 ft/min.

(d) When the ant reaches the stone the distance from the stone is 0. So, solve the equation $0=-2x+10$ for x: $0-10=-2x+10-10 \Rightarrow -10=-2x \Rightarrow \dfrac{-10}{-2}=\dfrac{-2x}{-2} \Rightarrow 5=x$

Checking Basic Concepts for Sections 3.3 & 3.4

1. The line crosses the x-axis at $x=-2$; therefore, the x-intercept is -2.

The line crosses the y-axis at $y=3$; therefore, the y-intercept is 3.

2. $2x-y=2 \Rightarrow 2x-y+y=2+y \Rightarrow 2x-2=2-2+y \Rightarrow y=2x-2$. Therefore:

When $x=-1$: $y=2(-1)-2 \Rightarrow y=-2-2 \Rightarrow y=-4$; When $x=0$: $y=2x-2=2(0)-2=-2$; When $x=1$: $y=2x-2=2(1)-2=0$; When $x=2$: $y=2x-2=2(2)-2=2$.

Therefore, the missing values in the table are -4, -2, 0, and 2. See Figure 2.

When $y=0$, $x=1$. Therefore the x-intercept is $x=1$.

When $x=0$, $y=-2$. Therefore, the y-intercept is $y=-2$.

x	-2	-1	0	1	2
y	-6	-4	-2	0	2

Figure 2

3. (a) To find the x-intercept, set $y=0$ and solve for x: $x-2y=6 \Rightarrow x-2(0)=6 \Rightarrow x=6$. Therefore, the x-intercept is $x=6$.

To find the y-intercept, set $x=0$ and solve for y: $x-2y=6 \Rightarrow 0-2(y)=6 \Rightarrow -2y=6 \Rightarrow$ $\dfrac{-2y}{-2}=\dfrac{6}{-2} \Rightarrow y=-3$. Therefore, the y-intercept is $y=-3$. See Figure 3a.

(b) $y = 2$ is the graph of a horizontal line. Therefore, the line does not cross the x-axis and does not have an x-intercept. The line crosses the y-axis at $y = 2$, and therefore its y-intercept is at $y = 2$. See Figure 3b.

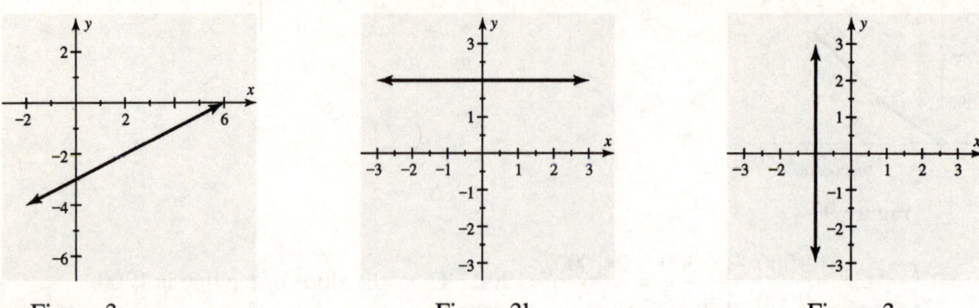

Figure 3a Figure 3b Figure 3c

(c) $x = -1$ is the graph of a vertical line. Therefore, the line crosses the x-axis at $x = -1$ and therefore its x-intercept is at $x = -1$. The line does not cross the y-axis, and therefore does not have a y-intercept. See Figure 3c.

4. A horizontal line passing through $(-2, 4)$ has the equation $y = 4$.

A vertical line passing through $(-2, 4)$ has the equation $x = -2$.

5. (a) $\frac{6-3}{2-(-2)} = \frac{3}{4}$ Therefore, the slope of the line is $\frac{3}{4}$.

(b) $\frac{3-3}{0-(-5)} = \frac{0}{5} = 0$ Therefore, the slope of the line is 0.

(c) $\frac{8-5}{1-1} = \frac{3}{0}$ Therefore, the slope of the line is undefined.

6. A point on the line is (2, 1) and another point is (0, 2). Thus, $\frac{2-1}{0-2} = \frac{1}{-2} = -\frac{1}{2}$.

Therefore, the slope of the line is $-\frac{1}{2}$.

7. See Figure 7.

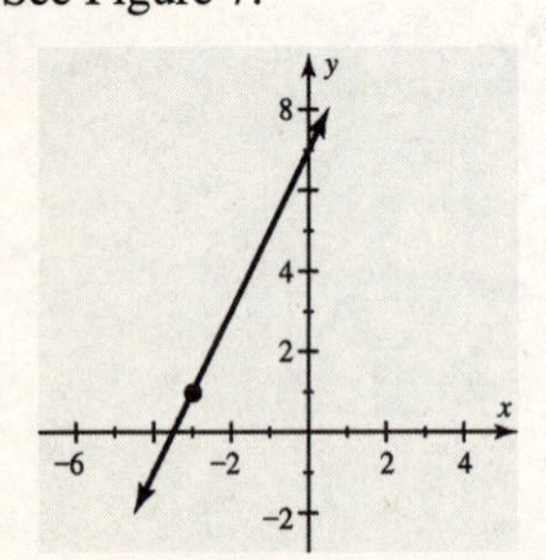

Figure 7

8. (a) $\frac{5-5}{1-0} = \frac{0}{1} = 0$. Thus $m_1 = 0$. $\frac{7-5}{2-1} = \frac{2}{1} = 2$. Thus $m_2 = 2$. $\frac{5-7}{5-2} = \frac{-2}{3} = -\frac{2}{3}$. Thus $m_3 = -\frac{2}{3}$.

(b) $m_1 = 0$: before the rain, the depth of water in the pond does not change.

$m_2 = 2$: while it rains, the depth of water in the pond increases by 2 feet for every hour of rain.

$m_3 = -\frac{2}{3}$: after it stops raining, the depth of water in the pond decreases by $-\frac{2}{3}$ feet per hour until it reaches its original depth.

(c) Initially the pond had a depth of 5 feet. For the first hour, there was no change in the depth of the pond. For the next hour, the depth of the pond increased at a rate of 2 feet per hour to a depth of 7 feet. Finally, the depth of the pond decreased for 3 hours at a rate of $\frac{2}{3}$ foot per hour until it was 5 feet deep.

9. (a) $\frac{22,300-18,100}{600-450} = \frac{4200}{150} = 28$. Therefore, the slope of the line is 28.

(b) For counties between 450 and 600 mi^2, the population increases at an average rate of 28 people/ mi^2.

(c) No. We do not know whether this trend continues.

3.5: Slope-Intercept Form

Concepts

1. $y = mx + b$
2. slope
3. y-intercept
4. horizontal
5. origin
6. $y = 2x + 5$
7. parallel
8. slope
9. perpendicular
10. reciprocals
11. f
12. d
13. a
14. b
15. e
16. c

Slope-Intercept Form

17. The graph of the line shows the y-intercept at -1. For every x-increase of 1, the y-value increases by 1, so the slope is 1. Therefore: $y = mx + b \Rightarrow y = x - 1$.

19. The graph of the line shows the y-intercept at 1. For every x-increase of 1, the y-value decreases by 2, so the slope is -2. Therefore: $y = mx + b \Rightarrow y = -2x + 1$.

21. The graph of the line shows the y-intercept at -2. For every x-increase of 2, the y-value increases by 1, so the slope is $\frac{1}{2}$. Therefore: $y = mx + b \Rightarrow y = \frac{1}{2}x - 2$.

23. The graph of the line shows the y-intercept at 0. For every x-increase of 1, the y-value decreases by 2, so the slope is -2. Therefore: $y = mx + b = -2x + 0 \Rightarrow y = -2x$.

25. The graph of the line shows the y-intercept at 2. For every x-increase of 4, the y-value increases by 3, so the slope is $\frac{3}{4}$. Therefore: $y = mx + b \Rightarrow y = \frac{3}{4}x + 2$.

27. See Figure 27. The slope-intercept form is $y = mx + b$, so with $m = 1$ and $b = 2$, $y = x + 2$.

29. See Figure 29. The slope-intercept form is $y = mx + b$, so with $m = 2$ and $b = -1$, $y = 2x - 1$.

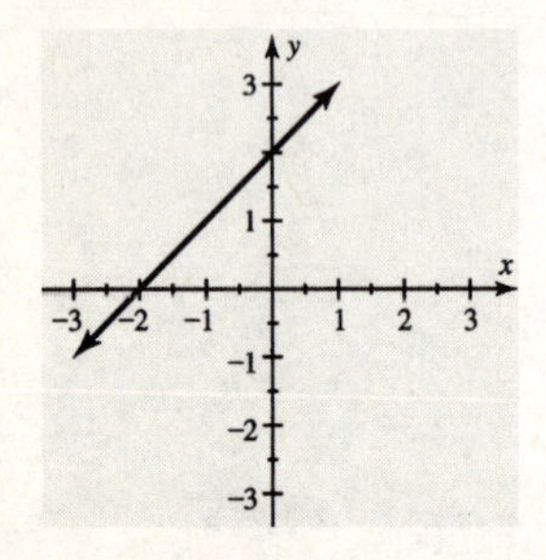

Figure 27

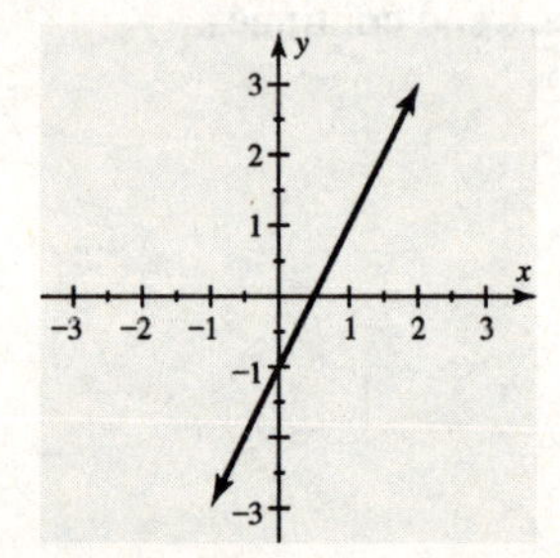

Figure 29

31. See Figure 31. The slope-intercept form is $y = mx + b$, so with $m = -\frac{1}{2}$ and $b = -2$, $y = -\frac{1}{2}x - 2$.

33. See Figure 33. $y = mx + b = \frac{1}{3}x + 0 \Rightarrow y = \frac{1}{3}x$.

35. See Figure 35. $y = mx + b \Rightarrow y = -2x + 1$.

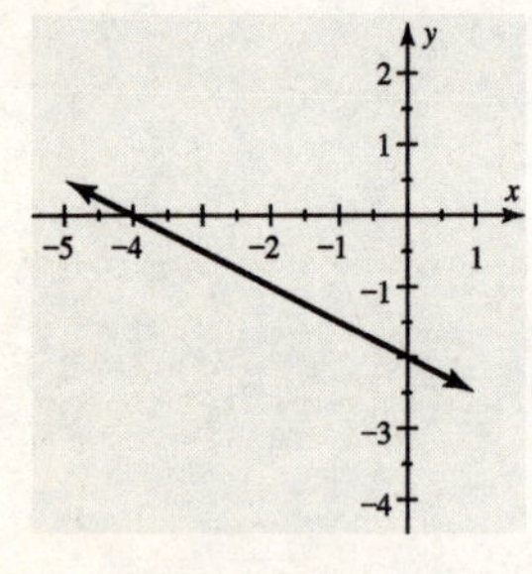

Figure 31

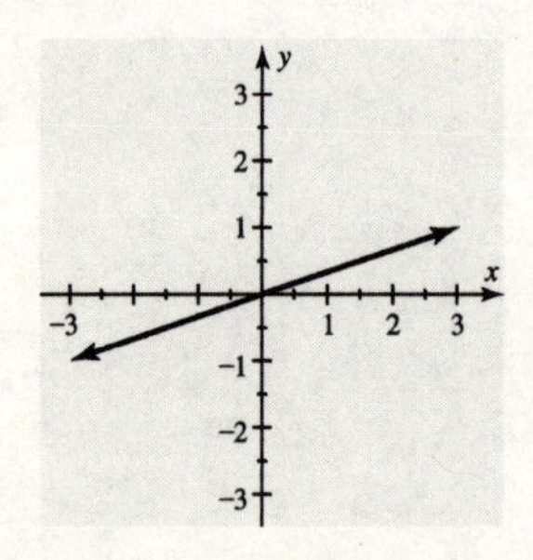

Figure 33

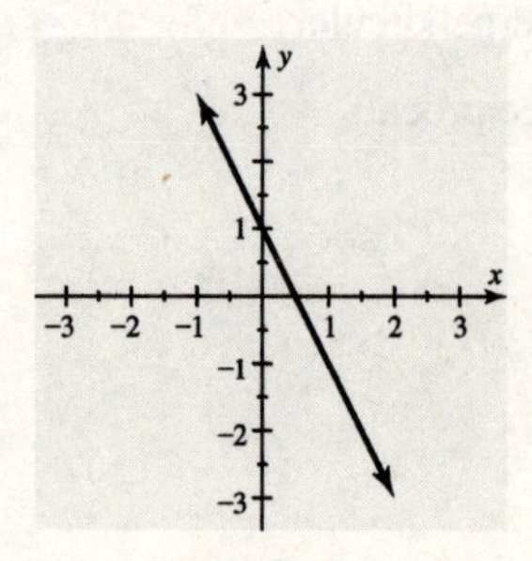

Figure 35

37. (a) To write the equation in slope-intercept form, solve for y:

$$x + y = 4 \Rightarrow x - x + y = 4 - x \Rightarrow y = 4 - x \Rightarrow y = -x + 4$$

(b) Because the slope-intercept form is $y = mx + b$, the slope is -1 and the y-intercept is 4.

39. (a) To write the equation in slope-intercept form, solve for y:

$$2x + y = 4 \Rightarrow 2x - 2x + y = 4 - 2x \Rightarrow y = 4 - 2x \Rightarrow y = -2x + 4$$

(b) Because the slope-intercept form is $y = mx + b$, the slope is -2 and the y-intercept is 4.

41. (a) To write the equation in slope-intercept form, solve for y: $x-2y=-4 \Rightarrow x-x-2y=-4-x \Rightarrow$

$$-2y=-4-x \Rightarrow -2y=-x-4 \Rightarrow \frac{-2y}{-2}=\frac{-x-4}{-2} \Rightarrow y=\frac{-x}{-2}-\frac{4}{-2} \Rightarrow y=\frac{1}{2}x+2$$

(b) Because the slope-intercept form is $y=mx+b$, the slope is $\frac{1}{2}$ and the y-intercept is 2.

43. (a) To write the equation in slope-intercept form, solve for y: $2x-3y=6 \Rightarrow 2x-2x-3y=6-2x \Rightarrow$

$$-3y=6-2x \Rightarrow -3y=-2x+6 \Rightarrow \frac{-3y}{-3}=\frac{-2x+6}{-3} \Rightarrow y=\frac{-2x}{-3}+\frac{6}{-3} \Rightarrow y=\frac{2}{3}x-2$$

(b) Because the slope-intercept form is $y=mx+b$, the slope is $\frac{2}{3}$ and the y-intercept is -2.

45. (a) To write the equation in slope-intercept form, solve for y: $x=4y-6 \Rightarrow x-4y=4y-4y-6 \Rightarrow$

$$x-4y=-6 \Rightarrow x-x-4y=-6-x \Rightarrow -4y=-x-6 \Rightarrow \frac{-4y}{-4}=\frac{-x}{-4}-\frac{6}{-4} \Rightarrow y=\frac{1}{4}x+\frac{3}{2}$$

(b) Because the slope-intercept form is $y=mx+b$, the slope is $\frac{1}{4}$ and the y-intercept is $\frac{3}{2}$.

47. (a) To write the equation in slope-intercept form, solve for y: $\frac{1}{2}x+\frac{3}{2}y=1 \Rightarrow \frac{1}{2}x-\frac{1}{2}x+\frac{3}{2}y=1-\frac{1}{2}x \Rightarrow$

$$\frac{3}{2}y=-\frac{1}{2}x+1 \Rightarrow \frac{2}{3}\left(\frac{3}{2}y\right)=\frac{2}{3}\left(-\frac{1}{2}x+1\right) \Rightarrow y=-\frac{1}{3}x+\frac{2}{3}$$

(b) Because the slope-intercept form is $y=mx+b$, the slope is $-\frac{1}{3}$ and the y-intercept is $\frac{2}{3}$.

49. See Figure 49.

51. See Figure 51.

53. See Figure 53.

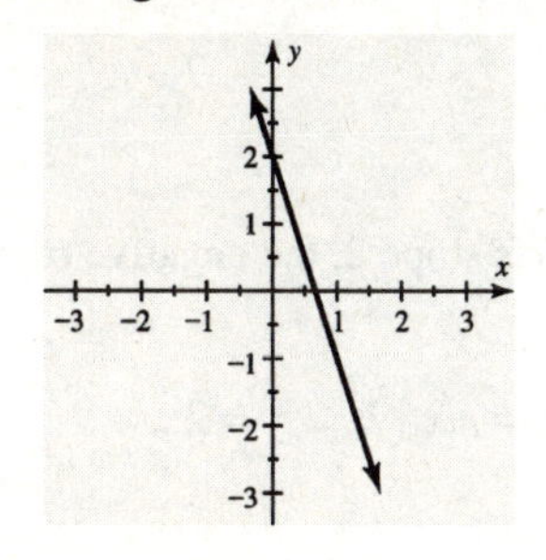

Figure 49

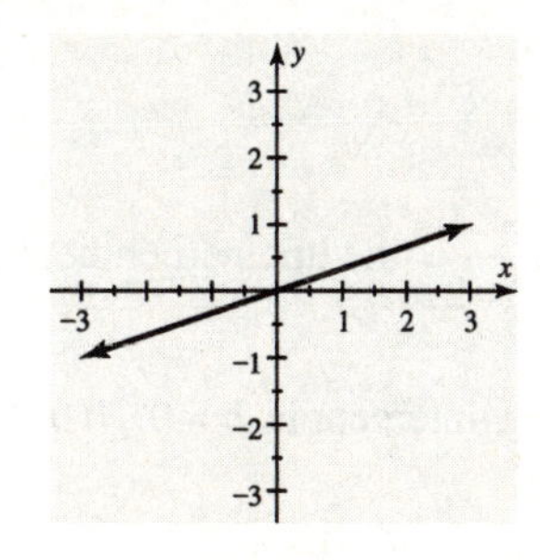

Figure 51

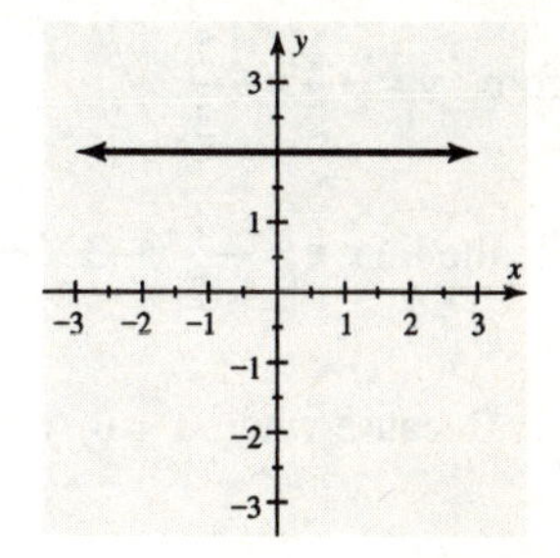

Figure 53

55. See Figure 55.

57. See Figure 57.

59. See Figure 59.

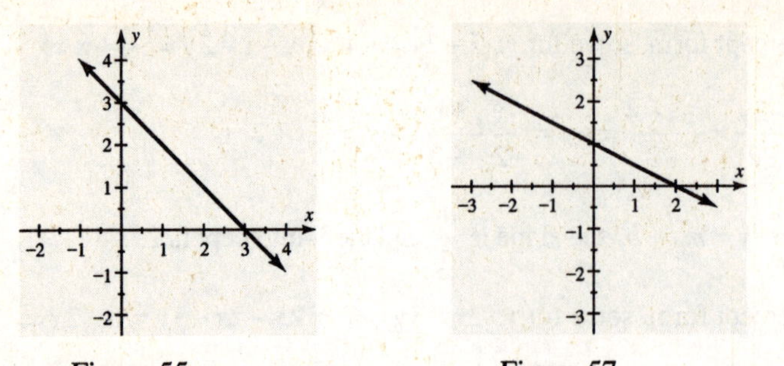

Figure 55 Figure 57 Figure 59

61. For every increase in x of 1, the y-value increases by 2. Therefore, the slope $m=2$. When $x=0,\ y=2$. Therefore the y-intercept $b=2$. Therefore: $y=mx+b \Rightarrow y=2x+2$.

63. For every increase in x of 2, the y-value increases by 2. Therefore, the slope $m=1$. When $x=0,\ y=-2$. Therefore the y-intercept $b=-2$. Therefore: $y=mx+b \Rightarrow y=x-2$.

Parallel and Perpendicular Lines

65. $y=\frac{4}{7}x+3$

67. Because the line parallel to $y=3x+1$ has slope 3, $m=3$.

Because when $x=0,\ y=0$, the y-intercept $b=0$. Therefore: $y=3x$.

69. Solve $2x+4y=5$ for y: $2x+4y=5 \Rightarrow 2x-2x+4y=5-2x \Rightarrow 4y=-2x+5 \Rightarrow$

$\frac{4y}{4}=\frac{-2x}{4}+\frac{5}{4} \Rightarrow y=-\frac{1}{2}x+\frac{5}{4}$.

Because the line parallel to $2x+4y=5$ has slope $-\frac{1}{2}$, $m=-\frac{1}{2}$. Then, to find the y-intercept b, replace x with 1 and y with 2: $y=-\frac{1}{2}x+b \Rightarrow 2=-\frac{1}{2}(1)+b \Rightarrow 2=-\frac{1}{2}+b \Rightarrow 2+\frac{1}{2}=-\frac{1}{2}+\frac{1}{2}+b \Rightarrow \frac{5}{2}=b$.

Therefore, $y=-\frac{1}{2}x+\frac{5}{2}$.

71. Because the line $y=-\frac{1}{2}x-3$` has slope $-\frac{1}{2}$, the line perpendicular to it has slope 2, the negative reciprocal of $-\frac{1}{2}$. Because when $x=0,\ y=0$, the y-intercept is $b=0$. Therefore, $y=mx+b \Rightarrow y=2x$.

73. Solve $x=-\frac{1}{3}y$ for y: $x=-\frac{1}{3}y \Rightarrow -3(x)=-3\left(-\frac{1}{3}y\right) \Rightarrow -3x=y \Rightarrow y=-3x$.

Because the line $y=-3x$ has slope -3, the line perpendicular to it has slope $\frac{1}{3}$, the negative reciprocal of -3. Then, to find the y-intercept b, replace y with 0 and x with -1: $0=\frac{1}{3}(-1)+b \Rightarrow 0=-\frac{1}{3}+b \Rightarrow$

$0+\frac{1}{3}=-\frac{1}{3}+\frac{1}{3}+b \Rightarrow \frac{1}{3}=b$. Therefore, $y=\frac{1}{3}x+\frac{1}{3}$.

Applications

75. (a) When $x = 0$, $y = 25$. Therefore, $25.

(b) It costs $0.25x$ for each additional mile x. Therefore, 25 cents.

(c) When $x = 0$, $y = 25$. Therefore the y-intercept is 25. This represents the fixed cost of renting a car.

(d) The slope is 0.25. This represents the cost per mile of renting the car.

77. (a) Because $b = 3.95$ and $m = 0.07$, $y = 0.07x + 3.95$. Set $x = 50$ and solve for y:

$y = 0.07(50) + 3.95 = 3.5 + 3.95 = 7.45$. Therefore, the charge is $7.45.

(b) $C = 0.07x + 3.95$

(c) Set $C = 8.64$ and solve for x: $8.64 = 0.07x + 3.95 \Rightarrow 8.64 - 3.95 = 0.07x \Rightarrow 4.69 = 0.07x \Rightarrow$

$\frac{4.69}{0.07} = \frac{0.07x}{0.07} \Rightarrow x = 67$ min.

79. (a) Because the y-value increases by $0.35 for every increase of 1 in the x-value, the slope $m = 0.35$. Because the fixed cost of $164.30 represents the y-value when $x = 0$, $b = 164.3$.

(b) The fixed cost of owning the car for one month.

3.6: Point-Slope Form

Concepts

1. One
2. One
3. $y = mx + b$
4. $y - y_1 = m(x - x_1)$ or $y = m(x - x_1) + y_1$.
5. 1; 3
6. distributive
7. Yes; every nonvertical line has one slope and one y-intercept.
8. No; it depends on the point used.

Point-Slope Form

9. Substitute -3 for x and 3 for y: $y - 1 = -\frac{2}{3}x \Rightarrow 3 - 1 = -\frac{2}{3}(-3) \Rightarrow 2 = \frac{6}{3} \Rightarrow 2 = 2$. Because this is a true statement, the point $(-3, 3)$ lies on the line.

11. Substitute 1 for x and 4 for y: $y - 3 = -(x - 1) \Rightarrow 4 - 3 = -(1 - 1) \Rightarrow 1 = 0$. Because this is not a true statement, the point $(1, 4)$ does not lie on the line.

13. Substitute 0 for x and 4 for y: $y = \frac{1}{2}(x + 4) + 2 \Rightarrow 4 = \frac{1}{2}(0 + 4) + 2 \Rightarrow 4 = \frac{1}{2}(4) + 2 \Rightarrow$ $4 = 2 + 2 \Rightarrow 4 = 4$. Because this is a true statement, the point $(0, 4)$ lies on the line.

15. The points (1, 2) and $(-3, -1)$ are on the line. First, determine the slope: $m=\frac{y_2-y_1}{x_2-x_1}=\frac{-1-2}{-3-1}=\frac{-3}{-4}=\frac{3}{4}$.

Then substitute the x and y values of the point (1, 2) into the point-slope

form: $y-y_1=m(x-x_1)\Rightarrow y-2=m(x-1)$. Then, substitute the value of the slope for m: $y-2=\frac{3}{4}(x-1)$.

17. The points $(3, -1)$ and $(-1, 1)$ are on the line. First, determine the slope:

$m=\frac{y_2-y_1}{x_2-x_1}=\frac{1-(-1)}{-1-3}=\frac{1+1}{-4}=\frac{2}{-4}=-\frac{1}{2}$.

Then substitute the x and y values of the point $(3, -1)$ and the value of the slope into the point-slope

form: $y-y_1=m(x-x_1)\Rightarrow y-(-1)=-\frac{1}{2}(x-3)\Rightarrow y+1=-\frac{1}{2}(x-3)$.

19. $y-y_1=m(x-x_1)\Rightarrow y-1=4(x-(-3))\Rightarrow y-1=4(x+3)$

21. $y-y_1=m(x-x_1)=y-(-3)=\frac{1}{2}(x-(-5))\Rightarrow y+3=\frac{1}{2}(x+5)$

23. $y-y_1=m(x-x_1)\Rightarrow y-30=1.5(x-2000)$

25. First, determine the slope: $m=\frac{y_2-y_1}{x_2-x_1}=\frac{-3-4}{-1-2}=\frac{-7}{-3}=\frac{7}{3}$. Then, insert the values of the first point and the value of the slope into the point-slope form: $y-y_1=m(x-x_1)\Rightarrow y-4=\frac{7}{3}(x-2)$.

27. First determine the slope: $m=\frac{y_2-y_1}{x_2-x_1}=\frac{-3-0}{0-5}=\frac{-3}{-5}=\frac{3}{5}$. Then, insert the values of the first point and the value of the slope into the point-slope form: $y-y_1=m(x-x_1)\Rightarrow y-0=\frac{3}{5}(x-5)\Rightarrow y=\frac{3}{5}(x-5)$.

29. First determine the slope: $m=\frac{y_2-y_1}{x_2-x_1}=\frac{65-15}{2000-1990}=\frac{50}{10}=5$. Then, insert the values of the first point and the value of the slope into the point-slope form: $y-y_1=m(x-x_1)\Rightarrow y-15=5(x-1990)$.

31. To convert to slope-intercept form, use the distributive property of multiplication and solve for y:

$y-4=3(x-2)\Rightarrow y-4=3x-6\Rightarrow y-4+4=3x-6+4\Rightarrow y=3x-2$.

33. To convert to slope-intercept form, use the distributive property of multiplication and solve for y:

$y+2=\frac{1}{3}(x+6)\Rightarrow y+2=\frac{1}{3}x+2\Rightarrow y+2-2=\frac{1}{3}x+2-2\Rightarrow y=\frac{1}{3}x$.

35. To convert to slope-intercept form, use the distributive property of multiplication and solve for y:

$y-\frac{3}{4}=\frac{2}{3}(x-1)\Rightarrow y-\frac{3}{4}=\frac{2}{3}x-\frac{2}{3}\Rightarrow y-\frac{3}{4}+\frac{3}{4}=\frac{2}{3}x-\frac{2}{3}+\frac{3}{4}\Rightarrow y=\frac{2}{3}x-\frac{8}{12}+\frac{9}{12}\Rightarrow y=\frac{2}{3}x+\frac{1}{12}$.

37. Use the distributive property of multiplication and solve for y: $y=-2(x-2)+5 \Rightarrow y=-2x+4+5 \Rightarrow$ $y=-2x+9$.

39. Use the distributive property of multiplication and solve for y:

$y=\frac{3}{5}(x-5)+1 \Rightarrow y=\frac{3}{5}x-3+1 \Rightarrow y=\frac{3}{5}x-2$.

41. $y=-16(x+1.5)+5 \Rightarrow y=-16x-24+5 \Rightarrow y=-16x-19$

43. Write the point-slope form in slope-intercept form:

$y-y_1=m(x-x_1) \Rightarrow y-y_1=mx-mx_1 \Rightarrow y-y_1+y_1=mx-mx_1+y_1$

$\Rightarrow y=mx-mx_1+y_1$. so the y-intercept is $-mx_1+y_1$.

45. First, find the point-slope form for the line: $y-y_1=m(x-x_1) \Rightarrow y-(-3)=-2(x-4)$.

Then, to find the slope-intercept form, use the distributive property of multiplication, then solve for y:

$y-(-3)=-2(x-4) \Rightarrow y+3=-2x+8 \Rightarrow y+3-3=-2x+8-3 \Rightarrow y=-2x+5$.

47. First, find the slope of the line: $m=\frac{y_2-y_1}{x_2-x_1}=\frac{-1-(-2)}{2-3}=\frac{-1+2}{-1}=\frac{1}{-1}=-1$. Then, find the point-slope form for the line by inserting the values of the point $(3,\ -2)$ and the value of the slope into the point-slope form: $y-y_1=m(x-x_1) \Rightarrow y-(-2)=-1(x-3) \Rightarrow y+2=-1(x-3)$. Then, use the distributive property of multiplication and solve for y: $y+2=-x+3 \Rightarrow y+2-2=-x+3-2 \Rightarrow y=-x+1$.

49. The points $(3,\ 0)$ and $\left(0,\ \frac{1}{3}\right)$ are on the line. First find the slope of the line:

$m=\frac{y_2-y_1}{x_2-x_1}=\frac{\frac{1}{3}-0}{0-3}=\frac{\frac{1}{3}}{-3}=\frac{1}{3}\left(-\frac{1}{3}\right)=-\frac{1}{9}$. Then, insert the values of the point $(3,\ 0)$ and the value of the slope into the point-slope form: $y-y_1=m(x-x_1) \Rightarrow y-0=-\frac{1}{9}(x-3)$.

Then, use the distributive property of multiplication and solve for y: $y=-\frac{1}{9}x+\frac{1}{3}$.

51. The line parallel to $y=2x-1$ has slope 2. Insert the values of the point $(2,\ -3)$ and the value of the slope into the point-slope form: $y-y_1=m(x-x_1) \Rightarrow y-(-3)=2(x-2) \Rightarrow y+3=2(x-2)$. Then, convert to slope-intercept form: $y+3=2x-4 \Rightarrow y+3-3=2x-4-3 \Rightarrow y=2x-7$.

53. The line perpendicular to $y=-\frac{1}{2}x+3$ has slope 2. Insert the values of the point $(6,\ -3)$ and the value of the slope into the point-slope form: $y-y_1=m(x-x_1) \Rightarrow y-(-3)=2(x-6) \Rightarrow y+3=2(x-6)$. Then, convert to slope-intercept form: $y+3=2x-12 \Rightarrow y+3-3=2x-12-3 \Rightarrow y=2x-15$.

55. As x increases by 1, y decreases by 2. Thus, the slope is -2. Insert the values of the point $(1, -3)$ and the value of the slope into the point slope form: $y - y_1 = m(x - x_1) \Rightarrow y - (-3) = -2(x-1) \Rightarrow y + 3 = -2(x-1)$. Then, convert to slope-intercept form: $y + 3 = -2x + 2 \Rightarrow y + 3 - 3 = -2x + 2 - 3 \Rightarrow y = -2x - 1$.

57. As x increases by 2, y increases by 1. Thus, the slope is $\frac{1}{2}$. Insert the values of the point $(-1, -3)$ and the value of the slope into the point slope form: $y - y_1 = m(x - x_1) \Rightarrow y - (-3) = \frac{1}{2}(x - (-1)) \Rightarrow y + 3 = \frac{1}{2}(x+1)$.

Then, convert to slope-intercept form: $y + 3 = \frac{1}{2}x + \frac{1}{2} \Rightarrow y + 3 - 3 = \frac{1}{2}x + \frac{1}{2} - 3 \Rightarrow y = \frac{1}{2}x - \frac{5}{2}$.

Graphical Interpretation

59. (a) As x increases, the y-value decreases. Because y represents the distance from home, the person is traveling toward home.

(b) After 1 hour, the person is 250 miles from home; after 4 hours, the person is 100 miles from home.

(c) Because in 3 hours the person travels 150 miles, the driver is traveling at 50 mph.

(d) First, find the slope: $m = \frac{y_2 - y_1}{x_2 - x_1} = \frac{100 - 250}{4 - 1} = \frac{-150}{3} = -50$. Then, $y = -50x + b$. Because when $x = 0$, $y = 300$, the y-intercept b is 300. Then, $y = -50x + 300$. The car is traveling toward home at 50 mph.

61. (a) Because y increases as x increases, water is entering the tank. After 4 minutes, 300 gallons.

(b) When $x = 0$, $y = 100$. therefore, the y-intercept is 100. This means that the tank held 100 gallons of water initially.

(c) First, find the slope: $m = \frac{y_2 - y_1}{x_2 - x_1} = \frac{200 - 100}{2 - 0} = \frac{100}{2} = 50$.

Then: $y = mx + b \Rightarrow y = 50x + 100$; the amount of water is increasing at a rate of 50 gal/min.

(d) The tank will be full after 8 minutes.

63. (a) Find the slope: $m = \frac{y_2 - y_1}{x_2 - x_1} = \frac{15 - 40}{6 - 1} = \frac{-25}{5} = -5°$ F/hr.

(b) One point of the graph of the line is $(1, 40)$ and another point is $(2, 35)$. First, find the point-slope form of the line: $y - y_1 = m(x - x_1) \Rightarrow y - 40 = -5(x-1)$. Then convert to slope-intercept form: $T = -5x + 5 + 40 \Rightarrow T = -5x + 45$. The temperature is decreasing at a rate of 5°F per hour.

(c) To find the x-intercept, use the slope-intercept form of the line and set $T = 0$: $T = -5x + 45 \Rightarrow 0 = -5x + 45 \Rightarrow 5x = 45 \Rightarrow x = 9$. Therefore, at 9 A.M. the temperature was 0°F.

(d) See Figure 63.

(e) At 4 A.M., the temperature was 25°F.

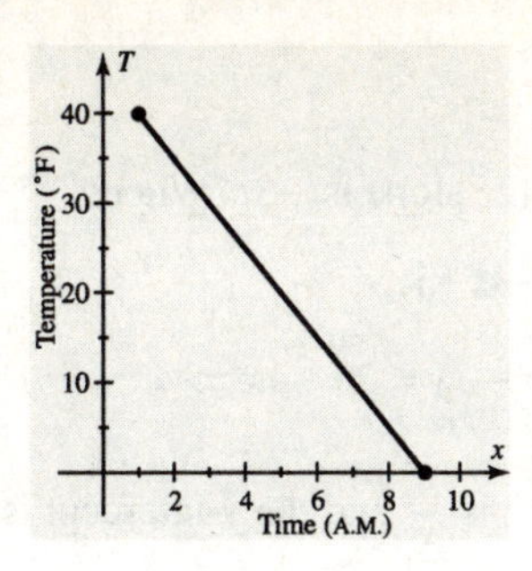

Figure 63

Applications

65. (a) First, substitute 1995 for

x: $y = 0.032(x-1995)+1.13 = 0.032(1995-1995)+1.13 = 0.032(0)+1.13 = 1.13.$

Thus, in 1995 there were 1,130,000 inmates.

Then, substitute 2005 for

x: $y = 0.032(x-1995)+1.13 = 0.032(2005-1995)+1.13 = 0.032(10)+1.13 = 0.32+1.13 = 1.45.$

Thus, in 2005 there were 1,450,000 inmates.

(b) One point on the graph of the line is (1995, 1.13) and another point is (2005, 1.45). Thus:

$m = \frac{y_2 - y_1}{x_2 - x_1} = \frac{1.45-1.13}{2005-1995} = \frac{0.32}{10} = 0.032.$ Therefore, the number of inmates increased on average by 32,000/yr.

67. (a) In 2002, average tuition and fees were \$4081; in 2004, average tuition and fees were \$5133.

(b) First, find the slope: $m = \frac{y_2 - y_1}{x_2 - x_1} = \frac{5133-4081}{2004-2002} = \frac{1052}{2} = 526.$ Then, substitute the values of the first point and the value of the slope into the point-slope form: $y - y_1 = m(x - x_1) \Rightarrow$

$y - 4081 = 526(x - 2002)$; tuition and fees increased on average by \$526/yr.

(c) From the graph, tuition and fees in 2005 were about \$5700. Substitute 2005 for

x: $y - 4081 = 526(2005-2002) \Rightarrow y - 4081 = 526(3) \Rightarrow y - 4081 = 1578 \Rightarrow$

$y - 4081 + 4081 = 1578 + 4081 \Rightarrow y = 5659.$ Thus, in 2005, tuition and fees would be about \$5659.

69. (a) The slope-intercept form is $y = mx + b.$ Solve for b by setting y equal to 39, m equal to 4.9, and x equal to 2004:

$39 = 4.9(2004) + b \Rightarrow 39 = 9819.6 + b \Rightarrow 39 - 9819.6 = 9819.6 - 9819.6 + b \Rightarrow -9780.6 = b.$

Therefore, $y = 4.9x - 9780.6.$

(b) Substitute 2010 for x: $y = 4.9(2010) - 9780.6 = 9849 - 9780.6 \Rightarrow y = 68.4$ million.

Checking Basic Concepts for Sections 3.5 & 3.6

1. For every x-value increase of 1, the y-value decreases by 3. Therefore, the slope is -3. When $x=0, y=1$. Thus, the y-intercept is 1. Therefore: $y=mx+b \Rightarrow y=-3x+1$.

2. To convert to slope-intercept form, solve for y: $4x-5y=20 \Rightarrow 4x-4x-5y=20-4x \Rightarrow$

   $-5y=-4x+20 \Rightarrow \frac{-5y}{-5}=\frac{-4x}{-5}+\frac{20}{-5} \Rightarrow y=\frac{4}{5}x-4$. Therefore, the slope is $\frac{4}{5}$ and the y-intercept is -4.

3. See Figure 3.

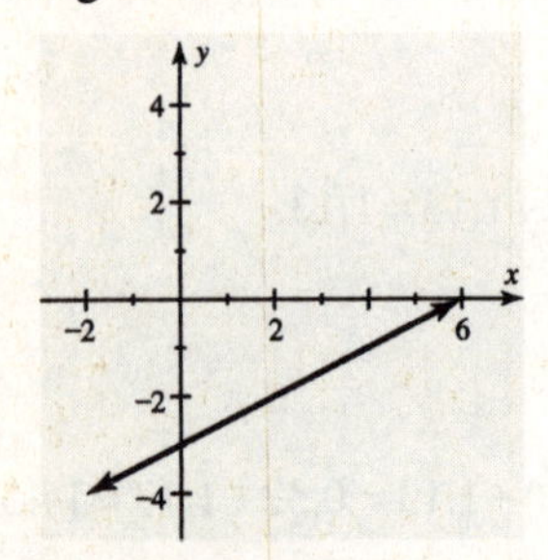

Figure 3

4. (a) The slope m is 3 and the y-intercept b is -2. Therefore: $y=mx+b \Rightarrow y=3x-2$.

   (b) The slope of the line $y=\frac{2}{3}x$ is $\frac{2}{3}$. Therefore, the slope of the line perpendicular to $y=\frac{2}{3}x$ has slope $-\frac{3}{2}$. Because for every 2 units of increase in run the rise decreases by 3, the y-intercept b exists when $x=0$. Thus, $b=0$. Then: $y=mx+b \Rightarrow y=-\frac{3}{2}x+0 \Rightarrow y=-\frac{3}{2}x$.

   (c) First, find the slope: $m=\frac{y_2-y_1}{x_2-x_1}=\frac{3-(-4)}{-2-1}=\frac{3+4}{-2-1}=\frac{7}{-3}=-\frac{7}{3}$. Then insert the values of the point $(1, -4)$ and the value of the slope into the point-slope form: $y-y_1=m(x-x_1) \Rightarrow$

   $y-(-4)=-\frac{7}{3}(x-1) \Rightarrow y+4=-\frac{7}{3}(x-1)$. Then, convert to slope-intercept form:

   $y+4=-\frac{7}{3}x+\frac{7}{3} \Rightarrow y+4-4=-\frac{7}{3}x+\frac{7}{3}-4 \Rightarrow y=-\frac{7}{3}x+\frac{7}{3}-\frac{12}{3} \Rightarrow y=-\frac{7}{3}x-\frac{5}{3}$.

5. $y-y_1=m(x-x_1) \Rightarrow y-3=-2(x-(-1)) \Rightarrow y-3=-2(x+1)$

6. Use the distributive property of multiplication and solve for y:

   $y+3=-2(x-2) \Rightarrow y+3=-2x+4 \Rightarrow y+3-3=-2x+4-3 \Rightarrow y=-2x+1$.

7. First, use the points $(-3,-3)$ and $(-1,1)$ to find the slope: $m=\frac{y_2-y_1}{x_2-x_1}=\frac{1-(-3)}{-1-(-3)}=\frac{1+3}{-1+3}=\frac{4}{2}=2$. Use the point-slope form to write an equation for the line then convert to slope-intercept form:

$y - y_1 = m(x - x_1) \Rightarrow y - (-3) = 2(x - (-3)) \Rightarrow y + 3 = 2(x + 3) \Rightarrow y + 3 = 2x + 6 \Rightarrow y + 3 - 3 = 2x + 6 - 3$

$\Rightarrow y = 2x + 3$

8. (a) First, find the slope: $m = \frac{y_2 - y_1}{x_2 - x_1} = \frac{12 - 36}{3 - 1} = \frac{-24}{2} = -12$. Then, insert the values for the point $(1, 36)$ and the value of the slope into the point-slope form: $y - y_1 = m(x - x_1) \Rightarrow y - 36 = -12(x - 1)$. Then convert to slope-intercept form: $y - 36 = -12x + 12 \Rightarrow y - 36 + 36 = -12x + 12 + 36 \Rightarrow y = -12x + 48$.

   (b) Because the slope is -12, the bicyclist gets 12 miles closer to home every hour ridden. Therefore the bicyclist is traveling at 12 mph.

   (c) The y-value will equal 0 when the rider reaches home Therefore, substitute 0 for y in the slope-intercept form and solve for x: $y = -12x + 48 \Rightarrow 0 = -12x + 48 \Rightarrow 0 - 48 = -12x + 48 - 48 \Rightarrow$

   $-48 = -12x \Rightarrow \frac{-48}{-12} = \frac{-12x}{-12} \Rightarrow 4 = x \Rightarrow x = 4$. Therefore, the rider will arrive home at 4 pm.

   (d) At noon, $x = 0$. Substitute 0 for x in the slope-intercept form and solve for y: $y = -12x + 48 \Rightarrow$

   $y = -12(0) + 48 \Rightarrow y = 48$. Therefore, at noon the rider was 48 miles from home.

9. (a) At noon, $t = 0$. Substitute 0 for t in the equation and solve for S: $S = 2t + 5 = 2(0) + 5 \Rightarrow S = 5$. Therefore, 5 inches of snow fell by noon.

   (b) Because the slope equals 2, the snow fell at a rate of 2 inches per hour in the afternoon.

   (c) The S-intercept is 5. This represents the total inches of snow that fell before noon.

   (d) The slope is 2. This represents that the rate of snowfall was 2 inches per hour.

3.7: Introduction to Modeling

Concepts

1. abstraction
2. predict
3. linear
4. exact
5. approximate
6. constant
7. constant rate of change
8. initial amount
9. f
10. d
11. a
12. c
13. e
14. b

Modeling Linear Data

15. For the linear equation $y = 2x + 2$, the y-intercept is 2 and the slope is 2. The ordered pairs in the table are modeled exactly by the linear equation.

17. For the linear equation $y = -4x$, the y-intercept is 0 and the slope is -4. The ordered pairs in the table are not modeled exactly by the linear equation.

19. For the linear equation $y = 1.4x - 4$, the y-intercept is -4 and the slope is 1.4. The ordered pairs in the table are not modeled exactly by the linear equation.

21. Because the line goes through the points on the graph, the linear model is exact. To find the equation of the line, first find the slope: $m = \frac{y_2 - y_1}{x_2 - x_1} = \frac{6-4}{4-3} = \frac{2}{1} = 2$. Then to find the y-intercept b, insert the values of the point (4, 6) and the value of the slope into the slope-intercept form and solve for b: $y = mx + b \Rightarrow$ $6 = 2(4) + b \Rightarrow 6 = 8 + b \Rightarrow 6 - 8 = 8 - 8 + b \Rightarrow -2 = b \Rightarrow b = -2$. Thus, the equation of the line is $y = 2x - 2$.

23. Because the line does not go through all the points on the graph, the linear model is approximate. To find the equation of the line, first find the slope: $m = \frac{y_2 - y_1}{x_2 - x_1} = \frac{6-4}{2-1} = \frac{2}{1} = 2$. Then to find the y-intercept b, insert the values of the point (2, 6) and the value of the slope into the slope-intercept form and solve for b: $y = mx + b \Rightarrow$ $6 = 2(2) + b \Rightarrow 6 = 4 + b \Rightarrow 6 - 4 = 4 - 4 + b \Rightarrow 2 = b \Rightarrow b = 2$. Thus, the equation of the line is $y = 2x + 2$.

25. Because the line does not go through all the points on the graph, the linear model is approximate. The equation of the line is $y = 2$ because the slope is 0 and because the y-intercept is at $y = 2$.

27. The initial value is 40, so the y-intercept b is 40. The rate of increase is 5 per minute, so the slope is 5. Thus: $y = mx + b \Rightarrow y = 5x + 40$.

29. The initial value is -5, so the y-intercept b is -5. The rate of decrease is 20 per day, so the slope is -20. Thus: $y = mx + b \Rightarrow y = -20x - 5$.

31. The initial value is 8, so the y-intercept b is 8. Because y remains constant, the slope is 0. Thus: $y = mx + b \Rightarrow y = 0(x) + 8 \Rightarrow y = 8$.

33. (a) See Figure 33a. A line could pass through all five points.

    (b) See Figure 33b.

    (c) Because y decreases by 2 for every increase in x, the slope m is -2. Because when $x = 0$, $y = 4$, the y-intercept b is 4. Thus: $y = mx + b \Rightarrow y = -2x + 4$.

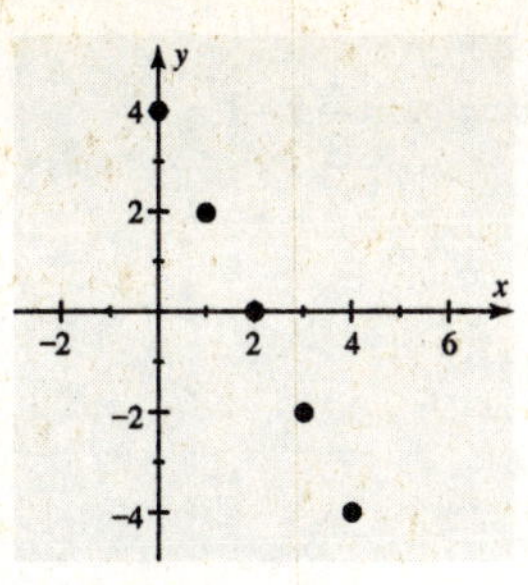

Figure 33a

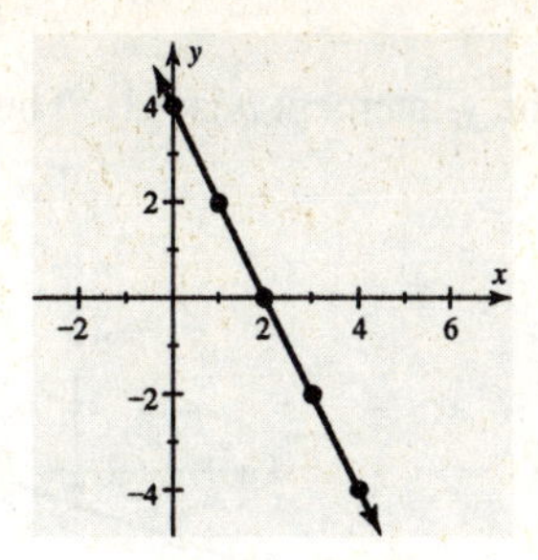

Figure 33b

35. (a) See Figure 35a. A line could not pass through all five points.

(b) See Figure 35b.

(c) To find the equation of the line, first find the slope: $m=\dfrac{y_2-y_1}{x_2-x_1}=\dfrac{0-4}{0-(-2)}=\dfrac{-4}{2}=-2.$

Then, the line passes through the origin, so the y-intercept b is 0. Thus: $y=mx+b\Rightarrow y=-2x+0\Rightarrow$ $y=-2x.$

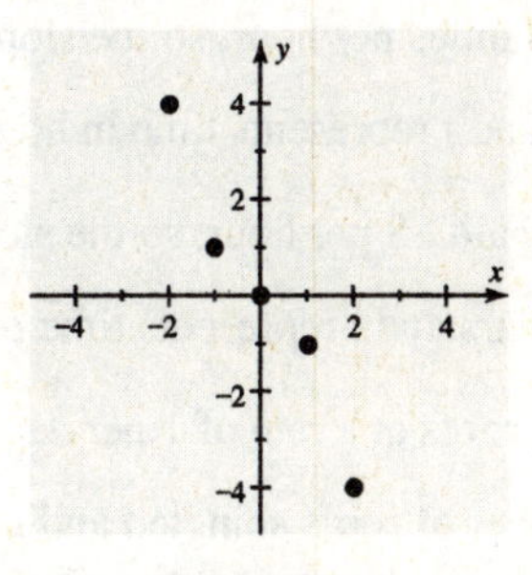

Figure 35a

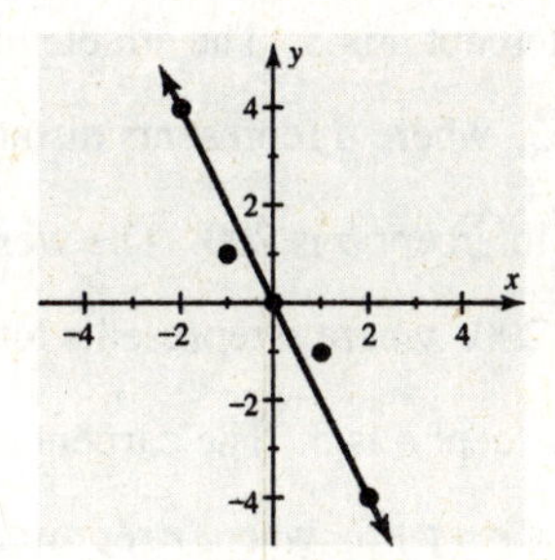

Figure 35b

37. (a) See Figure 37a. A line could pass through all five points.

(b) See Figure 37b.

(c) To find the equation of the line, first find the slope: $m=\dfrac{y_2-y_1}{x_2-x_1}=\dfrac{-2-0}{0-(-4)}=\dfrac{-2}{4}=-\dfrac{1}{2}.$

Then, when $x=0,\ y=-2,$ the y-intercept b is -2. Thus: $y=mx+b\Rightarrow y=-\dfrac{1}{2}x-2.$

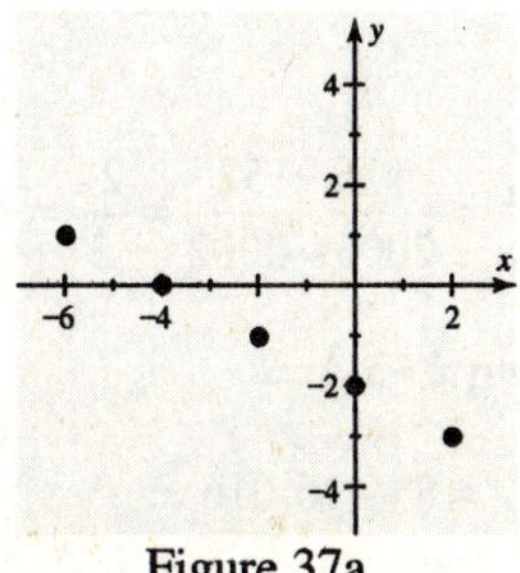

Figure 37a

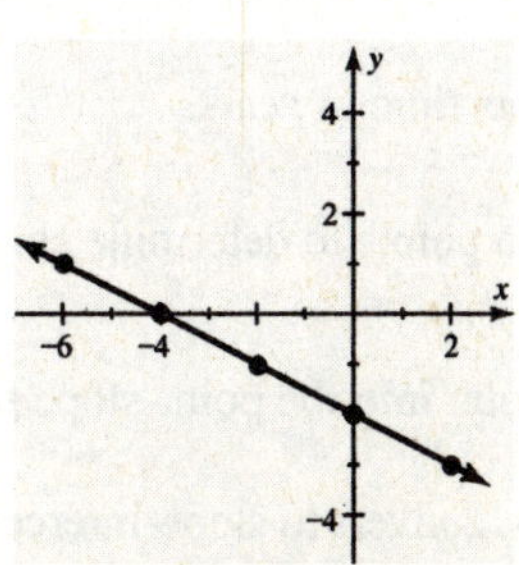

Figure 37b

39. (a) See Figure 39a. A line could not pass through all five points.

(b) See Figure 39b.

(c) To find the equation of the line, first find the slope: $m=\dfrac{y_2-y_1}{x_2-x_1}=\dfrac{0-(-1)}{3-0}=\dfrac{1}{3}.$

Then, when $x=0,\ y=-1,$ the y-intercept b is -1. Thus: $y=mx+b \Rightarrow y=\frac{1}{3}x-1.$

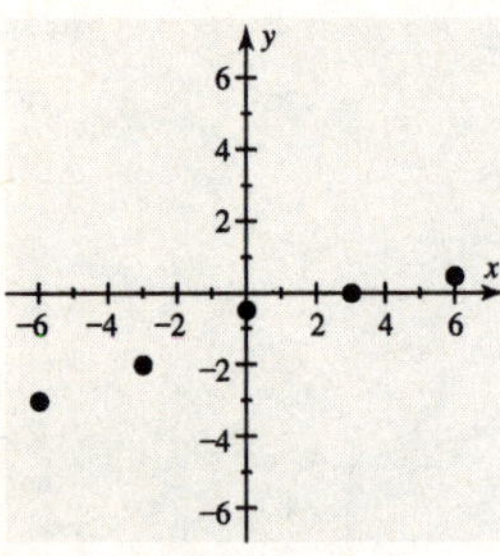

Figure 39a

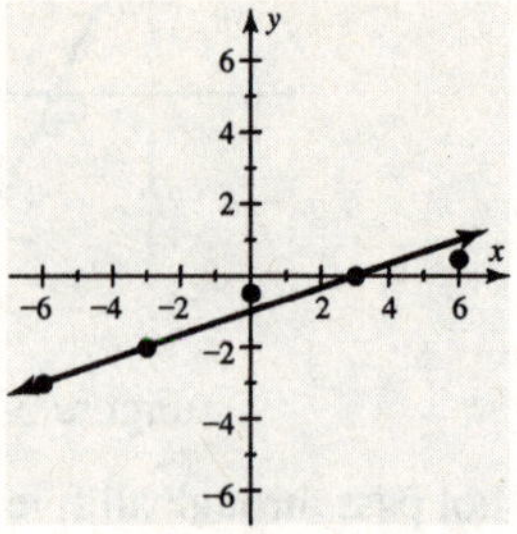

Figure 39b

Applications

41. The initial value is 200, so the g-intercept b is 200. The barrel is being filled at a rate of 5 gallons per minute, so the slope m is 5. Thus: $g=mt+b \Rightarrow g=5t+200,$ where g represents gallons of water and t represents time in minutes.

43. The initial value is 5, so the d-intercept b is 5. The athlete is jogging at 6 miles per hour, so the slope m is 6. Thus: $d=mt+b \Rightarrow d=6t+5,$ where d represents distance in miles and t represents time in hours.

45. The initial value is 200, so the p-intercept b is 200. The worker is being paid \$8 per hour, so the slope m is 8. Thus: $p=mt+b \Rightarrow p=8t+200,$ where p represents total pay in dollars and t represents time in hours.

47. The initial value is 5, so the r-intercept b is 5. The carpenter is shingling roofs at a rate of 1 per day, so the slope m is 1. Thus: $r=mt+b \Rightarrow r=t+5,$ where r represents total number of roofs shingled and t represents time in days.

49. (a) A point on the graph of the volume of the glacier is (1912, 5) and another point is (2002, 1). Because there is assumed to be a constant melt rate, to find the yearly rate find the slope of the line connecting the two points: $m=\frac{y_2-y_1}{x_2-x_1}=\frac{1-5}{2002-1912}=\frac{-4}{90}=-\frac{2}{45}$ acres per year.

(b) The slope m is $-\frac{2}{45}$. The initial value b is 5. Thus: $A=mt+b \Rightarrow A=-\frac{2}{45}t+5,$ where A represents acres of glacier and t represents time in years.

51. (a) First, use the first and last data points to determine slope: $m=\frac{y_2-y_1}{x_2-x_1}=\frac{176-152}{2005-2002}=\frac{24}{3}=8.$ Then, insert the values of the first point into the point-slope form: $y-y_1=m(x-x_1) \Rightarrow$

$y-152=8(x-2002)$. Then, convert to slope-intercept form: $y-152=8x-16{,}016 \Rightarrow$

$y-152+152=8x-16{,}016+152 \Rightarrow y=8x-15{,}864$

(b) Set $x=2009$ and solve for y: $y=8(2009)-15{,}864=16{,}072-15{,}864=208.$ Thus, the prison population in 2009 will be about 208,000.

53. (a) See Figure 53a.

(b) See Figure 53b.

(c) $m = \dfrac{y_2 - y_1}{x_2 - x_1} = \dfrac{120 - 60}{6 - 3} = \dfrac{60}{3} = 20.$ For every gallon of gasoline, the car travels 20 miles.

(d) Because when $x = 0,\ y = 0,$ the y-intercept b is 0. Thus, $y = 20x.$

(e) $y = 20x = 20(7) = 140$ miles.

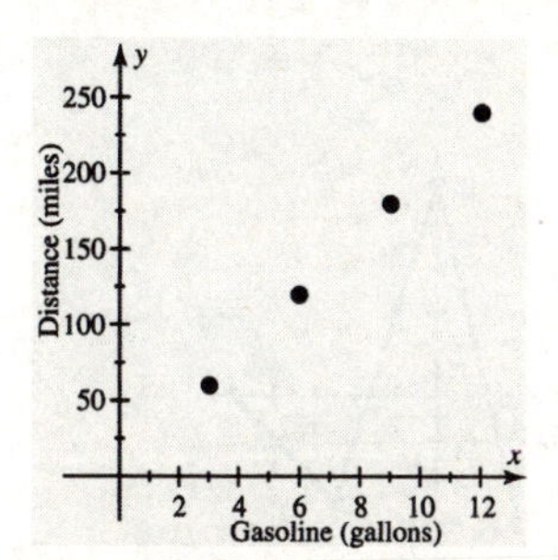

Figure 53a

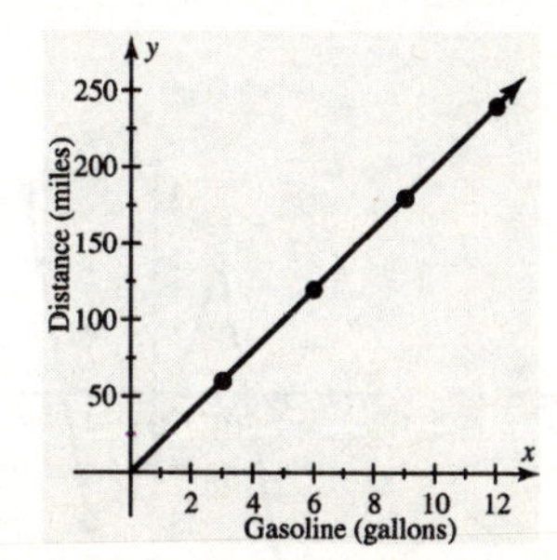

Figure 53b

Checking Basic Concepts for Section 3.7

1. The ordered pairs in the table indicate that the y-intercept is 10 and the slope is -5. Thus, the line $y = -5x + 10$ models exactly the ordered pairs.

2. Because the line does not pass through each point on the graph, the linear model is approximate. For each increase in run of 1, the value of the rise increases by 1, so the slope m is 1. When $x = 0,\ y = -1,$ so the y-intercept b is -1. Thus: $y = mx + b \Rightarrow y = x - 1.$

3. (a) The initial value b is 50 and the slope m is 10. Thus: $y = 10x + 50.$

   (b) The initial value b is 200 and the slope m is -2. Thus: $y = -2x + 200.$

4. (a) See Figure 4a. A line could pass through all four points.

   (b) See Figure 4b.

   (c) The y-intercept b is 1 and the slope m is $\dfrac{1-2}{0-(-2)} = \dfrac{-1}{2} = -\dfrac{1}{2}.$ Thus: $y = -\dfrac{1}{2}x + 1.$

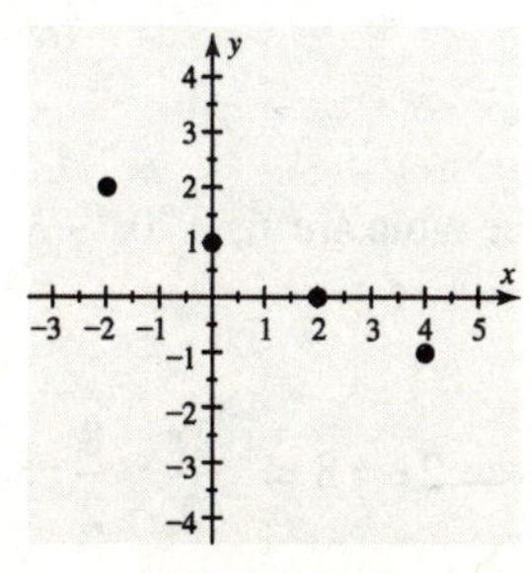

Figure 4a

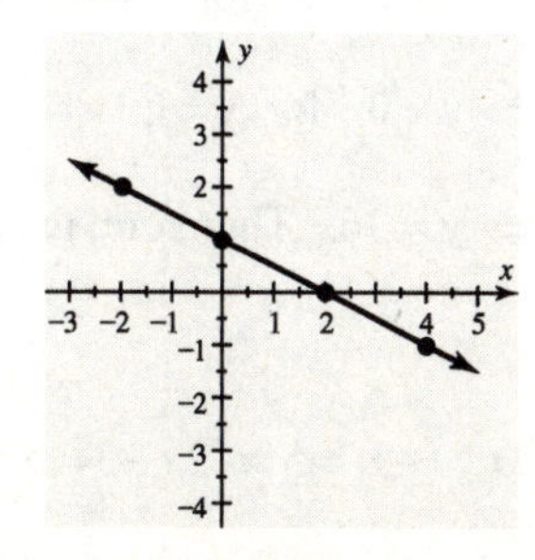

Figure 4b

5. (a) The slope m is 0.075 and the y-intercept is 0. Thus, $T = 0.075x.$

   (b) 2005 is 60 years after 1945, so substitute 60 for x and solve for T. $T = 0.075\ (60) = 4.5$. The temperature increase is 4.5° F.

Chapter 3 Review Exercises

Section 3.1

1. $(-2,\ 0)$: none; $(-1,\ 2)$: Quadrant II; $(0,\ 0)$: none; $(1,\ -2)$: Quadrant IV; $(1,\ 3)$: Quadrant I

2. See Figure 2.

3. See Figure 3.

4. See Figure 4.

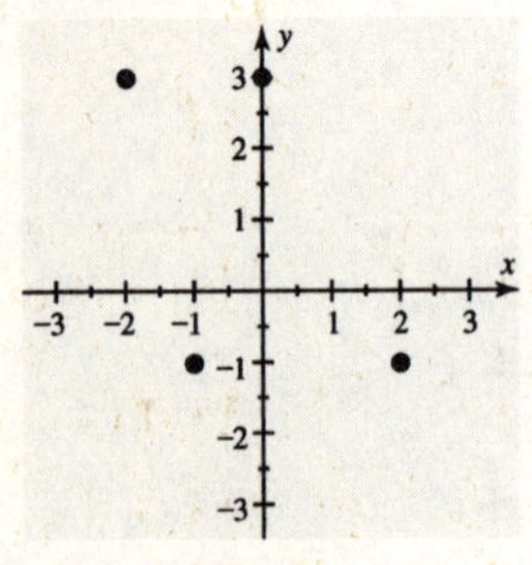

Figure 2

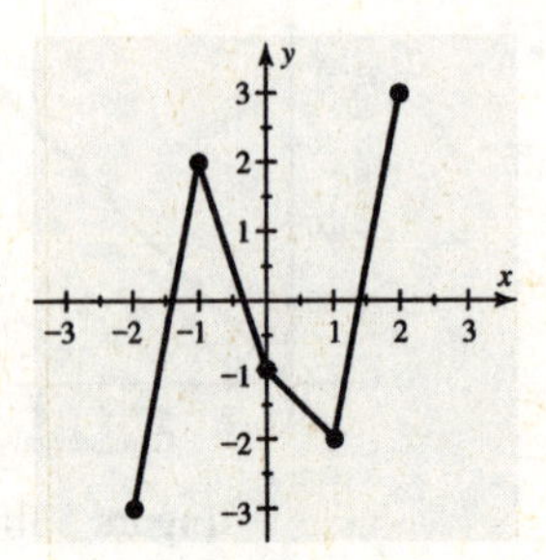

Figure 3

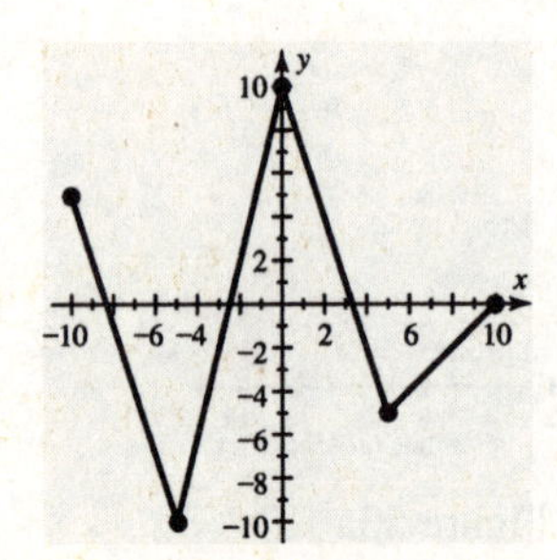

Figure 4

Section 3.2

5. Substitute 6 for x and 3 for y: $y = x - 3 \Rightarrow 3 = 6 - 3 \Rightarrow 3 = 3$. Because this is a true statement, the ordered pair (6, 3) is a solution.

6. Substitute -2 for x and 1 for y: $y = 5 - 2x \Rightarrow 1 = 5 - 2(-2) \Rightarrow 1 = 5 + 4 \Rightarrow 1 = 9$. Because this is not a true statement, the ordered pair $(-2,\ 1)$ is not a solution.

7. Substitute -1 for x and 6 for y: $3x - y = 3 \Rightarrow 3(-1) - 6 = 3 \Rightarrow -3 - 6 = 3 \Rightarrow -9 = 3$. Because this is not a true statement, the ordered pair $(-1,\ 6)$ is not a solution.

8. Substitute -4 for x and -3 for y: $\frac{1}{2}x + 2y = -8 \Rightarrow \frac{1}{2}(-4) + 2(-3) = -8 \Rightarrow -2 - 6 = -8 \Rightarrow -8 = -8$.

   Because this is a true statement, the ordered pair $(-4,\ -3)$ is a solution.

9. For $x = -2$: $y = -3x = -3(-2) \Rightarrow y = 6$; for $x = -1$: $y = -3x = -3(-1) \Rightarrow y = 3$;

   for $x = 0$: $y = -3x = -3(0) \Rightarrow y = 0$; for $x = 1$: $y = -3x = -3(1) \Rightarrow y = -3$;

   for $x = 2$: $y = -3x = -3(2) \Rightarrow y = -6$. Therefore, the missing values in the table are 6, 3, 0, -3, and -6.

   See Figure 9.

10. For $y = -3$: $2x + y = 5 \Rightarrow 2x + (-3) = 5 \Rightarrow 2x - 3 = 5 \Rightarrow 2x + 3 - 3 = 5 + 3 \Rightarrow 2x = 8 \Rightarrow \frac{2x}{2} = \frac{8}{2} \Rightarrow x = 4$;

    for $y = -1$: $2x + y = 5 \Rightarrow 2x + (-1) = 5 \Rightarrow 2x - 1 = 5 \Rightarrow 2x - 1 + 1 = 5 + 1 \Rightarrow$

    $2x = 6 \Rightarrow \frac{2x}{2} = \frac{6}{2} \Rightarrow x = 3$; for $y = 0$: $2x + y = 5 \Rightarrow 2x + 0 = 5 \Rightarrow 2x = 5 \Rightarrow \frac{2x}{2} = \frac{5}{2} \Rightarrow x = 2.5$;

    for $y = 1$: $2x + y = 5 \Rightarrow 2x + 1 = 5 \Rightarrow 2x + 1 - 1 = 5 - 1 \Rightarrow 2x = 4 \Rightarrow \frac{2x}{2} = \frac{4}{2} \Rightarrow x = 2$;

for $y=3$: $2x+y=5 \Rightarrow 2x+3=5 \Rightarrow 2x+3-3=5-3 \Rightarrow 2x=2 \Rightarrow \frac{2x}{2}=\frac{2}{2} \Rightarrow x=1.$

Therefore, the missing values in the tables are 4, 3, 2.5, 2, and 1. See Figure 10.

11. See Figure 11.

x	−2	−1	0	1	2
y	6	3	0	−3	−6

Figure 9

x	4	3	2.5	2	1
y	−3	−1	0	1	3

Figure 10

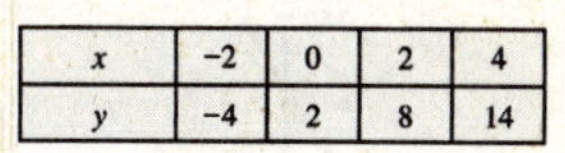

x	−2	0	2	4
y	−4	2	8	14

Figure 11

12. See Figure 12.

13. See Figure 13.

14. See Figure 14.

x	1	2	3	4
y	6	5	4	3

Figure 12

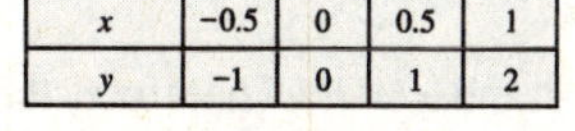

x	−0.5	0	0.5	1
y	−1	0	1	2

Figure 13

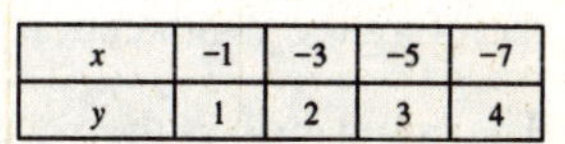

x	−1	−3	−5	−7
y	1	2	3	4

Figure 14

15. See Figure 15.

16. See Figure 16.

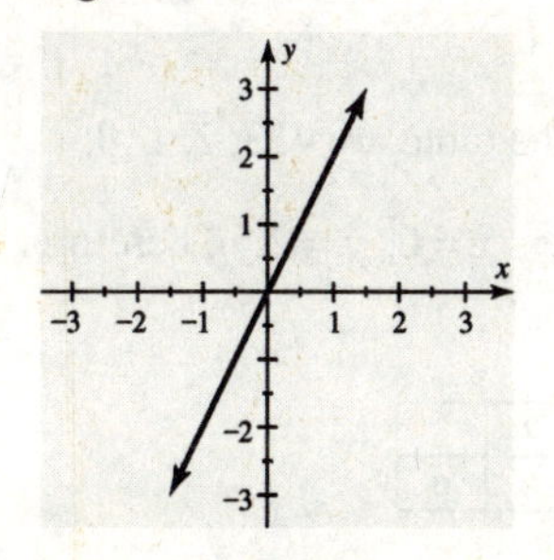

Figure 15

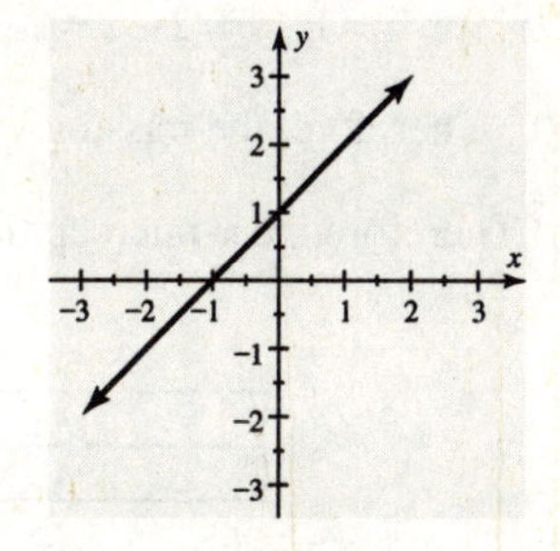

Figure 16

17. See Figure 17.

18. See Figure 18.

19. See Figure 19.

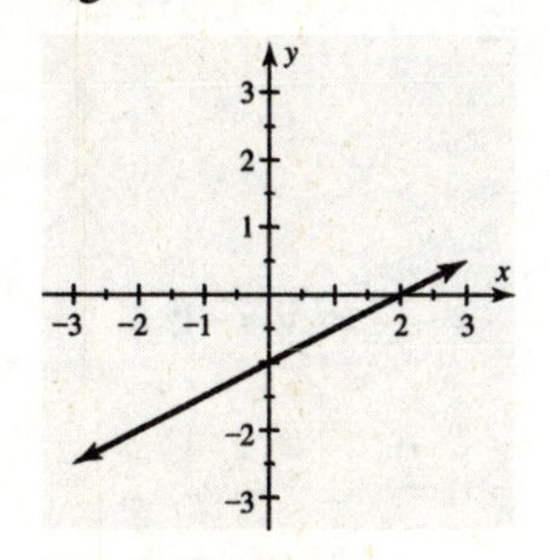

Figure 17

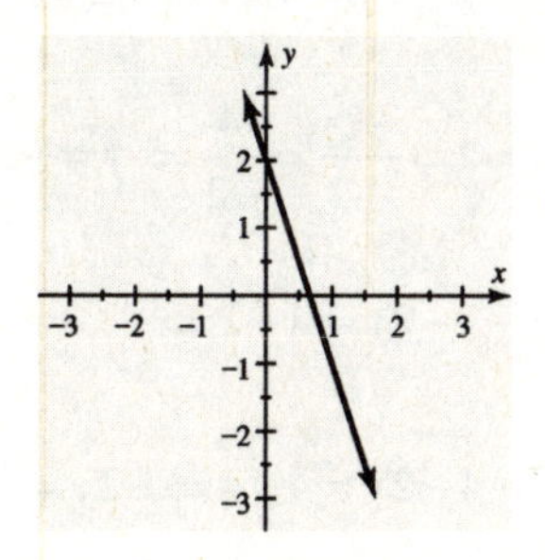

Figure 18

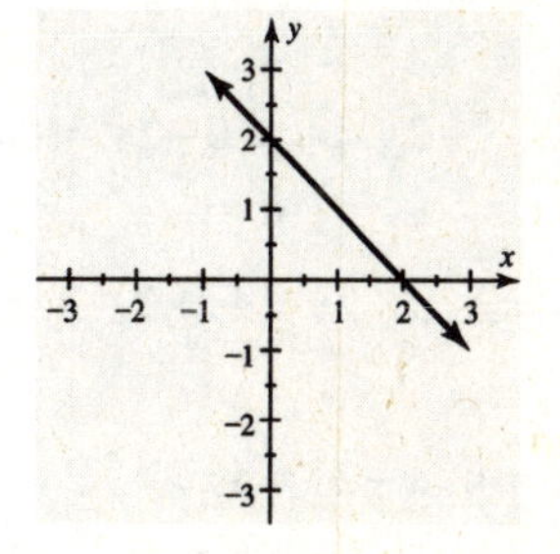

Figure 19

20. See Figure 20.

21. See Figure 21.

22. See Figure 22.

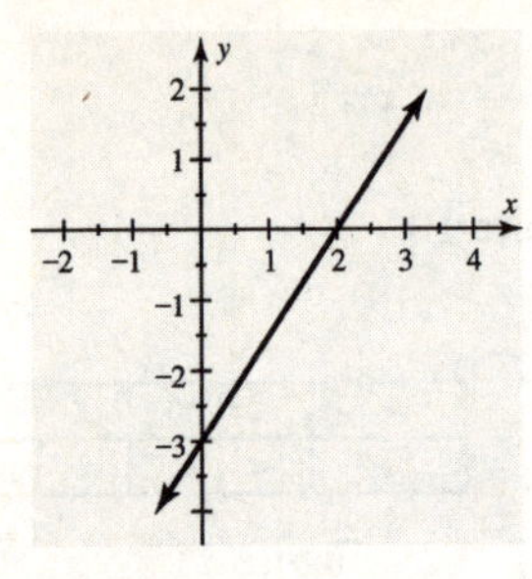

Figure 20

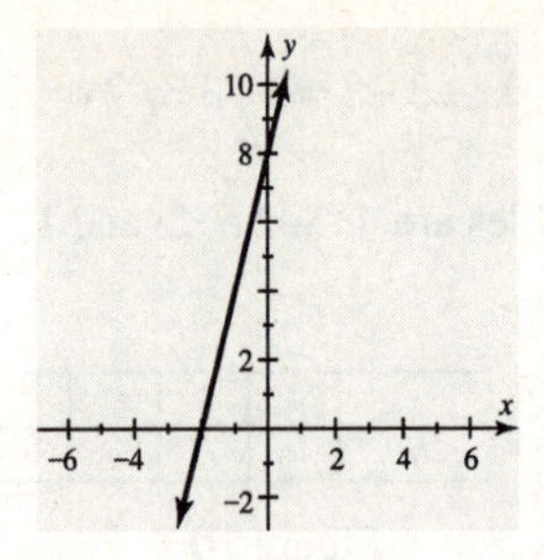

Figure 21

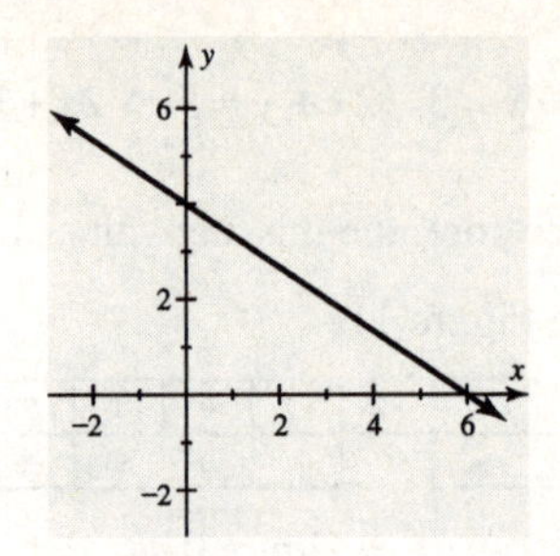

Figure 22

Section 3.3

23. The line crosses the x-axis at $x = 3$, therefore the x-intercept is 3. The line crosses the y-axis at $y = -2$, therefore the y-intercept is -2.

24. The graph crosses the x-axis at $x = -2$ and $x = 2$. Therefore, the x-intercepts are $2,\ -2$. The graph crosses the y-axis at $y = -4$, therefore, the y-intercept is -4.

25. For $x = -2$: $y = 2 - x = 2 - (-2) \Rightarrow y = 4$; for $x = -1$: $y = 2 - x = 2 - (-1) = 2 + 1 \Rightarrow y = 3$; for $x = 0$: $y = 2 - x = 2 - 0 \Rightarrow y = 2$; for $x = 1$: $y = 2 - x = 2 - 1 \Rightarrow y = 1$; for $x = 2$: $y = 2 - x = 2 - 2 \Rightarrow y = 0$. Therefore, the missing values in the table are 4, 3, 2, 1 ,0. See Figure 25. When $y = 0$, $x = 2$. Therefore the x-intercept is 2. When $x = 0$, $y = 2$. Therefore, the y-intercept is 2.

x	−2	−1	0	1	2
y	4	3	2	1	0

Figure 25

x	−4	−2	0	2	4
y	−4	−3	−2	−1	0

Figure 26

26. For $x = -4$: $x - 2y = 4 \Rightarrow -4 - 2y = 4 \Rightarrow -4 + 4 - 2y = 4 + 4 \Rightarrow -2y = 8 \Rightarrow \dfrac{-2y}{-2} = \dfrac{8}{-2} \Rightarrow y = -4$; for $x = -2$: $x - 2y = 4 \Rightarrow -2 - 2y = 4 \Rightarrow -2 + 2 - 2y = 4 + 2 \Rightarrow -2y = 6 \Rightarrow \dfrac{-2y}{-2} = \dfrac{6}{-2} \Rightarrow y = -3$; for $x = 0$: $x - 2y = 4 \Rightarrow 0 - 2y = 4 \Rightarrow -2y = 4 \Rightarrow \dfrac{-2y}{-2} = \dfrac{4}{-2} \Rightarrow y = -2$;

for $x = 2$: $x - 2y = 4 \Rightarrow 2 - 2y = 4 \Rightarrow 2 - 2 - 2y = 4 - 2 \Rightarrow -2y = 2 \Rightarrow \dfrac{-2y}{-2} = \dfrac{2}{-2} \Rightarrow y = -1$;

for $x = 4$: $x - 2y = 4 \Rightarrow 4 - 2y = 4 \Rightarrow 4 - 4 - 2y = 4 - 4 \Rightarrow -2y = 0 \Rightarrow \dfrac{-2y}{-2} = \dfrac{0}{-2} \Rightarrow y = 0$.

Therefore, the missing values in the table are $-4,\ -3,\ -2,\ -1,\ 0$. See Figure 26. When $y = 0$, $x = 4$. Therefore, the x-intercept is 4. When $x = 0$, $y = -2$. Therefore, the y-intercept is -2.

27. To find the x-intercept, set $y = 0$: $2x - 3y = 6 \Rightarrow 2x - 3(0) = 6 \Rightarrow 2x = 6 \Rightarrow \dfrac{2x}{2} = \dfrac{6}{2} \Rightarrow x = 3$. Therefore, the x-intercept is 3.

To find the y-intercept, set $x=0$: $2x-3y=6 \Rightarrow 2(0)-3y=6 \Rightarrow -3y=6 \Rightarrow \frac{-3y}{-3}=\frac{6}{-3} \Rightarrow$

$y=-2$. Therefore, the y-intercept is -2. See Figure 27.

28. Set $y=0$ and solve for x: $5x-y=5 \Rightarrow 5x-0=5 \Rightarrow 5x=5 \Rightarrow \frac{5x}{5}=\frac{5}{5} \Rightarrow x=1$. Therefore, the x-intercept is 1. Then set $x=0$ and solve for y: $5x-y=5 \Rightarrow 5(0)-y=5 \Rightarrow -y=5 \Rightarrow$

$-1(-y)=-1(5) \Rightarrow y=-5$. Therefore, the y-intercept is -5. See Figure 28.

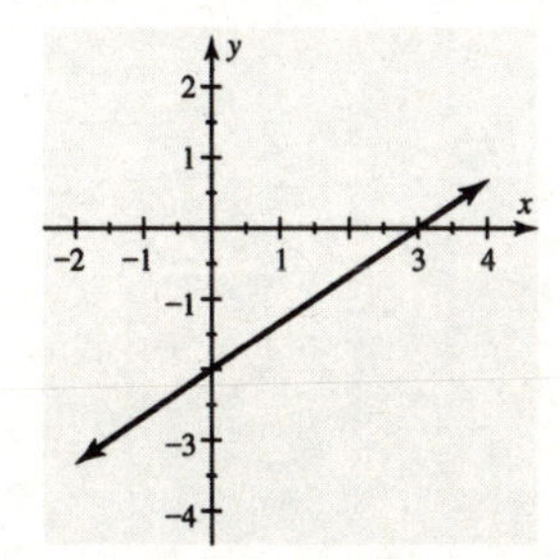

Figure 27

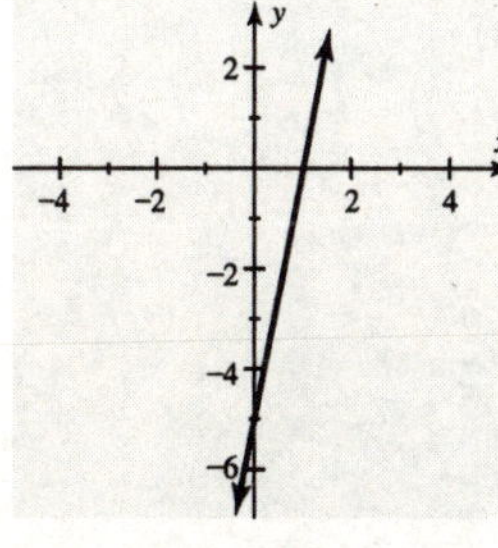

Figure 28

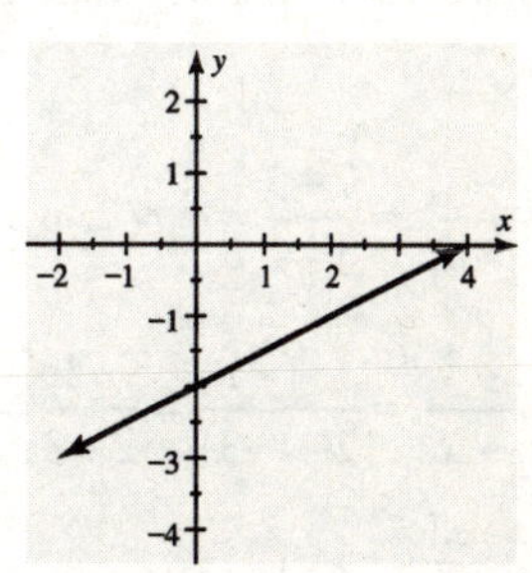

Figure 29

29. Set $y=0$ and solve for x: $0.1x-0.2y=0.4 \Rightarrow 0.1x-0.2(0)=0.4 \Rightarrow 0.1x=0.4 \Rightarrow \frac{0.1x}{0.1}=\frac{0.4}{0.1} \Rightarrow$

$x=4$. Therefore the x-intercept is 4. Then set $x=0$ and solve for y: $0.1x-0.2y=0.4 \Rightarrow$

$0.1(0)-0.2y=0.4 \Rightarrow -0.2y=0.4 \Rightarrow \frac{-0.2y}{-0.2}=\frac{0.4}{-0.2} \Rightarrow y=-2$. Therefore, the y-intercept is -2.

See Figure 29.

30. Set $y=0$ and solve for x: $\frac{x}{2}+\frac{y}{3}=1 \Rightarrow \frac{x}{2}+\frac{0}{3}=1 \Rightarrow \frac{x}{2}=1 \Rightarrow 2\left(\frac{x}{2}\right)=2(1) \Rightarrow x=2$.

Therefore the x-intercept is 2. Then set $x=0$ and solve for y: $\frac{x}{2}+\frac{y}{3}=1 \Rightarrow \frac{0}{2}+\frac{y}{3}=1 \Rightarrow \frac{y}{3}=1 \Rightarrow$

$3\left(\frac{y}{3}\right)=3(1) \Rightarrow y=3$. Therefore, the y-intercept is 3. See Figure 30.

31. (a) See Figure 31a.

(b) See Figure 31b.

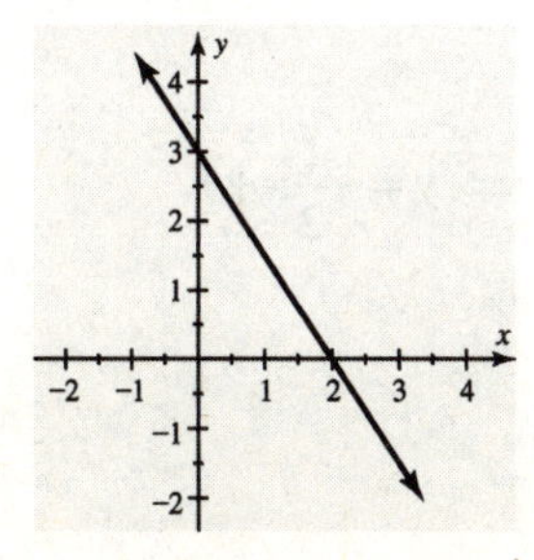

Figure 30

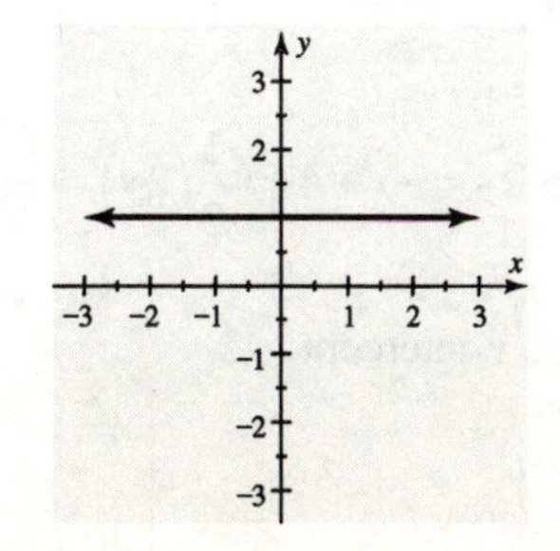

Figure 31a

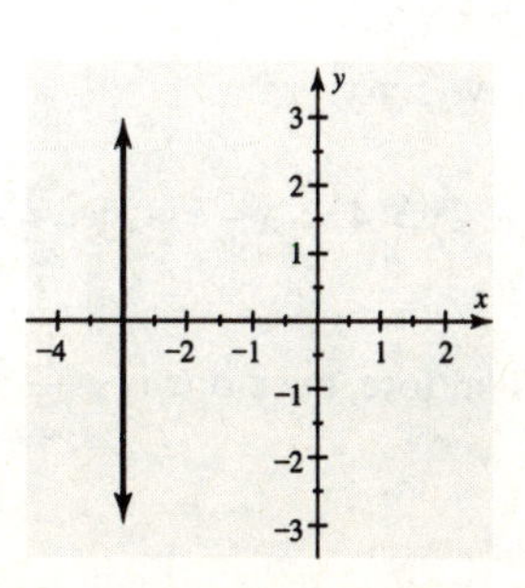

Figure 31b

32. $x=-1$; $y=1$

33. Because the y-value is constant, the slope $m = 0$. Because when $x = 0$, $y = 1$, the y-intercept $b = 1$.

Thus: $y = mx + b \Rightarrow y = 0x + 1 \Rightarrow y = 1$.

34. Because the x-value is a constant 3, the graph is a vertical line with slope m undefined. Thus, $x = 3$.

35. horizontal line: $y = 3$; vertical line: $x = -2$

36. (a) When $x = 0$, $y = 90$. Therefore, the y-intercept is 90. When $y = 0$, $x = 3$. Therefore, the x-intercept is 3.

(b) The driver is initially 90 miles from home; the driver arrives home after 3 hours.

Section 3.4

37. $m = \dfrac{y_2 - y_1}{x_2 - x_1} = \dfrac{7-3}{4-2} = \dfrac{4}{2} = 2.$

38. $m = \dfrac{y_2 - y_1}{x_2 - x_1} = \dfrac{-1-1}{2-(-3)} = \dfrac{-2}{2+3} = \dfrac{-2}{5} = -\dfrac{2}{5}.$

39. $m = \dfrac{y_2 - y_1}{x_2 - x_1} = \dfrac{1-1}{5-2} = \dfrac{0}{3} = 0.$

40. $m = \dfrac{y_2 - y_1}{x_2 - x_1} = \dfrac{10-6}{-5-(-5)} = \dfrac{4}{-5+5} = \dfrac{4}{0}.$ Thus, undefined.

41. The points (1, 3) and (0, 0) are on the graph. Insert these two points into the slope formula:

$m = \dfrac{y_2 - y_1}{x_2 - x_1} = \dfrac{0-3}{0-1} = \dfrac{-3}{-1} = 3.$

42. The points $(-2, 3)$ and (0, 2) are on the graph. Insert these two points into the slope formula:

$m = \dfrac{y_2 - y_1}{x_2 - x_1} = \dfrac{2-3}{0-(-2)} = \dfrac{-1}{0+2} = \dfrac{-1}{2} = -\dfrac{1}{2}.$

43. (a) See Figure 43.

(b) The slope is -2 and the y-intercept is 0.

44. (a) See Figure 44.

(b) The slope is 1 and the y-intercept is -1.

45. (a) See Figure 45.

(b) Solve for y:

$$x + 2y = 4 \Rightarrow x - x + 2y = 4 - x \Rightarrow 2y = -x + 4 \Rightarrow \frac{1}{2}(2y) = \frac{1}{2}(-x+4) \Rightarrow y = -\frac{1}{2}x + 2.$$

Therefore, the slope is $-\dfrac{1}{2}$ and the y-intercept is 2.

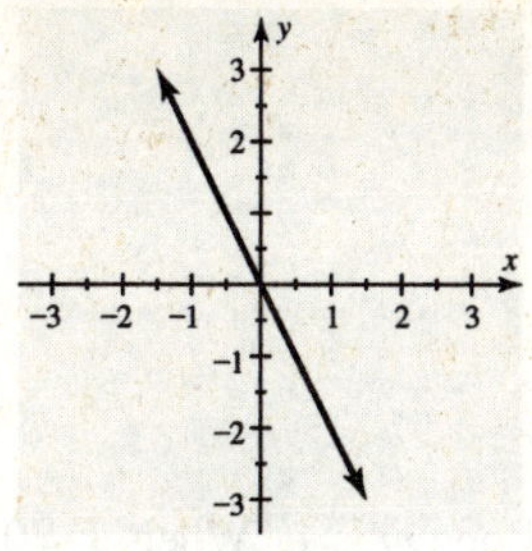

Figure 43

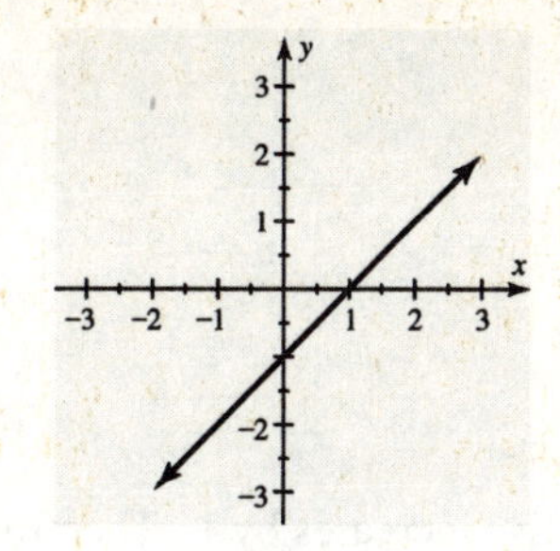

Figure 44

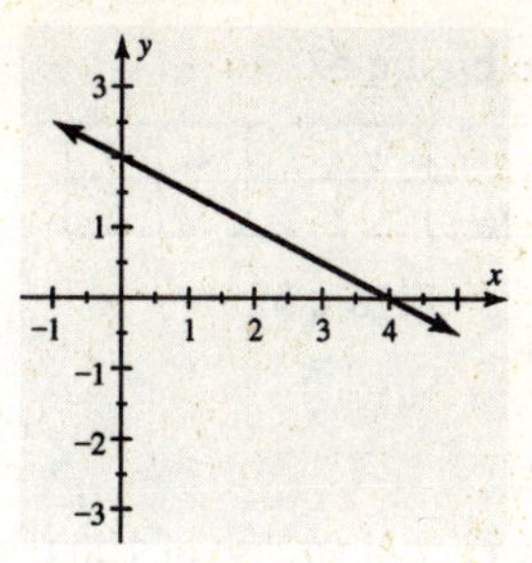

Figure 45

46. (a) See Figure 46.

(b) Solve for y:

$$2x-3y=-6 \Rightarrow 2x-2x-3y=-6-2x \Rightarrow -3y=-2x-6 \Rightarrow -\frac{1}{3}(-3y)=-\frac{1}{3}(-2x-6) \Rightarrow$$

$y=\frac{2}{3}x+2$. Therefore, the slope is $\frac{2}{3}$ and the y-intercept is 2.

47. See Figure 47.

48. See Figure 48.

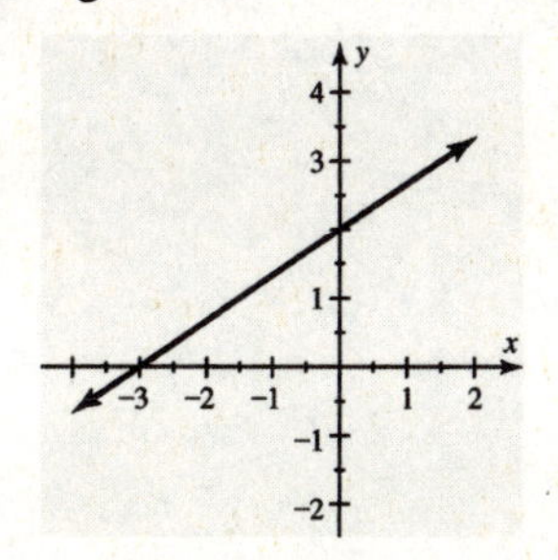

Figure 46

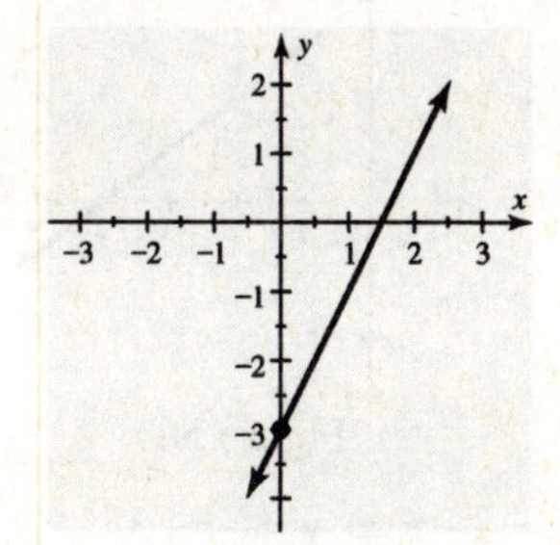

Figure 47

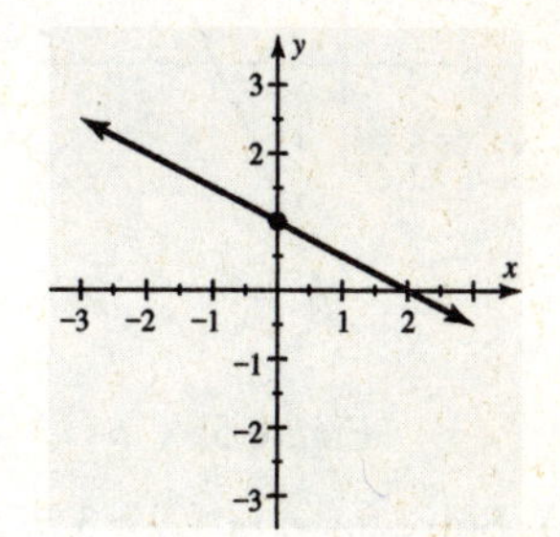

Figure 48

49. See Figure 49.

50. See Figure 50.

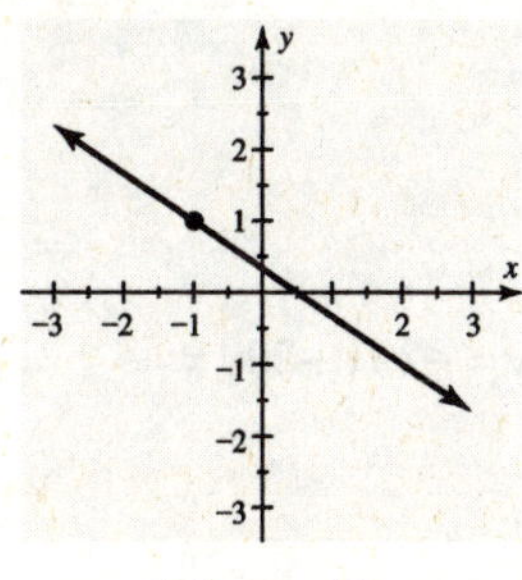

Figure 49

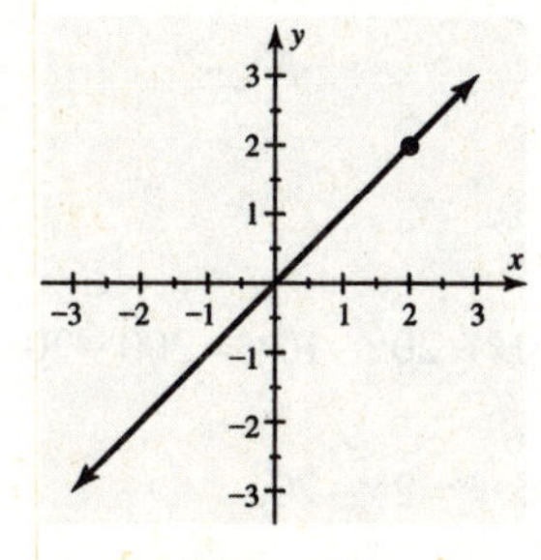

Figure 50

51. For every increase of 1 in the run, the rise increases by 2. Thus, the slope is 2. When $y=0$, $x=1$. Thus, the x-intercept is 1. When $x=0$, $y=-2$. Thus, the y-intercept is -2.

52. Because the slope is $\frac{1}{2}$, for every increase of 1 in the run, the rise increases by $\frac{1}{2}$. Thus, when $x=1$, $y=1+\frac{1}{2}=\frac{3}{2}$. When $x=2$, $y=\frac{3}{2}+\frac{1}{2}=\frac{4}{2}=2$. When $x=3$, $y=2+\frac{1}{2}=\frac{4}{2}+\frac{1}{2}=\frac{5}{2}$.

See Figure 52.

x	0	1	2	3
y	1	$\frac{3}{2}$	2	$\frac{5}{2}$

Figure 52

Section 3.5

53. For every increase of 1 in the run, the rise increases by 1. Therefore, the slope m is 1. When $x=0,\ y=1$. Therefore, the y-intercept is 1. Thus, $y=mx+b \Rightarrow y=1x+1 \Rightarrow y=x+1$.

54. For every increase of 1 in the run, the rise decreases by 2. Therefore, the slope m is -2. When $x=0,\ y=2$. Therefore, the y-intercept is 2. Thus, $y=mx+b \Rightarrow y=-2x+2$.

55. See Figure 55. $y=mx+b \Rightarrow y=2x-2$

56. See Figure 56. $y=mx+b \Rightarrow y=-\frac{3}{4}x+3$

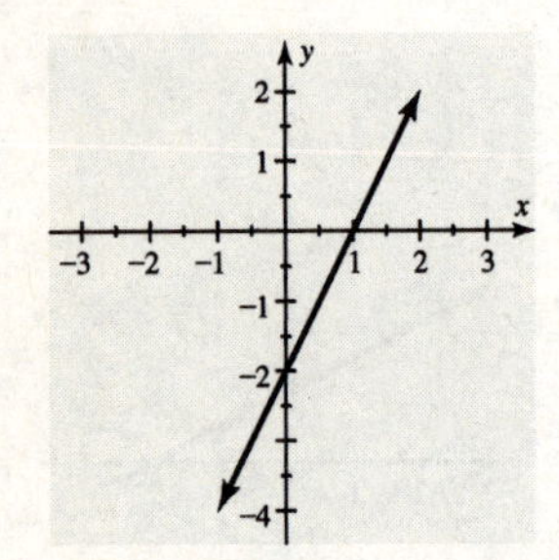

Figure 55

Figure 56

57. (a) Solve for y: $x+y=3 \Rightarrow x-x+y=3-x \Rightarrow y=-x+3$.

(b) The slope is -1 and the y-intercept is 3.

58. (a) Solve for y: $-3x+2y=-6 \Rightarrow -3x+3x+2y=-6+3x \Rightarrow 2y=3x-6 \Rightarrow$

$$\frac{1}{2}(2y)=\frac{1}{2}(3x-6) \Rightarrow y=\frac{3}{2}x-3.$$

(b) The slope is $\frac{3}{2}$ and the y-intercept is -3.

59. (a) Solve for y: $20x-10y=200 \Rightarrow 20x-20x-10y=200-20x \Rightarrow -10y=-20x+200 \Rightarrow$

$$-\frac{1}{10}(-10y)=-\frac{1}{10}(-20x+200) \Rightarrow y=2x-20.$$

(b) The slope is 2 and the y-intercept is -20.

60. (a) Solve for y: $5x-6y=30 \Rightarrow 5x-5x-6y=30-5x \Rightarrow -6y=-5x+30 \Rightarrow$

$$-\frac{1}{6}(-6y)=-\frac{1}{6}(-5x+30) \Rightarrow y=\frac{5}{6}x-5.$$

(b) The slope is $\frac{5}{6}$ and the y-intercept is -5.

61. See Figure 61.

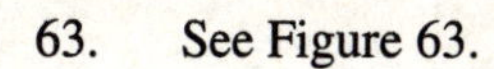

62. See Figure 62.

63. See Figure 63.

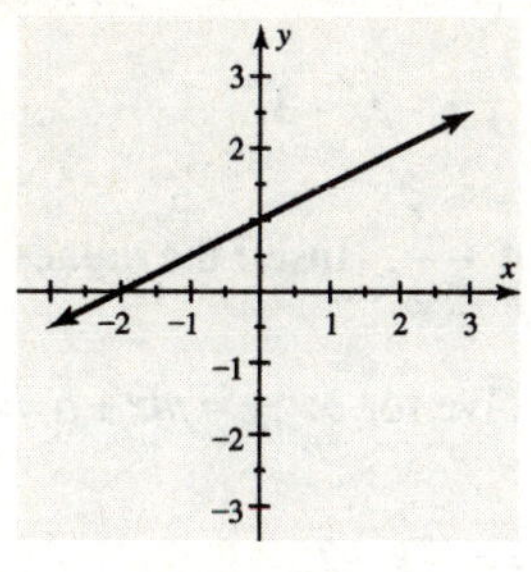

Figure 61

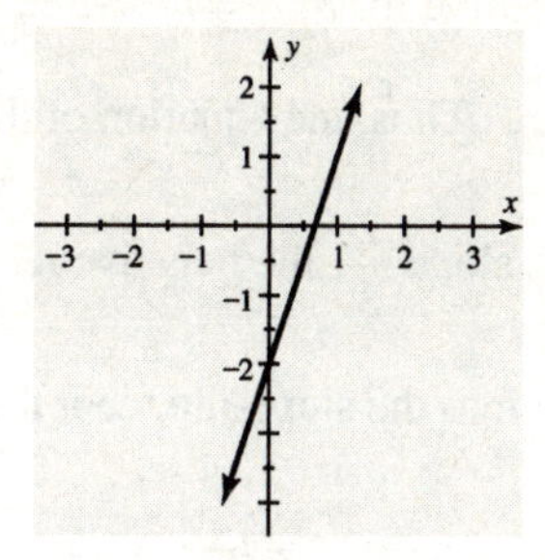

Figure 62

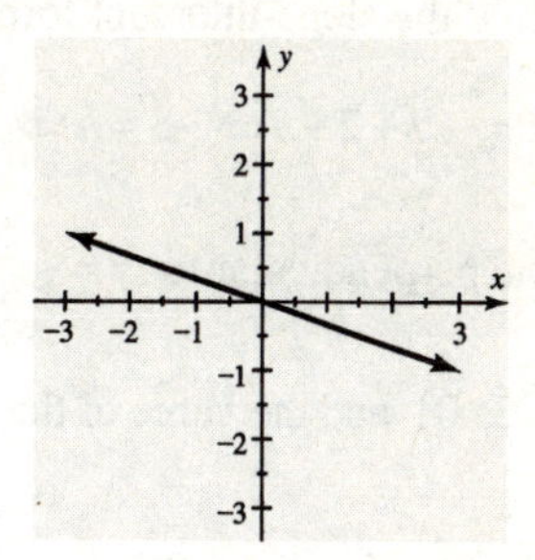

Figure 63

64. See Figure 64.

65. See Figure 65.

66. See Figure 66.

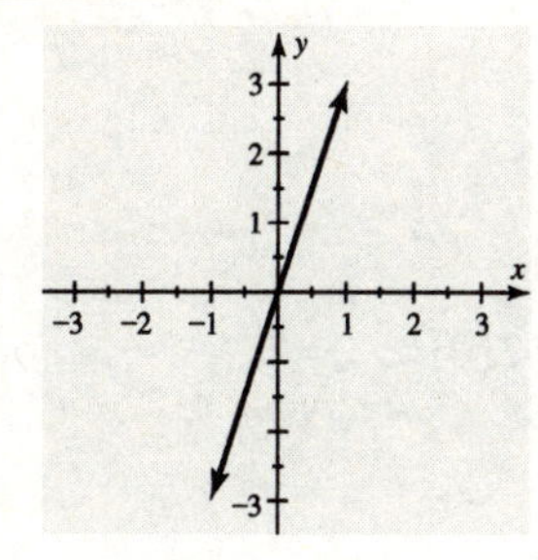

Figure 64

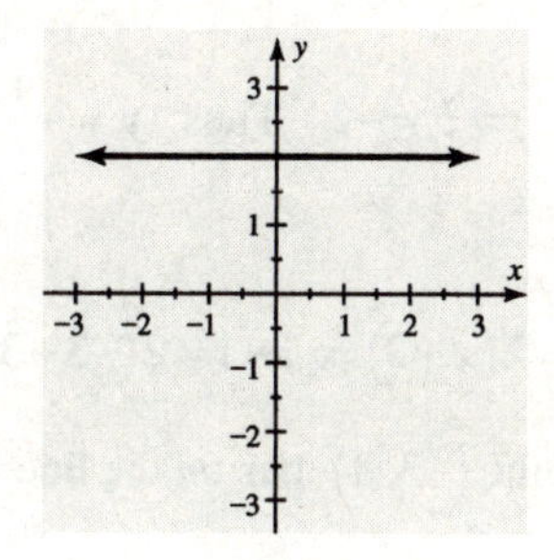

Figure 65

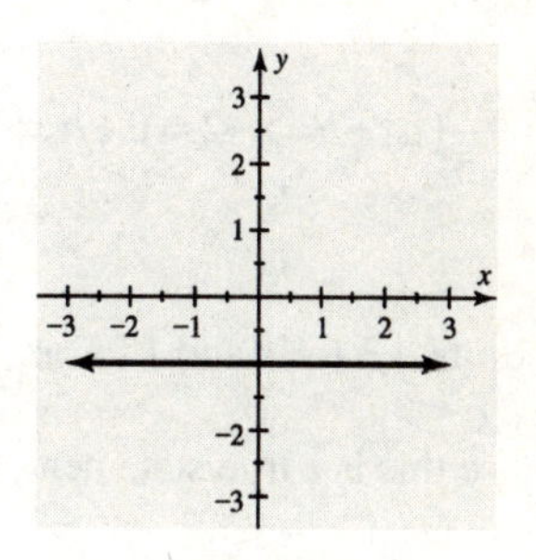

Figure 66

67. See Figure 67.

68. See Figure 68.

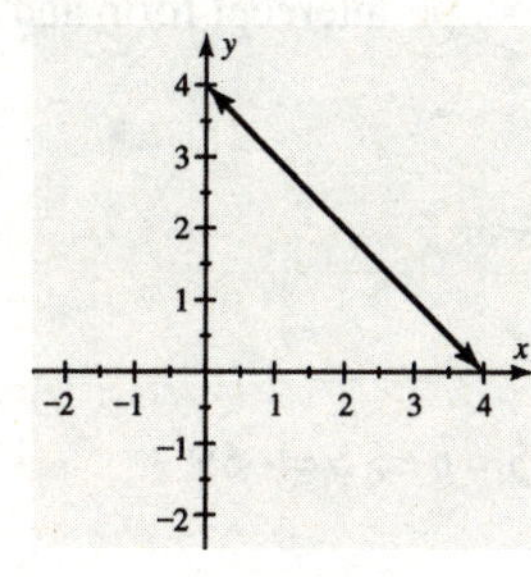

Figure 67

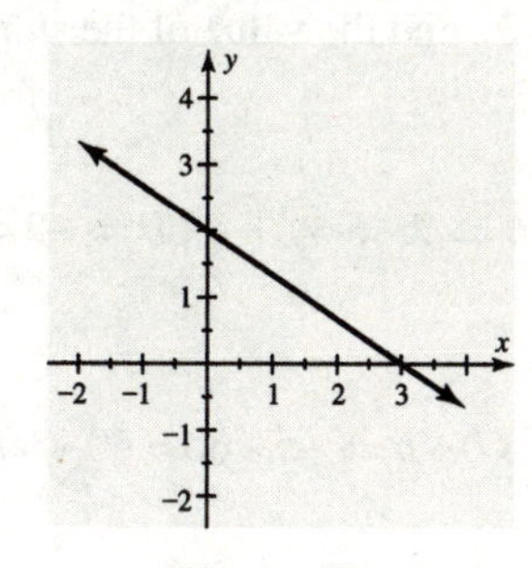

Figure 68

69. When the run increases by 1, the rise increases by 5. Thus, the slope m is 5. When $x=0$, $y=-5$. Therefore, the y-intercept b is -5. Thus, $y=mx+b \Rightarrow y=5x-5$.

70. When the run increases by 1, the rise decreases by 2. Thus, the slope m is -2. When $x=0$, $y=0$. Therefore, the y-intercept b is 0. Thus, $y=mx+b \Rightarrow y=-2x+0 \Rightarrow y=-2x$.

71. $y=mx+b \Rightarrow y=-\dfrac{5}{6}x+2$.

72. The line parallel to $y=-2x+1$ has slope -2. Insert the values of the point $(1,\ -5)$ and the value of the slope into the slope-intercept form and solve for b: $y=mx+b \Rightarrow -5=-\ 2(1)+b \Rightarrow -5=-\ 2+b \Rightarrow$ $-5+2=-\ 2+2+b \Rightarrow -3=b \Rightarrow b=-3$. Thus, the equation of the line is $y=-2x-3$.

73. The line perpendicular to $y=-\frac{3}{2}x$ has slope $\frac{2}{3}$, the negative reciprocal of $-\frac{3}{2}$. Insert the values of the point $(3,\ 0)$ and the value of the slope into the slope-intercept form and solve for b: $y=mx+b \Rightarrow$ $0=\frac{2}{3}(3)+b \Rightarrow 0=2+b \Rightarrow 0-2=2-2+b \Rightarrow -2=b \Rightarrow b=-2$. Thus, $y=\frac{2}{3}x-2$.

74. The line perpendicular to $y=5x-3$ has slope $-\frac{1}{5}$, the negative reciprocal of 5. Insert the values of the point $(0,\ -2)$ and the value of the slope into the slope-intercept form and solve for b: $y=mx+b \Rightarrow$ $-2=-\frac{1}{5}(0)+b \Rightarrow -2=0+b \Rightarrow -2=b \Rightarrow b=-2$. Thus, $y=-\frac{1}{5}x-2$.

Section 3.6

75. Substitute -3 for x and 1 for y: $y-1=2(x+3) \Rightarrow 1-1=2(-3+3) \Rightarrow 0=2(0) \Rightarrow 0=0$. Because this is a true statement, the point $(-3,\ 1)$ lies on the line.

76. Substitute 3 for x and -8 for y: $y=-3(x-1)+2 \Rightarrow -8=-3(3-1)+2 \Rightarrow -8=-3(2)+2 \Rightarrow$ $-8=-6+2 \Rightarrow -8=-4$. Because this is not a true statement, the point $(3,\ -8)$ is not on the line.

77. Substitute the values of the point $(1,\ 2)$ and the value of the slope into the slope-intercept form and solve for b:
$y=mx+b \Rightarrow 2=5(1)+b \Rightarrow 2=5+b \Rightarrow 2-5=5-5+b \Rightarrow -3=b \Rightarrow b=-3$.
Thus, $y=mx+b \Rightarrow y=5x-3$.

78. $y=mx+b \Rightarrow -5=20(3)+b \Rightarrow -5=60+b \Rightarrow -5-60=60-60+b \Rightarrow -65=b \Rightarrow b=-65$.
Thus, $y=mx+b \Rightarrow y=20x-65$.

79. Find the slope: $m=\frac{y_2-y_1}{x_2-x_1}=\frac{-1-1}{1-(-2)}=\frac{-2}{1+2}=\frac{-2}{3}=-\frac{2}{3}$. Then, insert the values of the point $(-2,\ 1)$ and the value of the slope into the point-slope form: $y-y_1=m(x-x_1) \Rightarrow$ $y-1=-\frac{2}{3}(x-(-2)) \Rightarrow$ $y-1=-\frac{2}{3}(x+2)$. Then, convert to slope-intercept form: $y-1=-\frac{2}{3}x-\frac{4}{3} \Rightarrow$ $y-1+1=-\frac{2}{3}x-\frac{4}{3}+1 \Rightarrow y=-\frac{2}{3}x-\frac{4}{3}+\frac{3}{3} \Rightarrow y=-\frac{2}{3}x-\frac{1}{3}$.

80. Find the slope: $m=\frac{y_2-y_1}{x_2-x_1}=\frac{30-(-30)}{40-20}=\frac{60}{20}=3$. Then, insert the values of the point $(20,-30)$ and the value of the slope into the point-slope form: $y-y_1=m(x-x_1)\Rightarrow y-(-30)=3(x-20)\Rightarrow$ $y+30=3(x-20)$. Then, convert to slope-intercept form: $y+30=3(x-20)\Rightarrow$ $y+30-30=3(x-20)-30\Rightarrow y=3x-60-30\Rightarrow y=3x-90$.

81. One point on the line is (3, 0) and another point is $(0,\ -4)$. Find the slope: $m=\frac{y_2-y_1}{x_2-x_1}=\frac{-4-0}{0-3}=\frac{-4}{-3}=\frac{4}{3}$.

Then, insert the values of the point $(3,\ 0)$ and the value of the slope into the point-slope form: $y-y_1=m(x-x_1)\Rightarrow\ y-0=\frac{4}{3}(x-3)$. Then, convert to slope-intercept form: $y=\frac{4}{3}x-4$.

82. One point on the line is $\left(\frac{1}{2},\ 0\right)$ and another point is $(0,\ -1)$. Find the slope:

$m=\frac{-1-0}{0-\frac{1}{2}}=\frac{-1}{-\frac{1}{2}}=-2(-1)=2.$

The y-intercept b is -1. Thus, $y=mx+b\Rightarrow y=2x-1$.

83. The line parallel to $y=2x$ has slope 2. Insert the values of the point $(5,\ 7)$ and the value of the slope into the point-slope form: $y-y_1=m(x-x_1)\Rightarrow\ y-7=2(x-5)$. Then, convert to slope-intercept form: $y-7=2x-10\Rightarrow\ y-7+7=2x-10+7\Rightarrow y=2x-10+7\Rightarrow y=2x-3$.

84. First, find the slope by converting to slope-intercept form: $y-4=\frac{3}{2}(x+1)\Rightarrow y-4=\frac{3}{2}x+\frac{3}{2}\Rightarrow$

$y-4+4=\frac{3}{2}x+\frac{3}{2}+4\Rightarrow y=\frac{3}{2}x+\frac{3}{2}+\frac{8}{2}\Rightarrow y=\frac{3}{2}x+\frac{11}{2}$. The line perpendicular to $y=\frac{3}{2}x+\frac{11}{2}$ has slope $-\frac{2}{3}$, the negative reciprocal of $\frac{3}{2}$. Insert the values of the point $(-1,\ 0)$ and the value of the slope into the point-slope form: $y-y_1=m(x-x_1)\Rightarrow y-0=-\frac{2}{3}(x-(-1))\Rightarrow y-0=-\frac{2}{3}(x+1)$. Then, convert to slope-intercept form: $y=-\frac{2}{3}x-\frac{2}{3}$.

85. Use the distributive property of multiplication and solve for y:

$y-2=3(x+1)\Rightarrow y-2=3x+3\Rightarrow y-2+2=3x+3+2\Rightarrow y=3x+5.$

86. Use the distributive property of multiplication and solve for y:

$y-9=\frac{1}{3}(x-6)\Rightarrow y-9=\frac{1}{3}x-2\Rightarrow y-9+9=\frac{1}{3}x-2+9\Rightarrow y=\frac{1}{3}x+7.$

87. Use the distributive property of multiplication:

$y=2(x+3)+5\Rightarrow y=2x+6+5\Rightarrow y=2x+11.$

88. Use the distributive property of multiplication:

$$y=-\frac{1}{4}(x-8)+1 \Rightarrow y=-\frac{1}{4}x+2+1 \Rightarrow y=-\frac{1}{4}x+3.$$

Section 3.7

89. Because a straight line would not pass through all the points, no.

90. Because the line does not pass through all the points, the linear model is approximate. When the run increases by 1, the rise decreases by 1. Therefore, the slope is -1. When $x=0$, $y=5$. Thus the y-intercept is 5. Therefore: $y=mx+b \Rightarrow y=-1x+5 \Rightarrow y=-x+5$.

91. Because the initial value is 40, $b=40$. Because y decreases at a rate of 2 pounds per minute, the slope is -2. Therefore: $y=mx+b \Rightarrow y=-2x+40$.

92. Because the initial value is 200, $b=200$. Because the rate of increase is 20 gallons per hour, the slope is 20. Therefore: $y=mx+b \Rightarrow y=20x+200$.

93. Because the initial value is 50, $b=50$. Because y remains constant, the slope is 0. Therefore: $y=50$.

94. Because the initial value is -20, $b=-20$. Because the rise is 5 feet per second, the slope is 5. Therefore: $y=5x-20$.

95. (a) See Figure 95a. The line could pass through all five points.

(b) See Figure 95b.

(c) The initial value is 10, and the slope is -4. Thus, $y=-4x+10$.

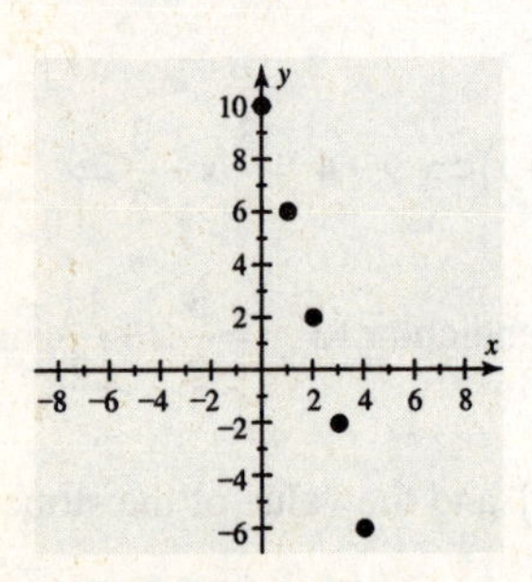

Figure 95a

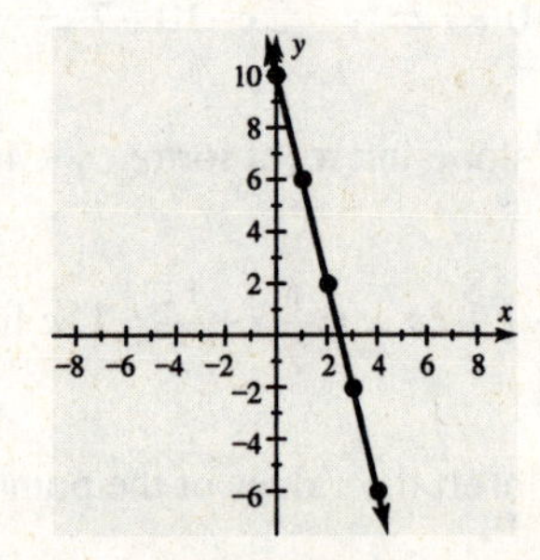

Figure 95b

96. (a) See Figure 96a. The line could not pass through all five points.

(b) See Figure 96b.

(c) $\frac{3-1}{0-(-4)}=\frac{2}{4}=\frac{1}{2}$, so the slope is $\frac{1}{2}$. The y-intercept is 3, so $y=\frac{1}{2}x+3$.

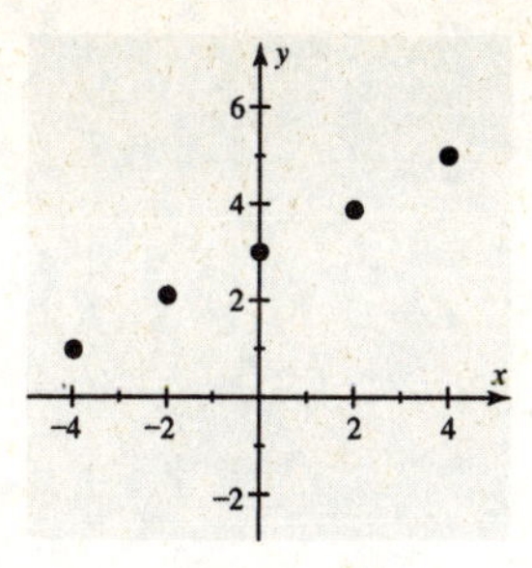

Figure 96a

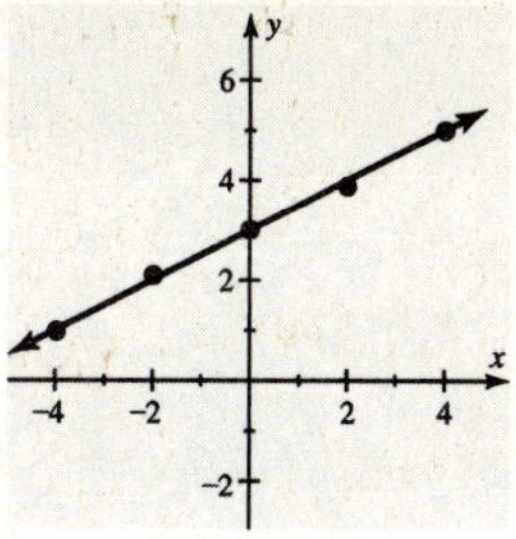

Figure 96b

Applications

97. (a) See Figure 97.

(b) The number of divorces increased significantly between 1960 and 1980, and then remined unchanged from 1980 to 2000.

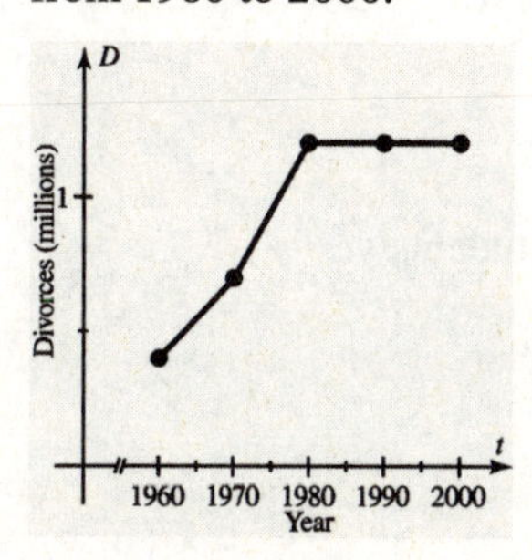

Figure 97

98. (a) $G = 100t$

(b) See Figure 98.

(c) When $y = 5$, $x = 50$. Therefore, 50 days.

99. (a) See Figure 99.

(b) v-intercept, 160; t-intercept, 5; the initial velocity was 160 ft/sec. and the velocity after 5 seconds was 0.

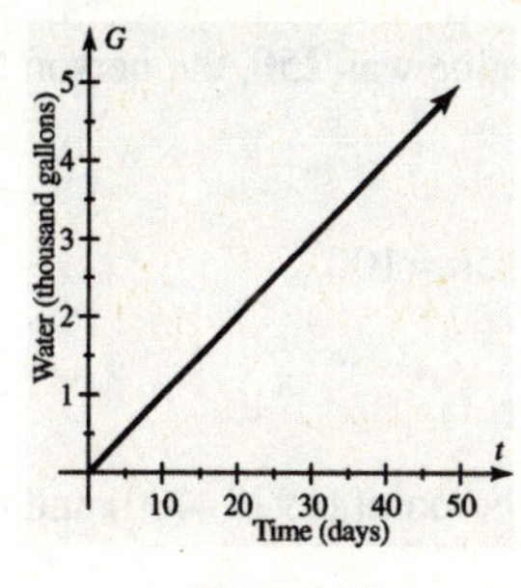

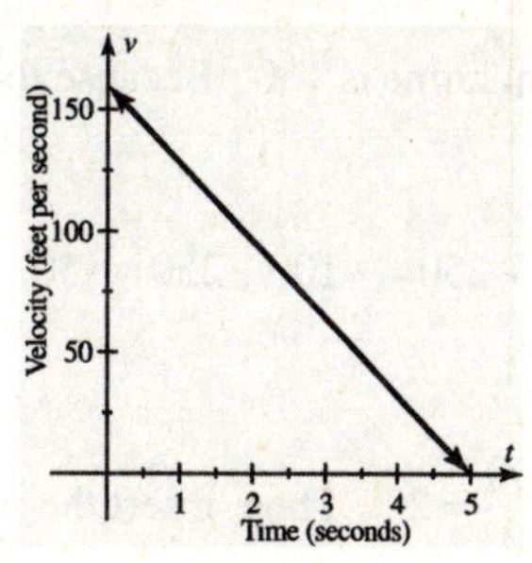

Figure 98

Figure 99

100. (a) $m_1 = 0$; $m_2 = -1500$; $m_3 = 500$

(b) $m_1 = 0$: The population remained unchanged. $m_2 = -1500$: The population decreased at a rate of 1500 insects per week. $m_3 = 500$: The population increased at a rate of 500 insects per week.

(c) For the first week the population did not change from the initial value of 4000. Over the next two weeks the population decreased at a rate of 1500 insects per week until it reached 1000. Finally, the population increased at a rate of 500 per week for two weeks, reaching 2000.

101. See Figure 101.

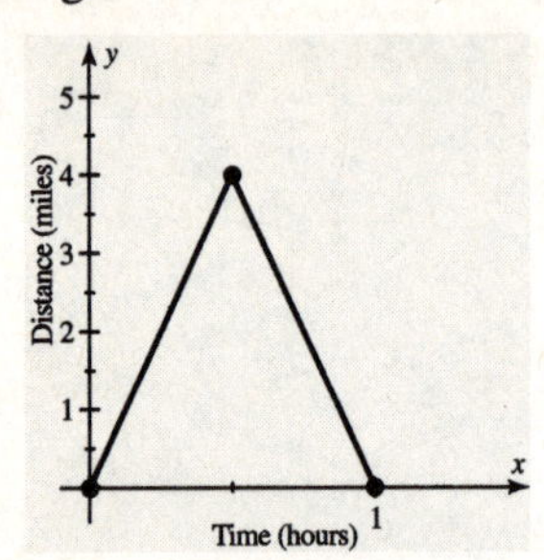

Figure 101

102. (a) $\frac{17,000-19,100}{2005-1985}=\frac{-2100}{20}=-105$

(b) The number of nursing homes decreased on average at a rate of 105/yr.

103. (a) $\frac{71-65}{2005-1995}=\frac{6}{10}=0.6$

(b) School enrollment is increasing on average at a rate of 0.6 million students/yr.

(c) $71+0.6(25)=71+15=86$ million students.

104. (a) Because the initial value is 35, \$35.

(b) Because the slope is 0.2, each additional mile costs 20¢.

(c) 35; the fixed cost of renting the car.

(d) The slope is 0.2; the cost of each mile driven.

105. (a) The person is driving toward home; the slope is negative.

(b) After 1 hour the car is 200 miles from home, after 3 hours the car is 100 miles from home.

(c) The initial value is 250, so the y-intercept b is 250. For each increase of 1 in run, the rise decreases by 50, so the slope is -50. Thus, $y=-50x+250$. The car is moving toward home at 50 mph.

(d) When time $=2$, the distance from home is 150. Because the initial value was 250, the person has traveled 100 miles.

Then: $y=-50x+250=-50(2)+250=-100+250=150$, so $250-150=100$.

106. (a) See Figure 106.

(b) Find the slope: $\frac{600-400}{1980-1970}=\frac{200}{10}=20$. Then, insert the values of the point (1970, 400) and the value of the slope into the point-slope form: $y-y_1=m(x-x_1)\Rightarrow y-400=20(x-1970)$. Then, convert to slope-intercept form: $y-400=20x-39,400\Rightarrow y-400+400=20x-39,400+400\Rightarrow$ $y=20x-39,000$; the number of icebergs increased at an average rate of 20 per year.

(c) Because you could draw a straight line through all the points on the graph, yes.

(d) $I=20(2005)-39,000=1100$ icebergs

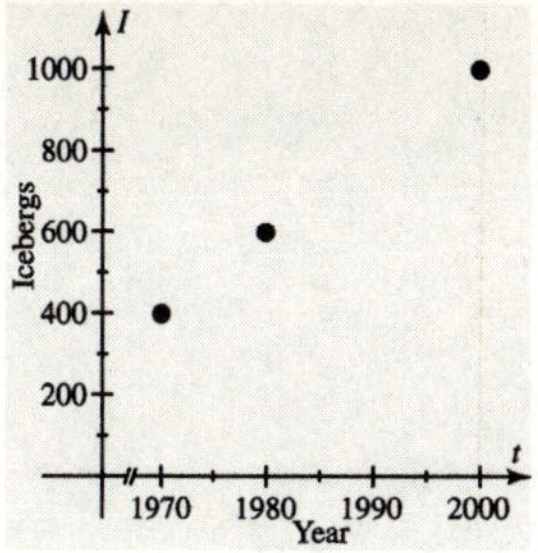

Figure 106

107. (a) Substitute 1998 for x: $D = -6.05(1998 - 1998) + 90.1 = -6.05(0) + 90.1 = 0 + 90.1 = 90.1 \Rightarrow$

$D = 90{,}100$. Then, substitute 2002 for x:

$D = -6.05(2002 - 1998) + 90.1 = -6.05(4) + 90.1 = -24.2 + 90.1 = 65.9 \Rightarrow D = 65{,}900$.

(b) $\dfrac{65.9 - 90.1}{2002 - 1998} = \dfrac{-24.2}{4} = -6.05$; deaths from pneumonia decreased at an average rate of 6050 per year.

108. (a) See Figure 108a.

(b) See Figure 108b.

(c) $\dfrac{200 - 40}{10 - 2} = \dfrac{160}{8} = 20$; The mileage is 20 miles per gallon.

(d) $y = 20x$; it is not an exact model.

(e) $y = 20(9) = 180$; about 180 miles.

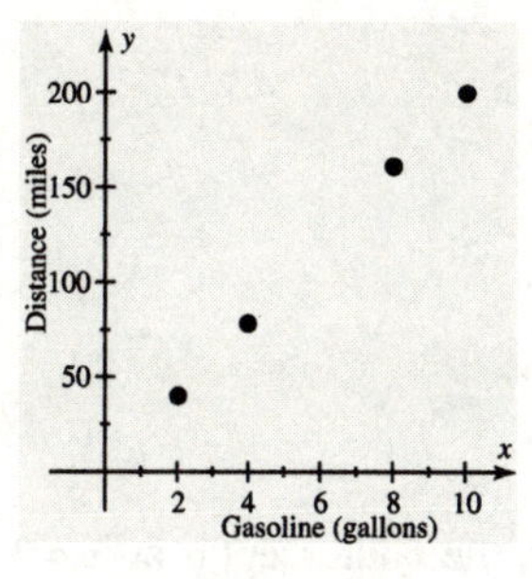

Figure 108a

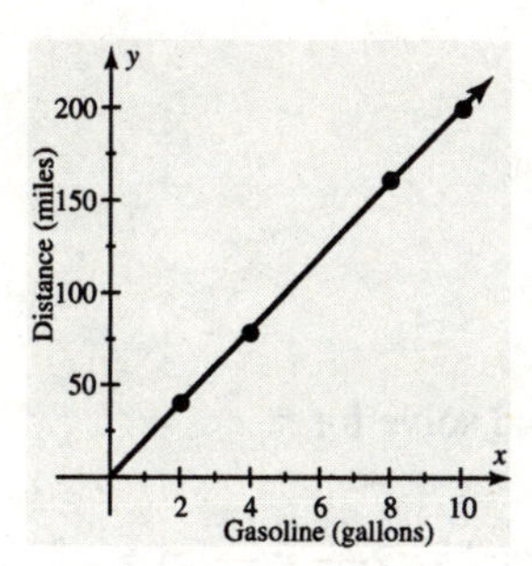

Figure 108b

Chapter 3 Test

1. $(-2, -2)$: Quadrant III; $(-2, 1)$: Quadrant II; $(0, -2)$: none; $(1, 0)$: none; $(2, 3)$: Quadrant I;

$(3, -1)$: Quadrant IV; $(3, 1)$: Quadrant I.

2. See Figure 2.

3. For $x = -2$: $y = 2x - 4 = 2(-2) - 4 = -4 - 4 \Rightarrow y = -8$.

For $x = -1$: $y = 2x - 4 = 2(-1) - 4 = -2 - 4 \Rightarrow y = -6$.

For $x = 0$: $y = 2x - 4 = 2(0) - 4 \Rightarrow y = -4$.

For $x = 1$: $y = 2x - 4 = 2(1) - 4 = 2 - 4 \Rightarrow y = -2$.

For $x = 2$: $y = 2x - 4 = 2(2) - 4 = 4 - 4 \Rightarrow y = 0$. See Figure 3.

The x-intercept is 2 and the y-intercept is -4.

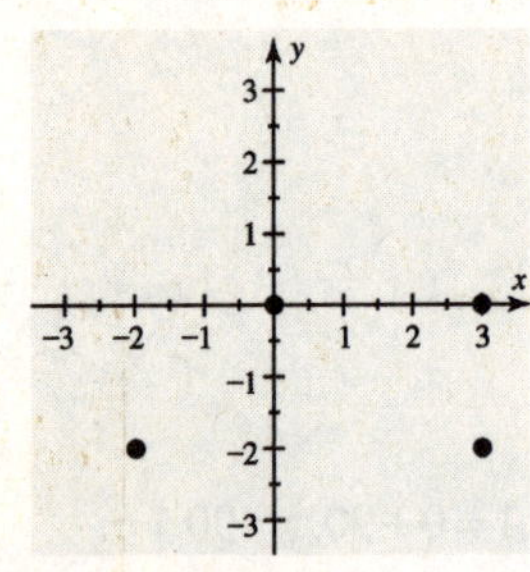

Figure 2

x	−2	−1	0	1	2
y	−8	−6	−4	−2	0

Figure 3

4. Let $x = 1$ and $y = -3$ in the equation $2x - y = 5$.

$$2(1) - (-3) \stackrel{?}{=} 5$$
$$2 + 3 \stackrel{?}{=} 5$$
$$5 = 5$$

Yes, the ordered pair $(1, -3)$ is a solution.

5. See Figure 5.

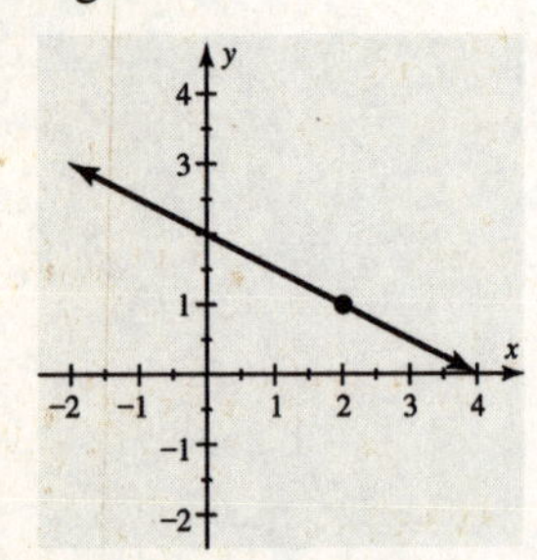

Figure 5

6. To find the x-intercept, set $y = 0$ and solve for x:

$5x - 3y = 15 \Rightarrow 5x - 3(0) = 15 \Rightarrow 5x = 15 \Rightarrow \dfrac{5x}{5} = \dfrac{15}{5} \Rightarrow x = 3$. Therefore, the x-intercept is at $x = 3$.

To find the y-intercept, set $x = 0$ and solve for y:

$5x - 3y = 15 \Rightarrow 5(0) - 3y = 15 \Rightarrow -3y = 15 \Rightarrow \dfrac{-3y}{-3} = \dfrac{15}{-3} \Rightarrow y = -5$. Therefore, the y-intercept is at $y = -5$.

7. See Figure 7.

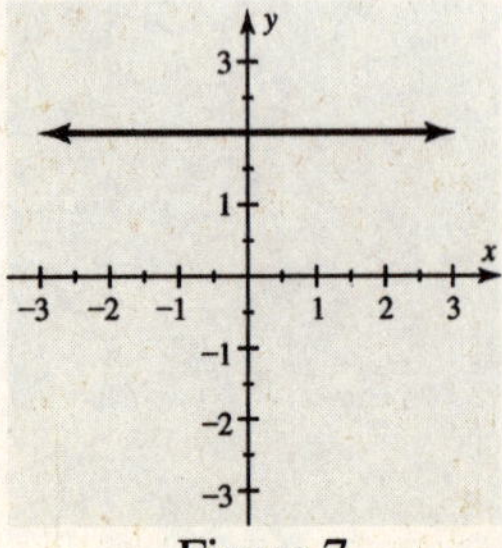

Figure 7

8. See Figure 8.

9. See Figure 9.

10. See Figure 10.

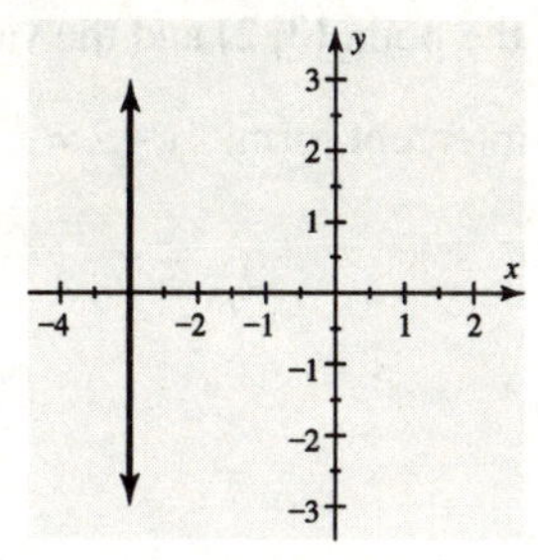

Figure 8

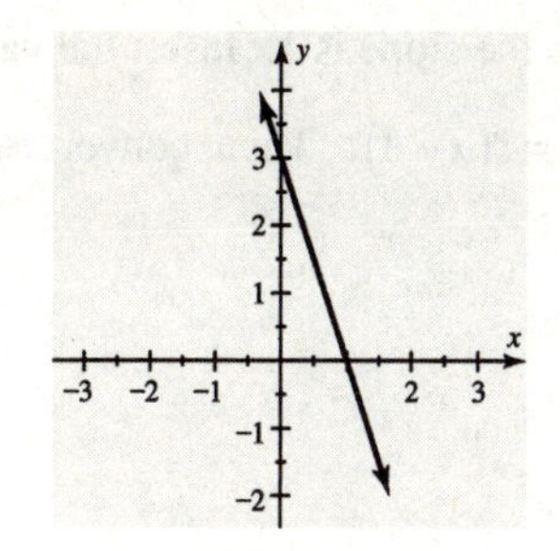

Figure 9

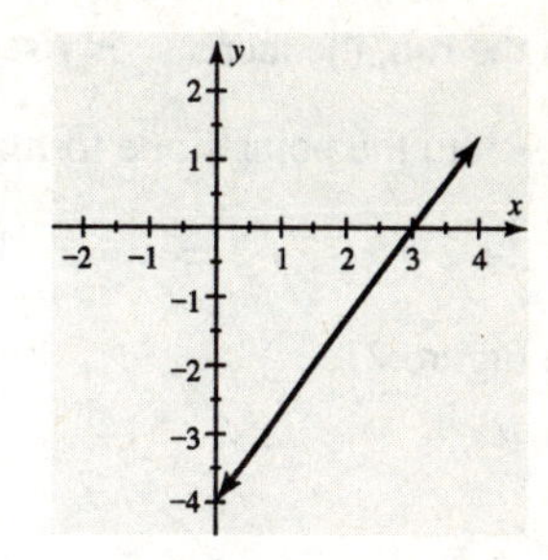

Figure 10

11. For every increase of 1 in the run, the rise decreases by 1, so the slope is −1. When $x=0,\ y=2,$ so the *y*-intercept is 2. Therefore, the equation of the line is $y=-x+2$.

12. Solve for *y*: $-4x+2y=1 \Rightarrow -4x+4x+2y=1+4x \Rightarrow 2y=4x+1 \Rightarrow \frac{2y}{2}=\frac{4x}{2}+\frac{1}{2} \Rightarrow\ y=2x+\frac{1}{2}$. The slope is 2 and the *y*-intercept is $\frac{1}{2}$.

13. The equation of a horizontal line through $(1,-5)$ is $y=-5$. The equation of a vertical line through $(1,-5)$ is $x=1$.

14. $m=\frac{y_2-y_1}{x_2-x_1}=\frac{1-3}{5-(-4)}=\frac{-2}{9}=-\frac{2}{9}$

15. $y=-\frac{4}{3}x-5$

16. The line parallel to $y=3x-1$ has slope 3. Insert the values of the point $(2,\ -5)$ and the value of the slope into the point-slope form: $y-y_1=m(x-x_1) \Rightarrow y-(-5)=3(x-2) \Rightarrow y+5=3(x-2)$. Then, convert to slope-intercept form: $y+5=3x-6 \Rightarrow y+5-5=3x-6-5 \Rightarrow y=3x-11$.

17. The line perpendicular to $y=\frac{1}{3}x$ has slope −3, the negative reciprocal to $\frac{1}{3}$. Insert the values of the point (1, 2) and the value of the slope into the point-slope form: $y-y_1=m(x-x_1) \Rightarrow y-2=-3(x-1)$. Then, convert to slope-intercept form: $y-2=-3x+3 \Rightarrow y-2+2=-3x+3+2 \Rightarrow y=-3x+5$.

18. Find the slope: $\frac{-1-2}{2-(-4)}=\frac{-3}{6}=-\frac{1}{2}$. Then, insert the values of the point $(-4,\ 2)$ and the value of the slope into the point-slope form: $y-2=-\frac{1}{2}(x-(-4)) \Rightarrow y-2=-\frac{1}{2}(x+4)$. Then, convert to slope-intercept form: $y-2=-\frac{1}{2}x-2 \Rightarrow y-2+2=-\frac{1}{2}x-2+2 \Rightarrow y=-\frac{1}{2}x$.

19. $y-3=\frac{1}{2}(x+4) \Rightarrow y-3=\frac{1}{2}x+2 \Rightarrow y-3+3=\frac{1}{2}x+2+3 \Rightarrow y=\frac{1}{2}x+5$

20. The linear model is approximate, because it does not go through all the points exactly. For every increase of 1 in the run, the increase in rise is 1, so the slope is 1. Insert the values of the point (3, 2) and the value of the slope into the point-slope form: $y-2=1(x-3)$. Then, convert to slope-intercept form: $y-2=x-3 \Rightarrow$ $y-2+2=x-3+2 \Rightarrow y=x-1$.

21. See Figure 21.

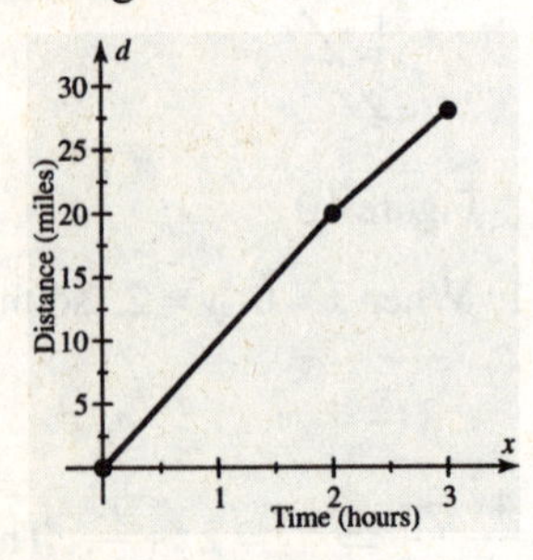

Figure 21

22. (a) $m_1=2;\ m_2=-9;\ m_3=2;\ m_4=5$

(b) $m_1=2$: The population increased at a rate of 2000 fish per year. $m_2=-9$: The population decreased at a rate of 9000 fish per year. $m_3=2$: The population increased at a rate of 2000 fish per year. $m_4=5$: The population increased at a rate of 5000 fish per year.

(c) For the first year the population increased from an initial value of 8000 to 10,000 at a rate of 2000 fish per year. During the second year the population dropped dramatically to 1000 at a rate of 9000 fish per year. Over the third year the population grew to 3000 at a rate of 2000 fish per year. Finally, over the fourth year, the population grew at a rate of 5000 fish per year to reach 8000.

23. $N=100x+2000$

Chapter 3 Extended and Discovery Exercises

1. (a) See Figure 1.

(b) Insert the values of the points (2007, 23.5) and (1993, 20.5) into the slope formula:

$\frac{23.5-20.5}{2007-1993}=\frac{3}{14}\approx 0.21$. Then insert the values of the point (1993, 20.5) and the value of the slope into the point-slope form:

$P-20.5=0.21(x-1993) \Rightarrow P-20.5+20.5=0.21(x-1993)+20.5$

$\Rightarrow P=0.21(x-1993)+20.5$. *Answers may vary.*

(c) Substitute 2010 for x and solve for P:

$P=0.21(2010-1993)+20.5 \Rightarrow P=0.21(17)+20.5$

$\Rightarrow P=3.57+20.5 \Rightarrow P=24.07 \Rightarrow P\approx 24.1$ percent.

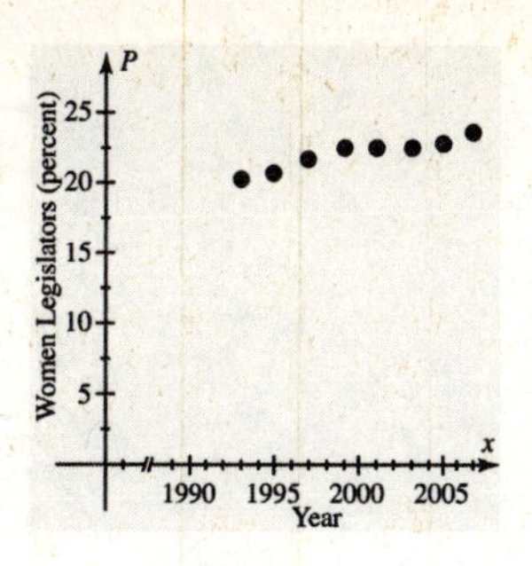

Figure 1

2. (a) See Figure 2.

(b) Find the slope: $\dfrac{281-179}{2000-1960}=\dfrac{102}{40}=2.55$. Then, insert the values of the point (1960, 179) and the value of the slope into the point-slope form: $P-179=2.55(x-1960)\Rightarrow$

$P-179+179=2.55(x-1960)+179\Rightarrow P=2.55(x-1960)+179$. *Answers may vary.*

(c) Substitute 2005 for x in the equation and solve for P: $P=2.55(2005-1960)+179\Rightarrow$

$P=2.55(45)+179\Rightarrow P\approx 294$ million.

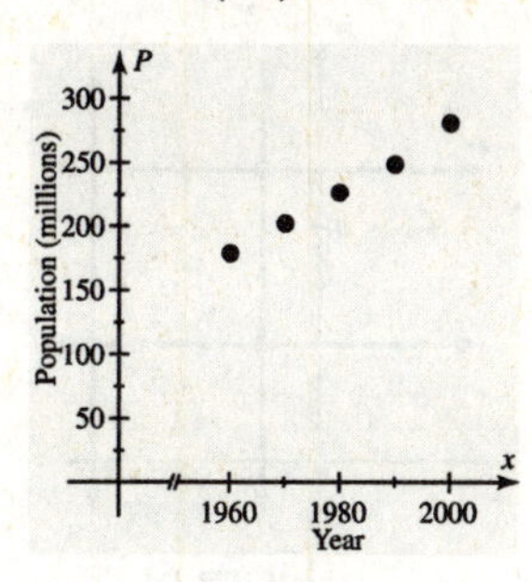

Figure 2

Creating Geometric Shapes

3. (a) The slope of the line connecting the points (0, 0) and (2, 3) is $\frac{3}{2}$, and the y-intercept is 0. Therefore the equation of the line is $y=\frac{3}{2}x$. The slope of the line connection the points (2, 3) and (3, 6) is 3. Therefore the point-slope form is: $y-3=3(x-2)$. The slope-intercept form is thus: $y-3+3=3x-6+3\Rightarrow\ y=3x-3$. The slope of the line connecting the points (3, 6) and (0, 0) is 2. The y-intercept is 0. Therefore the equation of the line is: $y=2x$.

(b) See Figure 3. A triangle is formed by the line segments connecting the three points.

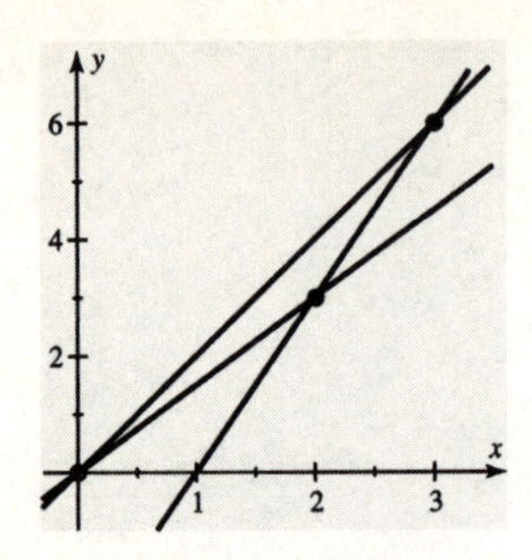

Figure 3

4. (a) $y=-x+1$

(b) See Figure 4. A parallelogram is formed.

5. (a) $y=\frac{1}{2}x$; $y=\frac{1}{2}x+3$; $y=-2x$; $y=-2x+10$

(b) See Figure 5. A rectangle is formed.

6. (a) (1, 5)

(b) $x=1$; $x=4$; $y=2$; $y=5$

(c) See Figure 6. A square is formed.

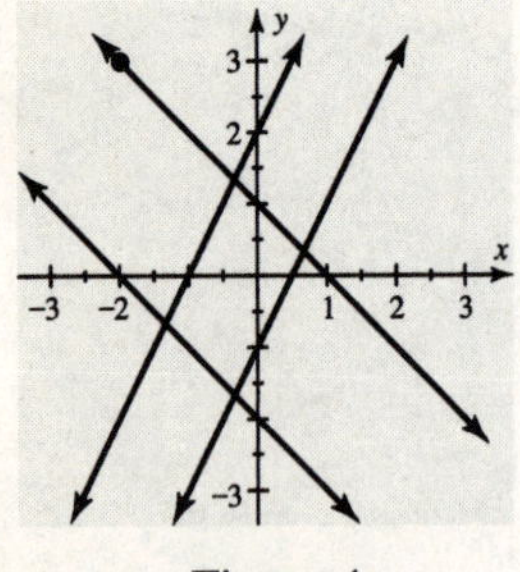

Figure 4

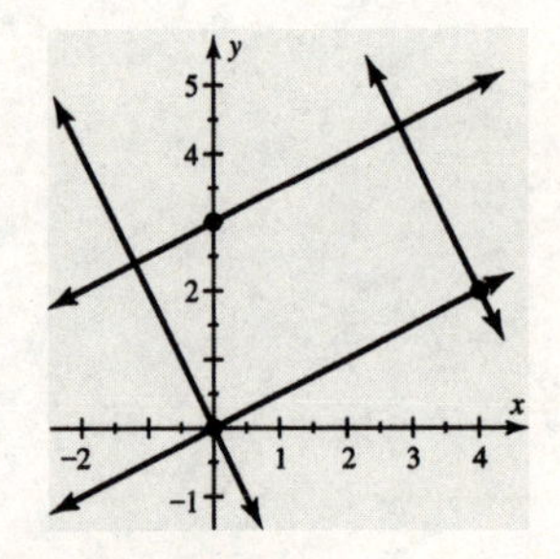

Figure 5

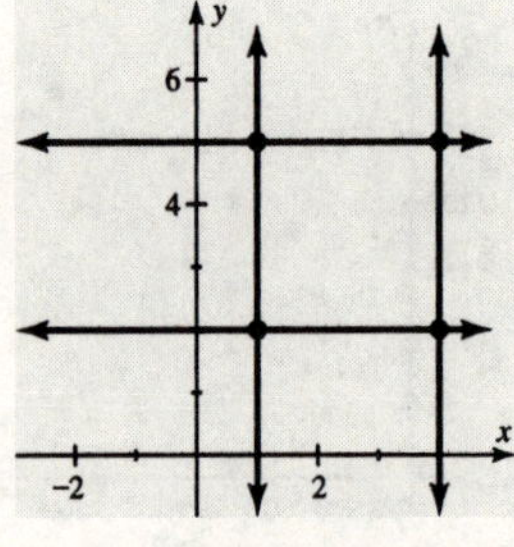

Figure 6

Chapters 1-3 Cumulative Review Exercises

1. Composite; $40=2\times2\times2\times5$
2. Prime
3. $n+10$
4. n^2-2
5. $\frac{4}{3}\cdot\frac{21}{24}=\frac{84}{72}=\frac{12\cdot7}{12\cdot6}=\frac{7}{6}$
6. $\frac{3}{4}\div\frac{9}{8}=\frac{3}{4}\cdot\frac{8}{9}=\frac{24}{36}=\frac{12\cdot2}{12\cdot3}=\frac{2}{3}$
7. $\frac{2}{3}+\frac{4}{3}=\frac{2+4}{3}=\frac{6}{3}=2$
8. $\frac{7}{10}-\frac{2}{15}=\frac{7\cdot3}{10\cdot3}-\frac{2\cdot2}{15\cdot2}=\frac{21}{30}-\frac{4}{30}=\frac{21-4}{30}=\frac{17}{30}$
9. $20-2\cdot3=20-(2\cdot3)=20-6=14$

10. $14-5-2=(14-5)-2=9-2=7$

11. $\dfrac{1+4}{1+2}=\dfrac{5}{3}$

12. $-3^2=-(3^2)=-(9)=-9$

13. Rational

14. Irrational, because $\sqrt{3}$ cannot be expressed as a fraction.

15. See Figure 15.

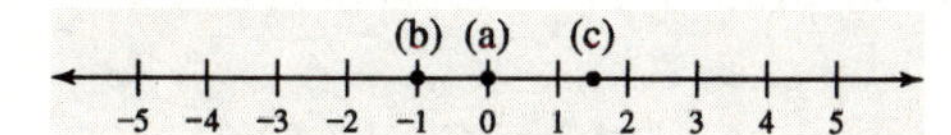

Figure 15

16. $|3-5|=|-2|=2$

17. $-12\div\left(-\dfrac{2}{3}\right)=-\dfrac{12}{1}\cdot\left(-\dfrac{3}{2}\right)=\dfrac{36}{2}=18$

18. $-\dfrac{2x}{5y}\div\left(\dfrac{x}{10y}\right)=-\dfrac{2x}{5y}\cdot\dfrac{10y}{x}=-\dfrac{20xy}{5xy}=-\dfrac{20}{5}\cdot\dfrac{xy}{xy}=-4\cdot 1=-4$

19. $3+4x-2+3x=3-2+4x+3x=1+7x=7x+1$

20. $2(x-1)-(x+2)=2x-2-x-2=2x-x-2-2=x-4$

21. $x+5=2\Rightarrow x+5-5=2-5\Rightarrow x=-3$. To check the solution, replace x with -3 in the original equation: $x+5=2\Rightarrow -3+5=2\Rightarrow 2=2$. The solution checks.

22. $\dfrac{1}{3}z=7\Rightarrow 3\left(\dfrac{1}{3}z\right)=3(7)\Rightarrow z=21$. Replace z with 21 in the original equation:

$\dfrac{1}{3}z=7\Rightarrow\dfrac{1}{3}(21)=7\Rightarrow\dfrac{21}{3}=7\Rightarrow 7=7$. The solution checks.

23. $3t-5=1\Rightarrow 3t-5+5=1+5\Rightarrow 3t=6\Rightarrow\dfrac{3t}{3}=\dfrac{6}{3}\Rightarrow t=2$. Replace t with 2 in the original equation: $3t-5=1\Rightarrow 3(2)-5=1\Rightarrow 6-5=1\Rightarrow 1=1$. The solution checks.

24. $2(x-3)=-6-x\Rightarrow 2x-6=-6-x\Rightarrow 2x+x-6=-6-x+x\Rightarrow 3x-6=-6$

$\Rightarrow 3x-6+6=-6+6\Rightarrow 3x=0\Rightarrow\dfrac{3x}{3}=\dfrac{0}{3}\Rightarrow x=0$. Replace x with 0 in the original equation:

$2(x-3)=-6-x\Rightarrow 2(0-3)=-6-0\Rightarrow 2(-3)=-6\Rightarrow -6=-6$. The solution checks.

25. When $x=-2$: $6-2x=6-2(-2)=6-(-4)=6+4=10$;

when $x=-1$: $6-2x=6-2(-1)=6-(-2)=6+2=8$;

when $x=0$: $6-2x=6-2(0)=6-0=6$; when $x=1$: $6-2x=6-2(1)=6-2=4$;

when $x = 2$: $6-2x = 6-2(2) = 6-4 = 2$; thus, the missing values in the table are 10, 8, 6, 4, and 2.

See Figure 25. From the table, $6-2x = 4$, when $x = 1$.

x	−2	−1	0	1	2
y	10	8	6	4	2

Figure 25

26. $2n+2 = n-5 \Rightarrow 2n-n+2 = n-n-5 \Rightarrow n+2 = -5 \Rightarrow n+2-2 = -5-2 \Rightarrow n = -7$

27. Let n represent the lowest integer. Then: $n+(n+1)+(n+2)+(n+3) = -98 \Rightarrow$

$n+n+1+n+2+n+3 = -98 \Rightarrow 4n+6 = -98 \Rightarrow 4n+6-6 = -98-6 \Rightarrow$

$4n = -104 \Rightarrow \frac{4n}{4} = -\frac{104}{4} \Rightarrow n = -26.$ Thus, the four integers are -26, -25, -24 and -23.

28. $d = rt \Rightarrow 80 = 10t \Rightarrow \frac{80}{10} = \frac{10t}{10} \Rightarrow 8 = t \Rightarrow t = 8$ hours.

29. Area of a rectangle is width multiplied by length, $A = LW$. Replace W with 18 and L with 40 and solve for A:

$A = LW \Rightarrow 18 \cdot 40 = 720 \text{ ft}^2.$

30. Area of a triangle is $\frac{1}{2}$ times base times height, $A = \frac{1}{2}bh$. Replace b with 8 and h with 5 and solve for A:

$A = \frac{1}{2}bh \Rightarrow A = \frac{1}{2}(8)(5) = \frac{1}{2}(40) = 20 \text{ ft}^2.$

31. $A = \frac{1}{2}bh \Rightarrow 2A = 2\left(\frac{1}{2}bh\right) \Rightarrow 2A = bh \Rightarrow \frac{2A}{h} = \frac{bh}{h} \Rightarrow \frac{2A}{h} = b \Rightarrow b = \frac{2A}{h}$

32. $P = 2w+2l \Rightarrow P-2w = 2w-2w+2l \Rightarrow P-2w = 2l \Rightarrow \frac{P-2w}{2} = \frac{2l}{2} \Rightarrow \frac{P-2w}{2} = l \Rightarrow l = \frac{P-2w}{2}$

33. $x < 2$

34. When $x = 0$: $2x-3 = 2(0)-3 = 0-3 = -3$; when $x = 1$: $2x-3 = 2(1)-3 = 2-3 = -1$;

when $x = 2$: $2x-3 = 2(2)-3 = 4-3 = 1$; when $x = 3$: $2x-3 = 2(3)-3 = 6-3 = 3$;

when $x = 4$: $2x-3 = 2(4)-3 = 8-3 = 5$; thus, the missing values in the table are -3, -1, 1, 3 and 5.

See Figure 34. From the table, $2x-3 \le 1$, when $x \le 2$.

x	0	1	2	3	4
$2x-3$	−3	−1	1	3	5

Figure 34

35. $3-6x < 3 \Rightarrow 3-3-6x < 3-3 \Rightarrow -6x < 0 \Rightarrow \frac{-6x}{-6} > \frac{0}{-6} \Rightarrow x > 0; \{x|x > 0\}$

36. $2x \le 1-(2x-1) \Rightarrow 2x \le 1-2x+1 \Rightarrow 2x+2x \le 1-2x+2x+1 \Rightarrow 4x \le 2 \Rightarrow \frac{4x}{4} \le \frac{2}{4} \Rightarrow x \le \frac{1}{2}; \left\{x \middle| x \le \frac{1}{2}\right\}$

37. $(-2,\ -3)$: Quadrant III; $(0,\ 3)$: none; $(2,\ -2)$: Quadrant IV; $(2,\ 1)$: Quadrant I

38. See Figures 38a and 38b.

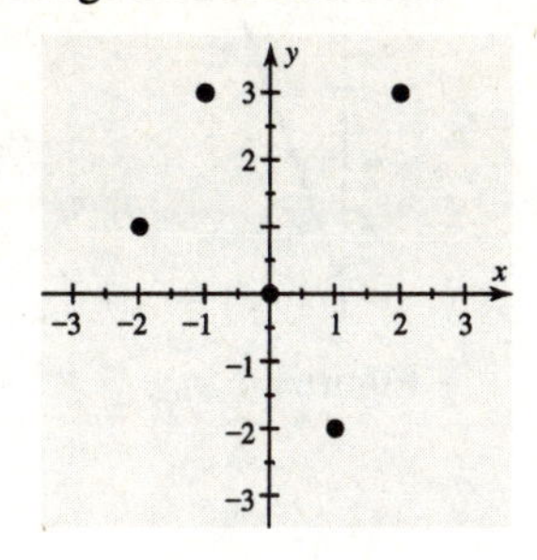

Figure 38a

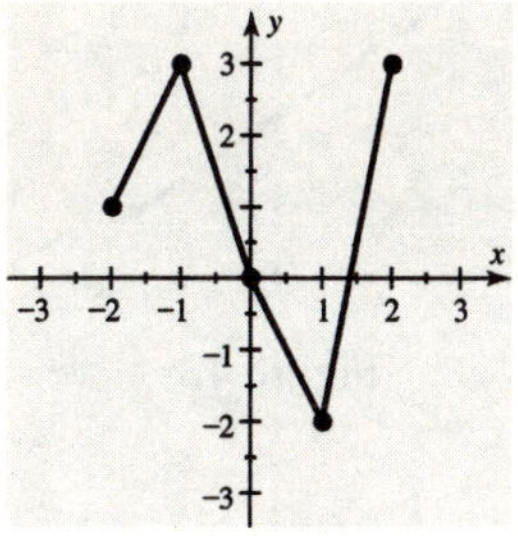

Figure 38b

39. Replace x with 5 and y with -1: $x-3y=8 \Rightarrow 5-3(-1)=8 \Rightarrow 5+3=8 \Rightarrow 8=8$. Because this is a true statement, $(5,\ -1)$ is a solution.

40. Solve the equation for y: $x+2y=4 \Rightarrow x-x+2y=4-x \Rightarrow 2y=4-x \Rightarrow \dfrac{2y}{2}=\dfrac{4}{2}-\dfrac{x}{2} \Rightarrow$

$y=2-\dfrac{x}{2}$. Then, when $x=-2$: $y=2-\dfrac{x}{2} \Rightarrow y=2-\dfrac{-2}{2}=2-(-1)=2+1=3$;

when $x=-1$: $y=2-\dfrac{x}{2} \Rightarrow y=2-\left(\dfrac{-1}{2}\right)=2+\dfrac{1}{2}=\dfrac{5}{2}=2.5$;

when $x=0$: $y=2-\dfrac{x}{2} \Rightarrow y=2-\dfrac{0}{2}=2-0=2$; when $x=1$: $y=2-\dfrac{x}{2} \Rightarrow y=2-\dfrac{1}{2}=\dfrac{3}{2}=1.5$;

when $x=2$: $y=2-\dfrac{x}{2} \Rightarrow y=2-\dfrac{2}{2}=2-1=1$. Thus, the missing values in the table are 3, 2.5, 2, 1.5, and 1. See Figure 40.

41. See Figure 41.

x	−2	−1	0	1	2
y	3	2.5	2	1.5	1

Figure 40

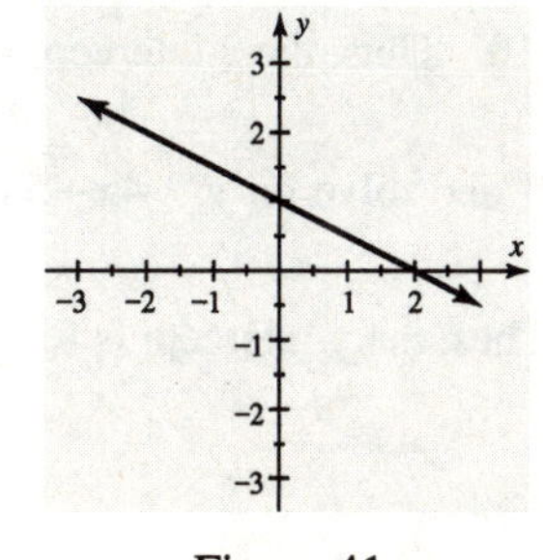

Figure 41

42. See Figure 42.

43. See Figure 43.

44. See Figure 44.

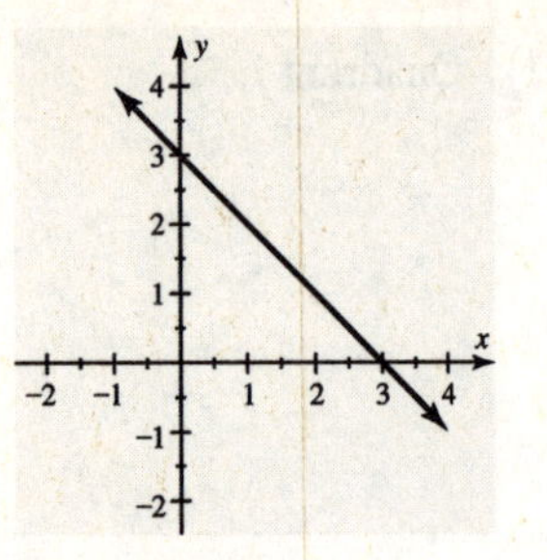

Figure 42

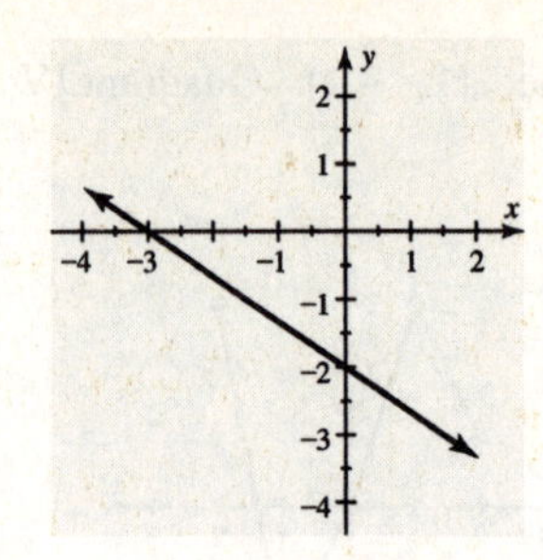

Figure 43

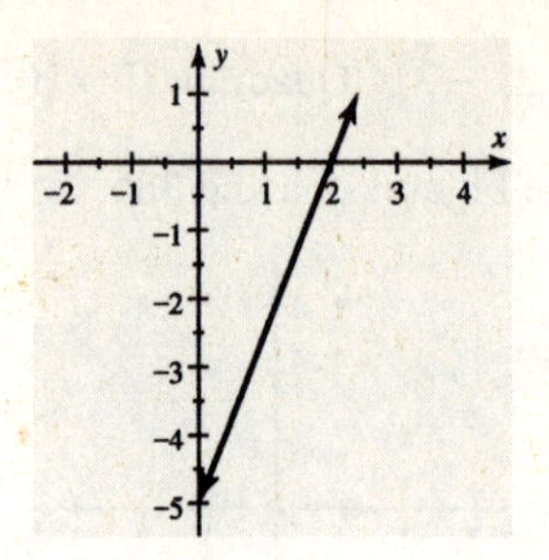

Figure 44

45. See Figure 45.

46. See Figure 46.

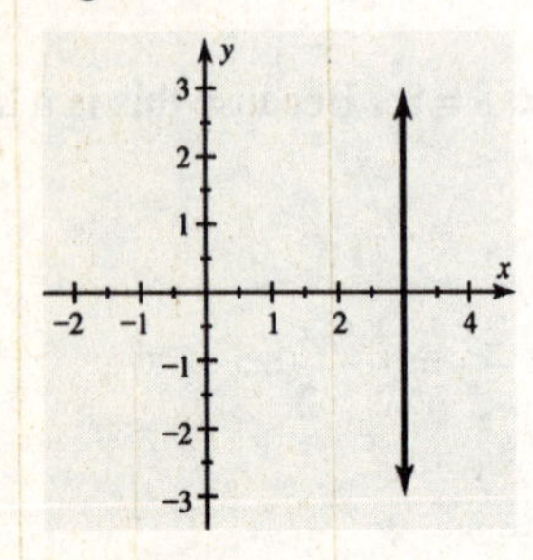

Figure 45

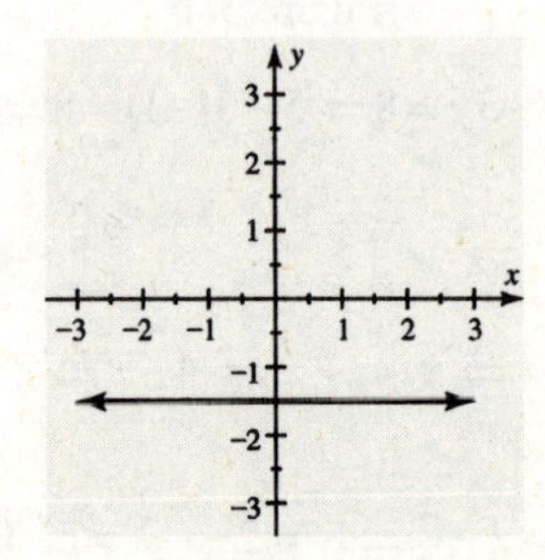

Figure 46

47. The x-intercept is $\frac{3}{2}$ and the y-intercept is -3. When the run increases by 1, the rise increases by 2, so the slope is 2. Thus, the equation of the line is $y = 2x - 3$.

48. The x-intercept is 2 and the y-intercept is 1. When the run increases by 2, the rise decreases by 1, so the slope is $-\frac{1}{2}$. Thus, the equation of the line is $y = -\frac{1}{2}x + 1$.

49. To find the x-intercept, set $y = 0$ and solve for x: $-4x + 5y = 40 \Rightarrow -4x + 5(0) = 40 \Rightarrow -4x + 0 = 40 \Rightarrow$

$-4x = 40 \Rightarrow \frac{-4x}{-4} = \frac{40}{-4} \Rightarrow x = -10$. Thus, the x-intercept is -10.

To find the y-intercept, set $x = 0$ and solve for y: $-4x + 5y = 40 \Rightarrow -4(0) + 5y = 40 \Rightarrow 0 + 5y = 40 \Rightarrow$

$5y = 40 \Rightarrow \frac{5y}{5} = \frac{40}{5} \Rightarrow y = 8$. Thus, the y-intercept is 8.

50. $x = \frac{3}{2}$; $y = -2$

51. See Figure 51.

52. See Figure 52.

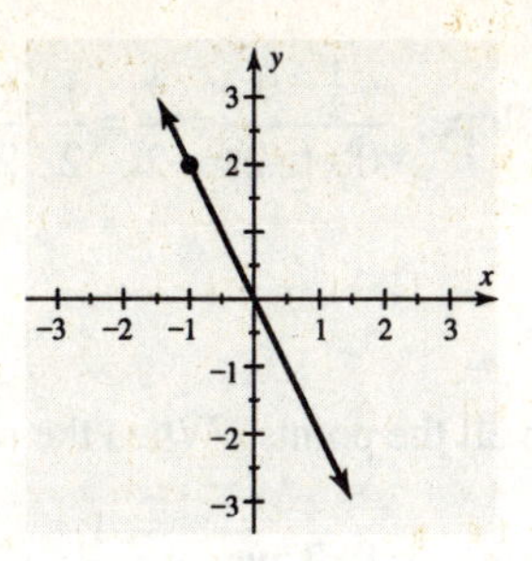

Figure 51

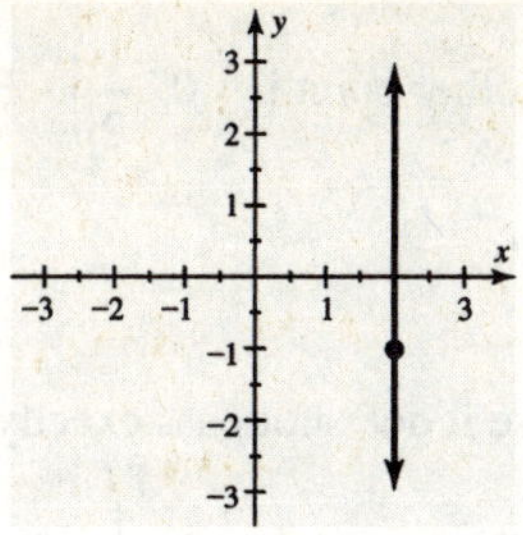

Figure 52

53. When x increases by 1, y increases by 3; therefore the slope is 3. When $x=0,\ y=-3$; therefore the y-intercept is -3. Thus, the slope-intercept form of the line is $y=3x-3$.

54. Solve for y: $3x-5y=15 \Rightarrow 3x-3x-5y=15-3x \Rightarrow -5y=-3x+15 \Rightarrow$

$\dfrac{-5y}{-5}=\dfrac{-3x}{-5}+\dfrac{15}{-5} \Rightarrow y=\dfrac{3}{5}x-3.$ See Figure 54.

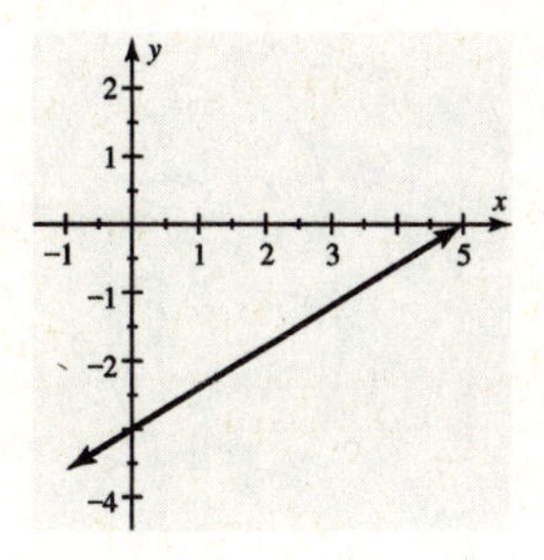

Figure 54

55. The line parallel to $y=-\dfrac{1}{3}x-5$ has slope $-\dfrac{1}{3}$. Insert the values of the point $(-3,\ 8)$ and the value of the slope into the point-slope form: $y-8=-\dfrac{1}{3}(x-(-3)) \Rightarrow y-8=-\dfrac{1}{3}(x+3)$. Then, convert to slope-intercept form: $y-8+8=-\dfrac{1}{3}x-1+8 \Rightarrow y=-\dfrac{1}{3}x+7.$

56. First, solve for y: $3x-2y=6 \Rightarrow 3x-3x-2y=6-3x \Rightarrow -2y=-3x+6 \Rightarrow \dfrac{-2y}{-2}=\dfrac{-3x}{-2}+\dfrac{6}{-2} \Rightarrow$

$y=\dfrac{3}{2}x-3$. The line perpendicular to $y=\dfrac{3}{2}x-3$ has slope $-\dfrac{2}{3}$, the negative reciprocal of $\dfrac{3}{2}$. Insert the values of the point $(0,\ -3)$ and the value of the slope into the point-slope form: $y-(-3)=-\dfrac{2}{3}(x-0) \Rightarrow y+3+(-3)=-\dfrac{2}{3}x-3 \Rightarrow y=-\dfrac{2}{3}x-3.$

57. First, find the slope: $\dfrac{-3-3}{2-(-1)}=\dfrac{-6}{2+1}=\dfrac{-6}{3}=-2$. Then, insert the values of the point $(-1,\ 3)$ and the value of the slope into the point-slope form: $y-3=-2(x-(-1) \Rightarrow y-3=-2(x+1) \Rightarrow$

$y-3=-2x-2 \Rightarrow y-3+3=-2x-2+3 \Rightarrow y=-2x+1.$

58. One point on the line is $(-2,\ 0)$ and another point is $\left(0,\ \dfrac{1}{2}\right)$. Find the slope: $\dfrac{\frac{1}{2}-0}{0-(-2)}=\dfrac{\frac{1}{2}}{2}=\dfrac{1}{2}\cdot\dfrac{1}{2}=\dfrac{1}{4}$.

The y-intercept is $\dfrac{1}{2}$. Thus, $y=\dfrac{1}{4}x+\dfrac{1}{2}$.

59. The linear model is approximate because it does not pass exactly through all the points. When the run increases by 2, the rise increases by 1. Thus, the slope is $\dfrac{1}{2}$. The y-intercept is 1. Thus, $y=\dfrac{1}{2}x+1$.

60. $y=5x+100$

61. $I=5000x+20{,}000$

62. (a) See Figure 62a. A line could pass though all four points.

(b) See Figure 62b.

(c) The y-intercept is 5 and the slope is -2. Thus, $y=-2x+5$.

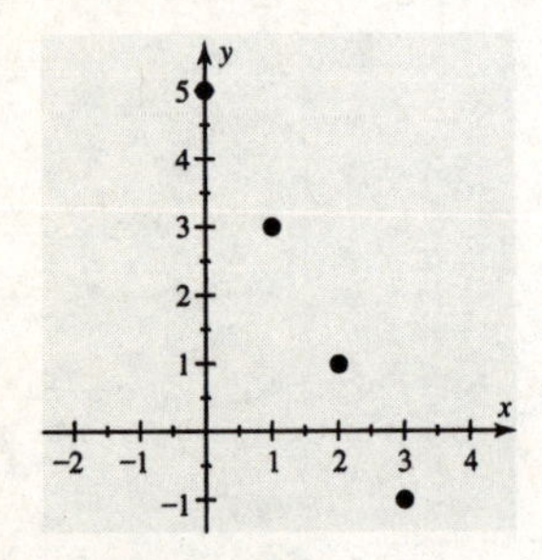

Figure 62a

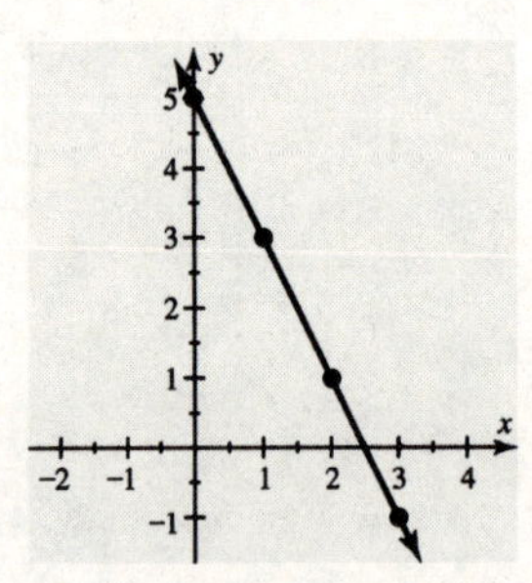

Figure 62b

63. $C=8x$

64. 9 out of 10, or $\dfrac{9}{10}$, of the portion of mail consisting of first-class mail and periodicals were first class mail.

Thus, $\dfrac{9}{10}\cdot\dfrac{11}{20}=\dfrac{99}{200}$ represents the fraction of all mail that was first-class mail.

65. (a) The slope of the line of the equation $I=807.4x-1{,}587{,}300$ is 807.4. Thus, the yearly average increase is \$807.40.

(b) Set I equal to 19,400 and solve for x: $19{,}400=807.4x-1{,}587{,}300 \Rightarrow$

$19{,}400+1{,}587{,}300=807.4x-1{,}587{,}300+1{,}587{,}300 \Rightarrow 1{,}606{,}700=807.4x \Rightarrow$

$\dfrac{1{,}606{,}700}{807.4}=\dfrac{807.4x}{807.4} \Rightarrow 1990=x \Rightarrow x=1990.$

66. $131-127=4$. Divide 4 by 127 to find the percent increase: $\dfrac{4}{127}\approx 0.031\approx 3.1\%$

67. Let x represent the amount of money invested at 3%. Then, $2x$ represents the amount of money invested at 4%. Thus, $0.03x+0.04(2x)=110 \Rightarrow 0.03x+0.08x=110 \Rightarrow 0.11x=110 \Rightarrow \dfrac{0.11x}{0.11}=\dfrac{110}{0.11} \Rightarrow x=1000.$

Thus, \$1000 is invested at 3% and \$2000 is invested at 4%.

68. (a) $\frac{27.2-22.4}{2000-1970}=\frac{4.8}{30}=0.16$

(b) Participation increased at an average rate of 0.16 million students/year.

(c) 2005 is 5 additional years beyond 2000. Since the slope is 0.16, $5\cdot 0.16=0.80$ additional students.

Because the number in 2000 was 27.2, $27.2+0.80=28$ million.

69. (a) Replace x with 200: $C=0.3(200)+25=60+25=\$85$.

(b) Because the initial value is 25, it costs \$25 to rent the car but not drive it.

(c) Because the slope is 0.3, it costs 30¢ to drive the car 1 additional mile.

Critical Thinking Solutions for Chapter 3

Section 3.1

- *Answers may vary.*

Section 3.4

- See Figure 4. The slopes of the two segments are 10 and -5, which are the two velocities of the runner. The second slope is negative because the distance between the runner and home is decreasing.

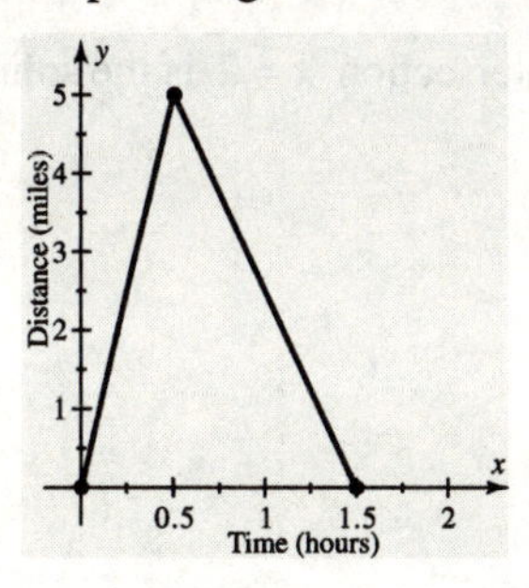

Figure 4

Section 3.6

- A positive slope indicates that water is entering the pool. A negative slope indicates water is leaving the pool.

Chapter 4: Systems of Linear Equations in Two Variables

4.1: Solving Systems of Linear Equations Graphically and Numerically

Concepts

1. ordered
2. intersection-of-graphs
3. one
4. The intersection is the solution $\Rightarrow$ $(11, 9)$.
5. Note $2(0)+3(0) \neq 8$ *and* $5(0)-4(0) \neq -3 \Rightarrow (0, 0)$ is not a solution to the system.
6. table
7. the same
8. intercepts

Solving Systems of Equations

9. Graph the system $y = 2,\ y = 2x$. See Figure 9. The intersection $x = 1$ is the solution.

11. Graph the system $y = 2,\ y = 4 - x \Rightarrow y = -x + 4$. See Figure 11. The intersection $x = 2$ is the solution.

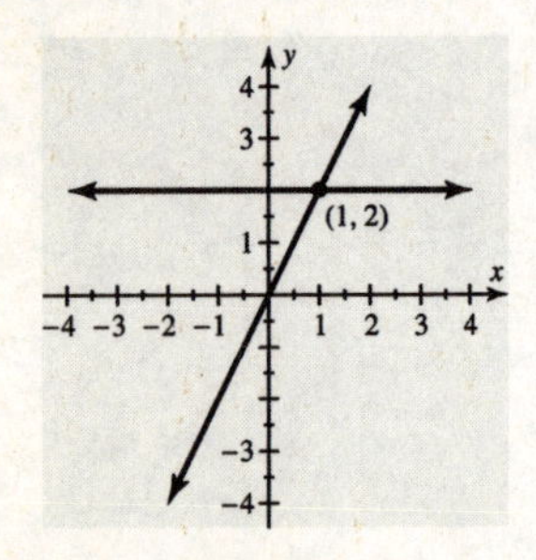

Figure 9

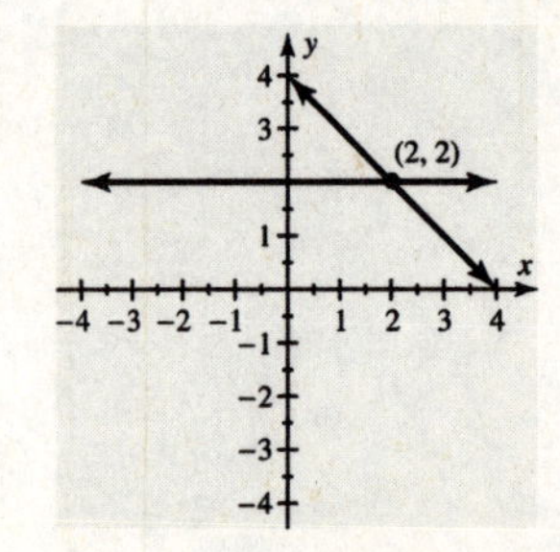

Figure 11

13. Graph the system $y = 2,\ y = -\dfrac{1}{2}x + 1$. See Figure 13. The intersection $x = -2$ is the solution.

15. Graph the system $y = 2,\ y = 2x + y = 6 \Rightarrow y = -2x + 6$. See Figure 15. The intersection $x = 2$ is the solution.

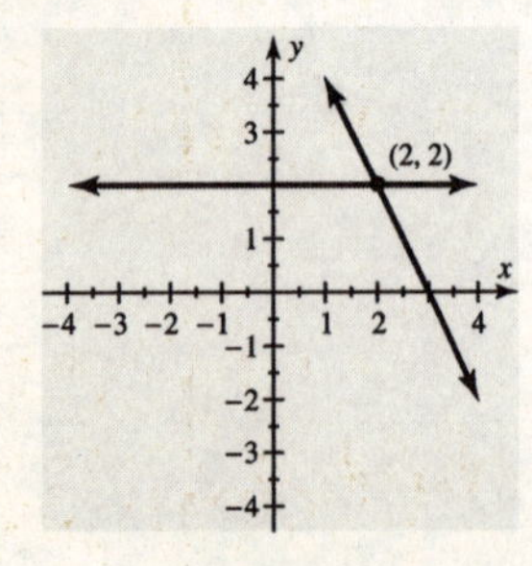

Figure 15

17. For the system $x + y = 2$ and $x - y = 0$, $(1, 1)$ is the solution because $(1)+(1) = 2$ and $(1)-(1) = 0$.

19. For the system $2x+3y=-5$ and $4x-5y=23$, $(2,-3)$ is the solution because $2(2)+3(-3)=4+(-9)=-5$ and $4(2)-5(-3)=8-(-15)=8+15=23$.

21. For the system $-5x+5y=-10$ and $4x+9y=8$, $(2,\ 0)$ is the solution because $-5(2)+5(0)=-10+0=$ -10 and $4(2)+9(0)=8+0=8$.

23. The intersection and solution is $(2,1)$, $y=1$ and $2+1=3$.

25. The intersection and solution is $(3,2)$, $3+2=5$ and $3-2=1$.

27. The intersection and solution is $(-1,1)$, $-(-1)+2(1)=1+2=3$ and $2(-1)+3(1)=-2+3=1$.

29. Since $y=4$ for both equations when $x=2$, the solution is $(2,4)$.

31. Since $y=1$ for both equations when $x=3$, the solution is $(3,1)$.

33. See Figure 33. $Y_1=Y_2=3$ when $x=1$, the solution is $(1,3)$.

x	0	1	2	3
$y=x+2$	2	3	4	5
$y=4-x$	4	3	2	1

Figure 33

35. (a) Graph the system. See Figure 35a. The intersection and solution is $(-1,\ 1)$.

(b) See Figure 35b. Since $y=1$ for both equations, when $x=-1$ the solution is $(-1,1)$.

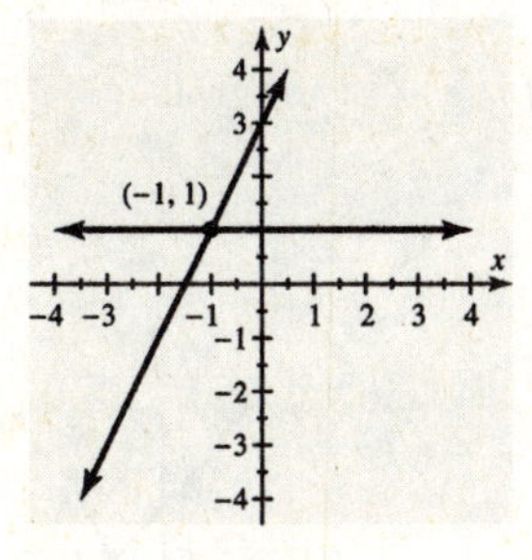

Figure 35a

x	−2	−1	0	1
$y=2x+3$	−1	1	3	5
$y=1$	1	1	1	1

Figure 35b

37. (a) Graph the system. See Figure 37a. The intersection and solution is $(3,1)$.

(b) See Figure 37b. Since $y=1$ for both equations, when $x=3$ the solution is $(3,1)$.

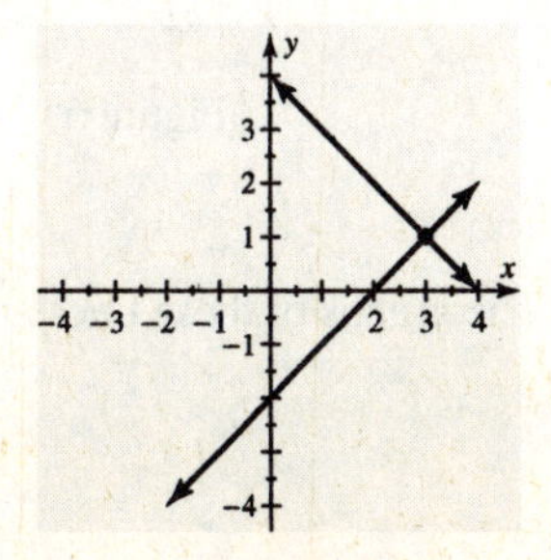

Figure 37a

x	2	3	4	5
$y=4-x$	2	1	0	−1
$y=x-2$	0	1	2	3

Figure 37b

39. (a) Graph the system. See Figure 39a. The intersection and solution is $(1,\ 3)$.

(b) See Figure 39b. Since $y = 3$ for both equations, when $x = 1$ the solution is $(1,\ 3)$.

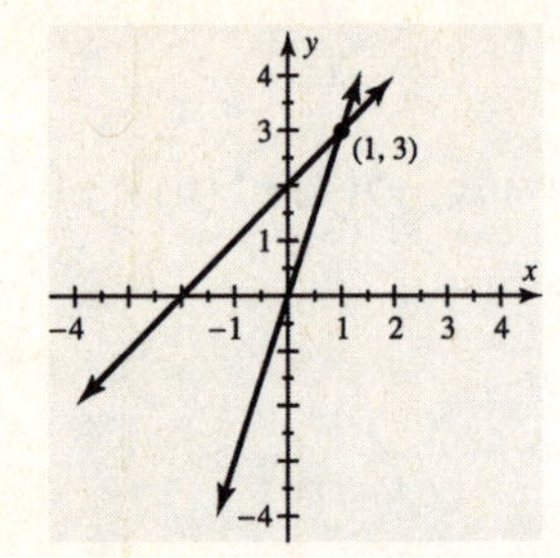

Figure 39a

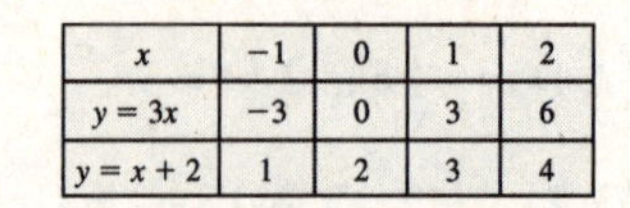

x	-1	0	1	2
$y = 3x$	-3	0	3	6
$y = x + 2$	1	2	3	4

Figure 39b

41. Graph the system. See Figure 41. The intersection and solution is $(-1,\ -2)$.

43. Graph the system. See Figure 43. The intersection and solution is $(1,\ -1)$.

45. Graph the system. See Figure 45. The intersection and solution is $(1,\ 2)$.

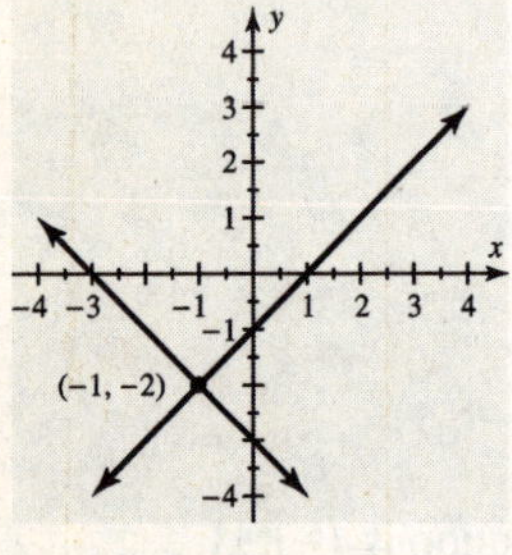

Figure 41

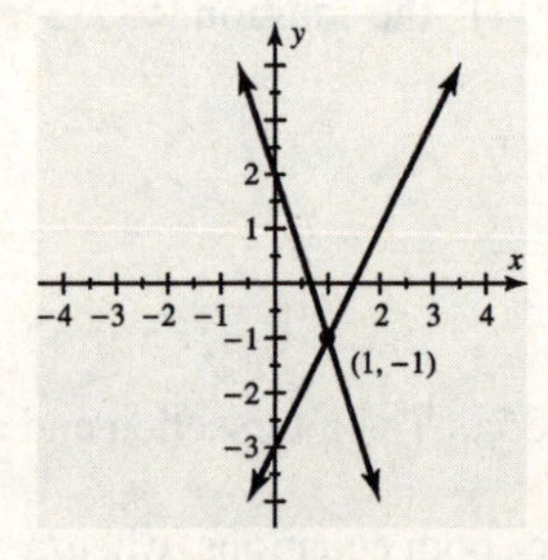

Figure 43

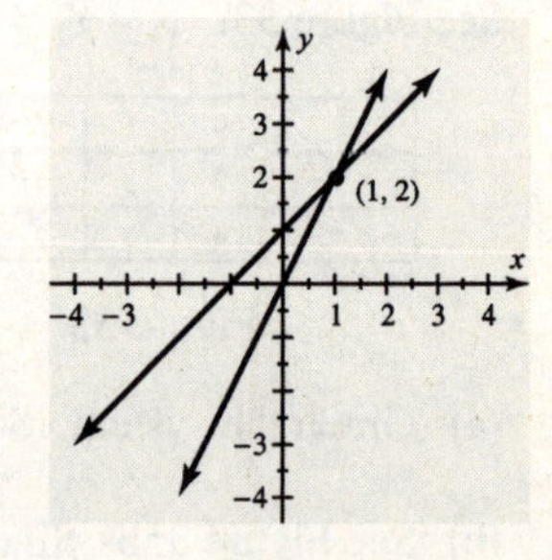

Figure 45

47. Graph the system. See Figure 47. The intersection and solution is $(3,\ 2)$.

49. Graph the system. See Figure 49. The intersection and solution is $(3,\ 1)$.

51. Graph the system. See Figure 51. The intersection and solution is $(1,\ 2)$.

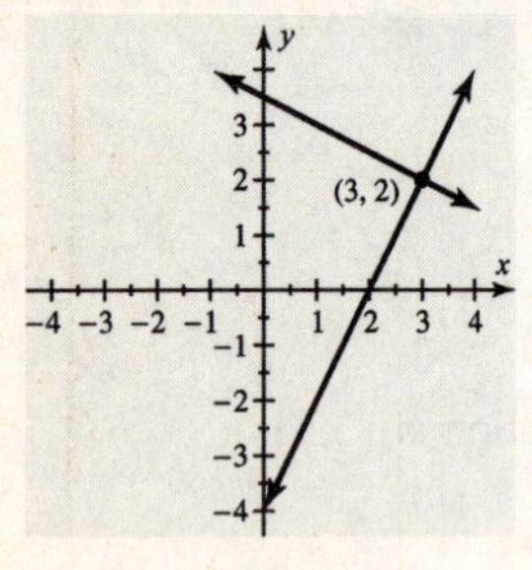

Figure 47

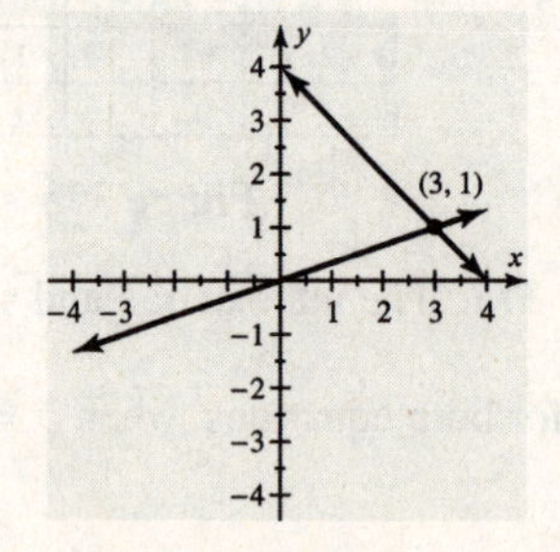

Figure 49

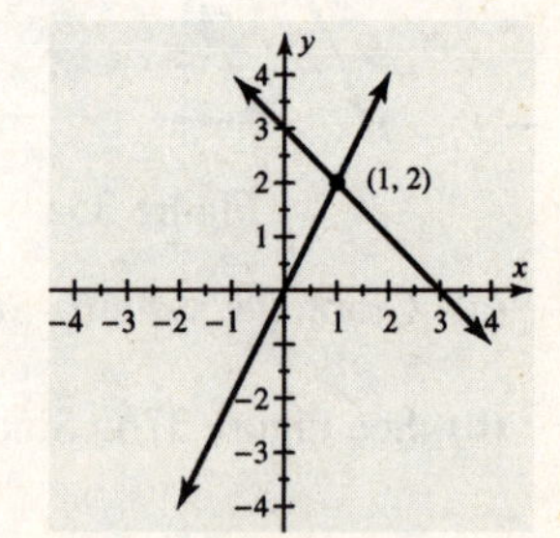

Figure 51

53. (a) Let x and y represent the numbers. Then $x + y = 4$ and $x - y = 0$.

(b) Write each equation in slope-intercept form: $y = -x + 4$ and $y = x$. The graphs of these lines are shown in the figure, and they intersect at the point $(2, 2)$. See Figure 53.

The unknown numbers are 2 and 2.

55. (a) Let x and y represent the numbers. Then $2x + y = 7$ and $x - y = 2$.

(b) Write each equation in slope-intercept form: $y = -2x + 7$ and $y = x - 2$. The graphs of these lines are shown in the figure, and they intersect at the point $(3, 1)$.

See Figure 55.

The unknown numbers are 3 and 1.

The unknown numbers are 1 and 4.

57. (a) Let x and y represent the numbers. Then $x = 3y$ and $x - y = 4$.

(b) Write each equation in slope-intercept form: $y = \frac{1}{3}x$ and $y = x - 4$. The graphs of these lines are shown in the figure, and they intersect at the point $(6, 2)$.

See Figure 57.

The unknown numbers are 6 and 2.

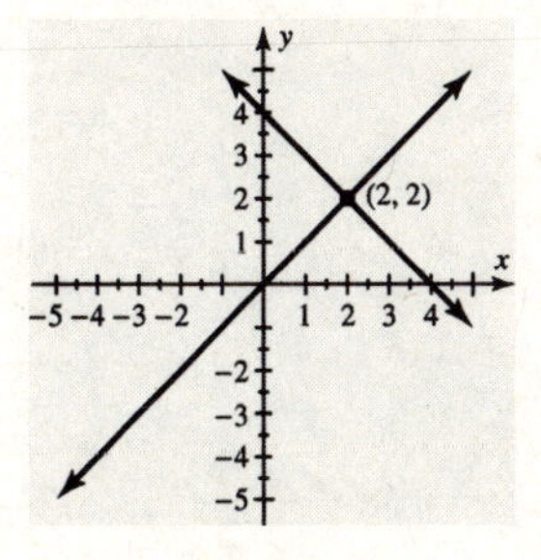

Figure 53

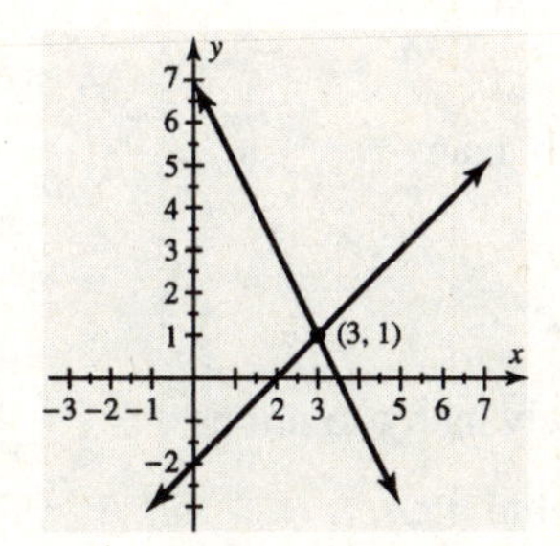

Figure 55

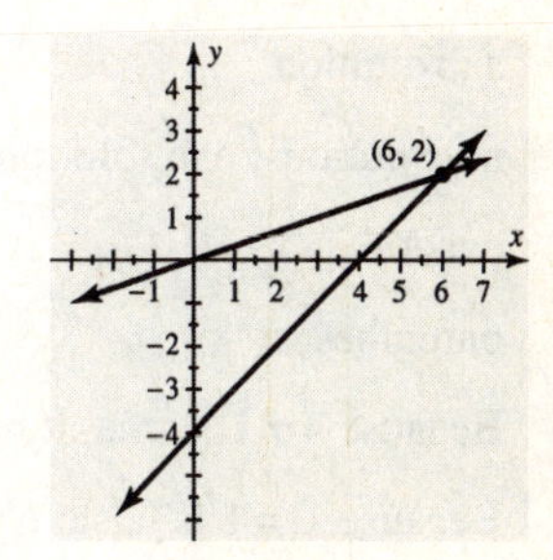

Figure 57

Applications

59. (a) $x =$ miles, $C =$ cost, then $C = 0.5x + 50$.

(b) Graph $C = 0.5x + 50$ and $C = 80$. See Figure 59b. The intersection and solution $(60, 80) \Rightarrow 60$ miles.

(c) See Figure 59c. Since $C = 80$ for both equations when $x = 60$, the solution is $(60, 80) \Rightarrow 60$ miles.

61. (a) $x = \%$ rock music, $y = \%$ country music, then $x + y = 36$ and $x = 2y$.

(b) Graph $x + y = 36$ and $x = 2y$. See Figure 61. The intersection and solution is $(24, 12)$.

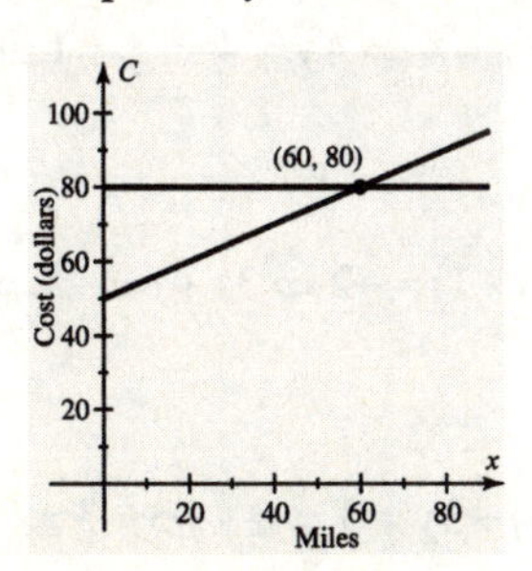

Figure 59b

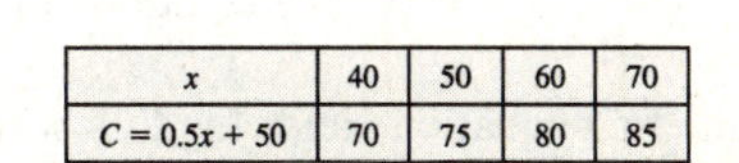

x	40	50	60	70
$C = 0.5x + 50$	70	75	80	85

Figure 59c

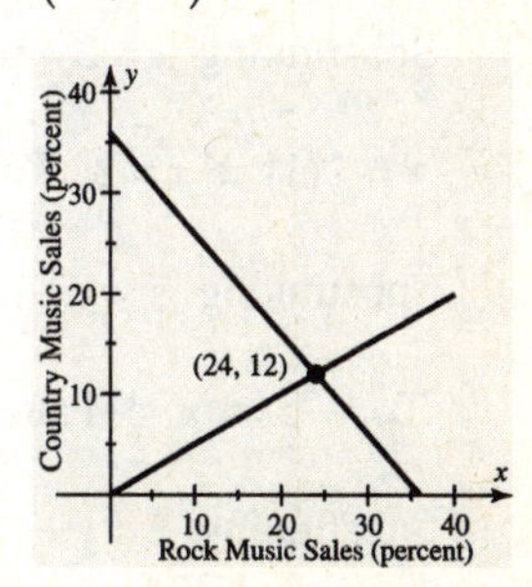

Figure 61

63. (a) Let x = length, y = width, then $2x+2y=28$ and $x=y+4$ or $x-y=4$.

(b) Graph $2x+2y=28$ and $x-y=4$. See Figure 63. The intersection and solution is $(9,\ 5) \Rightarrow 9$ in.×5 in.

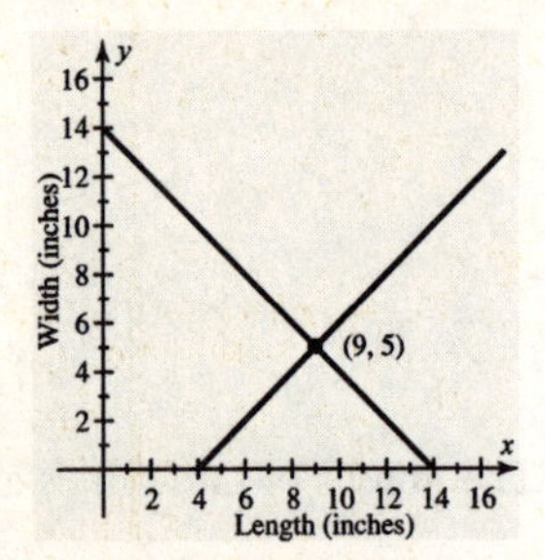

Figure 63

4.2: Solving Systems of Linear Equations by Substitution

Concepts

1. substitution
2. no solutions; one solution; infinitely many
3. exact
4. parentheses
5. Because $1=1$ is true, it has infinitely many solutions.
6. Because $0=1$ is false, it has no solutions.
7. consistent
8. inconsistent
9. independent
10. dependent

Solving Systems of Equations

11. Substituting $y=2x$ into the first equation yields the following: $x+(2x)=9 \Rightarrow 3x=9 \Rightarrow x=3$ and so $y=2(3) \Rightarrow y=6$. The solution is $(3,\ 6)$.

13. Substituting $x=2y$ into the first equation yields the following: $(2y)+2y=4 \Rightarrow 4y=4 \Rightarrow y=1$ and so $x=2(1) \Rightarrow x=2$. The solution is $(2,\ 1)$.

15. Substituting $y=x+1$ into the first equation yields the following: $2x+(x+1)=-2 \Rightarrow 3x+1=-2 \Rightarrow$ $3x=-3 \Rightarrow x=-1$ and so $y=-1+1 \Rightarrow y=0$. The solution is $(-1,\ 0)$.

17. Substituting $x=y+3$ into the first equation yields the following: $(y+3)+3y=3 \Rightarrow 4y+3=3 \Rightarrow$ $4y=0 \Rightarrow y=0$ and so $x=0+3 \Rightarrow x=3$. The solution is $(3,\ 0)$.

19. Substituting $y=2x-1$ into the first equation yields the following: $3x+2(2x-1)=\frac{3}{2} \Rightarrow$

$3x+4x-2=\frac{3}{2} \Rightarrow 7x-2=\frac{3}{2} \Rightarrow 7x=\frac{7}{2} \Rightarrow x=\frac{1}{2}$ and so $y=2\left(\frac{1}{2}\right)-1 \Rightarrow y=0$.

The solution is $\left(\frac{1}{2},\ 0\right)$.

21. Substituting $x=2-\frac{1}{2}y$ into the first equation yields the following: $2\left(2-\frac{1}{2}y\right)-3y=-12\Rightarrow$

$4-y-3y=-12\Rightarrow 4-4y=-12\Rightarrow -4y=-16\Rightarrow y=4$ and so $x=2-\frac{1}{2}(4)\Rightarrow x=2-2\Rightarrow$

$x=0$. The solution is $(0,\ 4)$.

23. Note that $3x-y=1\Rightarrow -y=-3x+1\Rightarrow y=3x-1$, substituting $y=3x-1$ into the first equation yields the following: $2x-3(3x-1)=-4\Rightarrow 2x-9x+3=-4\Rightarrow -7x+3=-4\Rightarrow -7x=-7\Rightarrow x=1$ and so $y=3(1)-1\Rightarrow y=2$. The solution is $(1,\ 2)$.

25. Note that $x-5y=26\Rightarrow x=5y+26$, substituting $x=5y+26$ into the second equation yields the following: $2(5y+26)+6y=-12\Rightarrow 10y+52+6y=-12\Rightarrow 16y+52=-12\Rightarrow 16y=-64\Rightarrow$

$y=-4$ and so $x=5(-4)+26\Rightarrow x=6$. The solution is $(6,\ -4)$.

27. Note that $y-3z=13\Rightarrow y=3z+13$, substituting $y=3z+13$ into the first equation yields the following:

$\frac{1}{2}(3z+13)-z=5\Rightarrow \frac{3}{2}z+\frac{13}{2}-z=5\Rightarrow \frac{1}{2}z+\frac{13}{2}=5\Rightarrow \frac{1}{2}z=-\frac{3}{2}\Rightarrow z=-3$ and so

$y=3(-3)+13\Rightarrow y=-9+13\Rightarrow y=4$. The solution is $(4,-3)$.

29. Note that $r+60t=-29\Rightarrow r=-60t-29$, substituting $r=-60t-29$ into the first equation yields the following: $10(-60t-29)-20t=20\Rightarrow -600t-290-20t=20\Rightarrow -620t-290=20\Rightarrow$

$-620t=310\Rightarrow t=-\frac{1}{2}$ and so $r=-60\left(-\frac{1}{2}\right)-29\Rightarrow r=30-29\Rightarrow r=1$. The solution is $\left(1,-\frac{1}{2}\right)$.

31. Note that $3x+2y=9\Rightarrow 3x=-2y+9\Rightarrow x=-\frac{2}{3}y+3$, substituting $x=-\frac{2}{3}y+3$ into the second

equation yields the following: $2\left(-\frac{2}{3}y+3\right)-3y=-7\Rightarrow -\frac{4}{3}y+6-3y=-7\Rightarrow -\frac{13}{3}y+6=-7\Rightarrow$

$-\frac{13}{3}y=-13\Rightarrow y=3$ and so $x=-\frac{2}{3}(3)+3\Rightarrow x=-2+3\Rightarrow x=1$. The solution is $(1,\ 3)$.

33. Note that $2a-3b=6\Rightarrow 2a=3b+6\Rightarrow a=\frac{3}{2}b+3$, substituting $a=\frac{3}{2}b+3$ into the second equation

yields the following: $-5\left(\frac{3}{2}b+3\right)+4b=-8\Rightarrow -\frac{15}{2}b-15+4b=-8\Rightarrow -\frac{7}{2}b-15=-8\Rightarrow$

$-\frac{7}{2}b=7\Rightarrow b=-2$ and so $a=\frac{3}{2}(-2)+3\Rightarrow a=-3+3\Rightarrow a=0$. The solution is $(0,\ -2)$.

35. Note that $2x-\frac{1}{2}y=3 \Rightarrow -\frac{1}{2}y=-2x+3 \Rightarrow y=4x-6,$ substituting $y=4x-6$ into the first equation yields the following: $-\frac{1}{2}x+3(4x-6)=5 \Rightarrow -\frac{1}{2}x+12x-18=5 \Rightarrow \frac{23}{2}x-18=5 \Rightarrow \frac{23}{2}x=23 \Rightarrow x=2$ and so $y=4(2)-6 \Rightarrow y=8-6 \Rightarrow y=2.$ The solution is $(2, 2)$.

37. Note that $-8a+2b=34 \Rightarrow 2b=8a+34 \Rightarrow b=4a+17,$ substituting $b=4a+17$ into the first equation yields the following: $3a+5(4a+17)=16 \Rightarrow 3a+20a+85=16 \Rightarrow 23a=-69 \Rightarrow a=-3$ and so $b=4(-3)+17 \Rightarrow b=-12+17 \Rightarrow b=5.$ The solution is $(-3, 5)$.

39. The lines intersect at one point, so there is one solution $\Rightarrow$ the system is consistent and the equations independent.

41. The lines coincide, so there are infinitely many solutions $\Rightarrow$ the system is consistent and the equations are dependent.

43. The lines are parallel, so there are no solutions $\Rightarrow$ the system is inconsistent.

45. Note that $x+y=4 \Rightarrow y=-x+4,$ substituting $y=-x+4$ into the second equation yields the following: $x+(-x+4)=2 \Rightarrow 4=2,$ which is false $\Rightarrow$ there are no solutions.

Graph the system. See Figure 45. The lines are parallel $\Rightarrow$ there are no solutions. The system is inconsistent.

47. Note that $x-2y=0 \Rightarrow x=2y,$ substituting $x=2y$ into the first equation yields the following: $2(2y)-y=3 \Rightarrow 3y=3 \Rightarrow y=1$ and so $x=2(1) \Rightarrow x=2.$ The solution is $(2, 1)$.

Graph the system. See Figure 47. The lines intersect at the solution (2, 1). The system has one solution $\Rightarrow$ the system is consistent and the equations are independent.

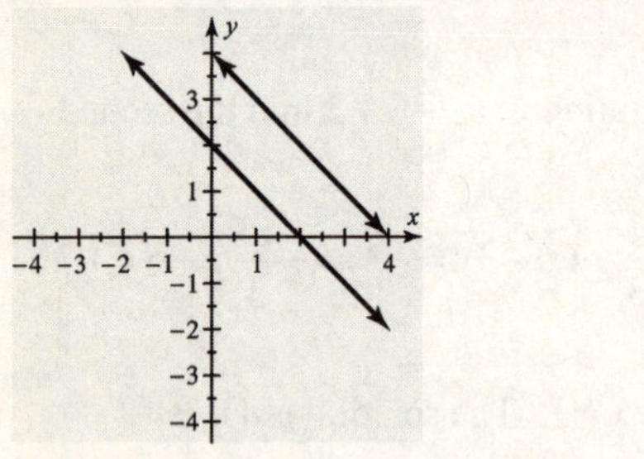

Figure 45

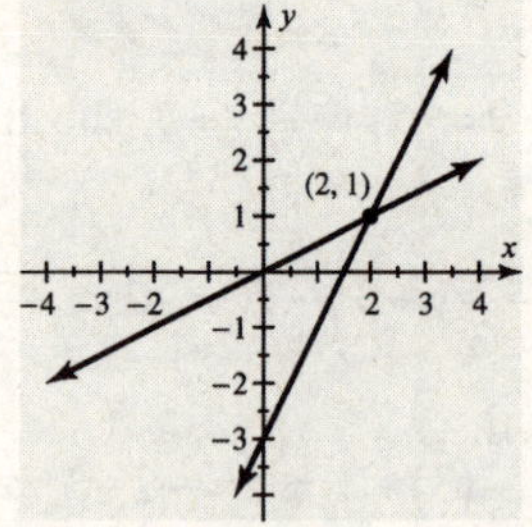

Figure 47

49. Note that $x-y=1 \Rightarrow x=y+1,$ substituting $x=y+1$ into the second equation yields the following: $2(y+1)-2y=2 \Rightarrow 2y+2-2y=2 \Rightarrow 2=2,$ which is true $\Rightarrow$ infinitely many solutions.

Graph the system. See Figure 49. The lines coincide $\Rightarrow$ infinitely many solutions. The system has infinitely many solutions $\Rightarrow$ the system is consistent and the equations dependent.

51. Substituting $x = 2y$ into the first equation yields the following: $(2y) - 2y = 4 \Rightarrow 0 = 4$ which is false $\Rightarrow$ no solution. Graph the system. See Figure 51. The lines are parallel $\Rightarrow$ no solutions. The system has no solutions $\Rightarrow$ the system is inconsistent.

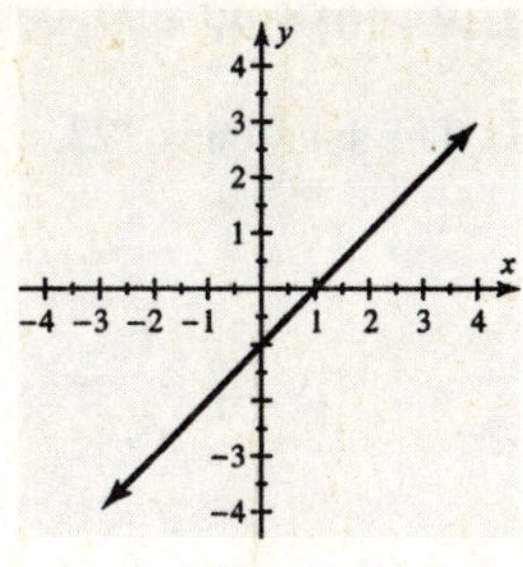

Figure 49

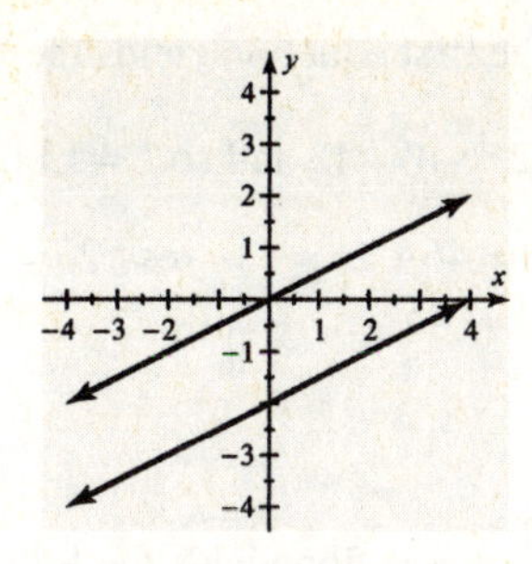

Figure 51

53. Note that $x + y = 7 \Rightarrow y = -x + 7$, substituting $y = -x + 7$ into the first equation yields the following: $x + (-x + 7) = 9 \Rightarrow 7 = 9$, which is false$\Rightarrow$ no solutions.

55. Note that $x - y = 4 \Rightarrow x = y + 4$, substituting $x = y + 4$ into the second equation yields the following: $2(y + 4) - 2y = 8 \Rightarrow 2y + 8 - 2y = 8 \Rightarrow 8 = 8$, which is true$\Rightarrow$ infinitely many solutions.

57. Note that $x + y = 4 \Rightarrow y = -x + 4$, substituting $y = -x + 4$ into the second equation yields the following: $x - (-x + 4) = 2 \Rightarrow 2x = 6 \Rightarrow x = 3$ and so $y = -(3) + 4 \Rightarrow y = 1$. The solution is (3, 1).

59. Note that $-x + y = -7 \Rightarrow y = x - 7$, substituting $y = x - 7$ into the first equation yields the following: $x - (x - 7) = 7 \Rightarrow 7 = 7$, which is true$\Rightarrow$ infinitely many solutions.

61. Note that $u - 2v = 5 \Rightarrow u = 2v + 5$, substituting $u = 2v + 5$ into the second equation yields the following: $2(2v + 5) - 4v = -2 \Rightarrow 4v + 10 - 4v = -2 \Rightarrow 10 = -2$, which is false$\Rightarrow$ no solutions.

63. Note that $r - 3t = -5 \Rightarrow r = 3t - 5$, substituting $r = 3t - 5$ into the first equation yields the following: $2(3t - 5) + 3t = 1 \Rightarrow 6t - 10 + 3t = 1 \Rightarrow 9t - 10 = 1 \Rightarrow 9t = 11 \Rightarrow t = \frac{11}{9}$ and so

$r = 3\left(\frac{11}{9}\right) - 5 \Rightarrow r = \frac{11}{3} - 5 \Rightarrow r = -\frac{4}{3}$. The solution is $\left(-\frac{4}{3}, \frac{11}{9}\right)$.

65. Substituting $y = 5x$ into the second equation yields the following: $5x = -3x \Rightarrow 8x = 0 \Rightarrow x = 0$, and so $y = 5(0) \Rightarrow y = 0$. The solution is $(0,\ 0)$.

67. Note that $5a = 4 - b \Rightarrow a = \frac{4}{5} - \frac{1}{5}b$, substituting $a = \frac{4}{5} - \frac{1}{5}b$ into the second equation yields the following: $5\left(\frac{4}{5} - \frac{1}{5}b\right) = 3 - b \Rightarrow 4 - b = 3 - b \Rightarrow 4 = 3$, which is false$\Rightarrow$ no solutions.

69. Note that $2x + 4y = 0 \Rightarrow 2x = -4y \Rightarrow x = -2y$, substituting $x = -2y$ into the second equation yields the following: $3(-2y) + 6y = 5 \Rightarrow -6y + 6y = 5 \Rightarrow 0 = 5$, which is false $\Rightarrow$ no solutions.

71. They are a single line.

Applications

73. (a) $2L+2W=72$ and $L=W+10$

(b) Substituting $L=W+10$ into the first equation yields the following: $2(W+10)+2W=72 \Rightarrow$ $2W+20+2W=72 \Rightarrow 4W=52 \Rightarrow W=13$ and so $L=13+10 \Rightarrow L=23$. The solution is (23, 13). Checking: $2(23)+2(13)=46+26=72$, yes. $23=13+10$, yes.

75. (a) $x+y=90$ and $x=\frac{1}{2}y$

(b) Substituting $x=\frac{1}{2}y$ into the first equation yields the following: $\frac{1}{2}y+y=90 \Rightarrow \frac{3}{2}y=90 \Rightarrow y=60$ and so $x=\frac{1}{2}(60) \Rightarrow x=30$. The solution is (30, 60).

(c) Graph the system. See Figure 75. The lines intersect at the solution (30, 60).

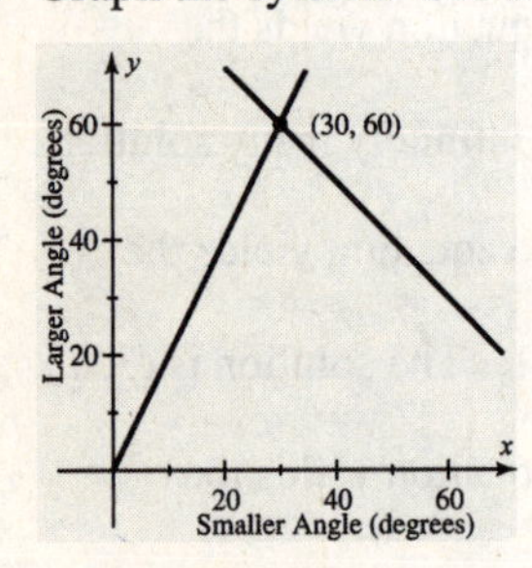

Figure 75

77. (a) $x-y=1.68$ and $y=0.98x$

(b) Substituting $y=0.98\,x$ into the first equation yields the following: $x-0.98x=1.68 \Rightarrow$ $0.02x=1.68 \Rightarrow x=84$ and so $84-y=1.68 \Rightarrow y=82.32$. The solution is $(84,\ 82.32)$.

79. Let $L=$ length and $W=$ width, then let $L=W+44$ and $2L+2W=288$, substituting $L=W+44$ into the second equation yields the following: $2(W+44)+2W=288 \Rightarrow 2W+88+2W=288 \Rightarrow$ $4W=200 \Rightarrow W=50$ and so $L=50+44 \Rightarrow L=94$. The solution is (94, 50) or 94 ft × 50 ft.

81. Let $x=$ larger number and $y=$ smaller number, then $x+y=70$ and $x=3y+2$, substituting $x=3y+2$ into the first equation yields the following: $(3y+2)+y=70 \Rightarrow 4y+2=70 \Rightarrow 4y=68 \Rightarrow y=17$ and so $x=3(17)+2 \Rightarrow x=51+2 \Rightarrow x=53$. The numbers are 17 and 53.

83. Let $x=$ speed of tugboat and $y=$ speed of current, then $15x-15y=120$ and $10x+10y=120$. Note that $10x+10y=120 \Rightarrow 10y=-10x+120 \Rightarrow y=-x+12$, substituting $y=-x+12$ into the first equation yields the following: $15x-15(-x+12)=120 \Rightarrow 15x+15x-180=120 \Rightarrow 30x=300 \Rightarrow x=10$ and so $y=-10+12 \Rightarrow y=2$. The tugboat speed is 10 mph and the current's speed is 2 mph.

85. Let x = liters of 20% solution and y = liters of 50% solution, then $x+y=10$ and $0.2x+0.5y=0.4(10)\Rightarrow$ $0.2x+0.5y=4$. Note that $x+y=10\Rightarrow y=-x+10$, substituting $y=-x+10$ into the second equation yields the following: $0.2x+0.5(-x+10)=4\Rightarrow 0.2x-0.5x+5=4\Rightarrow -0.3x=-1\Rightarrow x=3.33$ and so $3.33+y=10\Rightarrow y=6.67$. The amounts are: 3.33 liters of 20% solution and 6.67 liters of 50% solution.

87. Let $x = \text{mi}^2$ of Lake Superior and $y = \text{mi}^2$ of Lake Michigan, then $x + y = 54{,}000$ and $x = y+10{,}000$, substituting $x = y+10{,}000$ into the first equation yields the following: $(y+10{,}000)+y=54{,}000\Rightarrow$ $2y+10{,}000=54{,}000\Rightarrow 2y=44{,}000\Rightarrow y=22{,}000$ and so $x=22{,}000+10{,}000\Rightarrow x=32{,}000$. So Lake Superior has $32{,}000 \text{ mi}^2$ and Lake Michigan has $22{,}000 \text{ mi}^2$.

Checking Basic Concepts for Sections 4.1 & 4.2

1. (a) Graph the system $y=2$ and $y=1-\frac{1}{2}x$. See Figure 1a. The intersection and solution is $(-2,\ 2)\Rightarrow x=-2$.

   (b) Graph the system $y=2$ and $2x-3y=6$. See Figure 1b. The intersection and solution is $(6,\ 2)\Rightarrow x=6$.

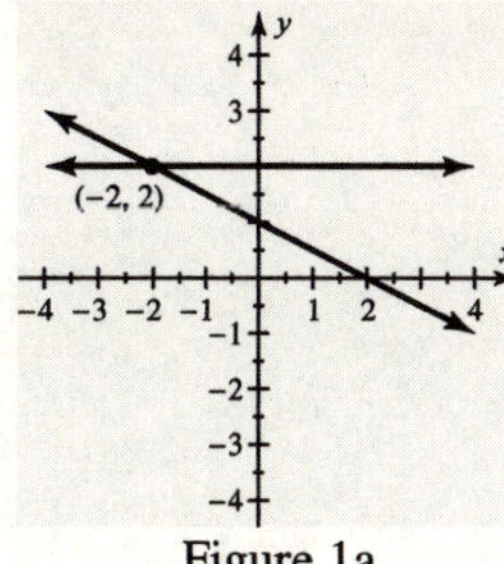

Figure 1a

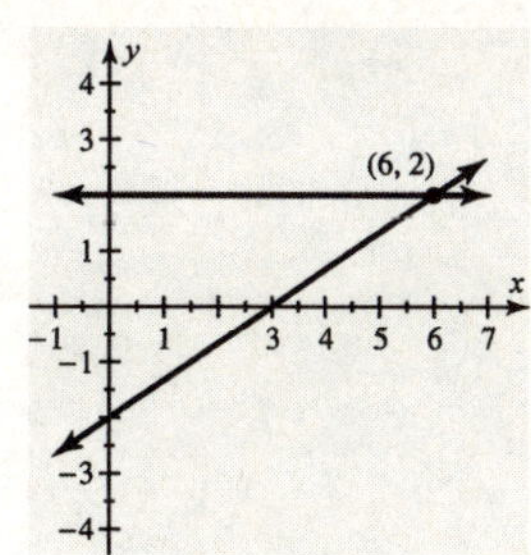

Figure 1b

2. (4, 2) is the solution because $2(4)-5(2)=8-10=-2$ and $3(4)+2(2)=12+4=16$.

3. Graph the system. See Figure 3. The intersection and solution is $(2,\ 1)$.

   Checking: $2-1=1$ (yes) and $2(2)+1=5$ (yes).

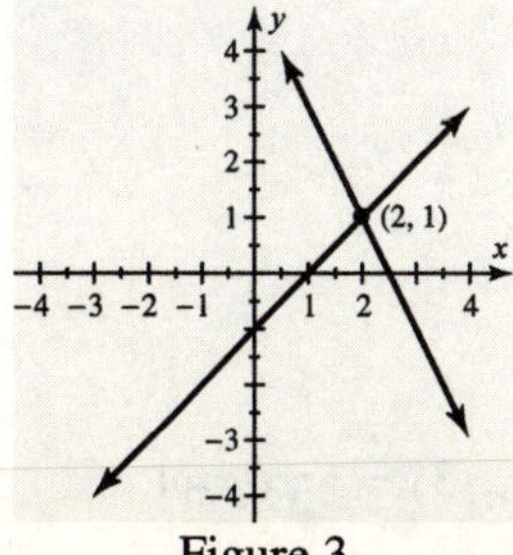

Figure 3

4. (a) Substituting $y=2-x$ into the first equation yields the following: $x+(2-x)=-1\Rightarrow 2=-1$ which is false $\Rightarrow$ no solution.

(b) Note that $-x+y=-2 \Rightarrow y=x-2$, substituting $y=x-2$ into the first equation yields the following: $4x-(x-2)=5 \Rightarrow 3x+2=5 \Rightarrow 3x=3 \Rightarrow x=1$ and so $y=1-2 \Rightarrow y=-1$. The solution is $(1,\ -1)$, one solution.

(c) Note that $x+2y=3 \Rightarrow x=-2y+3$, substituting $x=-2y+3$ into the second equation yields the following: $-(-2y+3)-2y=-3 \Rightarrow 2y-3-2y=-3 \Rightarrow -3=-3$, which is true $\Rightarrow$ infinitely many solutions.

5. (a) $x+y=300,\ 150x+200y=55{,}000$

(b) Note that $x+y=300 \Rightarrow y=-x+300$, substituting $y=-x+300$ into the second equation yields the following: $150x+200(-x+300)=55{,}000 \Rightarrow 150x-200x+60{,}000=55{,}000 \Rightarrow -50x=-5000 \Rightarrow$ $x=100$ and so $y=-100+300 \Rightarrow y=200$. The solution is (100, 200). Checking the answer $100+200=300$ (yes) and $150(100)+200(200)=15{,}000+40{,}000=55{,}000$ (true).

4.3: Solving Systems of Linear Equations by Elimination

Concepts

1. Substitution; elimination
2. addition
3. =
4. =
5. Add the equations.
6. Multiply the first equation by −3.
7. It has infinitely many solutions.
8. It has no solutions.

Using Elimination

9. Adding the two equations will eliminate the variable y.

$$\begin{array}{r} x-y=0 \\ \underline{x+y=2} \\ 2x=2 \end{array}$$

Thus, $x=1$. And so $1+y=2 \Rightarrow y=1$. The solution is (1, 1).

The result is supported by the graph's intersection point of (1, 1).

11. Adding the two equations will eliminate the variable y.

$$\begin{array}{r} 2x+3y=-1 \\ \underline{2x-3y=-7} \\ 4x=-8 \end{array}$$

Thus, $x=-2$. And so $2(-2)+3y=-1 \Rightarrow -4+3y=-1 \Rightarrow 3y=3 \Rightarrow y=1$.

The solution is $(-2,\ 1)$. The result is supported by the graph's intersection point of (−2, 1).

13. Multiplying the first equation by -1 and adding the two equations will eliminate both variables.

$$\begin{array}{r} -x-y=-3 \\ x+y=-1 \\ \hline 0=-4 \end{array}$$

This is false $\Rightarrow$ no solution. This result is supported by the graph's parallel lines which has no solution.

15. Multiplying the second equation by -2 and adding the two equations will eliminate both variables.

$$\begin{array}{r} 2x+2y=6 \\ -2x-2y=-6 \\ \hline 0=0 \end{array}$$

Which is true $\Rightarrow$ there are infinitely many solutions. This result is supported by the graph having two lines that coincide.

17. Adding the two equations will eliminate the variable y.

$$\begin{array}{r} x+y=7 \\ x-y=5 \\ \hline 2x=12 \end{array}$$

Thus, $x=6$. And so $6+y=7 \Rightarrow y=1$. The solution is $(6, 1)$.

19. Adding the two equations will eliminate the variable x.

$$\begin{array}{r} -x+y=5 \\ x+y=3 \\ \hline 2y=8 \end{array}$$

Thus, $y=4$. And so $x+4=3 \Rightarrow x=-1$. The solution is $(-1, 4)$.

21. Adding the two equations will eliminate the variable y.

$$\begin{array}{r} 2x+y=8 \\ 3x-y=2 \\ \hline 5x=10 \end{array}$$

Thus, $x=2$. And so $2(2)+y=8 \Rightarrow 4+y=8 \Rightarrow y=4$. The solution is $(2, 4)$.

23. Adding the two equations will eliminate the variable x.

$$\begin{array}{r} -2x+y=-3 \\ 2x-4y=0 \\ \hline -3y=-3 \end{array}$$

Thus, $y=1$. And so $-2x+1=-3 \Rightarrow -2x=-4 \Rightarrow x=2$. The solution is $(2, 1)$.

25. Multiplying the second equation by -1 and adding the two equations will eliminate the variable a.

$$\begin{array}{r} a+6b=2 \\ -a-3b=1 \\ \hline 3b=3 \end{array}$$

Thus, $b=1$. And so $a+6(1)=2 \Rightarrow a=-4$. The solution is $(-4, 1)$.

27. Multiplying the second equation by -1 and adding the two equations will eliminate the variable t.

$$\begin{array}{r} 3r-t=7 \\ -2r+t=-2 \\ \hline r=5 \end{array}$$

And so, $-2(5)+t=-2 \Rightarrow t=8$. The solution is $(5, 8)$.

29. Multiplying the second equation by -2 and adding the two equations will eliminate the variable v.

$$\begin{array}{r} 3u+2v=-16 \\ -4u-2v=18 \\ \hline -u=2 \end{array}$$

Thus $u=-2$. And so $3(-2)+2v=-16 \Rightarrow -6+2v=-16 \Rightarrow 2v=-10 \Rightarrow v=-5$.

The solution is $(-2, -5)$.

31. Multiplying the first equation by 2, the second equation by 5 and adding the two equations will eliminate the variable x.

$$\begin{array}{r} 10x-14y=10 \\ -10x+10y=-10 \\ \hline -4y=0 \end{array}$$

Thus $y=0$. And so $10x-14(0)=10 \Rightarrow 10x=10 \Rightarrow x=1$. The solution is $(1, 0)$.

33. Multiplying the second equation by -5, the first equation by 3 and adding the two equations will eliminate the variable x.

$$\begin{array}{r} 15x-9y=12 \\ -15x-10y=-50 \\ \hline -19y=-38 \end{array}$$

Thus $y=2$. And so $5x-3(2)=4 \Rightarrow 5x-6=4 \Rightarrow 5x=10 \Rightarrow x=2$.

The solution is $(2, 2)$.

35. Adding the two equations will eliminate the variable y.

$$\begin{array}{r} \frac{1}{2}x-y=3 \\ \frac{3}{2}x+y=5 \\ \hline 2x=8 \end{array}$$

Thus, $x=4$. And so $\frac{1}{2}(4)-y=3 \Rightarrow 2-y=3 \Rightarrow -y=1 \Rightarrow y=-1$.

The solution is $(4, -1)$.

37. Multiplying the first equation by 2 and adding the two equations will eliminate the variable x.

$$\begin{array}{r} -10x-20y=-44 \\ 10x+15y=35 \\ \hline -5y=-9 \end{array}$$

Thus $y=\frac{9}{5}$. And so $10x+15\left(\frac{9}{5}\right)=35 \Rightarrow 10x+27=35 \Rightarrow 10x=8 \Rightarrow x=\frac{4}{5}$.

The solution is $\left(\frac{4}{5}, \frac{9}{5}\right)$.

39. Since $y=2$, when $x=3$ for both equations $\Rightarrow$ the solution is $(3, 2)$.

41. Since $y=1$, when $x=0$ for both equations $\Rightarrow$ the solution is $(0, 1)$.

Using More Than One Method

43. (a) Adding the two equations will eliminate the variable y.

$$\begin{array}{r} 2x+y=5 \\ x-y=1 \\ \hline 3x=6 \end{array}$$

Thus, $x=2$. And so $2-y=1 \Rightarrow y=1$. The solution is $(2, 1)$.

(b) Graph the system. See Figure 43b. The intersection and solution is $(2, 1)$.

(c) See Figure 43c. Since $y=1$ when $x=2$ for both equations the solution is $(2, 1)$.

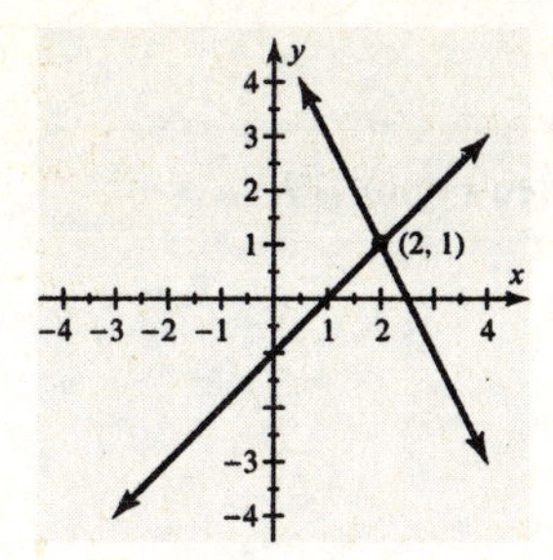

Figure 43b

x	0	1	2	3
$y=5-2x$	5	3	1	−1
$y=x-1$	−1	0	1	2

Figure 43c

45. (a) Multiplying the second equation by -1 and adding the two equations will eliminate the variable y.

$$\begin{aligned} 2x+y&=5\\ -x-y&=-1\\ \hline x&=4 \end{aligned}$$

And so $-4-y=-1 \Rightarrow -y=3 \Rightarrow y=-3$. The solution is $(4, -3)$.

(b) Graph the system. See Figure 45b. The intersection and solution is $(4, -3)$.

(c) See Figure 45c. Since $y=-3$ when $x=4$ for both equations the solution is $(4, -3)$.

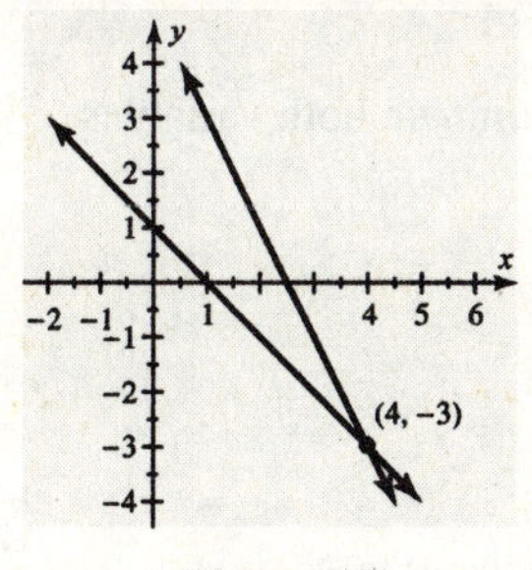

Figure 45b

x	2	3	4	5
$y=5-2x$	1	−1	−3	−5
$y=1-x$	−1	−2	−3	−4

Figure 45c

47. (a) Multiplying the second equation by 3 and adding the two equations will eliminate the variable x.

$$\begin{aligned} 6x+3y&=6\\ -6x+6y&=-6\\ \hline 9y&=0 \end{aligned}$$

Thus, $y=0$. And so $6x+3(0)=6 \Rightarrow 6x=6 \Rightarrow x=1$. The solution is $(1, 0)$.

(b) Graph the system. See Figure 47b. The intersection and solution is $(1, 0)$.

(c) See Figure 47c. Since $y=0$ when $x=1$ for both equations the solution is $(1, 0)$.

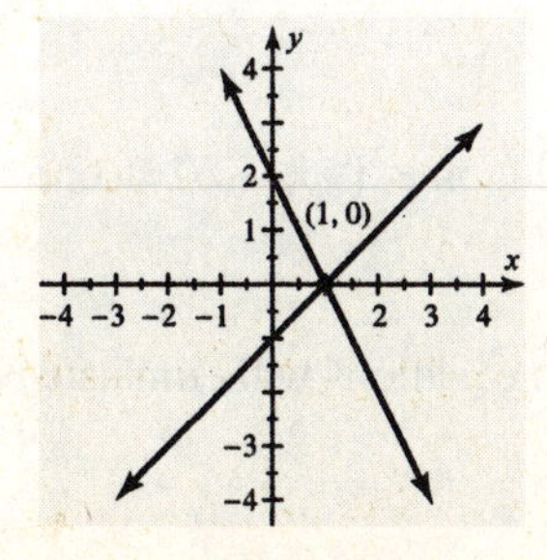

Figure 47b

x	−1	0	1	2
$y=2-2x$	4	2	0	−2
$y=x-1$	−2	−1	0	1

Figure 47c

Elimination and Other Types of Systems

49. Multiplying the second equation by 2 and adding the two equations will eliminate both variables.

$$\begin{array}{r} 2x-2y=4 \\ -2x+2y=-4 \\ \hline 0=0 \end{array}$$

This is always true $\Rightarrow$ infinitely many solutions. See Figure 49.

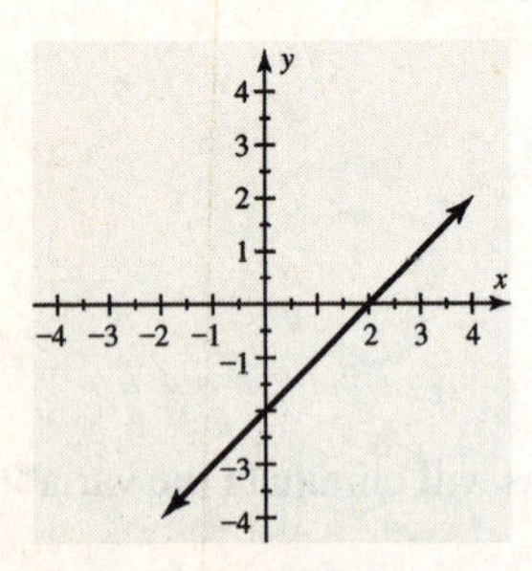

Figure 49

51. Adding the two equations will eliminate the variable y.

$$\begin{array}{r} x-y=0 \\ x+y=0 \\ \hline 2x=0 \end{array}$$

Thus, $x=0$. And so $0-y=0 \Rightarrow -y=0 \Rightarrow y=0$. The solution is $(0,\ 0) \Rightarrow$ one solution.

See Figure 51.

53. Multiplying the first equation by -1 and adding the two equations will eliminate both variables.

$$\begin{array}{r} -x+y=-4 \\ x-y=1 \\ \hline 0=-3 \end{array}$$

This is never true $\Rightarrow$ no solutions. See Figure 53.

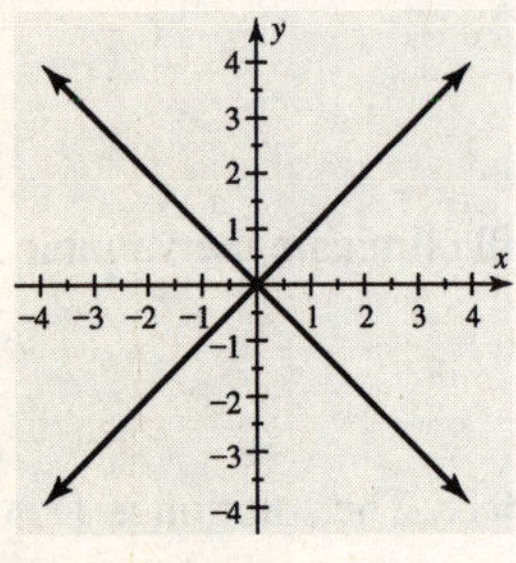

Figure 51

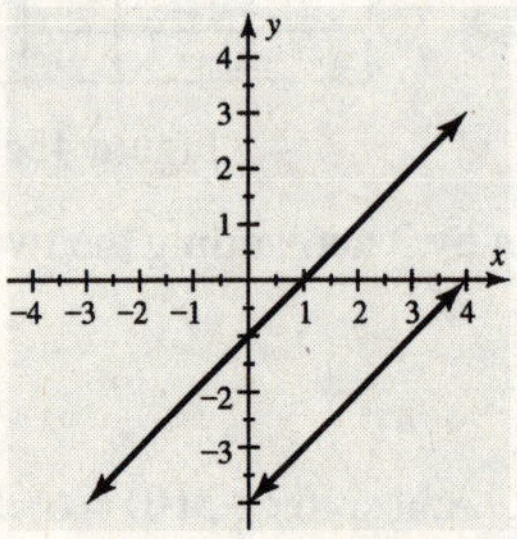

Figure 53

55. Multiplying the first equation by -1 and adding the two equations will eliminate the variable y.

$$\begin{array}{r} -x+y=-5 \\ 2x-y=4 \\ \hline x=-1 \end{array}$$

And so $-(-1)+y=-5 \Rightarrow 1+y=-5 \Rightarrow y=-6$. The solution is $(-1,\ -6) \Rightarrow$ one solution. See Figure 55.

57. Multiplying the first equation by 3, the second by -2 and adding the two equations will eliminate both variables.

$$\begin{array}{r} 12x-24y=\ \ 72 \\ -12x+24y=-72 \\ \hline 0=\ \ 0 \end{array}$$

This is always true $\Rightarrow$ infinitely many solutions. See Figure 57.

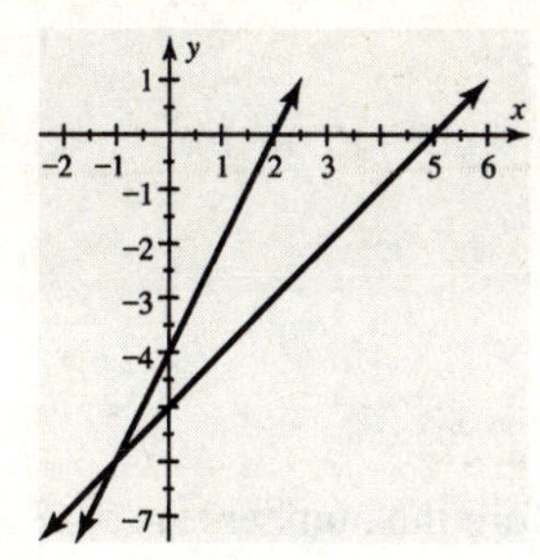

Figure 55

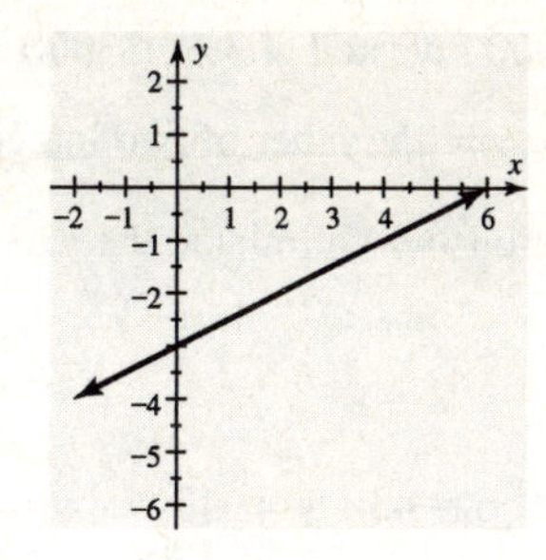

Figure 57

Applications

59. Let $x=$ skin cancer in men and let $y=$ skin cancer in women. Then $x+y=66{,}000$ and $x=y+10{,}000$. Note that $x=y+10{,}000 \Rightarrow x-y=10{,}000$. Adding the two equations will eliminate the variable y.

$$\begin{array}{r} x+y=66{,}000 \\ x-y=10{,}000 \\ \hline 2x=76{,}000 \end{array}$$

Thus $x=38{,}000$. And so $38{,}000-y=10{,}000 \Rightarrow -y=-28{,}000 \Rightarrow y=28{,}000$.

Therefore men: 38,000; Women: 28,000.

61. Let $x=$ minutes on a stationary bike and let $y=$ minutes on a stair climber. Then $x+y=30$ and $9x+11.5y=300$. Multiplying the first equation by -9 and adding the two equations will eliminate the variable x.

$$\begin{array}{r} -9x-9y=-270 \\ 9x+11.5y=300 \\ \hline 2.5y=30 \end{array}$$

Thus $y=12$. And so $x+12=30 \Rightarrow x=18$. Therefore bicycle: 18 minutes; stairclimber: 12 minutes.

63. Let $x=$ speed of riverboat and let $y=$ speed of the current. Then $8x+8y=64$ and $16x-16y=64$. Multiplying the first equation by 2 and adding the two equations will eliminate the variable y.

$$\begin{array}{r} 16x+16y=128 \\ 16x-16y=64 \\ \hline 32x=192 \end{array}$$

Thus $x=6$. And so $8(6)+8y=64 \Rightarrow 48+8y=64 \Rightarrow 8y=16 \Rightarrow y=2$.

Therefore current: 2 mph; boat: 6 mph.

65. Let $x=$ amount of money invested at 3% and let $y=$ amount of money invested at 5%. Then $x+y=5000$ and $0.03x+0.05y=210$. Multiplying the second equation by -20 and adding the two equations will eliminate the variable y.

$$\begin{aligned} x + y &= 5000 \\ -0.6x - y &= -4200 \\ \hline 0.4x &= 800 \end{aligned}$$

Thus $x = 2000$. And so $2000 + y = 5000 \Rightarrow y = 3000$.

Therefore \$2000 invested at 3%; \$3000 invested at 5%.

67. Let $x =$ one of two integers and let $y =$ the other of two integers. Then $x + y = -17$ and $x - y = -69$. Adding the two equations will eliminate the variable y.

$$\begin{aligned} x + y &= -17 \\ x - y &= -69 \\ \hline 2x &= -86 \end{aligned}$$

Thus $x = -43$. And so $-43 + y = -17 \Rightarrow y = 26$. Therefore the numbers are -43 and 26.

69. (a) The graphs intersection is $20 \text{ in.} \times 40 \text{ in.}$

(b) Note that $L = 2W \Rightarrow -2W + L = 0$, using this and $2W + 2L = 120$ and adding the two equations will eliminate the variable W.

$$\begin{aligned} -2W + L &= 0 \\ 2W + 2L &= 120 \\ \hline 3L &= 120 \end{aligned}$$

Thus $L = 40$. And so $2W + 2(40) = 120 \Rightarrow 2W + 80 = 120 \Rightarrow 2W = 40 \Rightarrow W = 20$.

Therefore $L = 40$ and $W = 20$ so $20 \text{ in.} \times 40 \text{ in.}$

4.4: Systems of Linear Inequalities

Concepts

1. All points below and including the line $y = k$
2. All points to the right of the line $x = k$
3. All points above and including the line $y = x$
4. test
5. dashed
6. solid
7. $Ax + By = C$
8. both inequalities
9. intersect
10. No

Solutions to Linear Inequalities

11. Yes, $3 > 2$ is true.

13. No, $0 \geq 2$ is false.

15. No, $4 \geq 5$ is false.

17. Yes, $0 < 3 - 1 \Rightarrow 0 < 2$ is true.

19. Yes, $-2 + 6 \leq 4 \Rightarrow 4 \leq 4$ is true.

21. No, $2(-1)+(-1)\geq -1 \Rightarrow -2+(-1)\geq -1 \Rightarrow -3\geq -1$ is false.

23. $x>1$

25. $y\geq 2$

27. $y<x$

29. $-x+y\leq 1$

31. See Figure 31.

33. See Figure 33.

35. See Figure 35.

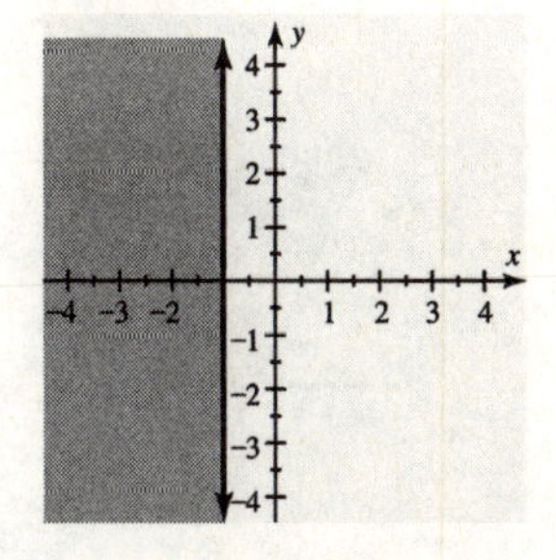

Figure 31

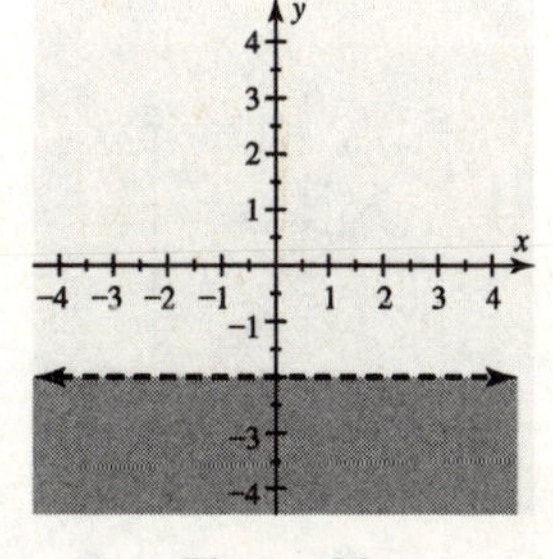

Figure 33

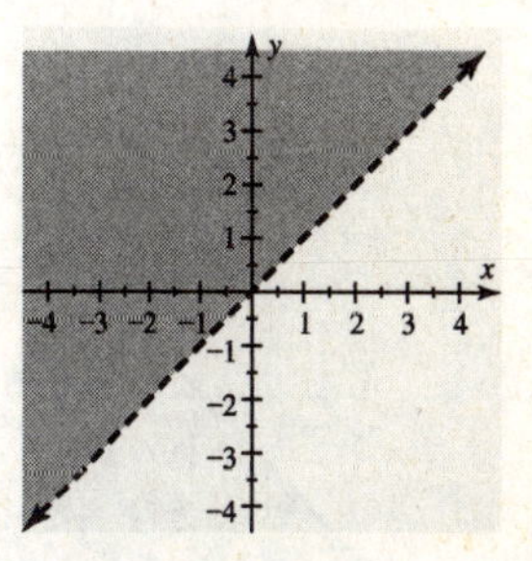

Figure 35

37. See Figure 37.

39. See Figure 39.

41. See Figure 41.

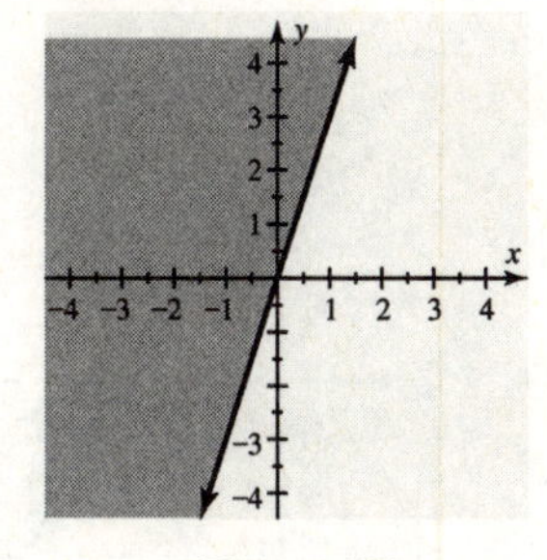

Figure 37

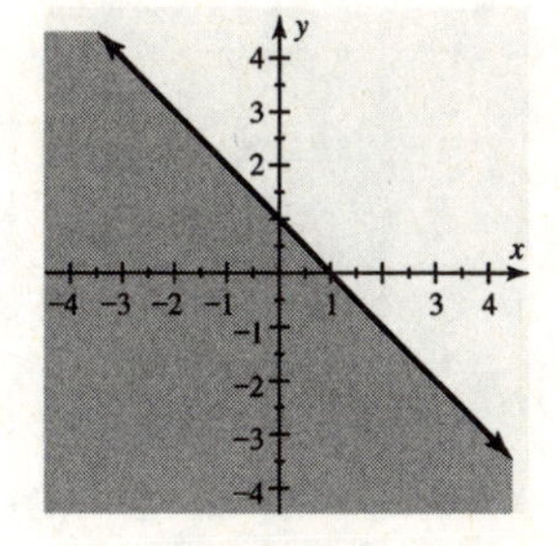

Figure 39

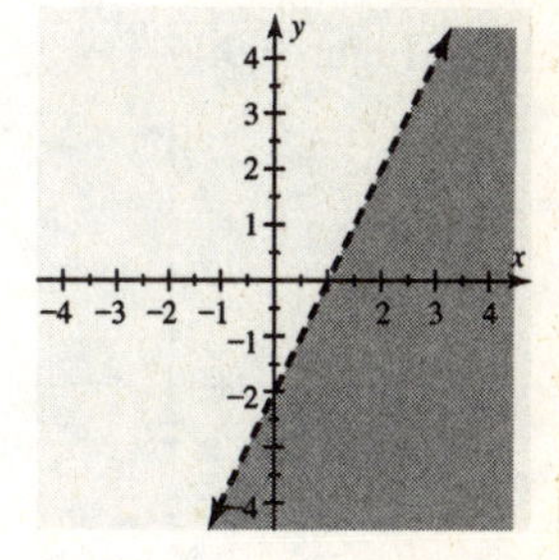

Figure 41

43. Yes, because $3-1<3 \Rightarrow 2<3$ is true and $3+1>3 \Rightarrow 4>3$ is true.

45. No, because $3(-2)-2(3)\geq 1 \Rightarrow -6-6\geq 1 \Rightarrow -12\geq 1$ is false.

47. Yes, because $4-2(-2)\geq 8 \Rightarrow 4+4\geq 8 \Rightarrow 8\geq 8$ is true and $-2(4)-5(-2)>0 \Rightarrow -8+10>0 \Rightarrow$ $2>0$ is true.

49. The region containing (1, 2), because $1\leq 2$ is true and $1+2\geq 2 \Rightarrow 3\geq 2$ is true.

51. The region containing (1, 0), because $1+0\leq 3 \Rightarrow 1\leq 3$ is true and $0\leq 2(1) \Rightarrow 0\leq 2$ is true.

53. See Figure 53.

55. See Figure 55.

57. See Figure 57.

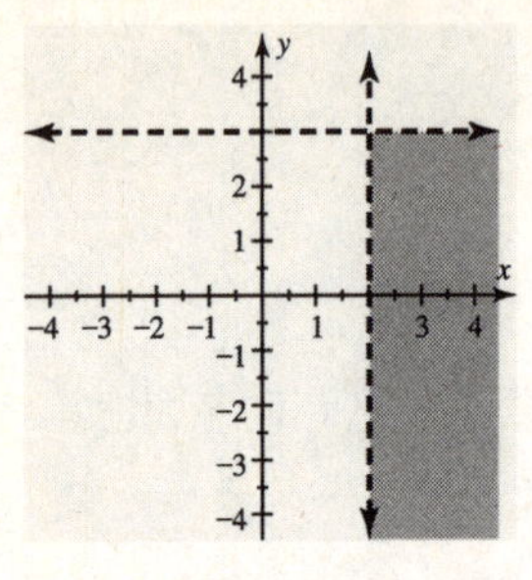

Figure 53

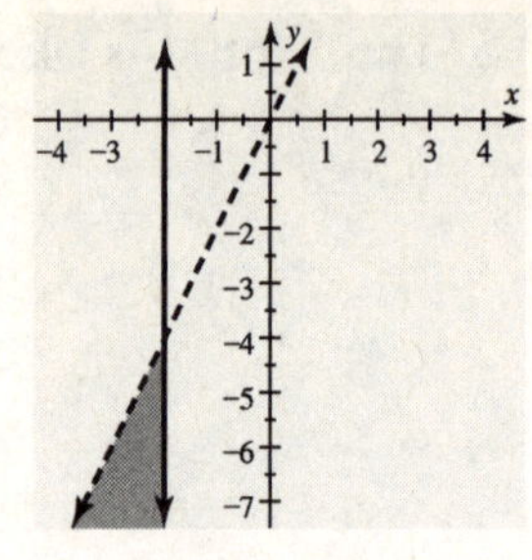

Figure 55

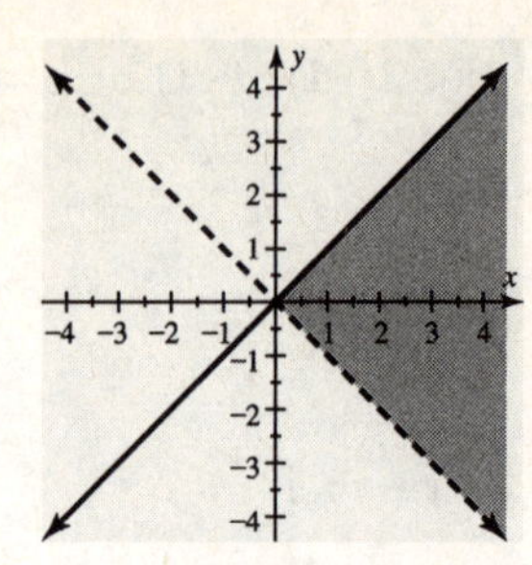

Figure 57

59. See Figure 59.

61. See Figure 61.

63. See Figure 63.

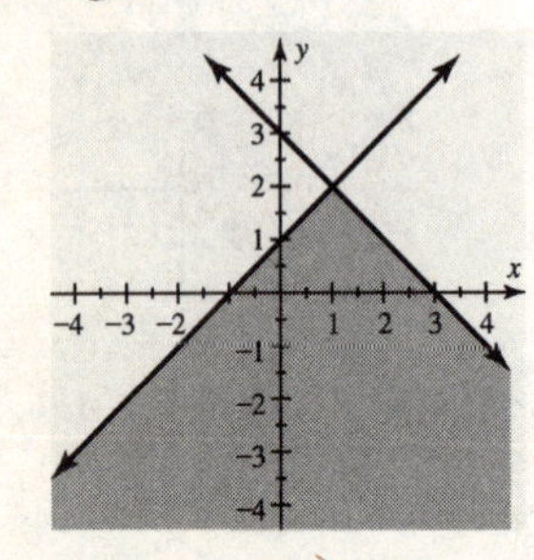

Figure 59

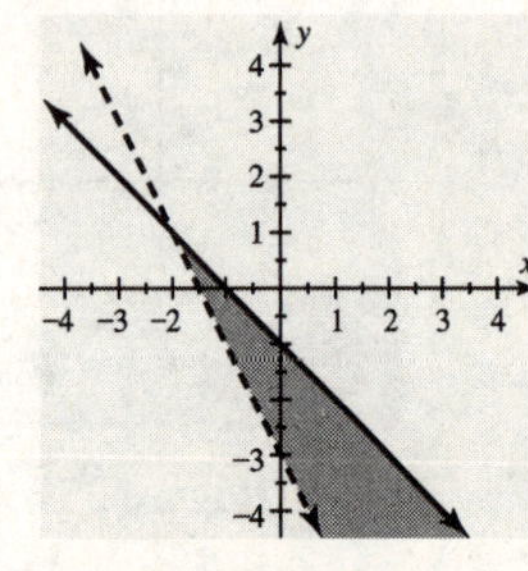

Figure 61

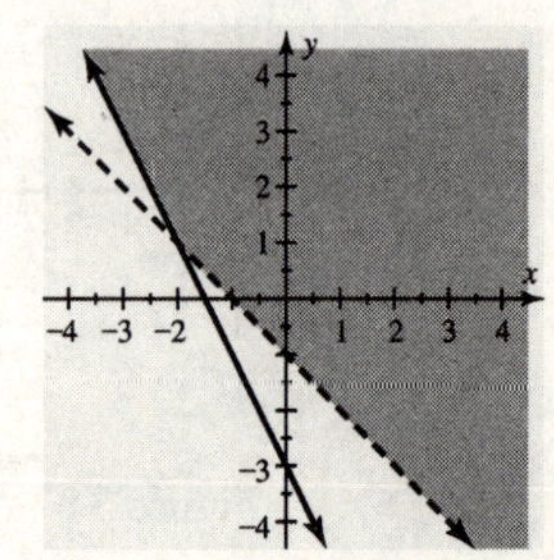

Figure 63

65. See Figure 65.

67. See Figure 67.

69. See Figure 69.

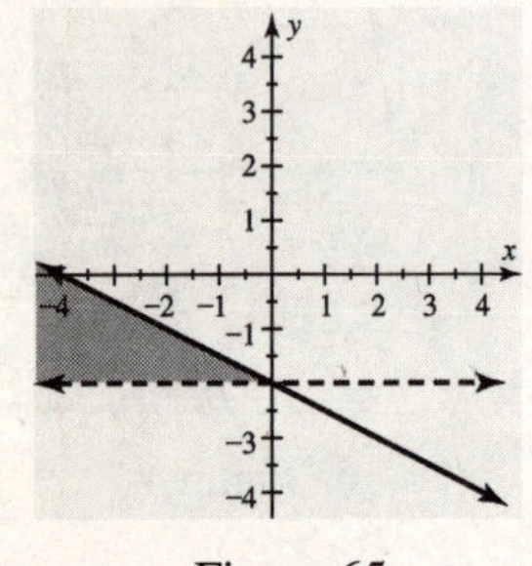

Figure 65

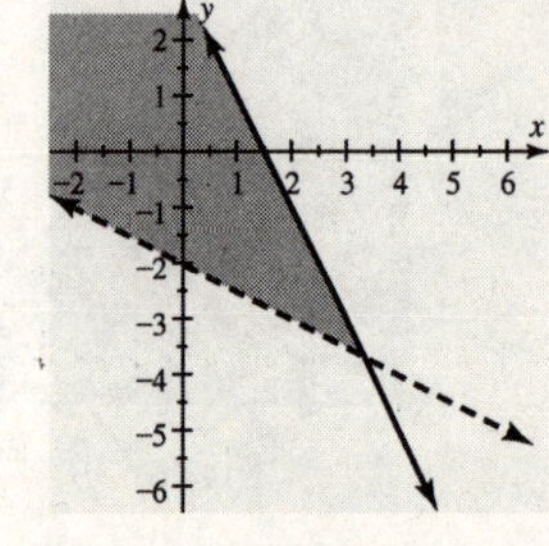

Figure 67

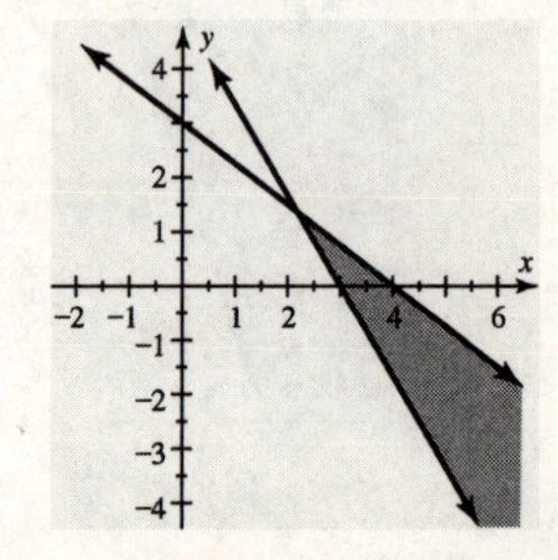

Figure 69

Applications

71. Graph the system, $R \geq 2P$ and $R + P < 90$. See Figure 71.

73. (a) $R = 220 - 20 \Rightarrow R = 200$ bpm for a 20 year old person and $R = 220 - 70 \Rightarrow R = 150$ bpm for a 70 year old person.

    (b) See Figure 73.

(c) The shaded region represents possible heart rates for ages 20 to 70.

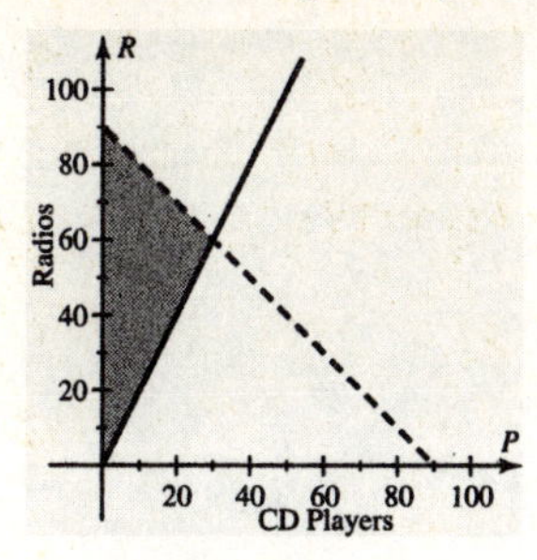

Figure 71

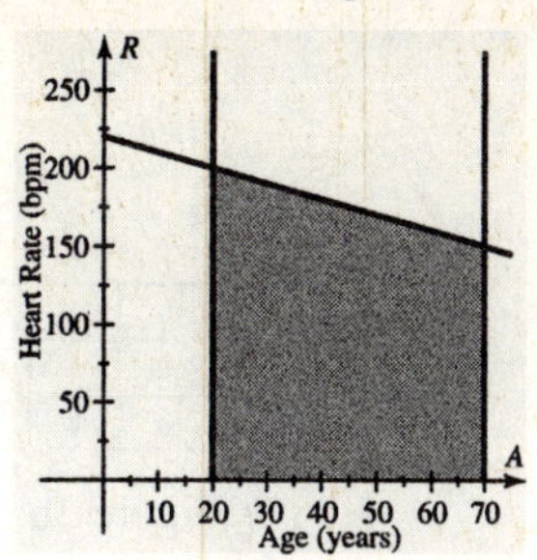

Figure 73

75. 150 to 200 lbs.

Checking Basic Concepts for Sections 4.3 & 4.4

1. Multiplying the second equation by -2 and adding the two equations will eliminate the variable x.

$$\begin{aligned} 2x+3y&=5 \\ -2x+14y&=12 \\ \hline 17y&=17 \end{aligned}$$

Thus $y=1$. And so $x-7(1)=-6 \Rightarrow x=1$. The solution is $(1, 1)$.

2. (a) Multiplying the first equation by -1 and adding the two equations will eliminate the variable x.

$$\begin{aligned} -x-y&=1 \\ x-2y&=2 \\ \hline -3y&=3 \end{aligned}$$

Thus $y=-1$. And so $x-2(-1)=2 \Rightarrow x+2=2 \Rightarrow x=0$. The solution is $(0, -1)$.

There is one solution.

(b) Adding the two equations will eliminate both variables.

$$\begin{aligned} 5x-6y&=4 \\ -5x+6y&=1 \\ \hline 0&=5 \end{aligned}$$

This is false, therefore there are no solutions.

(c) Multiplying the first equation by -2 and adding the two equations will eliminate both variables.

$$\begin{aligned} -2x+6y&=0 \\ 2x-6y&=0 \\ \hline 0&=0 \end{aligned}$$

This is true, therefore there are infinitely many solutions.

3. Substituting $y=2x$ into the first equation yields the following: $-2x+(2x)=0 \Rightarrow 0=0$. This is true $\Rightarrow$ infinitely many solutions. Graph the system. See Figure 3a. Line coincides $\Rightarrow$ infinitely many solutions. Numerically: See Figure 3b. Since for all values of x both equations produce the same solutions $\Rightarrow$ infinitely many solutions.

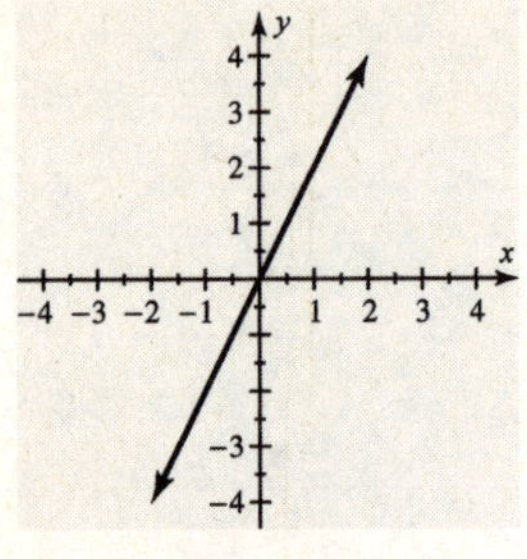

Figure 3a

x	0	1	2	3
$y=2x$	0	2	4	6
$-2x+y=0$	0	2	4	6

Figure 3b

4. (a) See Figure 4a.

(b) See Figure 4b.

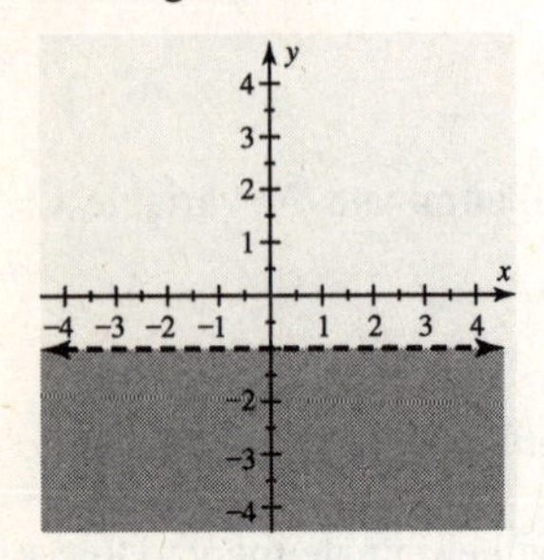

Figure 4a

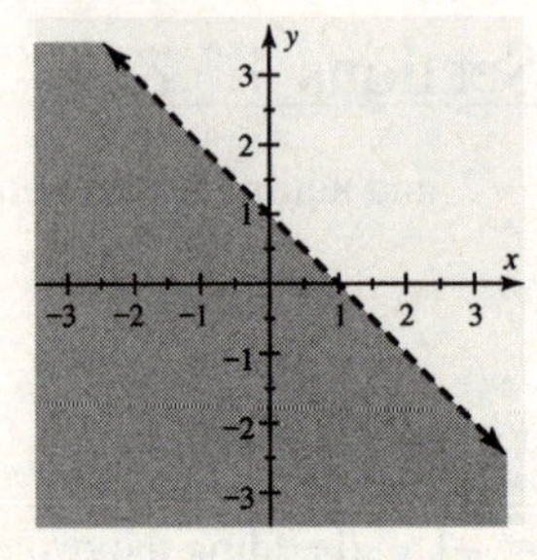

Figure 4b

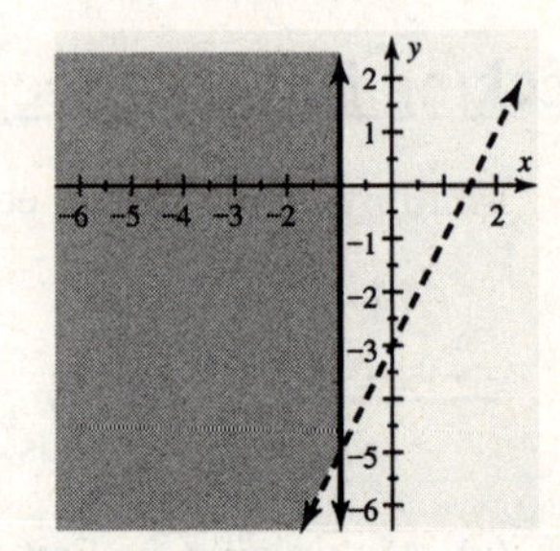

Figure 5

5. See Figure 5.

6. (a) $x+y=11$

$x=y+5$ or $x-y=5$.

(b) Substituting $x=y+5$ into the first equation yields the following: $(y+5)+y=11 \Rightarrow 2y+5=11 \Rightarrow 2y=6 \Rightarrow y=3$. And so $x+3=11 \Rightarrow x=8$. The solution is (8, 3) or New York had a population of 8 million people and Chicago had a population of 3 million people.

Chapter 4 Review Exercises

Section 4.1

1. Graph the system $y=3$ and $y=2x-3$. See Figure 1. The intersection and solution is $(3,3) \Rightarrow x=3$.

2. Graph the system $y=3$ and $y=\frac{3}{2}x$. See Figure 2. The intersection and solution is $(2, 3) \Rightarrow x=2$.

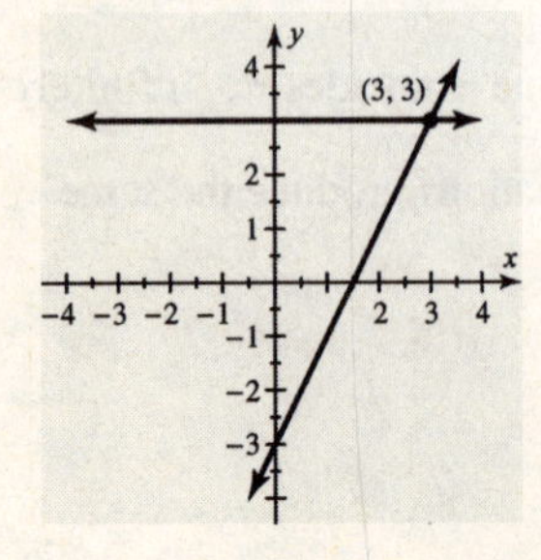

Figure 1

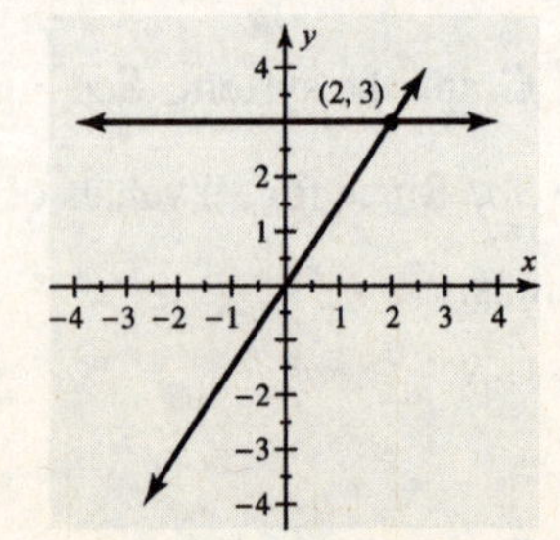

Figure 2

3. $(1, 2)$, because $1+2(2)=5 \Rightarrow 1+4=5$ is true and $1-(2)=-1$ is true.

4. $(5,\ 2)$, because $2(5)-2=8 \Rightarrow 10-2=8$ is true and $5+3(2)=11 \Rightarrow 5+6=11$ is true.

5. $(4,\ 3)$, because $\frac{1}{2}(4)=3-1 \Rightarrow 2=2$ is true and $2(4)=3(3)-1 \Rightarrow 8=9-1$ is true.

6. $(2,-4)$, because $5(2)-2(-4)=18 \Rightarrow 10+8=18$ is true and $-4=-2(2)$ is true.

7. The intersection and solution is (2, 2). Checking: $2=2$ is true and $-2(2)+2=-2 \Rightarrow -4+2=-2$ is true.

8. The intersection and solution is (1, 2). Checking: $1+2=3$ is true and $2=2(1)$ is true.

9. Since when $x=2,\ y=6$ for both equations $\Rightarrow$ the solution is $(2,\ 6)$.

10. Since when $x=1,\ y=1$ for both equations $\Rightarrow$ the solution is $(1,\ 1)$.

11. Graph the system. See Figure 11. The intersection and solution is $(4,-3)$.

12. Graph the system. See Figure 12. The intersection and solution is (1, 2).

13. Graph the system. See Figure 13. The intersection and solution is (1, 1).

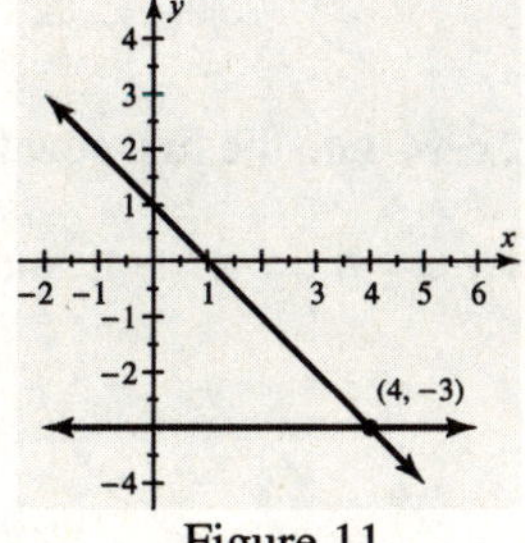

Figure 11

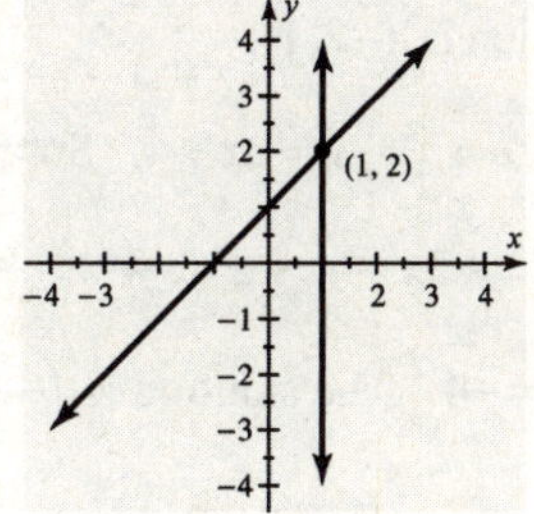

Figure 12

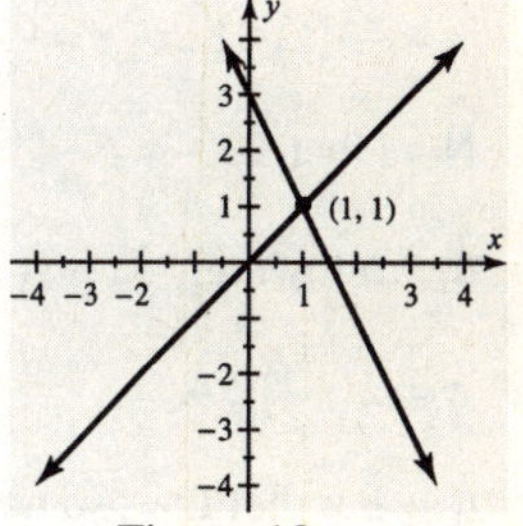

Figure 13

14. Graph the system. See Figure 14. The intersection and solution is (1, 2).

15. Graph the system. See Figure 15. The intersection and solution is (1, 1).

16. Graph the system. See Figure 16. The intersection and solution is $(-2,-1)$.

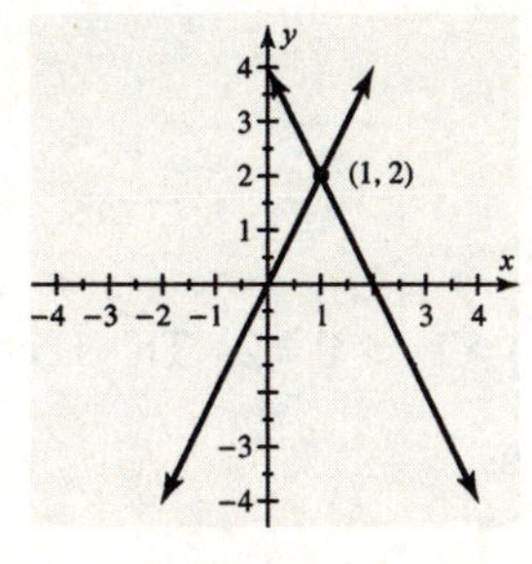

Figure 14

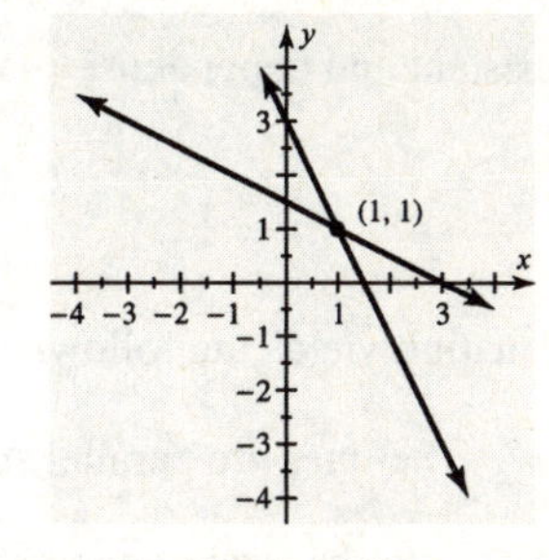

Figure 15

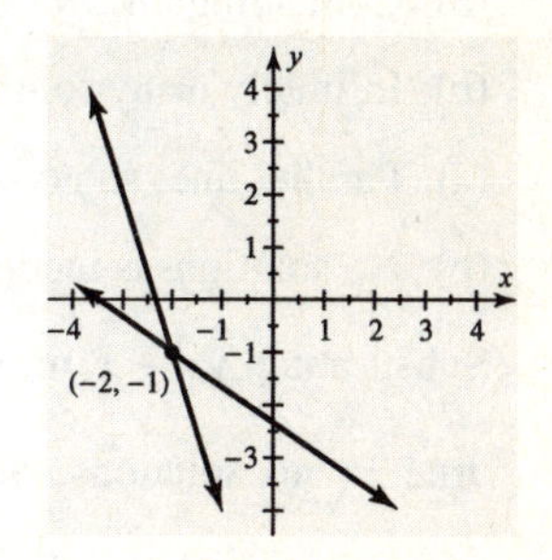

Figure 16

Section 4.2

17. Substituting $y=3x$ into the first equation yields the following: $x+(3x)=8 \Rightarrow 4x=8 \Rightarrow x=2$, and so $y=3(2) \Rightarrow y=6$. The solution is $(2,\ 6)$.

18. Substituting $y=-5x$ into the first equation yields the following: $x-2(-5x)=22 \Rightarrow x+10x=22 \Rightarrow 11x=22 \Rightarrow x=2$, and so $y=-5(2) \Rightarrow y=-10$. The solution is $(2,-10)$.

19. Note that $2x+y=5 \Rightarrow y=-2x+5$. Substituting $y=-2x+5$ into the second equation yields the following: $-3x+(-2x+5)=0 \Rightarrow -5x+5=0 \Rightarrow -5x=-5 \Rightarrow x=1$, and so $y=-2(1)+5 \Rightarrow y=3$. The solution is (1, 3).

20. Note that $x-y=-5 \Rightarrow x=y-5$. Substituting $x=y-5$ into the first equation yields the following: $3(y-5)-y=5 \Rightarrow 3y-15-y=5 \Rightarrow 2y=20 \Rightarrow y=10$, and so $x=10-5 \Rightarrow x=5$. The solution is (5, 10).

21. Note that $x+3y=1 \Rightarrow x=-3y+1$. Substituting $x=-3y+1$ into the second equation yields the following: $-2(-3y+1)+2y=6 \Rightarrow 6y-2+2y=6 \Rightarrow 8y-2=6 \Rightarrow 8y=8 \Rightarrow y=1$, and so $x=-3(1)+1 \Rightarrow x=-2$. The solution is $(-2,1)$.

22. Note that $2x-y=-4 \Rightarrow -y=-2x-4 \Rightarrow y=2x+4$. Substituting $y=2x+4$ into the first equation yields the following: $3x-2(2x+4)=-4 \Rightarrow 3x-4x-8=-4 \Rightarrow -x-8=-4 \Rightarrow -x=4 \Rightarrow x=-4$, and so $y=2(-4)+4 \Rightarrow y=-8+4 \Rightarrow y=-4$. The solution is $(-4,-4)$.

23. (a) Parallel lines so no solutions.
    (b) No solutions so inconsistent.

24. (a) An intersection so one solution.
    (b) One solution is consistent and independent.

25. (a) Coinciding lines so infinitely many solutions.
    (b) Infinitely many solutions is consistent and dependent.

26. (a) Parallel lines so no solutions.
    (b) No solutions is inconsistent.

27. Substituting $y=-x$ into the first equation yields the following: $x+(-x)=2 \Rightarrow 0=2$. This is not true $\Rightarrow$ no solutions. See Figure 27. The lines are parallel so no solutions.

28. Substituting $x+y=-2 \Rightarrow y=-x-2$ into the second equation yields the following: $x+(-x-2)=3 \Rightarrow -2=3$. This is not true $\Rightarrow$ no solutions. See Figure 28. The lines are parallel so no solutions.

29. Note that $x-2y=-2 \Rightarrow x=2y-2$. Substituting $x=2y-2$ into the first equation yields the following: $-(2y-2)+2y=2 \Rightarrow -2y+2+2y=2 \Rightarrow 2=2$. This is true $\Rightarrow$ infinitely many solutions. See Figure 29. The lines coincide, so there are infinitely many solutions.

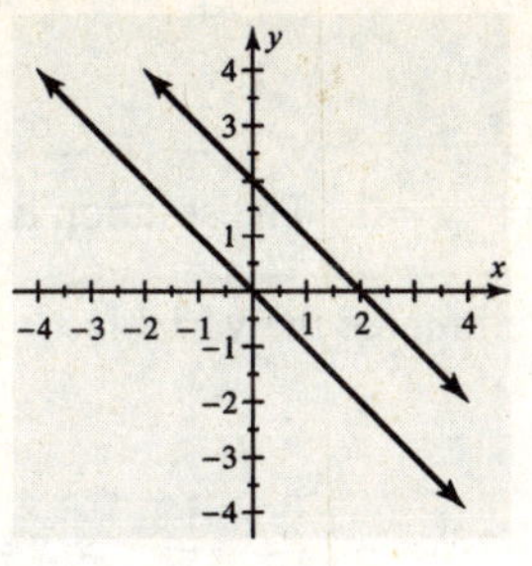

Figure 27

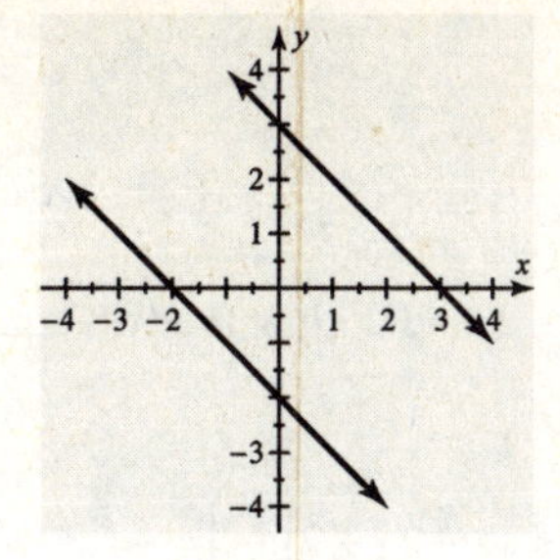

Figure 28

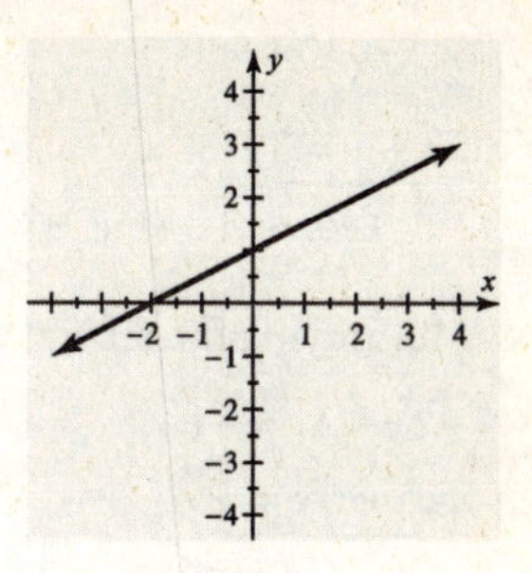

Figure 29

30. Note that $-x-y=-2 \Rightarrow -y=x-2 \Rightarrow y=-x+2$. Substituting $y=-x+2$ into the second equation yields the following: $2x-(-x+2)=1 \Rightarrow 2x+x-2=1 \Rightarrow 3x=3 \Rightarrow x=1$, and so $y=-1+2 \Rightarrow y=1$. The solution is $(1, 1)$. See Figure 30. The lines intersect at the solution (1, 1).

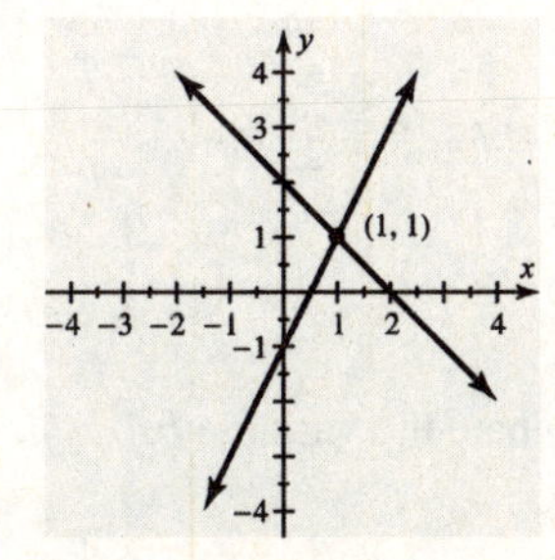

Figure 30

Section 4.3

31. The lines intersect at the solution (2, 1). Adding the two equations will eliminate the variable *y*.

$$\begin{array}{r} x+y=3 \\ \underline{x-y=1} \\ 2x=4 \end{array}$$

Thus $x=2$. And so $2+y=3 \Rightarrow y=1$. The solution is (2, 1).

32. The lines intersect at the solution $(-1, 2)$. Multiplying the second equation by -2 and adding the two equations will eliminate the variable *x*.

$$\begin{array}{r} 2x+3y=4 \\ \underline{-2x+4y=10} \\ 7y=14 \end{array}$$

Thus $y=2$. And so $2x+3(2)=4 \Rightarrow 2x+6=4 \Rightarrow 2x=-2 \Rightarrow x=-1$. The solution is $(-1, 2)$.

33. Adding the two equations will eliminate the variable *y*.

$$\begin{array}{r} x+y=10 \\ \underline{x-y=12} \\ 2x=22 \end{array}$$

Thus $x=11$. And so $11+y=10 \Rightarrow y=-1$. The solution is $(11,-1)$.

34. Adding the two equations will eliminate the variable *y*.

$$\begin{array}{r} 2x-y=2 \\ 3x+y=3 \\ \hline 5x=5 \end{array}$$

Thus $x=1$. And so $2(1)-y=2 \Rightarrow 2-y=2 \Rightarrow -y=0 \Rightarrow y=0$. The solution is (1, 0).

35. Multiplying the second equation by 2 and adding the two equations will eliminate the variable x.

$$\begin{array}{r} -2x+2y=-1 \\ 2x-6y=-6 \\ \hline -4y=-7 \end{array}$$

Thus $y=\frac{7}{4}$. And so $-2x+2\left(\frac{7}{4}\right)=-1 \Rightarrow -2x+\frac{14}{4}=-1 \Rightarrow -2x=-\frac{18}{4} \Rightarrow x=\frac{9}{4}$.

The solution is $\left(\frac{9}{4},\frac{7}{4}\right)$.

36. Multiplying the first equation by -1 and adding the two equations will eliminate the variable x.

$$\begin{array}{r} -2x+5y=0 \\ 2x+4y=9 \\ \hline 9y=9 \end{array}$$

Thus $y=1$. And so $2x+4(1)=9 \Rightarrow 2x+4=9 \Rightarrow 2x=5 \Rightarrow x=\frac{5}{2}$.

The solution is $\left(\frac{5}{2},1\right)$.

37. Multiplying the first equation by 2 and adding the two equations will eliminate the variable b.

$$\begin{array}{r} 4a+2b=6 \\ -3a-2b=-1 \\ \hline a=5 \end{array}$$

And so $4(5)+2b=6 \Rightarrow 20+2b=6 \Rightarrow 2b=-14 \Rightarrow b=-7$. The solution is $(5,-7)$.

38. Multiplying the first equation by -3 and adding the two equations will eliminate the variable a.

$$\begin{array}{r} -3a+9b=-6 \\ 3a+b=26 \\ \hline 10b=20 \end{array}$$

Thus $b=2$. And so $3a+2=26 \Rightarrow 3a=24 \Rightarrow a=8$. The solution is (8, 2).

39. Multiplying the first equation by 2, the second equation by 5 and adding the two equations will eliminate the variable r.

$$\begin{array}{r} 10r+6t=-2 \\ -10r-25t=-55 \\ \hline -19t=-57 \end{array}$$

Thus $t=3$. And so $10r+6(3)=-2 \Rightarrow 10r+18=-2 \Rightarrow 10r=-20 \Rightarrow r=-2$.

The solution is $(-2, 3)$.

40. Multiplying the first equation by 7, the second equation by 2 and adding the two equations will eliminate the variable t.

$$\begin{array}{r} 35r+14t=35 \\ 6r-14t=6 \\ \hline 41r=41 \end{array}$$

Thus $r=1$. And so $6(1)-14t=6 \Rightarrow 6-14t=6 \Rightarrow -14t=0 \Rightarrow t=0$.

The solution is $(1, 0)$.

41. (a) Adding the two equations will eliminate the variable y.

$$\begin{aligned} 3x + y &= 6 \\ x - y &= -2 \\ \hline 4x &= 4 \end{aligned}$$

Thus $x = 1$. And so $1 - y = -2 \Rightarrow -y = -3 \Rightarrow y = 3$. The solution is (1, 3).

(b) See Figure 41b. The intersection and solution is (1, 3).

(c) See Figure 41c. Since when $x = 1$, $y = 3$ for both equations the solution is (1, 3).

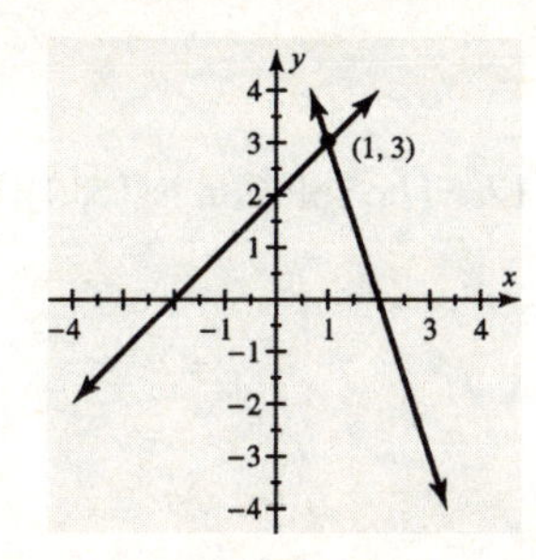

Figure 41b

x	−1	0	1	2
$y = 6 - 3x$	9	6	3	0
$y = x + 2$	1	2	3	4

Figure 41c

42. (a) Multiplying the second equation by 2 and adding the two equations will eliminate the variable x.

$$\begin{aligned} 2x + y &= 3 \\ -2x + 4y &= -8 \\ \hline 5y &= -5 \end{aligned}$$

Thus $y = -1$. And so $2x - 1 = 3 \Rightarrow 2x = 4 \Rightarrow x = 2$. The solution is $(2, -1)$.

(b) See Figure 42b. The intersection and solution is $(2, -1)$.

(c) See Figure 42c. Since when $x = 2$, $y = -1$ for both equations the solution is $(2, -1)$.

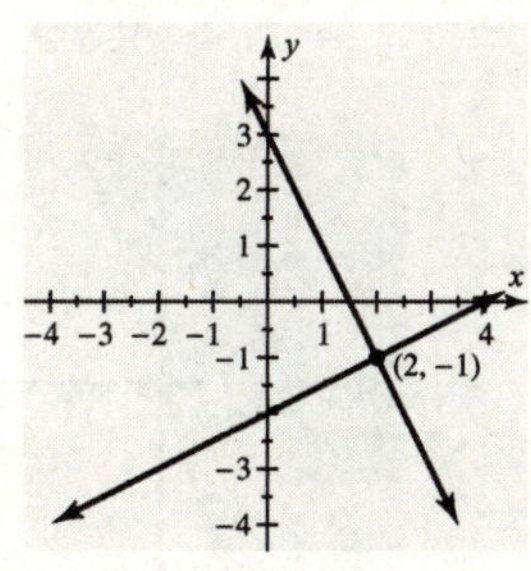

Figure 42b

x	−2	0	2	4
$y = 3 - 2x$	7	3	−1	−5
$y = (x - 4)/2$	−3	−2	−1	0

Figure 42c

43. Adding the two equations will eliminate both variables.

$$\begin{aligned} x - y &= 5 \\ -x + y &= -5 \\ \hline 0 &= 0 \end{aligned}$$

Since this is true, there are infinitely many solutions.

44. Multiplying the second equation by 3 and adding the two equations will eliminate both variables.

$$\begin{aligned} 3x - 3y &= 0 \\ -3x + 3y &= 0 \\ \hline 0 &= 0 \end{aligned}$$

Since this is true, there are infinitely many solutions.

45. Adding the two equations will eliminate both variables.

$$\begin{array}{r} -2x+y=3 \\ 2x-y=3 \\ \hline 0=6 \end{array}$$ Since this is false, there are no solutions.

46. Adding the two equations will eliminate the variable y.

$$\begin{array}{r} -2x+y=2 \\ 3x-y=3 \\ \hline x=5 \end{array}$$

And so $3(5)-y=3 \Rightarrow 15-y=3 \Rightarrow -y=-12 \Rightarrow y=12$. The solution is $(5,\ 12)$, therefore there is one solution.

Section 4.4

47. Yes, because $-3 \le 2$ is true.

48. No, because $-1 > -1$ is false.

49. No, because $1+2 < -2$ is false.

50. Yes, because $2(1)-3(-4) \ge 2 \Rightarrow 2+12 \ge 2$ is true.

51. $y > 1$

52. $y \le 2x+1$

53. See Figure 53.

54. See Figure 54.

55. See Figure 55.

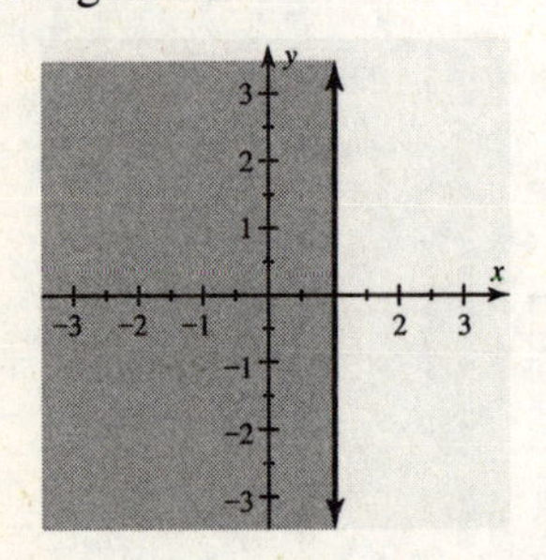

Figure 53

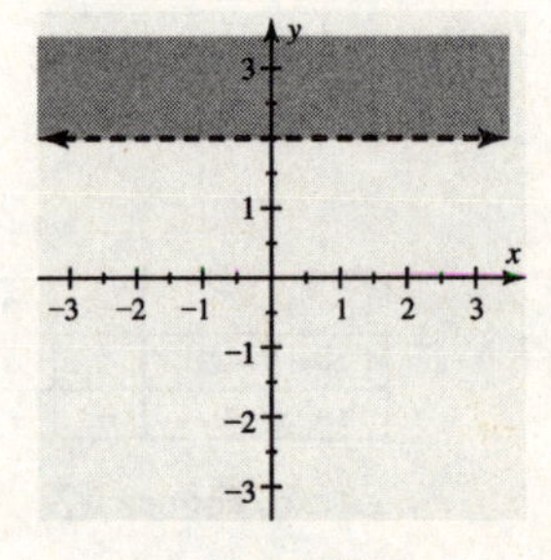

Figure 54

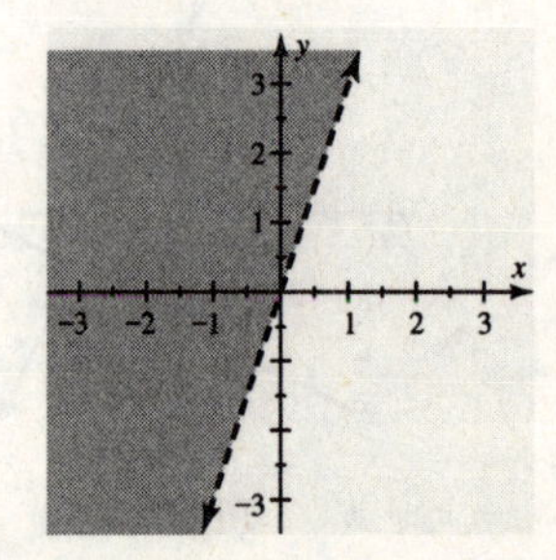

Figure 55

56. See Figure 56.

57. See Figure 57.

58. See Figure 58.

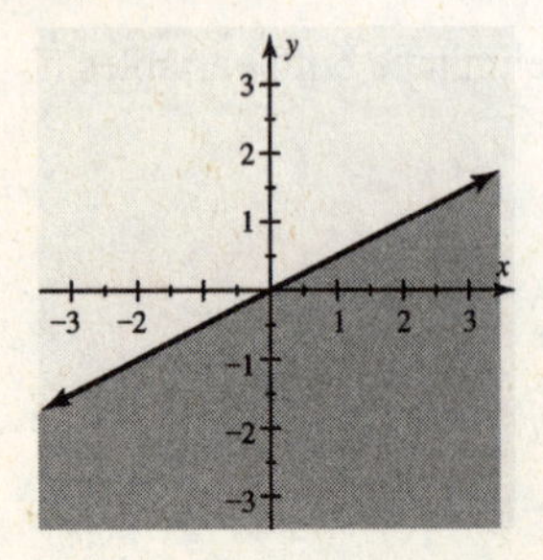

Figure 56

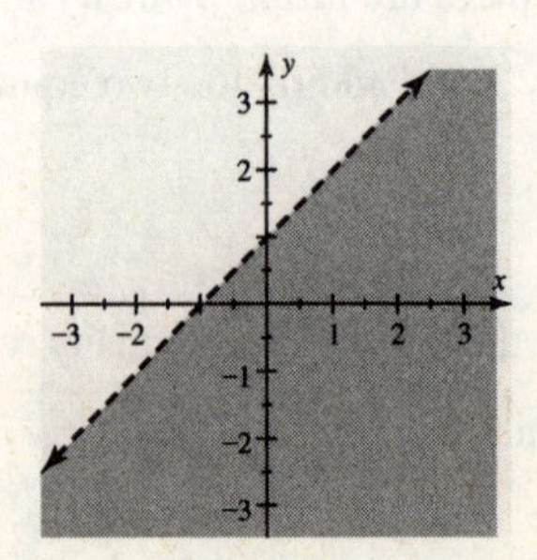

Figure 57

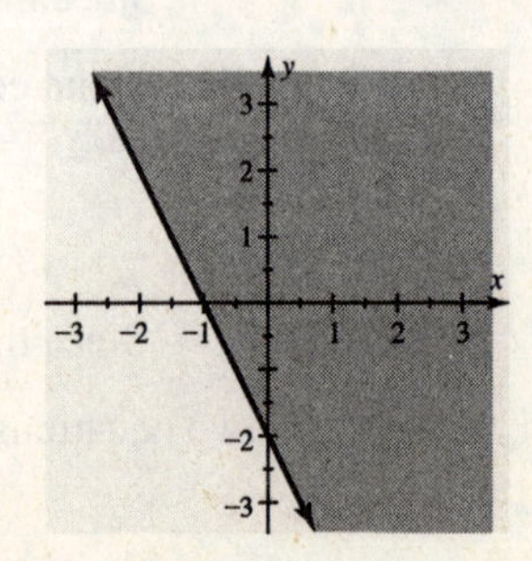

Figure 58

59. Yes, because $1-2(-2)>3 \Rightarrow 1+4>3$ is true and $2(1)+(-2)<3 \Rightarrow 0<3$ is true.

60. No, because $4-(-3)\geq 1 \Rightarrow 4+3\geq 1$ is true but $4(4)+3(-3)\leq 4 \Rightarrow 16-9\leq 4$ is false.

61. The region containing $(2,-2)$, because $-2\leq 1$ is true and $2(2)+(-2)\geq -1 \Rightarrow 4-2\geq -1$ is true.

62. The region containing $(1, 3)$, because $3\geq 1$ is true and $1+3\geq 2$ is true.

63. See Figure 63.

64. See Figure 64.

65. See Figure 65.

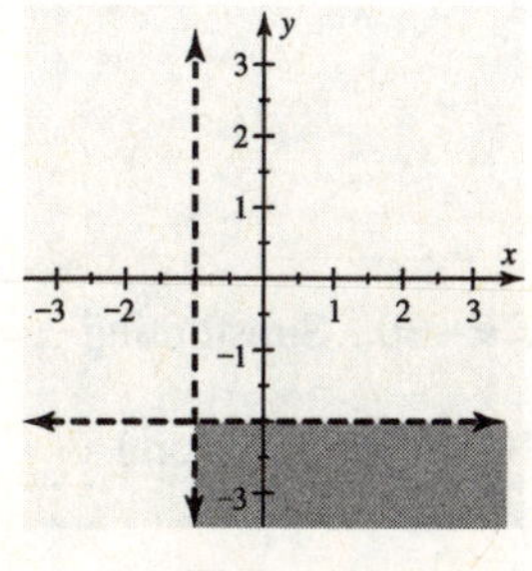

Figure 63

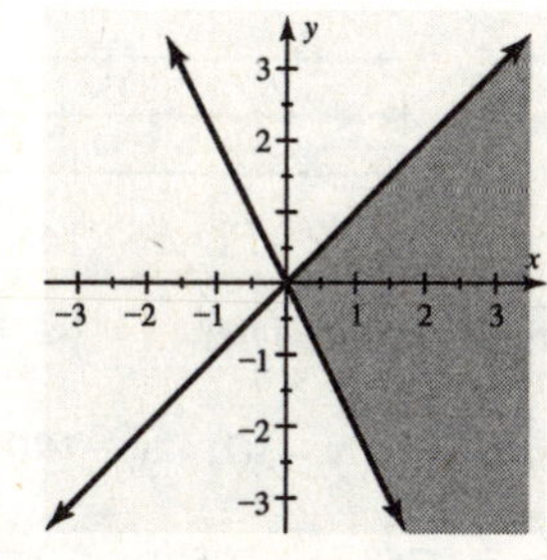

Figure 64

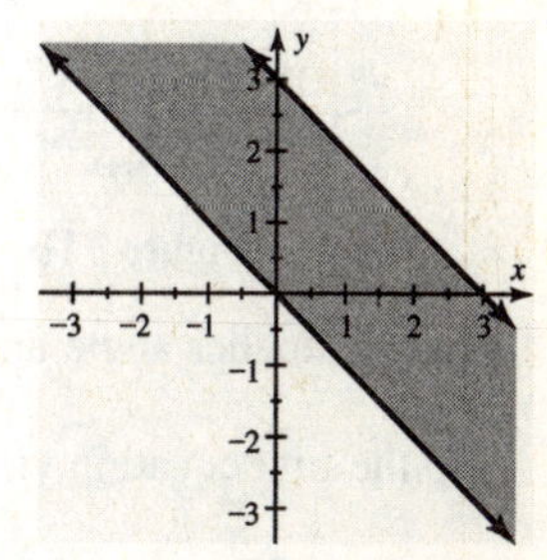

Figure 65

66. See Figure 66.

67. See Figure 67.

68. See Figure 68.

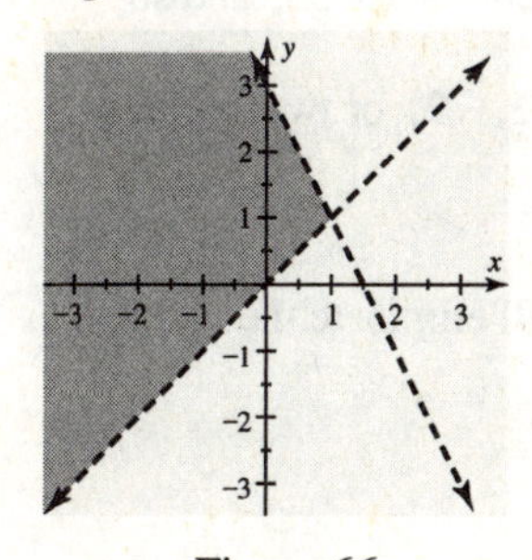

Figure 66

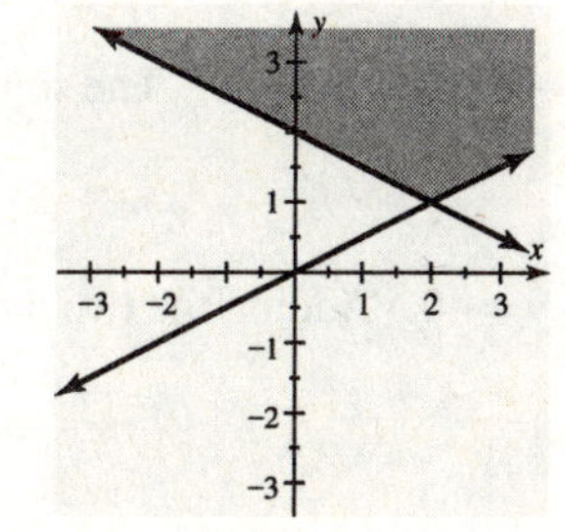

Figure 67

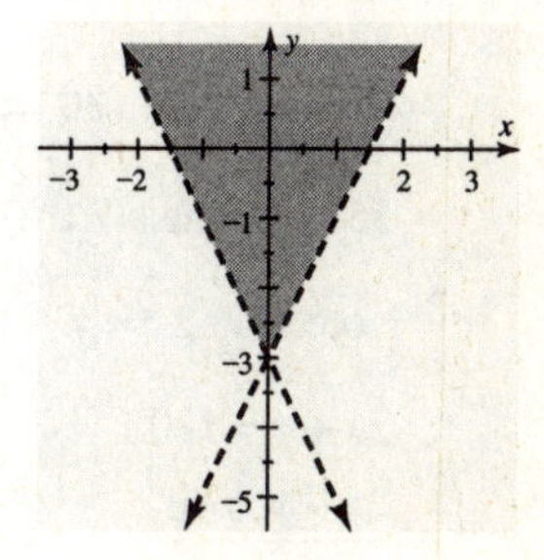

Figure 68

Applications

69. Let $x =$ number of motor vehicle deaths in 1912 and let $y =$ number of motor vehicle deaths in 2003. Then $13.75x = y$ and $y = x+39{,}525$. Substituting $13.75x = y$ into the second equation yields the following: $13.75x = x+39{,}525 \Rightarrow 12.75x = 39{,}525 \Rightarrow x = 3100$, and so $y = 3100+39{,}525 \Rightarrow y = 42{,}625$. Therefore 3100 deaths in 1912 and 42,625 deaths in 2003.

70. Let $x =$ lung cancer cases in men and let $y =$ lung cancer cases in women. Then $x+y = 185{,}000$ and $x = y+20{,}000$. Substituting $x = y+20{,}000$ into the first equation yields the following: $(y+20{,}000)+y = 185{,}000 \Rightarrow 2y = 165{,}000 \Rightarrow y = 82{,}500$, and so $x = 82{,}500+20{,}000 \Rightarrow x = 102{,}500$. Therefore men will have 102,500 cases of lung cancer reported and women will have 82,500 cases of lung cancer reported.

71. (a) $C = 0.2x + 40$

(b) See Figure 71b. The intersection and solution is (250, 90) or 250 miles.

(c) See Figure 71c. Since when

$x = 250,\ C = 90$ for both equations, the solution is $(250,\ 90) \Rightarrow$ 250 miles.

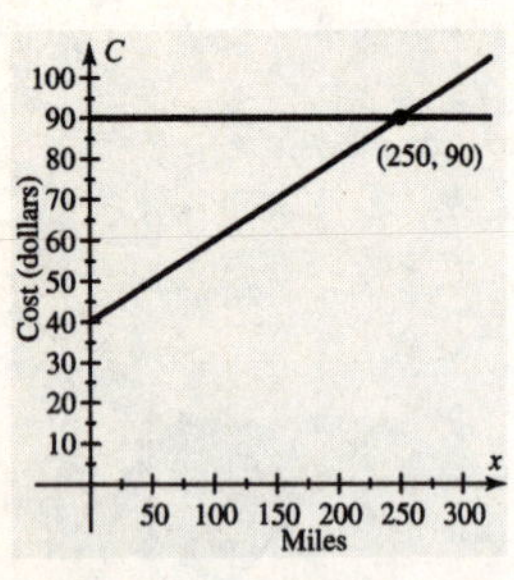

x	150	200	250	300
$C = 0.2x + 40$	70	80	90	100

Figure 71b Figure 71c

72. Let x = smaller angle and let y = larger angle. Then $x + y = 180$ and $x = y - 30$. Substituting $x = y - 30$ into the first equation yields the following: $(y - 30) + y = 180 \Rightarrow 2y = 210 \Rightarrow y = 105,$ and so $x = 105 - 30 \Rightarrow x = 75$. The angles are 75° and 105°.

73. (a) $2x + y = 180$ and $2x = y + 40$

(b) Note that $2x + y = 180 \Rightarrow y = -2x + 180$. Substituting $y = -2x + 180$ into the second equation yields the following: $2x = (-2x + 180) + 40 \Rightarrow 2x = -2x + 220 \Rightarrow 4x = 220 \Rightarrow x = 55,$ and so $y = -2(55) + 180 \Rightarrow y = -110 + 180 \Rightarrow y = 70$. The solution is $(55,\ 70)$ or two angles at 55° and one at 70°.

(c) Note that $2x = y + 40 \Rightarrow 2x - y = 40$. Adding the two equations will eliminate the variable y.

$$\begin{array}{r} 2x + y = 180 \\ \underline{2x - y = 40} \\ 4x = 220 \end{array}$$

Thus $x = 55$. And so $2(55) = y + 40 \Rightarrow 110 = y + 40 \Rightarrow y = 70$. The solution is (55, 70) or two angles at 55° and one angle at 70°.

74. Let x = garden's width and let y = garden's length. Then $2x + 2y = 88$ and $y = x + 4$. Substituting $y = x + 4$ into the first equation yields the following: $2x + 2(x + 4) = 88 \Rightarrow 2x + 2x + 8 = 88 \Rightarrow 4x + 8 = 88 \Rightarrow 4x = 80 \Rightarrow x = 20,$ and so $y = 20 + 4 \Rightarrow y = 24$. The solution is $(20,\ 24)$. The dimensions of the garden are 20 feet × 24 feet.

75. (a) Let x = number of \$80 rooms and let y = number of \$120 rooms. Then $x + y = 10$ and $80x + 120y = 920$.

(b) Multiplying the first equation by -80 and adding the two equations will eliminate the variable x.

$$\begin{array}{r} -80x-80y=-800 \\ 80x+120y=920 \\ \hline 40y=120 \end{array}$$

Thus $y=3$. And so $x+3=10 \Rightarrow x=7$. The solution is $(7, 3)$ or 7 \$80 rooms and 3 \$120 rooms.

76. Let $x=$ pounds of \$2 candy and let $y=$ pounds of \$3 candy. Then $x+y=18$ and $2x+3y=47$. Multiplying the first equation by -2 and adding the two equations will eliminate the variable x.

$$\begin{array}{r} -2x-2y=-36 \\ 2x+3y=47 \\ \hline y=11 \end{array}$$

And so $x+11=18 \Rightarrow x=7$. The solution is $(7, 11)$ or 7 pounds of \$2 candy and 11 pounds of \$3 candy.

77. Let $x=$ minutes on the stationary bike and let $y=$ minutes on the stair climber. Then $x+y=60$ and $9x+11y=590$. Multiplying the first equation by -9 and adding the two equations will eliminate the variable x.

$$\begin{array}{r} -9x-9y=-540 \\ 9x+11y=590 \\ \hline 2y=50 \end{array}$$

Thus $y=25$. And so $x+25=60 \Rightarrow x=35$. The solution is $(35, 25)$ or 35 minutes on the bike and 25 minutes on the stair climber.

78. Let $x=$ speed of the boat and let $y=$ speed of the current. Then $10x+10y=140$ and $14x-14y=140$. Multiplying the first equation by 7, the second by 5 and adding the two equations will eliminate the variable y.

$$\begin{array}{r} 70x+70y=980 \\ 70x-70y=700 \\ \hline 140x=1680 \end{array}$$

Thus $x=12$. And so $10(12)+10y=140 \Rightarrow 120+10y=140 \Rightarrow 10y=20 \Rightarrow y=2$. The solution is $(12, 2)$. The current is 2 mph.

79. (a) The intersection and solution is approximately 16 feet $\times$ 24 feet.

(b) Using the system $2L+2W=80$ and $L=\frac{3}{2}W$ and substituting $L=\frac{3}{2}W$ into the first equation yields the following: $2\left(\frac{3}{2}W\right)+2W=80 \Rightarrow 3W+2W=80 \Rightarrow 5W=80 \Rightarrow W=16$, and so $L=\frac{3}{2}(16) \Rightarrow L=24$. The dimensions are 16 feet $\times$ 24 feet.

80. Let $W=$ number of wheels made and let $T=$ number of trailers made. Then $W+T\le 30$ and $W\ge 2T$. Graph this system and shade. See Figure 80.

81. (a) $T = 150 - 0.7(20) \Rightarrow T = 150 - 14 \Rightarrow T = 136$ bpm for a 20 year old person and

$T = 150 - 0.7(60) \Rightarrow T = 150 - 42 \Rightarrow T = 108$ bpm for a 60 year old person.

(b) See Figure 81.

(c) The shaded region represents target heart rates above 70% of the maximum heart rate for ages 20 to 60.

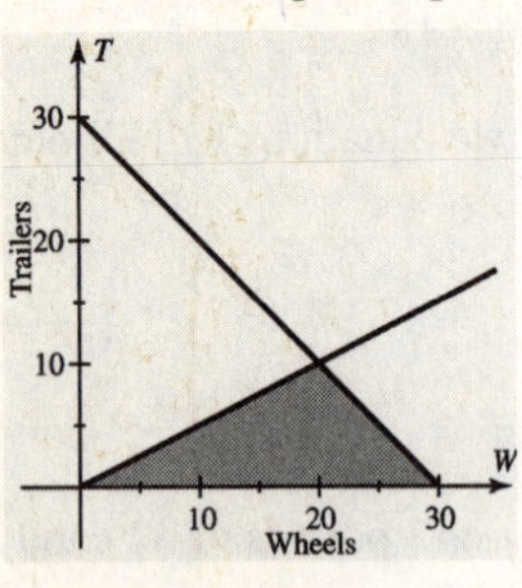

Figure 80

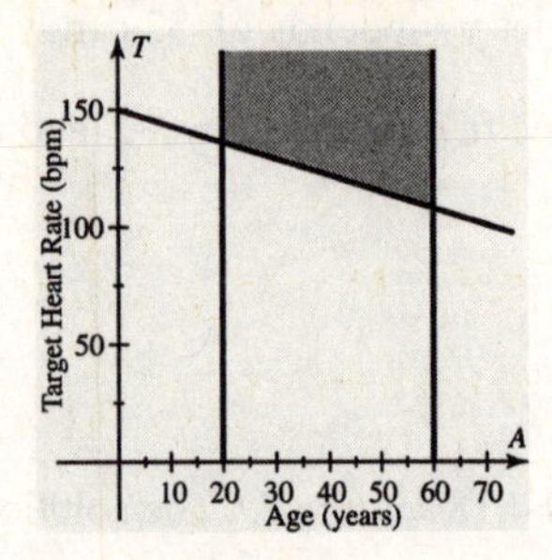

Figure 81

Chapter 4 Test

1. $(1, 2)$, because $3(1) + 2(2) = 7 \Rightarrow 3 + 4 = 7 \Rightarrow$ is true and $2(1) - 2 = 0$ is true.

2. The intersection and solution is $(-2, -1)$. Checking: $-2 + 4(-1) = -6 \Rightarrow -2 + (-4) = -6$ is true and $2(-2) + (-1) = -5 \Rightarrow -4 + (-1) = -5$.

3. Since when $x = -1$, $y = -2$ for both equations, the solution is $(-1, -2)$.

4. See Figure 4. The intersection and solution is $(-2, 3)$.

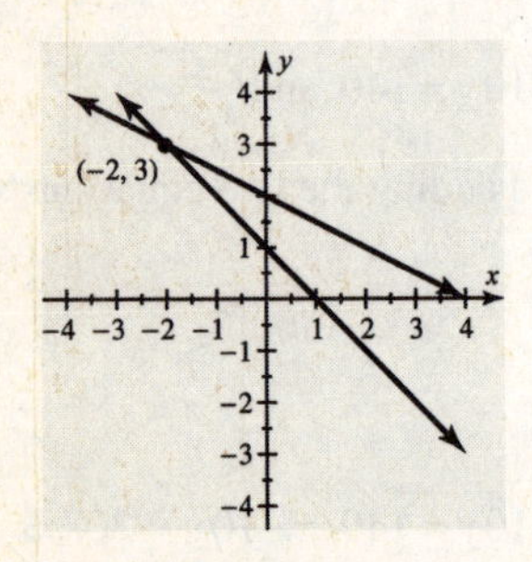

Figure 4

5. Substituting $y = 3x$ into the first equation yields the following: $3x + 2(3x) = 9 \Rightarrow 3x + 6x = 9 \Rightarrow 9x = 9 \Rightarrow x = 1$, and so $y = 3(1) \Rightarrow y = 3$. The solution is $(1, 3)$.

6. (a) Note that $x + 3y = 5 \Rightarrow x = -3y + 5$. Substituting $x = -3y + 5$ into the second equation yields the following: $3(-3y+5) - 2y = 4 \Rightarrow -9y + 15 - 2y = 4 \Rightarrow -11y + 15 = 4 \Rightarrow -11y = -11 \Rightarrow y = 1$, and so $x = -3(1) + 5 \Rightarrow x = -3 + 5 \Rightarrow x = 2$. The solution is $(2, 1) \Rightarrow$ one solution $\Rightarrow$ consistent.

(b) Note that $2x - y = -4 \Rightarrow -y = -2x - 4 \Rightarrow y = 2x + 4$. Substituting $y = 2x + 4$ into the first equation yields the following: $-x + \frac{1}{2}(2x+4) = 12 \Rightarrow -x + x + 2 = 12 \Rightarrow 2 = 12$. This is false $\Rightarrow$ no solutions $\Rightarrow$ inconsistent.

7. There is only one line, which indicates that the graphs are identical, or coincide.

   (a) There are infinitely many solutions.

   (b) The system is consistent and the equations are dependent.

8. The lines are parallel.

   (a) There are no solutions.

   (b) The system is inconsistent.

9. Adding the two equations will eliminate the variable y.

$$\begin{array}{r} x + 2y = 5 \\ \underline{3x - 2y = -17} \\ 4x = -12 \end{array}$$

Thus $x = -3$. And so $-3 + 2y = 5 \Rightarrow 2y = 8 \Rightarrow y = 4$. The solution is $(-3,\ 4)$.

10. Multiplying the second equation by 2 and adding the two equations will eliminate both variables.

$$\begin{array}{r} 2x - 2y = 3 \\ \underline{-2x + 2y = 10} \\ 0 = 13 \end{array}$$

This is false $\Rightarrow$ no solutions.

11. Multiply the first equation by 3 and then add.

$$\begin{array}{r} 3x - 6y = 9 \\ \underline{-3x + 6y = -9} \\ 0 = 0 \end{array}$$

The equation $0 = 0$ is always true, which indicates that the system has infinitely many solutions.

12. Multiply the first eqution by 2 and the second equation by 3 and then add.

$$\begin{array}{l} 8x + 6y = 10 \\ \underline{9x - 6y = -27} \\ 17x \quad = -17, \quad \text{or} \quad x = -1 \end{array}$$

Find y by substituting –1 for x in the first given equation, $4x + 3y = 5$.

$$4(-1) + 3y = 5 \Rightarrow -4 + 3y = 5 \Rightarrow -4 + 4 + 3y = 5 + 4 \Rightarrow 3y = 9 \Rightarrow \frac{3y}{3} = \frac{9}{3} \Rightarrow y = 3$$

The solution is $(-1, 3)$.

13. $y \leq -\dfrac{1}{2}x$

14. Point $(4, -3)$ is a solution because $2(4) + (-3) = 8 + (-3) = 5 > 3$ and $4 - (-3) = 7 \geq 7$.

15. See Figure 15.

16. See Figure 16.

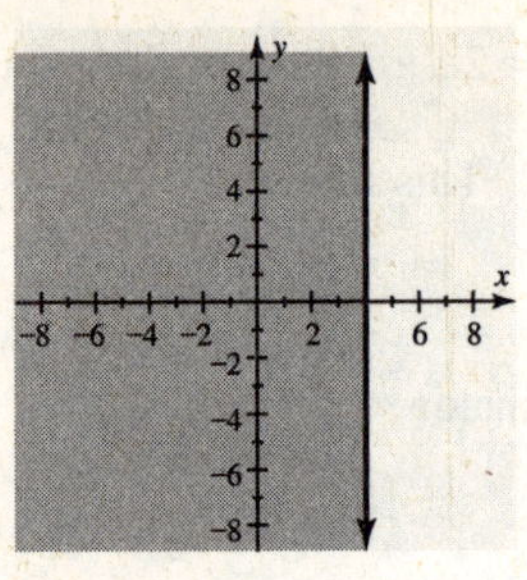

Figure 15

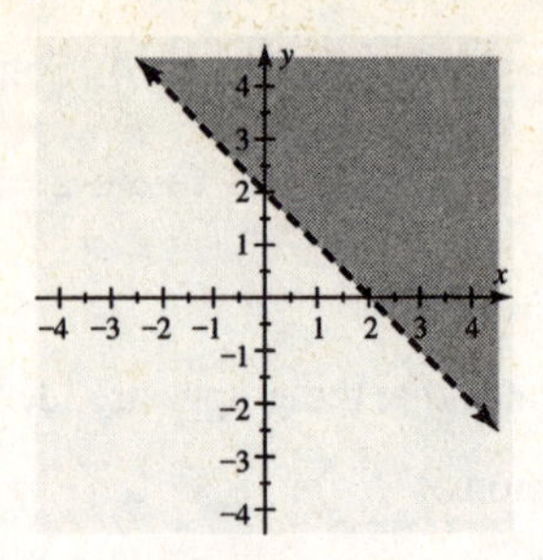

Figure 16

17. See Figure 17.

18. See Figure 18.

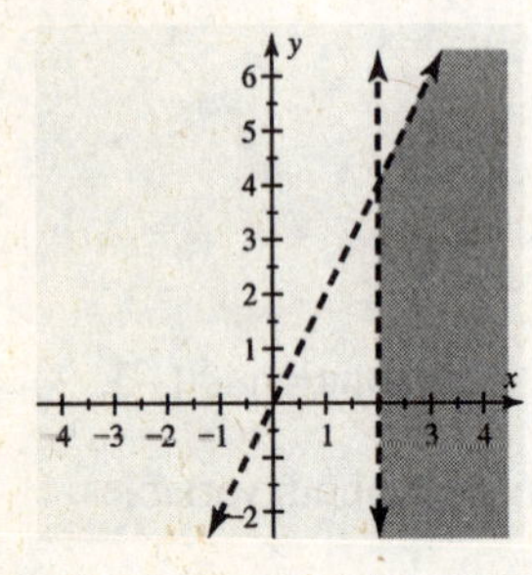

Figure 17

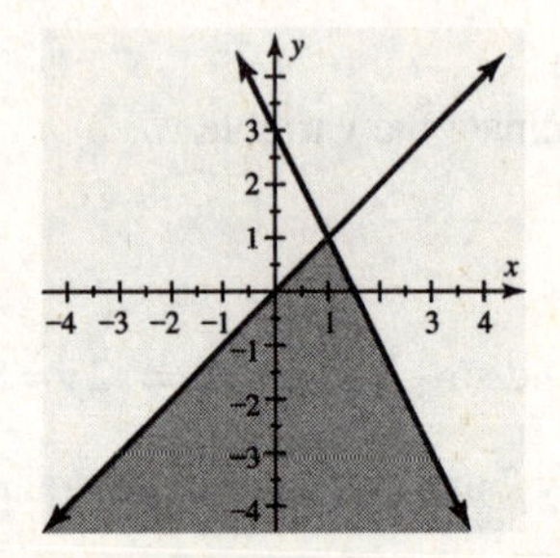

Figure 18

19. Let $x =$ taxes collected in 2003 and let $y =$ taxes collected in 2004. Then $x + y = 3.9$ and $y = x + 0.1$. Substituting $y = x + 0.1$ into the first equation yields the following: $x + (x + 0.1) = 3.9 \Rightarrow 2x + 0.1 = 3.9 \Rightarrow 2x = 3.8 \Rightarrow x = 1.9$, and so $y = 1.9 + 0.1 \Rightarrow y = 2.0$. The solution is (1.9, 2.0) or \$1.9 trillion collected in 2003 and \$2.0 trillion collected in 2004.

20. Let $x =$ hrs jogged at 6 mph and let $y =$ hrs jogged at 9 mph. Then $x + y = 1$ and $6x + 9y = 7$. Multiplying the first equation by -6 and adding the two equations will eliminate the variable x.

$$\begin{array}{r} -6x - 6y = -6 \\ 6x + 9y = 7 \\ \hline 3y = 1 \end{array}$$

Thus $y = \frac{1}{3}$. And so $x + \frac{1}{3} = 1 \Rightarrow x = \frac{2}{3}$. The solution is $\left(\frac{2}{3}, \frac{1}{3}\right)$ or $\frac{2}{3}$ hour at 6 miles per hour and $\frac{1}{3}$ hour at 9 miles per hour.

Chapter 4 Extended and Discovery Exercises

1. The forest has higher precipitation for higher temperatures than grasslands $\Rightarrow 7P - 5T \geq -70$.

2. The deserts have less precipitation for high temperatures than grasslands $\Rightarrow 35P - 3T \leq 140$.

3. The grassland is lower precipitation for higher temperatures than forests and higher than desert $\Rightarrow 7P - 5T \leq -70$ and $35P - 3T \geq 140$.

4. From the graph, grasslands would be the type of plant growth, and $7(14)-5(50)\le -70 \Rightarrow 98-250\le -70$ is true and $35(14)-3(50)\ge 140 \Rightarrow 490-150\ge 140$ is true $\Rightarrow$ yes.

5. -0.5; average the x-values -1 and 0 because the solution y-value 0 is half way between their corresponding y-values -1 and 1.

6. 0.5; average the x-values 0 and 1 because the solution y-value 5 is half way between their corresponding y-values 3 and 7.

7. 1.5; average the x-values 1 and 2 because the solution y-value 3.75 is half way between their corresponding y-values 3.5 and 4.

8. $\frac{1}{3}$; because 0 is one-third of the way between the y-values -1 and 2, you must choose a value one-third of the way between the corresponding x-values 0 and 1.

Chapters 1–4 Cumulative Review Exercises

1. $120 = 2\times 2\times 2\times 3\times 5$

2. (a) $2^3 \div \frac{5+7}{9-3} = 8\div\frac{12}{6} = 8\div 2 = 4$

   (b) $-\frac{2}{5}\cdot(5-25) = -\frac{2}{5}\cdot(-20) = \frac{-2}{5}\cdot\frac{-20}{1} = \frac{(-2)(-20)}{(5)(1)} = \frac{40}{5} = 8$

3. (a) -6.9 is a rational number.

   (b) $\sqrt{14}$ is an irrational number.

4. (a) $|-5| = 5$ and $-5 < 5$ so $-5 < |-5|$.

   (b) $|7| = 7$ and $|-1| = 1$ and $7 > 1$ so $|7| > |-1|$.

5. Associative (multiplication)

6. $30\cdot 102 = 30\cdot 100 + 30\cdot 2 = 3000 + 60 = 3060$

7. (a) $5x^2 - x^2 = (5-1)x^2 = 4x^2$

   (b) $3-2x+7x-5 = -2x+7x+3-5 = (-2+7)x+(3+-5) = 5x-2$

8. $5(2x+1) = 7+x \Rightarrow 10x+5 = 7+x \Rightarrow 10x-x+5 = 7+x-x$

   $\Rightarrow 9x+5 = 7 \Rightarrow 9x+5-5 = 7-5 \Rightarrow 9x = 2 \Rightarrow \frac{9x}{9} = \frac{2}{9} \Rightarrow x = \frac{2}{9}$

9. $1-(x+1) = x-1 \Rightarrow 1-x-1 = x-1 \Rightarrow -x = x-1 \Rightarrow -x-x = x-x-1 \Rightarrow -2x = -1 \Rightarrow \frac{-2x}{-2} = \frac{-1}{-2} \Rightarrow x = \frac{1}{2}$

10. First, solve for x: $2(5x+1) = 10x-3 \Rightarrow 10x+2 = 10x-3 \Rightarrow 10x-10x+2 = 10x-10x-3 \Rightarrow 2 = -3$. Because this statement is not true, the equation has no solutions.

11. Let x represent the smallest integer.

$$x+(x+1)+(x+2)+(x+3)=50 \Rightarrow 4x+6=50 \Rightarrow 4x+6-6=50-6 \Rightarrow 4x=44 \Rightarrow \frac{4x}{4}=\frac{44}{4} \Rightarrow x=11$$

Thus, the numbers are 11, 12, 13, and 14.

12. Convert feet to inches to keep the units consistent. Thus, $1 \text{ foot} = 1 \cdot 12 = 12$ inches. Then,

$A = LW = 36 \cdot 12 = 432 \text{ in}^2$. Or convert inches to feet to obtain $36 \text{ inches} = \frac{36}{12} = 3$ feet. Then,

$A = LW = 3 \cdot 1 = 3 \text{ ft}^2$.

13. The formula is given as $W = 3x - 7y$. To solve for x, proceed as follows:

$$W = 3x - 7y \Rightarrow W + 7y = 3x \Rightarrow \frac{W+7y}{3} = \frac{3x}{3} \Rightarrow x = \frac{W+7y}{3}$$

14. $3-(2x-7) \le 8x \Rightarrow 3-2x+7 \le 8x \Rightarrow 10-2x \le 8x \Rightarrow 10-2x+2x \le 8x+2x$

$$\Rightarrow 10 \le 10x \Rightarrow \frac{10}{10} \le \frac{10x}{10} \Rightarrow 1 \le x \Rightarrow x \ge 1$$

15. See Figure 15.

16. (a) See Figure 16a.

(b) See Figure 16b.

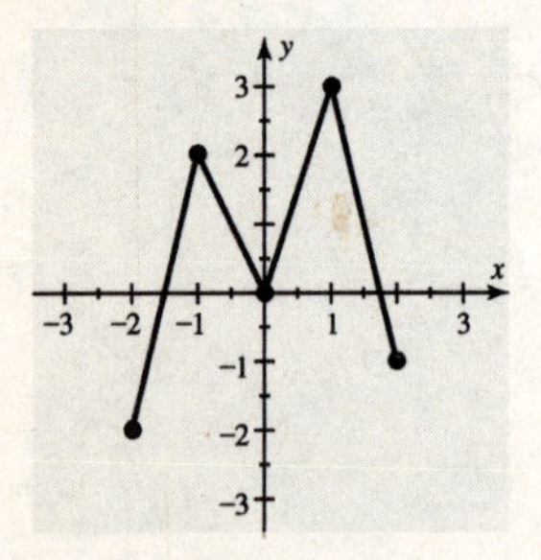

Figure 15

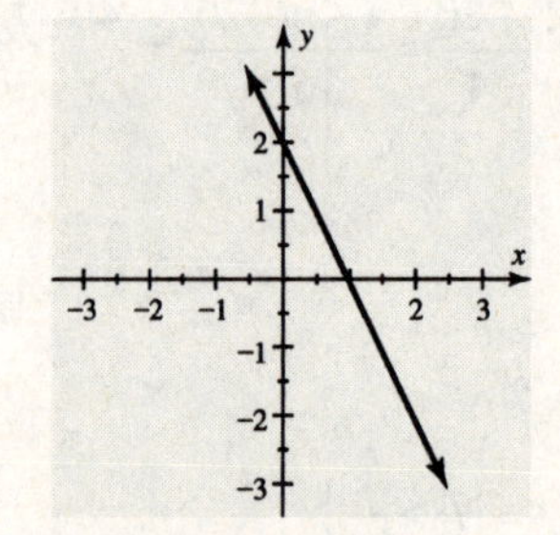

Figure 16a

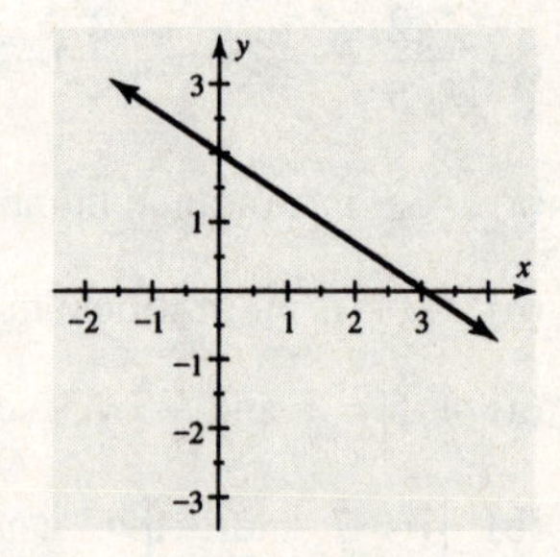

Figure 16b

17. To find the x-intercept, set $y = 0$ and solve for x: $4x - y = 8 \Rightarrow 4x - 0 = 8 \Rightarrow 4x = 8 \Rightarrow \frac{4x}{4} = \frac{8}{4} \Rightarrow x = 2$.

Therefore, the x-intercept is at $x = 2$. To find the y-intercept, set $x = 0$ and solve for

y: $4x - y = 8 \Rightarrow 4(0) - y = 8 \Rightarrow 0 - y = 8 \Rightarrow -y = 8 \Rightarrow y = -8$. Therefore, the y-intercept is at $y = -8$.

18. (a) See Figure 18a.

(b) See Figure 18b.

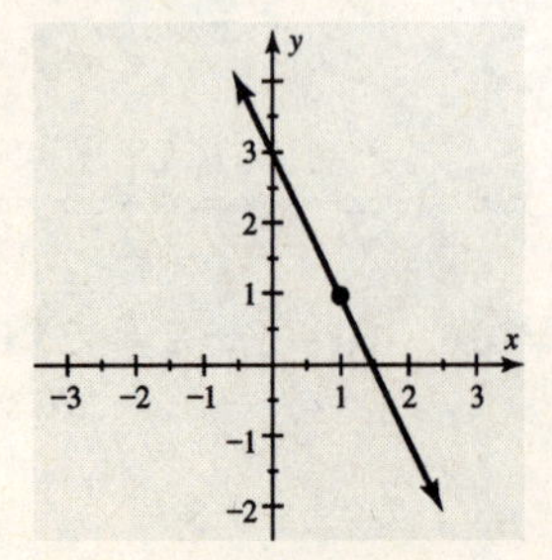

Figure 18a

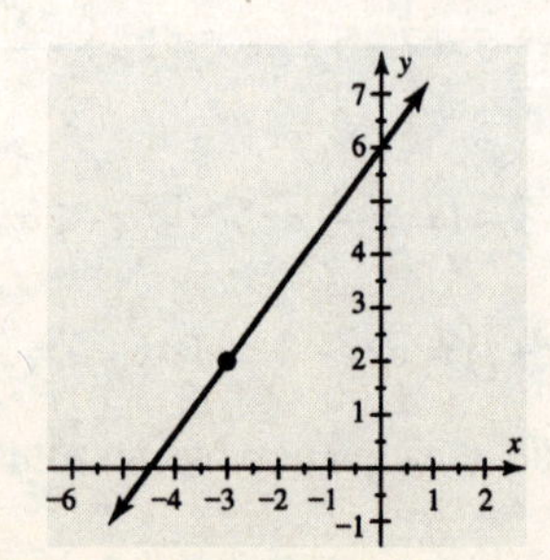

Figure 18b

19. (a) For every increase of 1 in the run, the rise decreases by $\frac{1}{2}$. Therefore, the slope m is $-\frac{1}{2}$. When $x=0,\ y=-1$. Therefore, the y-intercept is -1. Thus, $y=mx+b \Rightarrow y=-\frac{1}{2}x-1$.

(b) For every increase of 1 in the run, the rise increases by 4. Therefore, the slope m is 4. When $x=0,\ y=-3$. Therefore, the y-intercept is -3. Thus, $y=mx+b \Rightarrow y=4x-3$.

20. Find the slope: $m=\frac{y_2-y_1}{x_2-x_1}=\frac{4-2}{-4-4}=\frac{2}{-8}=-\frac{1}{4}$. Then, insert the values of the point (4, 2) and the value of the slope into the point-slope form:

$$y-y_1=m(x-x_1) \Rightarrow y-2=-\frac{1}{4}(x-4) \Rightarrow y-2=-\frac{1}{4}x+1 \Rightarrow y-2+2=-\frac{1}{4}x+1+2 \Rightarrow y=-\frac{1}{4}x+3$$

21. Solve the given equation for y: $2x-6y=7 \Rightarrow -2x+2x-6y=-2x+7 \Rightarrow -6y=-2x+7 \Rightarrow \frac{-6y}{-6}=\frac{-2x}{-6}+\frac{7}{-6}$

$\Rightarrow y=\frac{1}{3}x-\frac{7}{6}$. The slope of this line is $\frac{1}{3}$, so the slope of a line perpendicular to it is -3. Insert the values of the point (1, 1) and the value of the slope into the point-slope form:

$$y-y_1=m(x-x_1) \Rightarrow y-1=-3(x-1) \Rightarrow y-1=-3x+3 \Rightarrow y-1+1=-3x+3+1 \Rightarrow y=-3x+4.$$

22. (4, 4), because $3(4)-4=8 \Rightarrow 12-4=8$ is true and $2(4)+4=12 \Rightarrow 8+4=12$ is true.

23. Multiplying the second equation by –1 and adding the two equations will eliminate the variable x.

$$\begin{array}{r} x-y=4 \\ \underline{2x+y=-1} \\ 3x=3 \end{array}$$

Thus $x=1$. And so $1-y=4 \Rightarrow y=-3$.

The solution is $(1,-3)$.

24. Multiplying the first equation by 2 and adding the two equations will eliminate both variables.

$$\begin{array}{r} 4x+6y=8 \\ \underline{-4x-6y=7} \\ 0=15 \end{array}$$

Since this is false, there are no solutions.

25. Multiplying the first equation by 5 and adding the two equations will eliminate both variables.

$$\begin{array}{r} 15x-20y=40 \\ \underline{-15x+20y=-40} \\ 0=0 \end{array}$$

Since this is true, there are infinitely many solutions.

26. Multiplying the first equation by 3 and the second equation by 2 and adding the two equations will eliminate the variable y.

$$\begin{array}{r} 21x+6y=-9 \\ \underline{-10x-6y=-2} \\ 11x \qquad =-11 \end{array}$$

Thus $x=-1$. And so $7(-1)+2y=-3 \Rightarrow -7+2y=-3 \Rightarrow 2y=4 \Rightarrow y=2$. The solution is $(-1, 2)$.

27. See Figure 27.

28. See Figure 28.

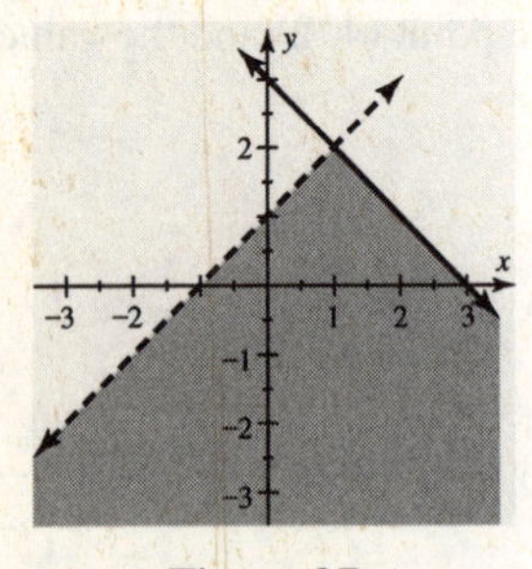

Figure 27

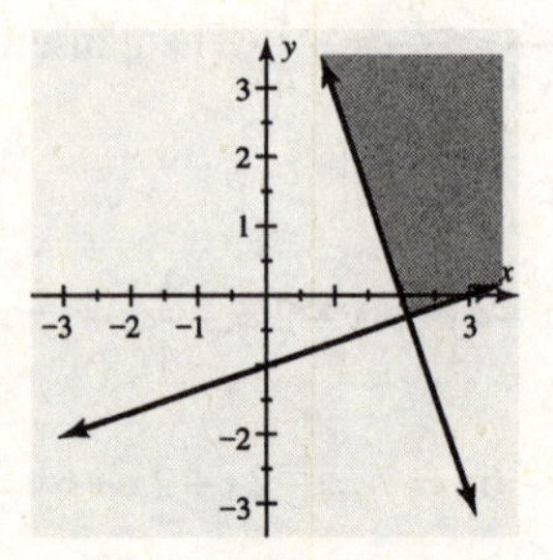

Figure 28

29. $F=16.5R$

30. The temperature change is the difference between the two temperatures: $83-(-11)=83+11=94°C$.

31. The sales tax \$17.15 is 7% of the cost of the camera, c. $17.15=0.07c \Rightarrow \dfrac{17.15}{0.07}=\dfrac{0.07c}{0.07} \Rightarrow 245=c$ The camera cost \$245.

32. The increase will be 9% of \$145 per credit, and $0.09(145)=13.05$. Thus, tuition is increasing by \$13.05 per credit so the new tuition will be $\$145+\$13.05=\$158.05$ per credit.

33. The truck uses 3 gallons per hour, so the slope m is –3. The initial amount is 30 gallons, so the y-intercept is $b=30$. Thus $y=mx+b \Rightarrow G=-3x+30$.

34. Let x represent the amount borrowed at 5%. Then the amount borrowed at 6% is $2400-x$. The total interest for one year will equal the sum of the interest for each loan. Therefore: $0.05x+0.06(2400-x)=132$

$\Rightarrow 0.05x+144-0.06x=132$

$\Rightarrow -0.01x+144=132$

$\Rightarrow -0.01x+144-144=132-144$

$\Rightarrow -0.01x=-12$

$\dfrac{-0.01x}{-0.01}=\dfrac{-12}{-0.01} \Rightarrow x=1200,\ 2400-x=1200.$

Therefore, the amount borrowed at 5% interest is \$1200 and the amount borrowed at 6% interest is \$1200.

Critical Thinking Solutions for Chapter 4

Section 4.1

- Parallel lines have no intersection, therefore there are no solutions.

Section 4.2

- Since it helps downstream and hurts upstream the current would be half the difference $\Rightarrow \dfrac{16-10}{2} = 3$ mph.

Section 4.4

- No, $(2,-2)$ is not a true solution for both inequalities; it does not satisfy $x+2y<-2$ because

  $2+2(-2)<-2 \Rightarrow -2<-2$ which is false.

- This region represents heavier weights and shorter heights. These ordered pairs correspond to people who weigh more than recommended.

Chapter 5: Polynomials and Exponents

5.1: Rules for Exponents

Concepts

1. base; exponent

2. 1

3. $\frac{1}{2} \cdot \frac{1}{2} \cdot \frac{1}{2} = \left(\frac{1}{2}\right)^3$

4. $x \cdot x \cdot x \cdot x = x^4$

5. $-1^2 = -(1 \cdot 1) = -1$

6. $(-1)^2 = (-1) \cdot (-1) = 1$

7. a^{m+n}

8. a^{mn}

9. $a^n b^n$

10. $\frac{a^n}{b^n}$

Properties of Exponents

11. $8^2 = 8 \cdot 8 = 64$

13. $(-2)^3 = (-2) \cdot (-2) \cdot (-2) = -8$

15. $-2^3 = -(2 \cdot 2 \cdot 2) = -8$

17. $6^0 = 1$

19. $2 \cdot 4^2 = 2 \cdot (4 \cdot 4) = 2 \cdot 16 = 32$

21. $1 + 5^2 = 1 + (5 \cdot 5) = 1 + 25 = 26$

23. $\frac{4^2}{2} = \frac{(4 \cdot 4)}{2} = \frac{16}{2} = 8$

25. $4 \cdot \frac{1}{2^3} = 4 \cdot \frac{1}{(2 \cdot 2 \cdot 2)} = 4 \cdot \frac{1}{8} = \frac{4}{8} = \frac{1}{2}$

27. $3 \cdot 3^2 = 3^1 \cdot 3^2 = 3^{1+2} = 3^3 = 27$

29. $4^2 \cdot 4^6 = 4^{2+6} = 4^8 = 65{,}536$

31. $2^3 \cdot 2^2 = 2^{3+2} = 2^5 = 32$

33. $x^3 \cdot x^6 = x^{3+6} = x^9$

35. $z^0z^4 = z^{0+4} = z^4$

37. $x^2x^2x^2 = x^{2+2+2} = x^6$

39. $4x^2 \cdot 5x^5 = 20x^{2+5} = 20x^7$

41. $3\left(-xy^3\right)\left(x^2y\right) = 3\left(-x^{1+2}y^{3+1}\right) = 3\left(-x^3y^4\right) = -3x^3y^4$

43. $\left(2^3\right)^2 = 2^{2\cdot3} = 2^6 = 64$

45. $\left(n^3\right)^4 = n^{3\cdot4} = n^{12}$

47. $x\left(x^3\right)^2 = x\left(x^{3\cdot2}\right) = x\left(x^6\right) = x^{1+6} = x^7$

49. $(-7b)^2 = (-7)\cdot(-7)\cdot(b\cdot b) = 49b^2$

51. $(ab)^3 = (a\cdot a\cdot a)(b\cdot b\cdot b) = a^3b^3$

53. $\left(2x^2\right)^2 = (2\cdot 2)\left(x^{2\cdot2}\right) = 4x^4$

55. $\left(-4b^2\right)^3 = (-4)\cdot(-4)\cdot(-4)\left(b^{2\cdot3}\right) = -64b^6$

57. $\left(x^2y^3\right)^7 = \left(x^{2\cdot7}y^{3\cdot7}\right) = x^{14}y^{21}$

59. $\left(y^3\right)^2\left(x^4y\right)^3 = \left(y^{3\cdot2}\right)\left(x^{4\cdot3}y^{1\cdot3}\right) = \left(y^6\right)\left(x^{12}y^3\right) = x^{12}y^{6+3} = x^{12}y^9$

61. $\left(a^2b\right)^2\left(a^2b^2\right)^3 = \left(a^{2\cdot2}b^{1\cdot2}\right)\left(a^{2\cdot3}b^{2\cdot3}\right) = \left(a^4b^2\right)\left(a^6b^6\right) = a^{4+6}b^{2+6} = a^{10}b^8$

63. $\left(\frac{1}{3}\right)^3 = \frac{(1\cdot1\cdot1)}{(3\cdot3\cdot3)} = \frac{1}{27}$

65. $\left(\frac{a}{b}\right)^5 = \frac{(a\cdot a\cdot a\cdot a\cdot a)}{(b\cdot b\cdot b\cdot b\cdot b)} = \frac{a^5}{b^5}$

67. $\frac{(x-y)^3}{3} = \frac{(x-y)^3}{3^3} = \frac{(x-y)^3}{3\cdot3\cdot3} = \frac{(x-y)^3}{27}$

69. $\left(\frac{5}{a+b}\right)^2 = \frac{5^2}{(a+b)^2} = \frac{5\cdot5}{(a+b)^2} = \frac{25}{(a+b)^2}$

71. $\left(\frac{2x}{5}\right)^3 = \frac{(2\cdot2\cdot2)\left(x^{1\cdot3}\right)}{(5\cdot5\cdot5)} = \frac{8x^3}{125}$

73. $\left(\frac{3x^2}{5y^4}\right)^3 = \frac{(3\cdot3\cdot3)\left(x^{2\cdot3}\right)}{(5\cdot5\cdot5)\left(y^{4\cdot3}\right)} = \frac{27x^6}{125y^{12}}$

75. $(x+y)(x+y)^3=(x+y)^{1+3}=(x+y)^4$

77. $(a+b)^2(a+b)^3=(a+b)^{2+3}=(a+b)^5$

79. $6(x^4y^6)^0=6(x^{4\cdot 0}y^{6\cdot 0})=6(x^0y^0)=6\cdot 1\cdot 1=6$

81. $a(a^2+2b^2)=a^{1+2}+2ab^2=a^3+2ab^2$

83. $3a^3(4a^2+2b)=3\cdot 4a^{3+2}+3\cdot 2a^3b=12a^5+6a^3b$

85. $(r+t)(rt)=(r^{1+1}t)+(r\,t^{1+1})=r^2t+rt^2$

87. For example, $a=3$, $b=1$, $m=1$, and $n=0$. $a^m\cdot b^n=3^1\cdot 1^0=3\cdot 1=3$ and $(ab)^{m+n}=(3\cdot 1)^{1+0}=3^1=3$.

Answers may vary.

Applications

89. $2x^2\cdot 5x^2=10x^{2+2}=10x^4$

91. $x\cdot 2x\cdot 4x=8x^{1+1+1}=8x^3$

93. $\pi(3x^2)^2=\pi(3\cdot 3)(x^{2\cdot 2})=\pi(9x^4)=9\pi x^4$

95. $1000(1+0.05)^3=1000(1.05)^3=1000(1.05\cdot 1.05\cdot 1.05)=1000(1.157625)\approx \$1157.63$

97. See Figure 97.

99. See Figure 99.

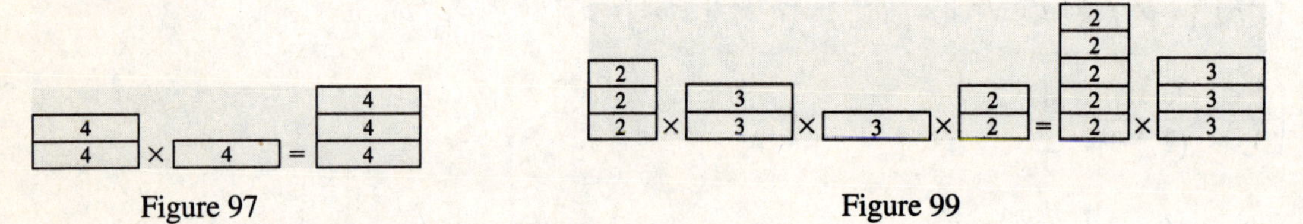

Figure 97 Figure 99

5.2: Addition and Subtraction of Polynomials

Concepts

1. monomial
2. −3
3. polynomial
4. degree
5. binomial
6. trinomial
7. 2; 3
8. like
9. opposite

10. vertically

Properties of Polynomials

11. Because $x^2 \Rightarrow$ the degree is 2; the coefficient is 3.

13. Because $ab = a^1b^1$ and $1+1=2 \Rightarrow$ the degree is 2; $-ab = -1ab \Rightarrow$ the coefficient is -1.

15. Because $rt = r^1t^1$ and $1+1=2 \Rightarrow$ the degree is 2; the coefficient is -5.

17. Because there are no variables the degree is 0; the coefficient is -6.

19. Yes it is a polynomial; 1 term $-x$; one variable x; $x^1 \Rightarrow$ degree is 1.

21. Yes it is a polynomial; 3 terms $4x^2, -5x$, and 9; one variable x; $x^2 \Rightarrow$ degree is 2.

23. Not a polynomial because it has a variable in the denominator.

25. Yes it is polynomial; 2 terms $3x^2y$ and $-xy^3$; two variables x and y; $xy^3 = x^1y^3$ and $1+3=4 \Rightarrow$ degree is 4.

27. Not a polynomial because it has negative exponents on the variables.

29. Yes it is a polynomial; 1 term -2^3a^4bc; three variables a, b, and c; -2^3a^4bc and $4+1+1=6 \Rightarrow$ degree is 6.

31. Yes, $5x + (-4x) = (5-4)x = 1x = x$.

33. Yes, $x^3 + (-6x^3) = [1 + (-6)]x^3 = -5x^3$.

35. No, x and y are not like terms.

37. Yes $\Rightarrow ab + ba = (1+1)ab = 2ab$.

39. Yes $\Rightarrow 7xy^2 + (-3xy^2) = (7-3)xy^2 = 4xy^2$.

Addition of Polynomials

41. $(3x+5) + (-4x+4) = 3x + (-4x) + 5 + 4 = (3-4)x + (5+4) = -x + 9$

43. $(3x^2+4x+1) + (x^2+4x-6) = 3x^2 + x^2 + 4x + 4x + 1 + (-6) = (3+1)x^2 + (4+4)x + (1-6) = 4x^2 + 8x - 5$

45. $(a^3-6) + (4a^3+7) = 4a^3 + a^3 + 7 + (-6) = (4+1)a^3 + (7-6) = 5a^3 + 1$

47. $(y^3+3y^2-5) + (3y^3+4y-4) = 3y^3 + y^3 + 3y^2 + 4y + (-5) + (-4) =$

$(3+1)y^3 + 3y^2 + 4y + (-5-4) = 4y^3 + 3y^2 + 4y - 9$

49. $(-xy+5) + (5xy-4) = 5xy + (-xy) + 5 + (-4) = (5-1)xy + (5-4) = 4xy + 1$

51. $(a^3b^2 + a^2b^3) + (a^2b^3 - a^3b^2) = a^3b^2 + (-a^3b^2) + a^2b^3 + a^2b^3 = (1-1)a^3b^2 + (1+1)a^2b^3 =$

$0a^3b^2 + 2a^2b^3 = 2a^2b^3$

53.
$$\begin{array}{r} 4x^2 - 2x + 1 \\ +5x^2 + 3x - 7 \\ \hline 9x^2 + x - 6 \end{array}$$

55.
$$\begin{array}{r} -x^2 + x + 0 \\ +2x^2 - 8x - 1 \\ \hline x^2 - 7x - 1 \end{array}$$

Subtraction of Polynomials

57. $-5x^2$

59. $-3a^2 + a - 4$

61. $2t^2 + 3t - 4$

63. $(3x+1)-(-x+3)=(3x+1)+(x-3)=(3+1)x+(1-3)=4x-2$

65. $\left(-x^2+6x+8\right)-\left(2x^2+x-2\right)=\left(-x^2+6x+8\right)+\left(-2x^2-x+2\right)=$

$(-1-2)x^2+(6-1)x+(8+2)=-3x^2+5x+10$

67. $\left(a^2-2a\right)-\left(4a^2+3a\right)=\left(a^2-2a\right)+\left(-4a^2-3a\right)=(1-4)a^2+(-2-3)a=-3a^2-5a$

69. $\left(z^3-2z^2-z\right)-\left(4z^2+5z+1\right)=\left(z^3-2z^2-z\right)+\left(-4z^2-5z-1\right)=$

$z^3+(-2-4)z^2+(-1-5)z-1=z^3-6z^2-6z-1$

71. $\left(4xy+x^2y^2\right)-\left(xy-x^2y^2\right)=\left(4xy+x^2y^2\right)+\left(-xy+x^2y^2\right)=(4-1)xy+(1+1)x^2y^2=3xy+2x^2y^2$

73. $\left(ab^2\right)-\left(ab^2+a^3b\right)=\left(ab^2\right)+\left(-ab^2-a^3b\right)=(1-1)ab^2-a^3b=0ab^2-a^3b=-a^3b$

75. $\left(x^2+2x-3\right)-\left(2x^2+7x+1\right)=$
$$\begin{array}{r} x^2 + 2x - 3 \\ +\left(-2x^2\right) - 7x - 1 \\ \hline -x^2 - 5x - 4 \end{array}$$

77. $\left(3x^3-2x\right)-\left(5x^3+4x+2\right)=$
$$\begin{array}{r} 3x^3 - 2x + 0 \\ +\left(-5x^3\right) - 4x - 2 \\ \hline -2x^3 - 6x - 2 \end{array}$$

Applications

79. (a) Let $t=0$, then $1.6t^2-28t+200=1.6\left(0^2\right)-28(0)+200=0-0+200=200$ bpm.

(b) Let $t=5$, then $1.6t^2-28t+200=1.6\left(5^2\right)-28(5)+200=1.6(25)-28(5)+200=$

$40-140+200=100$ bpm.

(c) It decreases quickly at first and then more slowly.

81. $z^2+z^2=(1+1)z^2=2z^2$; Let $z=10$ in., then $2(10^2)=2(100)=200$ in^2

83. $2x\cdot x+x\cdot x=2x^2+x^2$ or $3x^2$; Let $x=6$ feet, then $3x^2=3(6^2)=3(36)=108$ ft^2

85. $\pi x^2+\pi y^2$; Let $x=2$ feet and $y=3$ feet, then $\pi x^2+\pi y^2=\pi(2^2)+\pi(3^2)=\pi(4)+\pi(9)=13\pi$ft^2

87. (a) Slope $m=\dfrac{P_2-P_1}{t_2-t_1}$, so $m_1=\dfrac{5-4}{1987-1974}=\dfrac{1}{13}\Rightarrow m_1=0.077$; $m_2=\dfrac{6-5}{1999-1987}=\dfrac{1}{12}\Rightarrow$

$m_2=0.083$; $m_3=\dfrac{7-6}{2012-1999}=\dfrac{1}{13}\Rightarrow m_3=0.077$. A line is a reasonable estimate, but it is not exact.

(b) For the given years, the polynomial gives a reasonable estimate.

Checking Basic Concepts for Sections 5.1 & 5.2

1. (a) $-5^2=-(5\cdot 5)=-25$

(b) $3^2-2^3=(3\cdot 3)-(2\cdot 2\cdot 2)=9-8=1$

2. (a) $10^3\cdot 10^5=10^{3+5}=10^8$

(b) $(3x^2)(-4x^5)=-12x^{2+5}=-12x^7$

(c) $(a^3b)^2=a^{3\cdot 2}b^{1\cdot 2}=a^6b^2$

(d) $\left(\dfrac{x}{z^3}\right)^4=\dfrac{x^{1\cdot 4}}{z^{3\cdot 4}}=\dfrac{x^4}{z^{12}}$

3. 3 terms $5x^3y$, $-2x^2y$, and 5; 2 variables x and y; $x^3y=x^3y^1$, $3+1=4\Rightarrow$ the degree is 4.

4. (a) Let length $=2W$, width $=W$, and height $=H$, then $2W\cdot W\cdot H=2W^2H$.

(b) Let $W=12$ and $H=10$, then $2W^2H=2(12^2)(10)=2(144)(10)=2880$ in^3.

5. (a) $(2a^2+3a-1)+(a^2-3a+7)=2a^2+a^2+3a+(-3a)+(-1)+7=$

$(2+1)a^2+(3-3)a+(-1+7)=3a^2+0a+6=3a^2+6$

(b) $(4z^3+5z)-(2z^3-2z+8)=(4z^3+5z)+(-2z^3+2z-8)=(4-2)z^3+(5+2)z-8=2z^3+7z-8$

(c) $(x^2+2xy+y^2)-(x^2-2xy+y^2)=(x^2+2xy+y^2)+(-x^2+2xy-y^2)=$

$(1-1)x^2+(2+2)xy+(1-1)y^2=0x^2+4xy+0y^2=4xy$

6. Let $x=-2$, then $5x^2-7x=5(-2)^2-7(-2)=5(4)-7(-2)=20-(-14)=20+14=34$.

5.3: Multiplication of Polynomials

Concepts

1. The product rule
2. Distributive
3. One; two; three
4. term; term
5. No, $(x+y)(x+y)=x\cdot x+x\cdot y+y\cdot x+y\cdot y=x^2+xy+xy+y^2=x^2+2xy+y^2$
6. Vertically

Multiplication of Monomials

7. $x^2\cdot x^5=x^{2+5}=x^7$

9. $-3a\cdot 4a=(-3)(4)aa=-12a^{1+1}=-12a^2$

11. $4x^3\cdot 5x^2=(4)(5)x^3x^2=20x^{3+2}=20x^5$

13. $xy^2\cdot 4xy=(1)(4)xxy^2y=4x^{1+1}y^{2+1}=4x^2y^3$

15. $(-3xy^2)(4x^2y)=(-3)(4)xx^2y^2y=-12x^{1+2}y^{2+1}=-12x^3y^3$

Multiplication of Monomials and Polynomials

17. $3(x+4)=3\cdot x+3\cdot 4=3x+12$

19. $-5(9x+1)=-5\cdot 9x+(-5\cdot 1)=-45x-5$

21. $(4-z)z=4\cdot z+(-z\cdot z)=4z-z^2$

23. $-y(5+3y)=-y\cdot 5+(-y\cdot 3y)=-5y-3y^2$

25. $3x(5x^2-4)=3x\cdot 5x^2+3x\cdot(-4)=(3)(5)xx^2-12x=15x^{1+2}-12x=15x^3-12x$

27. $(6x-6)x^2=6x\cdot x^2+(-6)\cdot x^2=6x^{1+2}-6x^2=6x^3-6x^2$

29. $-8(4t^2+t+1)=-8\cdot 4t^2+(-8)(t)+(-8)(1)=-32t^2-8t-8$

31. $n^2(-5n^2+n-2)=(n^2)(-5n^2)+(n^2)(n)+(n^2)(-2)=-5n^{2+2}+n^{2+1}-2n^2=-5n^4+n^3-2n^2$

33. $xy(x+y)=xy\cdot x+xy\cdot y=x^{1+1}y+xy^{1+1}=x^2y+xy^2$

35. $x^2(x^2y-xy^2)=x^2\cdot x^2y+(x^2)(-xy^2)=x^{2+2}y-x^{2+1}y^2=x^4y-x^3y^2$

37. $-ab(a^3-2b^3)=-ab\cdot a^3+(-ab)(-2b^3)=-a^{1+3}b+2ab^{3+1}=-a^4b+2ab^4$

Multiplication of Polynomials

39. (a) Area $=x^2+3x$. See Figure 39.

    (b) $x(x+3)=x\cdot x+x\cdot 3=x^{1+1}+3x=x^2+3x$

41. (a) Area $= x^2 + 2x + 2x + 4 = x^2 + 4x + 4$. See Figure 41.

(b) $(x+2)(x+2) = x \cdot x + x \cdot 2 + 2 \cdot x + 2 \cdot 2 = x^{1+1} + 2x + 2x + 4 = x^2 + 4x + 4$

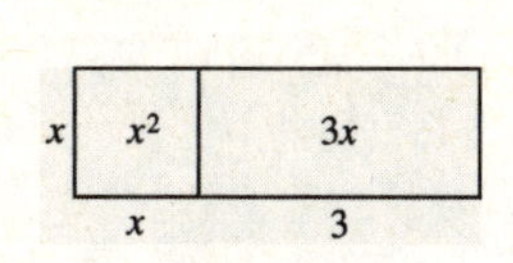

Figure 39

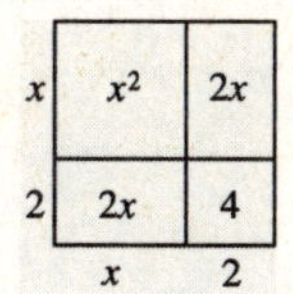

Figure 41

43. (a) Area $= x^2 + 3x + 6x + 18 = x^2 + 9x + 18$. See Figure 43.

(b) $(x+3)(x+6) = x \cdot x + x \cdot 6 + 3 \cdot x + 3 \cdot 6 = x^{1+1} + 6x + 3x + 18 = x^2 + 9x + 18$

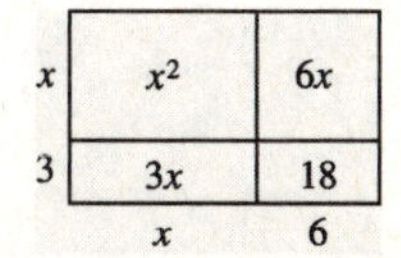

Figure 43

45. $(x+3)(x+5) = x \cdot x + x \cdot 5 + 3 \cdot x + 3 \cdot 5 = x^{1+1} + 5x + 3x + 15 = x^2 + 8x + 15$

47. $(x-8)(x-9) = x \cdot x + (x)(-9) + (-8)(x) + (-8)(-9) = x^{1+1} + (-9x) + (-8x) + 72 = x^2 - 17x + 72$

49. $(3z-2)(2z-5) = 3z \cdot 2z + (3z)(-5) + (-2)(2z) + (-2)(-5) = 6z^{1+1} + (-15z) + (-4z) + 10 = 6z^2 - 19z + 10$

51. $(8b-1)(8b+1) = 8b \cdot 8b + 8b \cdot 1 + (-1)(8b) + (-1)(1) = 64b^{1+1} + 8b - 8b - 1 = 64b^2 - 1$

53. $(10y+7)(y-1) = 10y \cdot y + (10y)(-1) + 7 \cdot y + (7)(-1) = 10y^{1+1} - 10y + 7y - 7 = 10y^2 - 3y - 7$

55. $(5-3a)(1-2a) = 5 \cdot 1 + (5)(-2a) + (-3a)(1) + (-3a)(-2a) = 5 + (-10a) + (-3a) + 6a^{1+1} = 5 - 13a + 6a^2$

57. $(1-3x)(1+3x) = 1 \cdot 1 + 1 \cdot 3x + (-3x)(1) + (-3x)(3x) = 1 + 3x - 3x - 9x^{1+1} = 1 - 9x^2$

59. $(x-1)(x^2+1) = x \cdot x^2 + x \cdot 1 + (-1)(x^2) + (-1)(1) = x^{1+2} + x - x^2 - 1 = x^3 - x^2 + x - 1$

61. $(x^2+4)(4x-3) = x^2 \cdot 4x + (x^2)(-3) + 4 \cdot 4x + (4)(-3) = 4x^{2+1} - 3x^2 + 16x - 12 = 4x^3 - 3x^2 + 16x - 12$

63. $(2n+1)(n^2+3) = 2n \cdot n^2 + 2n \cdot 3 + 1 \cdot n^2 + 1 \cdot 3 = 2n^{1+2} + 6n + n^2 + 3 = 2n^3 + n^2 + 6n + 3$

65. $(m+1)(m^2+3m+1) = m \cdot m^2 + m \cdot 3m + m \cdot 1 + 1 \cdot m^2 + 1 \cdot 3m + 1 \cdot 1 =$

$m^{1+2} + 3m^{1+1} + m + m^2 + 3m + 1 = m^3 + 3m^2 + m + m^2 + 3m + 1 = m^3 + 4m^2 + 4m + 1$

67. $(3x-2)(2x^2-x+4) = 3x \cdot 2x^2 + (3x)(-x) + 3x \cdot 4 + (-2)(2x^2) + (-2)(-x) + (-2)(4) =$

$6x^{1+2} - 3x^{1+1} + 12x - 4x^2 + 2x - 8 = 6x^3 - 3x^2 + 12x - 4x^2 + 2x - 8 = 6x^3 - 7x^2 + 14x - 8$

69. $(x+1)(x^2-x+1) = x \cdot x^2 + (x)(-x) + x \cdot 1 + 1 \cdot x^2 + (1)(-x) + 1 \cdot 1 =$

$x^{1+2} - x^{1+1} + x + x^2 - x + 1 = x^3 - x^2 + x + x^2 - x + 1 = x^3 + 1$

71. $(4b^2+3b+7)(b^2+3) = 4b^2 \cdot b^2 + 4b^2 \cdot 3 + 3b \cdot b^2 + 3b \cdot 3 + 7 \cdot b^2 + 7 \cdot 3 =$

$4b^{2+2}+12b^2+3b^{1+2}+9b+7b^2+21=4b^4+12b^2+3b^3+9b+7b^2+21=\ 4b^4+3b^3+19b^2+9b+21$

73.
$$\begin{array}{r} x^2-3x+1 \\ x+2 \\ \hline 2x^2-6x+2 \\ x^3-3x^2+x \\ \hline x^3-\ x^2-5x+2 \end{array}$$

75.
$$\begin{array}{r} a^2+2a+4 \\ a-2 \\ \hline -2a^2-4a-8 \\ a^3+2a^2+4a \\ \hline a^3 \qquad\qquad -8 \end{array}$$

77.
$$\begin{array}{r} 3x^2-x+1 \\ 2x^2+1 \\ \hline 3x^2-x+1 \\ 6x^4-2x^3+2x^2 \\ \hline 6x^4-2x^3+5x^2-x+1 \end{array}$$

79. Every term in the first polynomial must be multiplied by every term in the second polynomial, so there are $m \cdot n$ products, hence $m \cdot n$ terms.

Applications

81. (a) Let $h=$ height, $h+2=$ length, and $h-4=$ width, then $(h-4)(h+2)=$

$h \cdot h+h \cdot 2+(-4)(h)+(-4)(2)=h^{1+1}+2h-4h-8=h^2-2h-8 \Rightarrow (h)\left(h^2-2h-8\right)=$

$h \cdot h^2+(h)(-2h)+(h)(-8)=h^{1+2}-2h^{1+1}-8h=h^3-2h^2-8h.$

(b) $h^3-2h^2-8h=25^3-2(25)^2-8(25)=15{,}625-2(625)-200=15{,}625-1250-200=\ 14{,}175\text{ in}^3$

83. (a) $x(50-x)=x \cdot 50+(x)(-x)=50x-x^{1+1}=50x-x^2$

(b) $50x-x^2=50(25)-\left(25^2\right)=1250-625=625$

85. $(x+1)(x+1)=x \cdot x+x \cdot 1+1 \cdot x+1 \cdot 1=x^{1+1}+x+x+1=x^2+2x+1 \Rightarrow 1$ side $=$

x^2+2x+1. A cube has all sides the same $\Rightarrow 6$ sides $=(6)\left(x^2+2x+1\right)=6 \cdot x^2+6 \cdot 2x+6 \cdot 1=\ 6x^2+12x+6.$

87. (a) $t(64-16t)=t \cdot 64+(t)(-16t)=64t-16t^{1+1}=64t-16t^2$

(b) $64t-16t^2=64(2)-16(2)^2=128-16(4)=128-64=64;$

$t(64-16t)=2(64-16(2))=2(64-32)=2(32)=64$

(c) Yes; yes

5.4: Special Products

Concepts

1. $(a+b)(a-b)=(a)^2-(b)^2=a^2-b^2$

2. Yes; the product is a sum and a difference.

3. $(a+b)^2=(a)^2+2(a)(b)+(b)^2=a^2+2ab+b^2$

4. $(a-b)^2=(a)^2-2(a)(b)+(b)^2=a^2-2ab+b^2$

5. No; $(x+y)^2=x^2+2xy+y^2$

6. No; $(r-t)^2=r^2-2rt+t^2$

7. No; $(z+5)^3=(z+5)(z+5)^2$

8. $(a+b)^2$

Product of a Sum and Difference

9. $(x-3)(x+3)=(x)^2-(3)^2=x^2-9$

11. $(x+1)(x-1)=(x)^2-(1)^2=x^2-1$

13. $(4x-1)(4x+1)=(4x)^2-(1)^2=16x^2-1$

15. $(1+2a)(1-2a)=(1)^2-(2a)^2=1-4a^2$

17. $(2x+3y)(2x-3y)=(2x)^2-(3y)^2=4x^2-9y^2$

19. $(ab-5)(ab+5)=(ab)^2-(5)^2=a^2b^2-25$

21. $(ab+4)(ab-4)=(ab)^2-(4)^2=a^2b^2-16$

23. $(a^2-b^2)(a^2+b^2)=(a^2)^2-(b^2)^2=a^4-b^4$

25. $(x^3-y^3)(x^3+y^3)=(x^3)^2-(y^3)^2=x^6-y^6$

27. $101\cdot 99=(100+1)(100-1)=(100)^2-(1)^2=10,000-1=9999$

29. $23\cdot 17=(20+3)(20-3)=(20)^2-(3)^2=400-9=391$

31. $90\cdot 110=(100-10)(100+10)=(100)^2-(10)^2=10,000-100=9900$

Squaring Binomials

33. $(x+1)^2=(x)^2+2(x)(1)+(1)^2=x^2+2x+1$

35. $(a-2)^2=(a)^2-2(a)(2)+(2)^2=a^2-4a+4$

37. $(2x+3)^2=(2x)^2+2(2x)(3)+(3)^2=4x^2+12x+9$

39. $(3b+5)^2=(3b)^2+2(3b)(5)+(5)^2=9b^2+30b+25$

41. $\left(\frac{3}{4}a-4\right)^2=\left(\frac{3}{4}a\right)^2-2\left(\frac{3}{4}a\right)(4)+(4)^2=\frac{9}{16}a^2-6a+16$

43. $(1-b)^2=(1)^2-2(1)(b)+(b)^2=1-2b+b^2$

45. $(5+y^3)^2=(5)^2+2(5)(y^3)+(y^3)^2=25+10y^3+y^6$

47. $(a^2+b)^2=(a^2)^2+2(a^2)(b)+(b)^2=a^4+2a^2b+b^2$

Cubing Binomials

49. $(a+1)^3=(a+1)(a+1)^2=(a+1)(a^2+2a+1)=$

$a\cdot a^2+a\cdot 2a+a\cdot 1+1\cdot a^2+1\cdot 2a+1\cdot 1=a^3+2a^2+a+a^2+2a+1=a^3+3a^2+3a+1$

51. $(x-2)^3=(x-2)(x-2)^2=(x-2)(x^2-4x+4)=$

$x\cdot x^2+x\cdot(-4x)+x\cdot 4+(-2)\cdot x^2+(-2)(-4x)+(-2)4=$

$x^3+(-4x^2)+4x+(-2x^2)+8x+(-8)=x^3-6x^2+12x-8$

53. $(2x+1)^3=(2x+1)(2x+1)^2=(2x+1)(4x^2+4x+1)=$

$2x\cdot 4x^2+2x\cdot 4x+2x\cdot 1+1\cdot 4x^2+1\cdot 4x+1\cdot 1=8x^3+8x^2+2x+4x^2+4x+1=8x^3+12x^2+6x+1$

55. $(6u-1)^3=(6u-1)(6u-1)^2=(6u-1)(36u^2-12u+1)=$

$6u\cdot 36u^2+6u\cdot(-12u)+6u\cdot 1+(-1)\cdot 36u^2+(-1)\cdot(-12u)+(-1)\cdot 1=$

$216u^3+(-72u^2)+6u+(-36u^2)+12u+(-1)=216u^3-108u^2+18u-1$

57. $t(t+2)^3=t(t+2)(t+2)^2=t(t+2)(t^2+4t+4)=(t^2+2t)(t^2+4t+4)=$

$t^2\cdot t^2+t^2\cdot 4t+t^2\cdot 4+2t\cdot t^2+2t\cdot 4t+2t\cdot 4=\ t^4+4t^3+4t^2+2t^3+8t^2+8t=t^4+6t^3+12t^2+8t$

Multiplication of Polynomials

59. $4(5x+9)=4\cdot 5x+4\cdot 9=20x+36$

61. $(x-5)(x+7)=\ x\cdot x+x\cdot 7+(-5)(x)+(-5)(7)=x^2+2x-35$

63. $(3x-5)^2=(3x)^2-2(3x)(5)+(-5)^2=9x^2-30x+25$

65. $(5x+3)(5x+4)=5x\cdot 5x+5x\cdot 4+3\cdot 5x+3\cdot 4=25x^2+35x+12$

67. $(4b-5)(4b+5)=(4b)^2-(5)^2=16b^2-25$

69. $-5x(4x^2-7x+2)=(-5x)(4x^2)+(-5x)(-7x)+(-5x)(2)=-20x^3+35x^2-10x$

71. $(4-a)^3=(4-a)(4-a)^2=(4-a)(16-8a+a^2)=$

$4\cdot16+(4)(-8a)+4\cdot a^2+(-a)(16)+(-a)(-8a)+(-a)(a^2)=64-48a+12a^2-a^3$

73. $x(x+3)^2=x(x^2+6x+9)=x^3+6x^2+9x$

75. $(x+2)(x-2)(x+1)(x-1)=(x^2-4)(x^2-1)=x^4-x^2-4x^2+4$

$=x^4-5x^2+4$

77. $(a^n+b^n)(a^n-b^n)=(a^n)^2-(b^n)^2=a^{2\cdot n}-b^{2\cdot n}=a^{2n}-b^{2n}$

Applications

79. (a) $(x+2)(x+2)=(x)^2+2(x)(2)+(2)^2=x^2+4x+4$

(b) $x\cdot x+2\cdot x+x\cdot 2+2\cdot 2=x^2+2x+2x+4=x^2+4x+4$

81. (a) $(2x+3)(2x+3)=(2x)^2+2(2x)(3)+(3)^2=4x^2+12x+9$

(b) $2x\cdot 2x+3\cdot 2x+2x\cdot 3+3\cdot 3=4x^2+6x+6x+9=4x^2+12x+9$

83. (a) $6(x+5)^2=6(x^2+10x+25)=6x^2+60x+150$

(b) $(x+5)^3=(x+5)(x+5)^2=(x+5)(x^2+10x+25)=$

$x\cdot x^2+x\cdot 10x+x\cdot 25+5\cdot x^2+5\cdot 10x+5\cdot 25=x^3+15x^2+75x+125$

85. (a) $(1+x)^2=(1)^2+2(1)(x)+(x)^2=1+2x+x^2$

(b) $1+2x+x^2=1+2(0.10)+(0.10)^2=1+0.20+0.01=1.21$; the money increases by 1.21 times in 2 years if the interest rate is 10%.

87. (a) $(1-x)^2=(1)^2-2(1)(x)+(x)^2=1-2x+x^2$

(b) $1-2x+x^2=1-2(0.50)+(0.50)^2=1-1+0.25=0.25$; if the chance of rain on each day is 50%, then there is a 25% chance of no rain on either day.

89. (a) $(z+16)^2-(z)^2=(z^2+32z+256)-(z^2)=32z+256$

(b) $32z+256=32(60)+256=1920+256=2176$; the area of an 8-foot-wide sidewalk around a $60\,\text{ft}\times 60\,\text{ft}$ pool is $2176\ \text{ft}^2$.

Checking Basic Concepts for Sections 5.3 & 5.4

1. (a) $(-3xy^4)(5x^2y)=(-3)(5)xx^2y^4y=-15x^3y^5$

(b) $-x(6-4x)=(-x)(6)+(-x)(-4x)=-6x+4x^2$

(c) $3ab(a^2-2ab+b^2)=3ab\cdot a^2+(3ab)(-2ab)+3ab\cdot b^2=3a^3b-6a^2b^2+3ab^3$

2. (a) $(x+3)(4x-3)=x\cdot 4x+(x)(-3)+3\cdot 4x+(3)(-3)=4x^2+9x-9$

(b) $(x^2-1)(2x^2+2)=x^2\cdot 2x^2+x^2\cdot 2+(-1)(2x^2)+(-1)(2)=2x^4-2$

(c) $(x+y)(x^2-xy+y^2)=x\cdot x^2+(x)(-xy)+x\cdot y^2+y\cdot x^2+(y)(-xy)+y\cdot y^2=x^3+y^3$

3. (a) $(5x+2)(5x-2)=(5x)^2-(2)^2=25x^2-4$

(b) $(x+3)^2=(x)^2+2(x)(3)+(3)^2=x^2+6x+9$

(c) $(2-7x)^2=(2)^2-2(2)(7x)+(7x)^2=4-28x+49x^2$

(d) $(t+2)^3=(t+2)(t+2)^2=(t+2)(t^2+4t+4)=t\cdot t^2+t\cdot 4t+t\cdot 4+2\cdot t^2+2\cdot 4t+2\cdot 4=t^3+6t^2+12t+8$

4. (a) $4\pi(a+7)^2=4\pi(a^2+14a+49)=4\pi(a^2)+4\pi(14a)+4\pi(49)=4\pi a^2+56\pi a+196\pi$

(b) $4\pi a^2+56\pi a+196\pi=4\pi(8)^2+56\pi(8)+196\pi=256\pi+448\pi+196\pi=900\pi$

5. (a) $(m+5)^2=(m)^2+2(5)(m)+(5)^2=m^2+10m+25$

(b) $m\cdot m+m\cdot 5+5\cdot m+5\cdot 5=m^2+5m+5m+25=m^2+10m+25$

5.5: Integer Exponents and the Quotient Rule

Concepts

1. $a^{-n}=\dfrac{1}{a^n}$

2. $2^{-3}=\dfrac{1}{2^3}=\dfrac{1}{2\cdot 2\cdot 2}=\dfrac{1}{8}$

3. $\dfrac{1}{a^{-n}}=\dfrac{a^n}{1}=a^n$

4. $\dfrac{1}{2^{-3}}=\dfrac{2^3}{1}=2\cdot 2\cdot 2=8$

5. $\dfrac{a^m}{a^n}=a^{m-n}$

6. $\dfrac{2^5}{2^2}=2^{5-2}=2^3=2\cdot 2\cdot 2=8$

7. $\dfrac{a^{-n}}{b^{-m}}=\dfrac{b^m}{a^n}$

8. $\dfrac{2^{-3}}{5^{-2}}=\dfrac{5^2}{2^3}=\dfrac{5\cdot 5}{2\cdot 2\cdot 2}=\dfrac{25}{8}$

9. $\left(\dfrac{a}{b}\right)^{-n}=\left(\dfrac{b}{a}\right)^n$

10. $\left(\frac{3}{5}\right)^{-2} = \left(\frac{5}{3}\right)^{2} = \frac{5^2}{3^2} = \frac{5\cdot 5}{3\cdot 3} = \frac{25}{9}$

11. $1 \le |b| < 10$

12. no

Negative Exponents

13a. $4^{-1} = \frac{1}{4^1} = \frac{1}{4}$

13b. $\left(\frac{1}{3}\right)^{-2} = \left(\frac{3}{1}\right)^{2} = \frac{3^2}{1^2} = \frac{3\cdot 3}{1\cdot 1} = \frac{9}{1} = 9$

15a. $2^3 \cdot 2^{-2} = 2^{3+(-2)} = 2^1 = 2$

15b. $10^4 \cdot 10^{-2} = 10^{4+(-2)} = 10^2 = 10\cdot 10 = 100$

17a. $3^{-2}\cdot 3^{-1}\cdot 3^{-1} = 3^{-2+(-1)+(-1)} = 3^{-4} = \frac{1}{3^4} = \frac{1}{3\cdot 3\cdot 3\cdot 3} = \frac{1}{81}$

17b. $\left(2^3\right)^{-1} = 2^{(3)(-1)} = 2^{-3} = \frac{1}{2^3} = \frac{1}{2\cdot 2\cdot 2} = \frac{1}{8}$

19a. $\left(3^2 4^3\right)^{-1} = 3^{(2)(-1)}4^{(3)(-1)} = 3^{-2}4^{-3} = \frac{1}{3^2 4^3} = \frac{1}{(3\cdot 3)(4\cdot 4\cdot 4)} = \frac{1}{9\cdot 64} = \frac{1}{576}$

19b. $\frac{4^5}{4^2} = 4^{5-2} = 4^3 = 4\cdot 4\cdot 4 = 64$

21a. $\frac{1^9}{1^7} = 1^{9-7} = 1^2 = 1\cdot 1 = 1$

21b. $\frac{1}{4^{-3}} = 4^3 = 4\cdot 4\cdot 4 = 64$

23a. $\frac{5^{-2}}{5^{-4}} = 5^{-2-(-4)} = 5^2 = 5\cdot 5 = 25$

23b. $\left(\frac{2}{7}\right)^{-2} = \left(\frac{7}{2}\right)^{2} = \frac{7^2}{2^2} = \frac{7\cdot 7}{2\cdot 2} = \frac{49}{4}$

25a. $x^{-1} = \frac{1}{x}$

25b. $a^{-4} = \frac{1}{a^4}$

27a. $x^{-2}\cdot x^{-1}\cdot x = x^{-2+(-1)+1} = x^{-2} = \frac{1}{x^2}$

27b. $a^{-5}\cdot a^{-2}\cdot a^{-1} = a^{-5+(-2)+(-1)} = a^{-8} = \frac{1}{a^8}$

29a. $x^2y^{-3}x^{-5}y^6 = x^{2+(-5)}y^{-3+6} = x^{-3}y^3 = \frac{y^3}{x^3}$

29b. $(xy)^{-3} = x^{-3}y^{-3} = \frac{1}{x^3y^3}$

31a. $(2t)^{-4} = 2^{-4}t^{-4} = \frac{1}{2^4t^4} = \frac{1}{16t^4}$

31b. $(x+1)^{-7} = \frac{1}{(x+1)^7}$

33a. $\left(a^{-2}\right)^{-4} = a^{(-2)(-4)} = a^8$

33b. $\left(rt^3\right)^{-2} = r^{-2}t^{(3)(-2)} = r^{-2}t^{-6} = \frac{1}{r^2t^6}$

35a. $(ab)^2\left(a^2\right)^{-3} = \left(a^2b^2\right)\left(a^{(2)(-3)}\right) = \left(a^2b^2\right)\left(a^{-6}\right) = a^{2+(-6)}b^2 = a^{-4}b^2 = \frac{b^2}{a^4}$

35b. $\frac{x^4}{x^2} = x^{4-2} = x^2$

37a. $\frac{a^{10}}{a^{-3}} = a^{10-(-3)} = a^{13}$

37b. $\frac{4z}{2z^4} = \frac{4}{2}\cdot\frac{z}{z^4} = 2z^{1-4} = 2z^{-3} = \frac{2}{z^3}$

39a. $\frac{-4xy^5}{6x^3y^2} = \frac{-4}{6}\cdot\frac{xy^5}{x^3y^2} = \frac{-2}{3}x^{1-3}y^{5-2} = \frac{-2}{3}x^{-2}y^3 = -\frac{2y^3}{3x^2}$

39b. $\frac{x^{-4}}{x^{-1}} = x^{-4-(-1)} = x^{-3} = \frac{1}{x^3}$

41a. $\frac{10b^{-4}}{5b^{-5}} = \frac{10}{5}\cdot\frac{b^{-4}}{b^{-5}} = 2b^{-4-(-5)} = 2b$

41b. $\left(\frac{a}{b}\right)^3 = \frac{a^3}{b^3}$

43a. $\frac{1}{y^{-5}} = y^5$

43b. $\frac{4}{2t^{-3}} = \frac{4}{2}\cdot\frac{1}{t^{-3}} = 2\cdot t^3 = 2t^3$

45a. $\frac{1}{(xy)^{-2}} = \frac{1}{x^{-2}y^{-2}} = x^2y^2$

45b. $\frac{1}{\left(a^2b\right)^{-3}} = \frac{1}{a^{(2)(-3)}b^{-3}} = \frac{1}{a^{-6}b^{-3}} = a^6b^3$

47a. $\left(\frac{a}{b}\right)^{-2} = \left(\frac{b}{a}\right)^{2} = \frac{b^2}{a^2}$

47b. $\left(\frac{u}{4v}\right)^{-1} = \left(\frac{4v}{u}\right)^{1} = \frac{4v}{u}$

49. $\frac{a^n}{a^m} = a^{n-m} = a^{-1(m-n)} = (a^{m-n})^{-1} = \frac{1}{a^{m-n}}$

Scientific Notation

51. Thousand

53. Billion

55. Hundredth

57. Move the decimal point three places to the right, $2\times10^3 = 2000$.

59. Move the decimal point four places to the right, $4.5\times10^4 = 45{,}000$.

61. Move the decimal point three places to the left, $8\times10^{-3} = 0.008$.

63. Move the decimal point four places to the left, $4.56\times10^{-4} = 0.000456$.

65. Move the decimal point seven places to the right, $3.9\times10^7 = 39{,}000{,}000$.

67. Move the decimal point five places to the right, $-5\times10^5 = -500{,}000$.

69. Move the decimal point three places to the left, $2000 = 2\times10^3$.

71. Move the decimal point two places to the left, $567 = 5.67\times10^2$.

73. Move the decimal point seven places to the left, $12{,}000{,}000 = 1.2\times10^7$.

75. Move the decimal point three places to the right, $0.004 = 4\times10^{-3}$.

77. Move the decimal point four places to the right, $0.000895 = 8.95\times10^{-4}$.

79. Move the decimal point two places to the right, $-0.05 = -5\times10^{-2}$.

81. $\left(5\times10^3\right)\left(3\times10^2\right) = (5\cdot3)\times\left(10^3\cdot10^2\right) = 15\times10^{3+2} = 15\times10^5 = 1.5\times10^6$;

Move the decimal point six places to the right, $1.5\times10^6 = 1{,}500{,}000$.

83. $\left(-3\times10^{-3}\right)\left(5\times10^2\right) = (-3\cdot5)\times\left(10^{-3}\cdot10^2\right) = -15\times10^{-3+2} = -15\times10^{-1} = -1.5\times10^0 = -1.5$

85. $\frac{4\times10^5}{2\times10^2} = \frac{4}{2}\cdot\frac{10^5}{10^2} = 2\times10^{5-2} = 2\times10^3$;

Move the decimal point three places to the right, $2\times10^3 = 2000$.

87. $\frac{8\times10^{-6}}{4\times10^{-3}} = \frac{8}{4}\cdot\frac{10^{-6}}{10^{-3}} = 2\times10^{-6-(-3)} = 2\times10^{-3}$;

Move the decimal point three places to the left, $2\times10^{-3} = 0.002$.

Applications

89. (a) $(1.86\times10^5)(3.15\times10^7)=(1.86\cdot3.15)\times(10^5\cdot10^7)\approx5.859\times10^{5+7}\approx5.859\times10^{12}$ miles.

(b) $(5.859\times10^{12})(4.27)=(5.859\cdot4.27)\times(10^{12})\approx25\times10^{12}\approx2.5\times10^{13}$ miles.

91. $\frac{10^5\cdot\pi(\text{light years})}{1}\div\frac{2\times10^8(\text{years})}{1}\cdot\frac{5.859\times10^{12}}{1}\approx9.2\times10^9$ miles.

93. (a) Move the decimal point 13 places to the left, $12,460,000,000,000=1.246\times10^{13}$.

(b) $\frac{1.246\times10^{13}}{2.98\times10^8}=\frac{1.246}{2.98}\cdot\frac{10^{13}}{10^8}\approx0.41812\times10^{13-8}=0.41812\times10^5=\$41,812$

5.6: Division of Polynomials

Concepts

1. $\frac{a+b}{d}=\frac{a}{d}+\frac{b}{d}$

2. $\frac{a+b-c}{d}=\frac{a}{d}+\frac{b}{d}-\frac{c}{d}$

3. term

4. $\frac{5x^2+2x}{2x}=\frac{5x^2}{2x}+\frac{2x}{2x}=\frac{5x}{2}+1\neq5x^2+1$; No

5. $\frac{5x^2+2x}{2x}=\frac{5x^2}{2x}+\frac{2x}{2x}=\frac{5x}{2}+1\neq\frac{5x^2}{2x}$; No

6. $9\cdot4+1=37$

7. $(x+1)\cdot(2x^2-2x+1)+4=2x^3-x+5$

8. $0x^2$

Division by a Monomial

9. $\frac{6x^2}{3x}=\frac{6}{3}\cdot\frac{x^2}{x}=2x^{2-1}=2x$; Checking: $3x\cdot2x=(3\cdot2)(x\cdot x)=6x^2$

11. $\frac{z^4+z^3}{z}=\frac{z^4}{z}+\frac{z^3}{z}=z^{4-1}+z^{3-1}=z^3+z^2$; Checking: $(z)(z^3+z^2)=(z^3)(z)+(z^2)(z)=z^4+z^3$

13. $\frac{a^5-6a^3}{2a^3}=\frac{a^5}{2a^3}-\frac{6a^3}{2a^3}=\frac{1}{2}a^{5-3}-3a^{3-3}=\frac{a^2}{2}-3a^0=\frac{a^2}{2}-3$

Checking: $(2a^3)\left(\frac{a^2}{2}-3\right)=(2a^3)\left(\frac{a^2}{2}\right)-(2a^3)(3)=\frac{2a^{3+2}}{2}-6a^3=a^5-6a^3$

15. $\frac{4x-7x^4}{x^2}=\frac{4x}{x^2}-\frac{7x^4}{x^2}=4x^{1-2}-7x^{4-2}=4x^{-1}-7x^2=\frac{4}{x}-7x^2$

17. $\frac{9x^4-3x+6}{3x}=\frac{9x^4}{3x}-\frac{3x}{3x}+\frac{6}{3x}=3x^{4-1}-1+\frac{2}{x}=3x^3-1+\frac{2}{x}$

19. $\frac{12y^4-3y^2+6y}{3y^2}=\frac{12y^4}{3y^2}-\frac{3y^2}{3y^2}+\frac{6y}{3y^2}=4y^{4-2}-1+2y^{1-2}=4y^2-1+2y^{-1}=4y^2-1+\frac{2}{y}$

21. $\frac{15m^4-10m^3+20m^2}{5m^2}=\frac{15m^4}{5m^2}-\frac{10m^3}{5m^2}+\frac{20m^2}{5m^2}=3m^{4-2}-2m^{3-2}+4m^{2-2}=3m^2-2m+4$

23.

$$\begin{array}{r|l} & 2x+1 \\ x-2 & 2x^2-3x+1 \\ & \underline{2x^2-4x} \\ & \quad\quad x+1 \\ & \quad\quad \underline{x-2} \\ & \quad\quad\quad 3 \end{array}$$

The solution is: $2x+1+\frac{3}{x-2}$

Checking: $(x-2)(2x+1)+3=x\cdot 2x+x\cdot 1+(-2)(2x)+(-2)(1)+3=2x^2+x-4x-2+3=2x^2-3x+1$

25.

$$\begin{array}{r|l} & x+1 \\ x+1 & x^2+2x+1 \\ & \underline{x^2+\ x} \\ & \quad\quad x+1 \\ & \quad\quad \underline{x+1} \\ & \quad\quad\quad 0 \end{array}$$

The solution is: $x+1$

Checking: $(x+1)(x+1)=(x)^2+(2)(x)(1)+(1)^2=x^2+2x+1$

27.

$$\begin{array}{r|l} & x^2+1 \\ x-1 & x^3-x^2+x-2 \\ & \underline{x^3-x^2} \\ & \quad\quad\quad x-2 \\ & \quad\quad\quad \underline{x-1} \\ & \quad\quad\quad -1 \end{array}$$

The solution is: $x^2+1+\frac{-1}{x-1}$

Checking: $(x-1)(x^2+1)+(-1)=x\cdot x^2+x\cdot 1+(-1)(x^2)+(-1)(1)+(-1)=$

$x^3+x-x^2-1-1=x^3-x^2+x-2$

29.

$$\begin{array}{r} x^2 - x + 2 \\ 4x+1 \overline{\smash{\big)} 4x^3 - 3x^2 + 7x + 3} \\ \underline{4x^3 + x^2 } \\ -4x^2 + 7x \\ \underline{-4x^2 - x } \\ 8x + 3 \\ \underline{8x + 2} \\ 1 \end{array}$$

The solution is: $x^2 - x + 2 + \dfrac{1}{4x+1}$

31. $x-2\overline{\smash{\big)}x^3 - x + 2} \Rightarrow$

$$\begin{array}{r} x^2 + 2x + 3 \\ x-2 \overline{\smash{\big)} x^3 - 0x^2 - x + 2} \\ \underline{x^3 - 2x^2 } \\ 2x^2 - x \\ \underline{2x^2 - 4x } \\ 3x + 2 \\ \underline{3x - 6} \\ 8 \end{array}$$

The solution is:

$x^2 + 2x + 3 + \dfrac{8}{x-2}$

33. $x-1\overline{\smash{\big)}3x^3 + 2} \Rightarrow$

$$\begin{array}{r} 3x^2 + 3x + 3 \\ x-1 \overline{\smash{\big)} 3x^3 + 0x^2 + 0x + 2} \\ \underline{3x^3 - 3x^2 } \\ 3x^2 + 0x \\ \underline{3x^2 - 3x } \\ 3x + 2 \\ \underline{3x - 3} \\ 5 \end{array}$$

The solution is:

$3x^2 + 3x + 3 + \dfrac{5}{x-1}$

35. $x^2+1\overline{\smash{\big)}x^3 + 3x^2 + 1} \Rightarrow$

$$\begin{array}{r} x + 3 \\ x^2+1 \overline{\smash{\big)} x^3 + 3x^2 + 0x + 1} \\ \underline{x^3 + 0x^2 + x } \\ 3x^2 - x + 1 \\ \underline{3x^2 + 0x + 3} \\ -x - 2 \end{array}$$

The solution is:

$x + 3 + \dfrac{-x-2}{x^2+1}$

37. $x^2-x+1\overline{\smash{\big)}x^3 + 1} \Rightarrow$

$$\begin{array}{r} x + 1 \\ x^2-x+1 \overline{\smash{\big)} x^3 + 0x^2 + 0x + 1} \\ \underline{x^3 - x^2 + x } \\ x^2 - x + 1 \\ \underline{x^2 - x + 1} \\ 0 \end{array}$$

The solution is: $x + 1$

39. $x+2\overline{)x^3+8} \Rightarrow$

$$\begin{array}{r} x^2-2x+4 \\ x+2\overline{)x^3+0x^2+0x+8} \\ \underline{x^3+2x^2} \\ -2x^2+0x \\ \underline{-2x^2-4x} \\ 4x+8 \\ \underline{4x+8} \\ 0 \end{array}$$

The solution is: x^2-2x+4

41. They are the same.

Applications

43. $2x \cdot L = 8x^2 \Rightarrow L = \dfrac{8x^2}{2x} \Rightarrow L = 4x^{2-1} \Rightarrow L = 4x$

45. $2x^2 \cdot H = 2x^3 + 4x^2 \Rightarrow H = \dfrac{2x^3+4x^2}{2x^2} \Rightarrow H = \dfrac{2x^3}{2x^2} + \dfrac{4x^2}{2x^2} = x^{3-2} + 2x^{2-2} = x+2.$ See Figure 45.

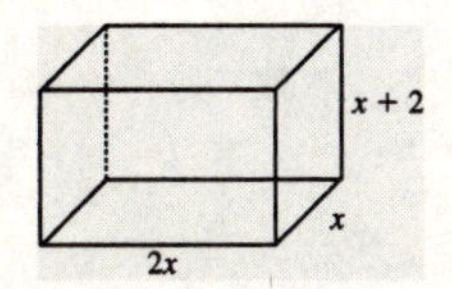

Figure 45

Checking Basic Concepts for Sections 5.5 & 5.6

1. (a) $9^{-2} = \dfrac{1}{9^2} = \dfrac{1}{9 \cdot 9} = \dfrac{1}{81}$

(b) $\dfrac{3x^{-3}}{6x^4} = \dfrac{1}{2}x^{-3-4} = \dfrac{1}{2}x^{-7} = \dfrac{1}{2x^7}$

(c) $\left(4ab^{-4}\right)^{-2} = 4^{-2}a^{-2}b^8 = \dfrac{b^8}{16a^2}$

2. (a) $\dfrac{1}{z^{-5}} = z^5$

(b) $\dfrac{x^{-3}}{y^{-6}} = \dfrac{y^6}{x^3}$

(c) $\left(\dfrac{3}{x^2}\right)^{-3} = \left(\dfrac{x^2}{3}\right)^3 = \dfrac{x^{2 \cdot 3}}{3^3} = \dfrac{x^6}{27}$

3. (a) Move the decimal point four places to the left: $45{,}000 = 4.5 \times 10^4$.

(b) Move the decimal point four places to the right: $0.000234 = 2.34 \times 10^{-4}$.

(c) Move the decimal point two places to the right: $0.01 = 1 \times 10^{-2}$.

4. (a) Move the decimal point four places to the right: $4.71 \times 10^4 = 47{,}100$.

(b) Move the decimal point three places to the left: $6\times10^{-3}=0.006$.

5. $\dfrac{25a^4-15a^3}{5a^3}=\dfrac{25a^4}{5a^3}-\dfrac{15a^3}{5a^3}=5a^{4-3}-3a^{3-3}=5a-3$

6. The quotient is: $3x+2$; The remainder is: -2

$$
\begin{array}{r}
3x+2 \\
x-1\overline{)\,3x^2-x-4} \\
\underline{3x^2-3x} \\
2x-4 \\
\underline{2x-2} \\
-2
\end{array}
$$

7. (a) Move the decimal point seven places to the left: $93{,}000{,}000=9.3\times10^7$.

(b) $\dfrac{9.3\times10^7}{1.86\times10^5}=\dfrac{9.3}{1.86}\cdot\dfrac{10^7}{10^5}\approx5\times10^{7-5}=5\times10^2=500$ seconds. (8 minutes 20 seconds)

Chapter 5 Review Exercises

Section 5.1

1. $5^3=5\cdot5\cdot5=125$

2. $-3^4=-(3\cdot3\cdot3\cdot3)=-81$

3. $4(-2)^0=4(1)=4$

4. $3+3^2-3^0=3+(3\cdot3)-1=3+9-1=11$

5. $\dfrac{-5^2}{5}=\dfrac{-(5\cdot5)}{5}=\dfrac{-25}{5}=-5$

6. $\left(\dfrac{-5}{5}\right)^2=\dfrac{(-5)(-5)}{(5)(5)}=\dfrac{25}{25}=1$

7. $6^2\cdot6^3=6^{2+3}=6^5$

8. $10^5\cdot10^7=10^{5+7}=10^{12}$

9. $z^4\cdot z^5=z^{4+5}=z^9$

10. $y^2\cdot y\cdot y^3=y^{2+1+3}=y^6$

11. $5x^2\cdot6x^7=5\cdot6\cdot x^{2+7}=30x^9$

12. $(ab^3)(a^3b)=a^{1+3}b^{3+1}=a^4b^4$

13. $(2^5)^2=2^{5\cdot2}=2^{10}$

14. $(m^4)^5=m^{4\cdot5}=m^{20}$

15. $(ab)^3=a^3b^3$

16. $\left(x^2y^3\right)^4 = x^{2\cdot 4}y^{3\cdot 4} = x^8y^{12}$

17. $(xy)^3(x^2y^4)^2 = (x^3y^3)(x^{2\cdot 2}y^{4\cdot 2}) = (x^3y^3)(x^4y^8) = x^{3+4}y^{3+8} = x^7y^{11}$

18. $\left(a^2b^9\right)^0 = 1$

19. $(r-t)^4(r-t)^5 = (r-t)^{4+5} = (r-t)^9$

20. $(a+b)^2(a+b)^4 = (a+b)^{2+4} = (a+b)^6$

21. $\left(\dfrac{3}{x-y}\right)^2 = \dfrac{3^2}{(x-y)^2} = \dfrac{3\cdot 3}{(x-y)^2} = \dfrac{9}{(x-y)^2}$

22. $\left(\dfrac{x+y}{2}\right)^3 = \dfrac{(x+y)^3}{2^3} = \dfrac{(x+y)^3}{2\cdot 2\cdot 2} = \dfrac{(x+y)^3}{8}$

Section 5.2

23. degree is 7; coefficient is 6

24. degree is 5; coefficient is -1

25. Yes, it is a polynomial; 1 term: $8y$; 1 variable: y; y^1 so it is a 1st degree polynomial.

26. Yes, it is a polynomial; 4 terms: $8x^3$, $-3x^2$, x and -5; 1 variables: x; x^3 so it is a 3rd degree polynomial.

27. Yes, it is a polynomial; 3 terms: a^2, $2ab$ and b^2; 2 variables: a and b; a^2 so it is a 2nd degree polynomial.

28. No, it is not a polynomial, because there are variables in the denominator.

29.
$$\begin{array}{r} 3x^2+4x+8 \\ +\,2x^2-5x-5 \\ \hline 5x^2-\;\;x+3 \end{array}$$

30. $-6x^2+3x+7$

31. $(4x-3)+(-x+7) = 4x+(-x)+(-3)+7 = 3x+4$

32. $\left(3x^2-1\right)-\left(5x^2+12\right) = \left(3x^2-1\right)+\left(-5x^2-12\right) = 3x^2+\left(-5x^2\right)+(-1)+(-12) = -2x^2-13$

33. $\left(x^2+5x+6\right)-\left(3x^2-4x+1\right) = \left(x^2+5x+6\right)+\left(-3x^2+4x-1\right) =$

$x^2+\left(-3x^2\right)+5x+4x+6+(-1) = -2x^2+9x+5$

34. $\left(a^3+4a^2\right)+\left(a^3-5a^2+7a\right) = a^3+a^3+4a^2+\left(-5a^2\right)+7a = 2a^3-a^2+7a$

35. $\left(xy+y^2\right)+\left(4y^2-4xy\right) = y^2+4y^2+xy+(-4xy) = 5y^2-3xy$

36. $\left(7x^2+2xy+y^2\right)-\left(7x^2-2xy+y^2\right) = \left(7x^2+2xy+y^2\right)+\left(-7x^2+2xy-y^2\right) =$

$7x^2+\left(-7x^2\right)+2xy+2xy+y^2+\left(-y^2\right) = 4xy$

Section 5.3

37. $-x^2 \cdot x^3 = -x^{2+3} = -x^5$

38. $-\left(r^2t^3\right)(rt) = -\left(r^{2+1}t^{3+1}\right) = -r^3t^4$

39. $-3(2t-5) = (-3)(2t) + (-3)(-5) = -6t + 15$

40. $2y(1-6y) = 2y \cdot 1 + 2y(-6y) = 2y - 12y^2$

41. $6x^3\left(3x^2+5x\right) = 6x^3 \cdot 3x^2 + 6x^3 \cdot 5x = 18x^{3+2} + 30x^{3+1} = 18x^5 + 30x^4$

42. $-x\left(x^2-2x+9\right) = (-x)\left(x^2\right) + (-x)(-2x) + (-x)(9) = -x^{1+2} + 2x^{1+1} - 9x = -x^3 + 2x^2 - 9x$

43. $-ab\left(a^2-2ab+b^2\right) = -ab \cdot a^2 + (-ab)(-2ab) + (-ab)\left(b^2\right) = -a^3b + 2a^2b^2 - ab^3$

44. $(a-2)(a+5) = a \cdot a + a \cdot 5 + (-2)(a) + (-2)(5) = a^2 + 5a - 2a - 10 = a^2 + 3a - 10$

45. $(8x-3)(x+2) = 8x \cdot x + 8x \cdot 2 + (-3)(x) + (-3)(2) = 8x^2 + 16x - 3x - 6 = 8x^2 + 13x - 6$

46. $(2x-1)(1-x) = 2x \cdot 1 + (2x)(-x) + (-1)(1) + (-1)(-x) = 2x - 2x^2 - 1 + x = -2x^2 + 3x - 1$

47. $\left(y^2+1\right)(2y+1) = y^2 \cdot 2y + y^2 \cdot 1 + 1 \cdot 2y + 1 \cdot 1 = 2y^3 + y^2 + 2y + 1$

48. $\left(y^2-1\right)\left(2y^2+1\right) = y^2 \cdot 2y^2 + y^2 \cdot 1 + (-1)\left(2y^2\right) + (-1)(1) = 2y^4 + y^2 - 2y^2 - 1 = 2y^4 - y^2 - 1$

49. $(z+1)(z^2-z+1) = z \cdot z^2 + (z)(-z) + z \cdot 1 + 1 \cdot z^2 + (1)(-z) + 1 \cdot 1 = z^3 - z^2 + z + z^2 - z + 1 = z^3 + 1$

50. $(4z-3)(z^2-3z+1) = 4z \cdot z^2 + (4z)(-3z) + 4z \cdot 1 + (-3)(z^2) + (-3)(-3z) + (-3)(1) =$

$4z^3 - 12z^2 + 4z - 3z^2 + 9z - 3 = 4z^3 - 15z^2 + 13z - 3$

51. (a) $z^2 + z$; See Figure 51.

(b) $z(z+1) = z \cdot z + z \cdot 1 = z^2 + z$

52. (a) $2x^2 + 4x$; See Figure 52.

(b) $2x(x+2) = 2x \cdot x + 2x \cdot 2 = 2x^2 + 4x$

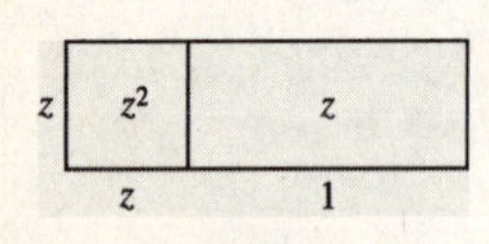

Figure 51

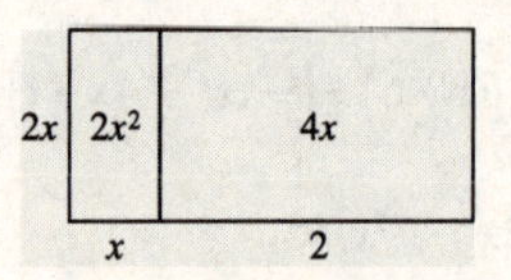

Figure 52

Section 5.4

53. $(z+2)(z-2) = (z)^2 - (2)^2 = z^2 - 4$

54. $(5z-9)(5z+9) = (5z)^2 - (9)^2 = 25z^2 - 81$

55. $(1-3y)(1+3y) = (1)^2 - (3y)^2 = 1 - 9y^2$

56. $(5x+4y)(5x-4y)=(5x)^2-(4y)^2=25x^2-16y^2$

57. $(rt+1)(rt-1)=(rt)^2-(1)^2=r^2t^2-1$

58. $(2m^2-n^2)(2m^2+n^2)=(2m^2)^2-(n^2)^2=4m^4-n^4$

59. $(x+1)^2=(x)^2+(2)(x)(1)+(1)^2=x^2+2x+1$

60. $(4x+3)^2=(4x)^2+(2)(4x)(3)+(3)^2=16x^2+24x+9$

61. $(y-3)^2=(y)^2-(2)(y)(3)+(3)^2=y^2-6y+9$

62. $(2y-5)^2=(2y)^2-(2)(2y)(5)+(5)^2=4y^2-20y+25$

63. $(4+a)^2=(4)^2+(2)(4)(a)+(a)^2=16+8a+a^2$

64. $(4-a)^2=(4)^2-(2)(4)(a)+(a)^2=16-8a+a^2$

65. $(x^2+y^2)^2=(x^2)^2+(2)(x^2)(y^2)+(y^2)^2=x^4+2x^2y^2+y^4$

66. $(xy-2)^2=(xy)^2-(2)(xy)(2)+(2)^2=x^2y^2-4xy+4$

67. $(z+5)^3=(z+5)(z+5)^2=(z+5)(z^2+10z+25)=$

$z\cdot z^2+z\cdot 10z+z\cdot 25+5\cdot z^2+5\cdot 10z+5\cdot 25=z^3+10z^2+25z+5z^2+50z+125=\ z^3+15z^2+75z+125$

68. $(2z-1)^3=(2z-1)(2z-1)^2=(2z-1)(4z^2-4z+1)=$

$2z\cdot 4z^2+(2z)(-4z)+2z\cdot 1+(-1)(4z^2)+(-1)(-4z)+(-1)(1)=8z^3-8z^2+2z-4z^2+4z-1=$

$8z^3-12z^2+6z-1$

69. $59\cdot 61=(60-1)(60+1)=60^2-1^2=3600-1=3599$

70. $22\cdot 18=(20+2)(20-2)=20^2-2^2=400-4=396$

Section 5.5

71. $9^{-1}=\frac{1}{9}$

72. $3^{-2}=\frac{1}{3^2}=\frac{1}{3\cdot 3}=\frac{1}{9}$

73. $4^3\cdot 4^{-2}=4^{3+(-2)}=4^1=4$

74. $10^{-6}\cdot 10^3=10^{-6+3}=10^{-3}=\frac{1}{10^3}=\frac{1}{10\cdot 10\cdot 10}=\frac{1}{1000}$

75. $\frac{1}{6^{-2}}=6^2=36$

76. $\dfrac{5^7}{5^9} = 5^{7-9} = 5^{-2} = \dfrac{1}{5^2} = \dfrac{1}{5 \cdot 5} = \dfrac{1}{25}$

77. $z^{-2} = \dfrac{1}{z^2}$

78. $y^{-4} = \dfrac{1}{y^4}$

79. $a^{-4} \cdot a^2 = a^{-4+2} = a^{-2} = \dfrac{1}{a^2}$

80. $x^2 \cdot x^{-5} \cdot x = x^{2+(-5)+1} = x^{-2} = \dfrac{1}{x^2}$

81. $(2t)^{-2} = 2^{-2}t^{-2} = \dfrac{1}{2^2 t^2} = \dfrac{1}{4t^2}$

82. $\left(ab^2\right)^{-3} = a^{-3}b^{(2)(-3)} = a^{-3}b^{-6} = \dfrac{1}{a^3 b^6}$

83. $(xy)^{-2}(x^{-2}y)^{-1} = (x^{-2}y^{-2})(x^{(-2)(-1)}y^{-1}) = x^{-2+2}y^{-2+(-1)} = y^{-3} = \dfrac{1}{y^3}$

84. $\dfrac{x^6}{x^2} = x^{6-2} = x^4$

85. $\dfrac{4x}{2x^4} = 2x^{1-4} = 2x^{-3} = \dfrac{2}{x^3}$

86. $\dfrac{20x^5y^3}{30xy^6} = \dfrac{2}{3}x^{5-1}y^{3-6} = \dfrac{2}{3}x^4y^{-3} = \dfrac{2x^4}{3y^3}$

87. $\left(\dfrac{a}{b}\right)^5 = \dfrac{a^5}{b^5}$

88. $\dfrac{4}{t^{-4}} = 4t^4$

89. $\left(\dfrac{x}{3}\right)^{-3} = \left(\dfrac{3}{x}\right)^3 = \dfrac{3^3}{x^3} = \dfrac{27}{x^3}$

90. $\dfrac{2}{(ab)^{-1}} = 2ab$

91. $\left(\dfrac{x}{y}\right)^{-2} = \left(\dfrac{y}{x}\right)^2 = \dfrac{y^2}{x^2}$

92. $\left(\dfrac{3u}{2v}\right)^{-1} = \left(\dfrac{2v}{3u}\right)^1 = \dfrac{2v}{3u}$

93. Move the decimal point two places to the right, $6 \times 10^2 = 600$.

94. Move the decimal point four places to the right, $5.24\times10^4 = 52,400$.

95. Move the decimal point three places to the left, $3.7\times10^{-3} = 0.0037$.

96. Move the decimal point two places to the left, $6.234\times10^{-2} = 0.06234$.

97. Move the decimal point four places to the left, $10,000 = 1\times10^4$.

98. Move the decimal point seven places to the left, $56,100,000 = 5.61\times10^7$.

99. Move the decimal point five places to the right, $0.000054 = 5.4\times10^{-5}$.

100. Move the decimal point three places to the right, $0.001 = 1\times10^{-3}$.

101. $(4\times10^2)(6\times10^4) = (4\cdot6)\times(10^2\cdot10^4) = 24\times10^6 = 2.4\times10^7$;

Move the decimal point seven places to the right, $2.4\times10^7 = 24,000,000$.

102. $\left(\dfrac{8\times10^3}{4\times10^4}\right) = \dfrac{8}{4}\times\dfrac{10^3}{10^4} = 2\times10^{3-4} = 2\times10^{-1}$;

Move the decimal point one place to the left, $2\times10^{-1} = 0.2$.

Section 5.6

103. $\dfrac{5x^2+3x}{3x} = \dfrac{5x^2}{3x} + \dfrac{3x}{3x} = \dfrac{5}{3}x^{2-1} + 1 = \dfrac{5}{3}x + 1$; Checking: $3x\left(\dfrac{5}{3}x+1\right) = 3x\cdot\dfrac{5}{3}x + 3x\cdot1 = 5x^2+3x$

104. $\dfrac{6b^4-4b^2+2}{2b^2} = \dfrac{6b^4}{2b^2} - \dfrac{4b^2}{2b^2} + \dfrac{2}{2b^2} = 3b^{4-2} - 2b^{2-2} + \dfrac{1}{b^2} = 3b^2 - 2 + \dfrac{1}{b^2}$;

Checking: $(2b^2)\left(3b^2-2+\dfrac{1}{b^2}\right) = 2b^2\cdot3b^2 + (2b^2)(-2) + 2b^2\cdot\dfrac{1}{b^2} = 6b^4-4b^2+2$

105.

$$\begin{array}{r|l} & 3x+2 \\ \hline x-1 & 3x^2 - x + 2 \\ & \underline{3x^2 - 3x} \\ & \quad 2x+2 \\ & \quad \underline{2x-2} \\ & \qquad 4 \end{array}$$

The solution is: $3x+2+\dfrac{4}{x-1}$

Checking: $(x-1)(3x+2)+4 = x\cdot3x + x\cdot2 + (-1)(3x) + (-1)(2) + 4 =$

$3x^2+2x-3x-2+4 = 3x^2-x+2$

106.

$$\begin{array}{r|l} & 3x-4 \\ \hline 3x+2 & 9x^2-6x-2 \\ & \underline{9x^2+6x} \\ & \quad -12x-2 \\ & \quad \underline{-12x-8} \\ & \qquad 6 \end{array}$$

The solution is: $3x-4+\dfrac{6}{3x+2}$

Checking: $(3x+2)(3x-4)+6=3x\cdot 3x+(3x)(-4)+2\cdot 3x+(2)(-4)+6=$

$9x^2-12x+6x-8+6=9x^2-6x-2$

107.

$$\begin{array}{r l}
 & x^2-3x-1 \\
4x+1\,\big) & 4x^3-11x^2-7x-1 \\
 & \underline{4x^3+\ \ x^2} \\
 & \quad -12x^2-7x \\
 & \quad \underline{-12x^2-3x} \\
 & \qquad\quad -4x-1 \\
 & \qquad\quad \underline{-4x-1} \\
 & \qquad\qquad 0
\end{array}$$

The solution is: x^2-3x-1

Checking: $(4x+1)(x^2-3x-1)=4x\cdot x^2+(4x)(-3x)+(4x)(-1)+1\cdot x^2+(1)(-3x)+(1)(-1)=$

$4x^3-12x^2-4x+x^2-3x-1=4x^3-11x^2-7x-1$

108.

$$\begin{array}{r l}
 & x^2 \\
2x-1\,\big) & 2x^3-x^2-1 \\
 & \underline{2x^3-x^2} \\
 & \qquad \underline{0-1} \\
 & \qquad\ \ -1
\end{array}$$

The solution is: $x^2+\dfrac{-1}{2x-1}$

Checking: $(x^2)(2x-1)+(-1)=x^2\cdot 2x+(x^2)(-1)+(-1)=2x^3-x^2-1$

109.

$$\begin{array}{r l}
 & x-1 \\
x^2+1\,\big) & x^3-x^2-\ \ x+1 \\
 & \underline{x^3+0x^2+\ \ x} \\
 & \quad -x^2-2x+1 \\
 & \quad \underline{-x^2+0x-1} \\
 & \qquad\quad -2x+2
\end{array}$$

The solution is: $x-1+\dfrac{-2x+2}{x^2+1}$

Checking: $(x-1)(x^2+1)+(-2x+2)=x\cdot x^2+x\cdot 1+(-1)(x^2)+(-1)(1)+(-2x+2)=$

$x^3+x-x^2-1-2x+2=x^3-x^2-x+1$

110.

$$\begin{array}{r} x^2+2x+5 \\ x^2+x+1\overline{)\,x^4+3x^3+8x^2+7x+5} \\ \underline{x^4+\ x^3+\ x^2} \qquad\qquad \\ 2x^3+7x^2+7x \\ \underline{2x^3+2x^2+2x} \\ 5x^2+5x+5 \\ \underline{5x^2+5x+5} \\ 0 \end{array}$$

The solution is: x^2+2x+5

Checking: $(x^2+x+1)(x^2+2x+5)=$

$x^2\cdot x^2+x^2\cdot 2x+x^2\cdot 5+x\cdot x^2+x\cdot 2x+x\cdot 5+1\cdot x^2+1\cdot 2x+1\cdot 5=$

$x^4+2x^3+5x^2+x^3+2x^2+5x+x^2+2x+5=x^4+3x^3+8x^2+7x+5$

Applications

111. (a) $t^2+60 \Rightarrow 0^2+60=60$ bpm

(b) $t^2+60 \Rightarrow 10^2+60=100+60=160$ bpm

(c) It increases.

112. $L\times W=2xy=1$ rectangle $\Rightarrow$ 3 rectangles $=3\cdot 2xy=6xy$; for $x=3$ ft, $y=4$ ft, $6xy=6(3)(4)=72$ ft^2

113. $5\cdot 3z=15z$, $5\cdot 2z=10z$, $2z\cdot(3z+2z)=6z^2+4z^2 \Rightarrow 15z+10z+6z^2+4z^2=10z^2+25z$; for

$z=6$ in. $\Rightarrow 10z^2+25z=10(6)^2+25(6)=10(36)+150=360+150=510$ in^2

114. $(x^2y)(x^2y)=x^{2+2}y^{1+1}=x^4y^2$

115. $P(1+0.06)^3=P(1.06)^3=P(1.191016)$; let $P=\$700 \Rightarrow \$700(1.191016)=\$833.71$.

116. $\frac{4}{3}\pi(x+2)^3=\frac{4}{3}\pi(x+2)(x+2)^2=\frac{4}{3}\pi(x+2)(x^2+4x+4)=$

$\frac{4}{3}\pi(x\cdot x^2+x\cdot 4x+x\cdot 4+2\cdot x^2+2\cdot 4x+2\cdot 4)=\frac{4}{3}\pi(x^3+4x^2+4x+2x^2+8x+8)=$

$\frac{4}{3}\pi(x^3+6x^2+12x+8)=\frac{4}{3}\pi\cdot x^3+\frac{4}{3}\pi\cdot 6x^2+\frac{4}{3}\pi\cdot 12x+\frac{4}{3}\pi\cdot 8=\frac{4}{3}\pi x^3+8\pi x^2+16\pi x+\frac{32}{3}\pi$

117. (a) $t(96-16t)=96t-16t^2$

(b) $t=2 \Rightarrow 96t-16t^2=96(2)-16(2)^2=192-16(4)=192-64=128$, after 2 seconds the ball is 128 feet high.

118. (a) If $P=2L+2W$ then $1200=2L+2W \Rightarrow 1200-2L=2W \Rightarrow 600-L=W$. Now $A=L\cdot W$ so

$A=L\cdot(600-L) \Rightarrow A=600L-L^2$.

(b) $L=50 \Rightarrow 600L-L^2=600(50)-(50)^2=30{,}000-2500=27{,}500$. A rectangular building with a perimeter of 1200 ft and a side of length 50 ft has an area of $27{,}500\text{ ft}^2$.

119. (a) $(x+5)(x+5)=(x)^2+2(x)(5)+(5)^2=x^2+10x+25$

(b) $x\cdot x+5\cdot x+x\cdot 5+5\cdot 5=x^2+5x+5x+25=x^2+10x+25$

120. (a) $(x+4)(x+4)-(x-4)(x-4)=x^2+8x+16-\left(x^2-8x+16\right)=16x$

(b) $x=100 \Rightarrow 16x=16(100)=1600$

121. 2.19 trillion $=2.19\times10^{12}$, 249 million $=2.49\times10^{8}$

$$\frac{2.19\times10^{12}}{2.49\times10^{8}}=\frac{2.19}{2.49}\times\frac{10^{12}}{10^{8}}=0.8795\times10^{12-8}=0.8795\times10^{4}=8.795\times10^{3}\approx \$8{,}795/\text{person}.$$

122. 233 million $=2.33\times10^{8}$;

$(2.23)\left(2.33\times10^{8}\right)=(2.23)(2.33)\times10^{8}=5.1959\times10^{8}\approx 519{,}590{,}000$ gal or 5.1959×10^{8} gal

Chapter 5 Test

1. Yes, it is a polynomial; 3 terms $5x^2, -3xy, -7y^3$; 2 variables x and y; $y^3 \Rightarrow$ 3rd degree.

2. x^3-4x+8

3. $(-3x+4)+(7x+2)=(-3x)+(7x)+4+2=4x+6$

4. $(y^2-2y+6)-(4y^3+5)=(y^3-2y+6)+(-4y^3-5)=y^3+(-4y^3)+(-2y)+6+(-5)=-3y^3-2y+1$

5. $\left(5x^2-x+3\right)-\left(4x^2-2x+10\right)=\left(5x^2-x+3\right)+\left(-4x^2+2x-10\right)=$

$5x^2+\left(-4x^2\right)+(-x)+2x+3+(-10)=x^2+x-7$

6. $\left(a^3+5ab\right)+\left(3a^3-3ab\right)=a^3+3a^3+5ab+(-3ab)=4a^3+2ab$

7. (a) $-4^2+10=-(4\cdot4)+10=-16+10=-6$

(b) $8^{-2}=\dfrac{1}{8^2}=\dfrac{1}{8\cdot8}=\dfrac{1}{64}$

(c) $\dfrac{1}{2^{-3}}=2^3=2\cdot2\cdot2=8$

(d) $-3x^0=-3(1)=-3$

8. $6y^4\cdot4y^7=6\cdot4y^{4+7}=24y^{11}$

9. $(a^2b^3)^2(ab^2)=a^{2\cdot2}b^{3\cdot2}ab^2=a^{4+1}b^{6+2}=a^5b^8$

10. $x^7\cdot x^{-3}=x^{7+(-3)}=x^4$

11. $(a^{-1}b^{2})^{-3} = a^{(-1)(-3)}b^{(2)(-3)} = a^{3}b^{-6} = \dfrac{a^{3}}{b^{6}}$

12. $ab\left(a^{2}-b^{2}\right) = ab\cdot a^{2} + ab\left(-b^{2}\right) = a^{3}b - ab^{3}$

13. $\left(\dfrac{3a^{2}}{2b^{-3}}\right)^{-2} = \dfrac{3^{-2}a^{(2)(-2)}}{2^{-2}b^{(-3)(-2)}} = \dfrac{2^{2}}{3^{2}a^{4}b^{6}} = \dfrac{4}{9a^{4}b^{6}}$

14. $\dfrac{12xy^{4}}{6x^{2}y} = 2x^{1-2}y^{4-1} = 2x^{-1}y^{3} = \dfrac{2y^{3}}{x}$

15. $\left(\dfrac{2}{a+b}\right)^{4} = \dfrac{2^{4}}{(a+b)^{4}} = \dfrac{16}{(a+b)^{4}}$

16. $3x^{2}\left(4x^{3}-6x+1\right) = 3x^{2}\cdot 4x^{3} + \left(3x^{2}\right)(-6x) + 3x^{2}\cdot 1 = 12x^{5} - 18x^{3} + 3x^{2}$

17. $(z-3)(2z+4) = z\cdot 2z + z\cdot 4 + (-3)(2z) + (-3)(4) = 2z^{2} + 4z - 6z - 12 = 2z^{2} - 2z - 12$

18. $\left(7y^{2}-3\right)\left(7y^{2}+3\right) = \left(7y^{2}\right)^{2} - (3)^{2} = 49y^{4} - 9$

19. $(3x-2)^{2} = (3x)^{2} - (2)(3x)(2) + (2)^{2} = 9x^{2} - 12x + 4$

20. $(m+3)^{3} = (m+3)(m+3)^{2} = (m+3)(m^{2}+6m+9) =$

$m\cdot m^{2} + m\cdot 6m + m\cdot 9 + 3\cdot m^{2} + 3\cdot 6m + 3\cdot 9 = m^{3} + 6m^{2} + 9m + 3m^{2} + 18m + 27 = m^{3} + 9m^{2} + 27m + 27$

21. $(y+2)(y^{2}-2y+3) = (y+2)(y^{2}) + (y+2)(-2y) + (y+2)(3)$

$= y\cdot y^{2} + 2\cdot y^{2} + y(-2y) + 2(-2y) + y\cdot 3 + 2\cdot 3$

$= y^{3} + 2y^{2} + (-2y^{2}) + (-4y) + 3y + 6$

$= y^{3} - y + 6$

22. $78\cdot 82 = (80-2)(80+2) = (80)^{2} - (2)^{2}$

$= 6400 - 4 = 6396$

23. Move the decimal point three places to the left, $6.1\times 10^{-3} = 0.0061$.

24. Move the decimal point three places to the left, $5410 = 5.41\times 10^{3}$.

25. $\dfrac{9x^{3}-6x^{2}+3x}{3x^{2}} = \dfrac{9x^{3}}{3x^{2}} - \dfrac{6x^{2}}{3x^{2}} + \dfrac{3x}{3x^{2}} = 3x^{3-2} - 2x^{2-2} + \dfrac{1}{x} = 3x - 2 + \dfrac{1}{x}$

26.
$$\begin{array}{r} x^2 - x + 1 \\ x+2\overline{)x^3 + x^2 - x + 1} \\ \underline{x^3 + 2x^2} \\ -x^2 - x \\ \underline{-x^2 - 2x} \\ x+1 \\ \underline{x+2} \\ -1 \end{array}$$

The solution is: $x^2 - x + 1 + \frac{-1}{x+2}$

27. (a) $20t$

(b) $2t + 2000$

(c) $(20t) - (2t + 2000) = 20t - 2t - 2000 = 18t - 2000$; profit from selling t tickets.

28. $(3x)(2x) + (3x)(2x) = 6x^2 + 6x^2 = 12x^2$; For $x = 10$ feet $\Rightarrow 12x^2 = 12(10)^2 = 12(100) = 1200 \text{ ft}^2$

29. $(3x)(x+3)(x+6) = (3x)(x \cdot x + x \cdot 6 + 3 \cdot x + 3 \cdot 6) = (3x)\left(x^2 + 9x + 18\right) =$

$3x \cdot x^2 + 3x \cdot 9x + 3x \cdot 18 = 3x^3 + 27x^2 + 54x$

30. (a) $t(88 - 16t) = t \cdot 88 - t \cdot 16t = 88t - 16t^2$

(b) $t = 3 \Rightarrow 88t - 16t^2 = 88(3) - 16(3)^2 = 264 - 16(9) = 264 - 144 = 120$

After 3 seconds the ball is 120 feet high.

Chapter 5 Extended and Discovery Exercises

1. Conjecture: Make the power of the 10's the same and you can add or subtract the lead numbers and multiply by that power of 10. $\left(4 \times 10^3\right) + \left(3 \times 10^3\right) = (4+3) \times 10^3 = 7 \times 10^3$; Answer checks.

2. Conjecture: Make the power of the 10's the same and you can add or subtract the lead numbers and multiply by that power of 10. $\left(5 \times 10^{-2}\right) - \left(2 \times 10^{-2}\right) = (5-2) \times 10^{-2} = 3 \times 10^{-2}$; Answer checks.

3. Conjecture: Make the power of the 10's the same and you can add or subtract the lead numbers and multiply by that power of 10. $\left(1.2 \times 10^4\right) - \left(3 \times 10^3\right) = \left(12 \times 10^3\right) - \left(3 \times 10^3\right) = (12-3) \times 10^3 = 9 \times 10^3$; Answer checks.

4. Conjecture: Make the power of the 10's the same and you can add or subtract the lead numbers and multiply by that power of 10. $\left(2 \times 10^2\right) + \left(6 \times 10^1\right) = \left(2 \times 10^2\right) + \left(0.6 \times 10^2\right) = (2+0.6) \times 10^2 = 2.6 \times 10^2$; Answer checks.

5. Conjecture: Make the power of the 10's the same and you can add or subtract the lead numbers and multiply by that power of 10. $\left(2 \times 10^{-1}\right) + \left(4 \times 10^{-2}\right) = \left(2 \times 10^{-1}\right) + \left(0.4 \times 10^{-1}\right) = (2+0.4) \times 10^{-1} = 2.4 \times 10^{-1}$; Answer checks.

6. Conjecture: Make the power of the 10's the same and you can add or subtract the lead numbers and multiply by that power of 10. $(2\times10^{-3})-(5\times10^{-2})=(0.2\times10^{-2})-(5\times10^{-2})=(0.2-5)\times10^{-2}=-4.8\times10^{-2}$; Answer checks.

7. (a) The length of the box is $30-2x$, the width is $20-2x$, and the height is x. If $V=L\times W\times H$ then,

$$V=(30-2x)(20-2x)(x)=(600-60x-40x+4x^2)(x)=(4x^2-100x+600)(x)=4x^3-100x^2+600x$$

(b) Let $x=4$, then $4(4)^3-100(4)^2+600(4)=4(64)-100(16)-600(4)=256-1600+2400=1056\text{ in}^3$

8. (a) The length and width would be $25-2x$ and the height would be x. Therefore the base is equal to $(25-2x)^2=625-100x+4x^2$ and the four sides are each $(25-2x)(x)=25x-2x^2$. 4 sides = $100x-8x^2$. Adding, $625-100x+4x^2+100x-8x^2=625-4x^2$.

(b) Let $x=3$, then $625-4(3)^2=625-4(9)=625-36=589\text{ in}^2$.

9. See Figure 9a & 9b. No, not equal for all values of x; $3x(4-5x)=12x-15x^2$

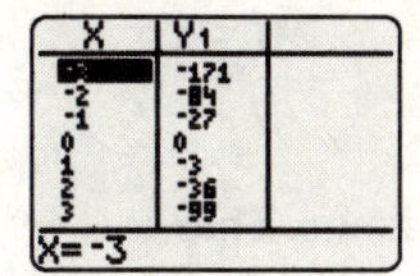
Figure 9a

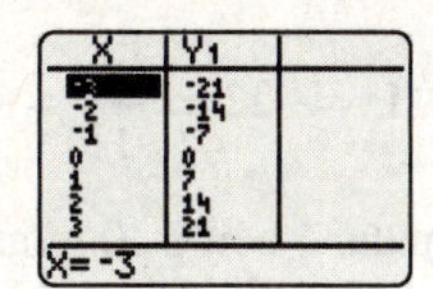
Figure 9b

10. See Figure 10a & 10b. No, not equal for all values of x; $(x-1)^2=x^2-2x+1$

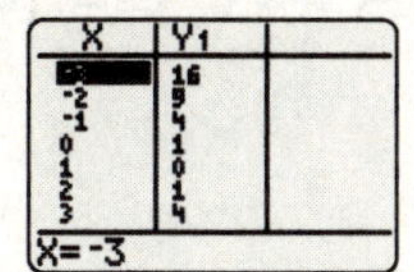
Figure 10a

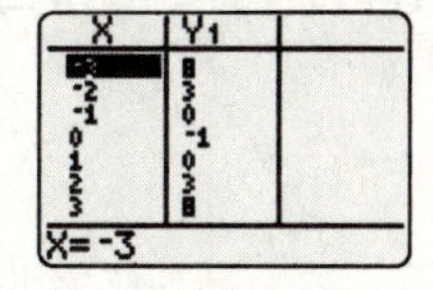
Figure 10b

11. See Figure 11a & 11b. Yes, equal for all values of x.

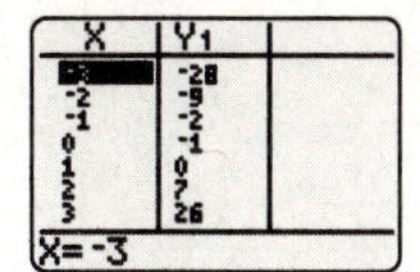
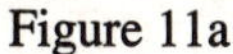
Figure 11a

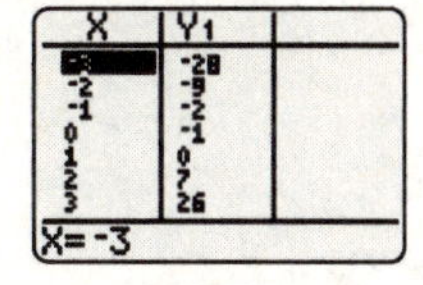
Figure 11b

12. See Figure 12a & 12b. No, not equal for all values of x; $(x-2)^3=(x-2)(x-2)^2=$

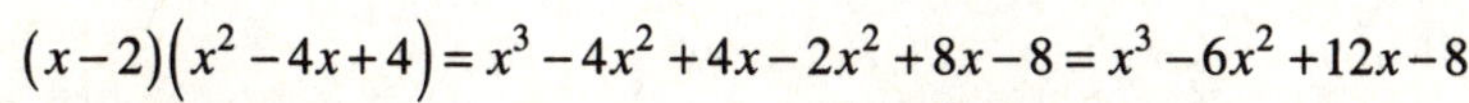
$(x-2)(x^2-4x+4)=x^3-4x^2+4x-2x^2+8x-8=x^3-6x^2+12x-8$

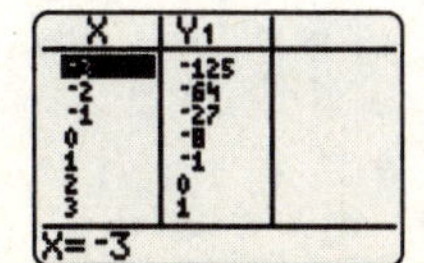
Figure 12a

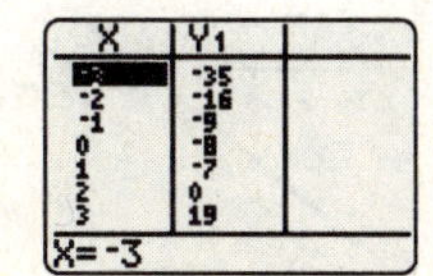
Figure 12b

Chapters 1–5 Cumulative Review Exercises

1. (a) $18-2\cdot 5=18-10=8$

   (b) $42\div 7+2=6+2=8$

2. (a) $21-(-8)=21+8=29$

   (b) $-\frac{7}{3}\div\left(-\frac{14}{9}\right)=-\frac{7}{3}\cdot -\frac{9}{14}=\frac{7\cdot 9}{3\cdot 14}=\frac{7\cdot 3\cdot 3}{3\cdot 2\cdot 7}=\frac{3}{2}$

3. (a) $(x-3)+x=4+x \Rightarrow x+x-3=4+x \Rightarrow 2x-3=4+x$

   $\Rightarrow 2x-x-3=4+x-x \Rightarrow x-3=4 \Rightarrow x-3+3=4+3$

   $\Rightarrow x=7$

   (b) $2(5x-4)=1+10x \Rightarrow 10x-8=1+10x \Rightarrow 10x-10x-8=1+10x-10x \Rightarrow -8=1$. Because this statement is not true, the equation has no solutions.

4. (a) $2+6x=2(3x+1) \Rightarrow 2+6x=6x+2 \Rightarrow 2+6x=2+6x$. Because this statement is true for any value for x the equation has infinitely many solutions.

   (b) $11x-9=-31 \Rightarrow 11x-9+9=-31+9 \Rightarrow 11x=-22 \Rightarrow \frac{11x}{11}=\frac{-22}{11} \Rightarrow x=-2$

5. Use the distance (d) = rate $(r)\times$time (t) formula. 4 hours 30 minutes equals 4.5 hours. Thus,

   $d=rt \Rightarrow 306=r(4.5) \Rightarrow \frac{306}{4.5}=\frac{4.5r}{4.5} \Rightarrow 68=r$, or 68 mph.

6. Let x represent the smallest number. $x+(x+1)+(x+2)=-114 \Rightarrow 3x+3=-114 \Rightarrow 3x+3-3=-114-3$

   $\Rightarrow 3x=-117 \Rightarrow \frac{3x}{3}=\frac{-117}{3} \Rightarrow x=-39$

   Thus, the numbers are –39, –38, and –37.

7. (a) $42\%=\frac{42}{100}=\frac{21\cdot 2}{50\cdot 2}=\frac{21}{50}$

   (b) $0.076=\frac{76}{1000}=\frac{19\cdot 4}{250\cdot 4}=\frac{19}{250}$

8. $Wx=4x+Y \Rightarrow Wx-4x=4x-4x+Y \Rightarrow (W-4)x=Y$

   $\Rightarrow \frac{(W-4)x}{W-4}=\frac{Y}{W-4} \Rightarrow x=\frac{Y}{W-4}$

9. 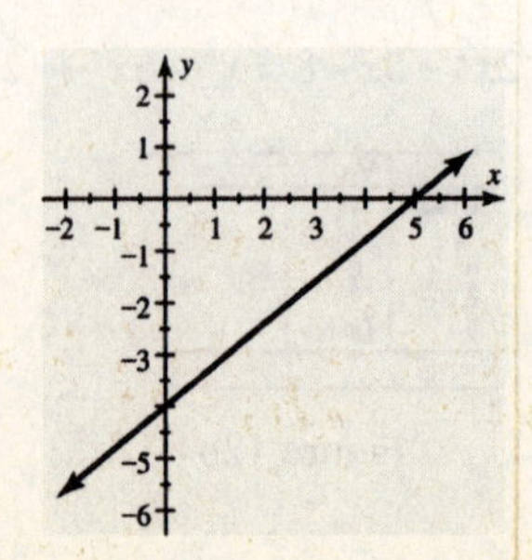

10.

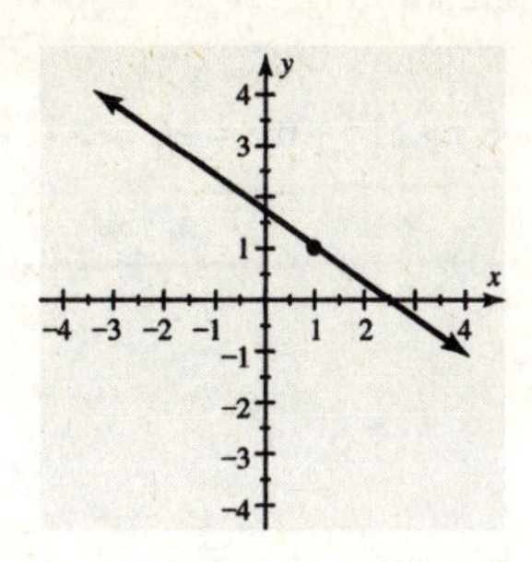

11. For every increase of 1 in the run, the rise decreases by 2. Therefore, the slope m is –2. When $x=0,\ y=1$. Therefore, the y-intercept is 1. Thus $y=mx+b \Rightarrow y=-2x+1$.

12. To find the x-intercept, set $y=0$: $2y=3x-6 \Rightarrow 2(0)=3x-6 \Rightarrow 0=3x-6 \Rightarrow 3x=6 \Rightarrow \frac{3x}{3}=\frac{6}{3} \Rightarrow x=2$. Therefore, the x-intercept is 2.

To find the y-intercept, set $x=0$: $2y=3x-6 \Rightarrow 2y=3(0)-6 \Rightarrow 2y=-6 \Rightarrow \frac{2y}{2}=\frac{-6}{2} \Rightarrow y=-3$. Therefore, the y-intercept is –3.

13. $3x-6y=7 \Rightarrow -6y=-3x+7 \Rightarrow y=\frac{1}{2}x-\frac{7}{6}$. The line parallel to $y=\frac{1}{2}x-\frac{7}{6}$ has slope $\frac{1}{2}$. Insert the values of the point (2, –3) and the value of the slope into the slope-intercept form and solve for b: $y=mx+b \Rightarrow -3=\frac{1}{2}(2)+b \Rightarrow -3=1+b \Rightarrow -3-1=1-1+b \Rightarrow -4=b$. Thus, $y=\frac{1}{2}x-4$.

14. Find the slope: $m=\frac{y_2-y_1}{x_2-x_1}=\frac{4-(-5)}{1-(-2)}=\frac{4+5}{1+2}=\frac{9}{3}=3$. Then, insert the values of the point (1, 4) and the value of the slope into the point-slope form: $y-y_1=m(x-x_1) \Rightarrow y-4=3(x-1)$. Then, convert to slope-intercept from: $y-4=3(x-1) \Rightarrow y-4=3x-3 \Rightarrow y-4+4=3x-3+4 \Rightarrow y=3x+1$.

15. Multiplying the first equation by –2 and adding the two equations will eliminate both variables.

$$\begin{array}{r} -8x-6y=12 \\ \underline{8x+6y=12} \\ 0=24 \end{array}$$

Since this is false, there are no solutions.

16. Note that $x-3y=5 \Rightarrow x=3y+5$. Substituting $x=3y+5$ into the second equation yields the following:

$3(3y+5)+y=5 \Rightarrow 9y+15+y=5 \Rightarrow 10y+15=5 \Rightarrow 10y=-10 \Rightarrow \frac{10y}{10}=\frac{-10}{10} \Rightarrow y=-1$, and so $x=3(-1)+5 \Rightarrow x=2$. The solution is (2, –1).

17. Multiplying the first equation by 3 and adding the two equations will eliminate both variables.

$$\begin{array}{r} 3x+12y=-24 \\ \underline{-3x-12y=\ \ 24} \\ 0=0 \end{array}$$

Since this is true, there are infinitely many solutions.

18. Note that $x-5y=30 \Rightarrow x=5y+30$. Substituting $x=5y+30$ into the second equation yields the following:

$2(5y+30)+y=-6 \Rightarrow 10y+60+y=-6 \Rightarrow 11y+60=-6 \Rightarrow 11y=-66 \Rightarrow \frac{11y}{11}=\frac{-66}{11} \Rightarrow y=-6$, and so

$x=5(-6)+30=0$. The solution is $(0, -6)$.

19. See Figure 19.

20. See Figure 20.

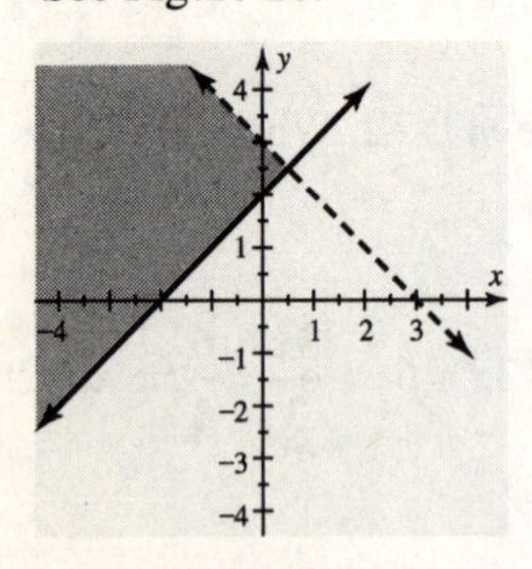

Figure 19

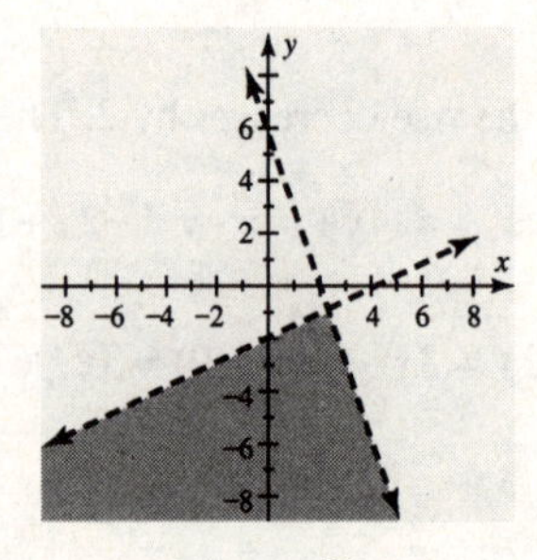

Figure 20

21. (a) $3x^2 \cdot 5x^3 = (3\cdot 5)x^{2+3} = 15x^5$

(b) $(x^3y)^2(x^4y^5) = (x^{3\cdot 2}y^2)(x^4y^5) = x^{6+4}y^{2+5} = x^{10}y^7$

22. (a) $(5x^2-3x+4)-(3x^2-2x+1) = 5x^2+(-3x^2)+(-3x)+2x+4+(-1) = 2x^2-x+3$

(b) $(7a^3-4a^2-5)+(5a^3+4a^2+a) = 7a^3+5a^3+(-4a^2)+4a^2+a+(-5) = 12a^3+a-5$

23. (a) $(2x+3)(x-7) = 2x\cdot x+2x(-7)+3\cdot x+3(-7) = 2x^2-14x+3x-21 = 2x^2-11x-21$

(b) $(y+3)(y^2-3y-1) = y\cdot y^2+y(-3y)+y(-1)+3\cdot y^2+3(-3y)+3(-1)$

$= y^3-3y^2-y+3y^2-9y-3 = y^3-10y-3$

(c) $(4x+7)(4x-7) = (4x)^2-(7)^2 = 16x^2-49$

(d) $(5a+3)^2 = (5a)^2+2(5a)(3)+(3)^2 = 25a^2+30a+9$

24. (a) $x^{-5}\cdot x^3\cdot x = x^{-5+3+1} = x^{-1} = \frac{1}{x}$

(b) $\left(\frac{2}{x^3}\right)^{-3} = \left(\frac{x^3}{2}\right)^3 = \frac{x^{3\cdot 3}}{2^3} = \frac{x^9}{2\cdot 2\cdot 2} = \frac{x^9}{8}$

(c) $\frac{3x^2y^{-1}}{6x^{-2}y} = \frac{3}{6}x^{2-(-2)}y^{-1-1} = \frac{1}{2}x^4y^{-2} = \frac{x^4}{2y^2}$

(d) $(xy^{-2})^3(x^{-2}y)^{-2} = (x^3y^{-2\cdot 3})(x^{(-2)(-2)}y^{-2}) = (x^3y^{-6})(x^4y^{-2}) = x^{3+4}y^{-6+(-2)} = x^7y^{-8} = \frac{x^7}{y^8}$

25. Move the decimal point 10 places to the left, $24{,}000{,}000{,}000 = 2.4\times 10^{10}$.

26. Move the decimal point 7 places to the left, $4.71\times 10^{-7} = 0.000000471$.

27. (a) $\dfrac{8x^3-2x}{2x}=\dfrac{8x^3}{2x}-\dfrac{2x}{2x}=4x^{3-1}-1=4x^2-1$

(b)
$$\begin{array}{r} 2x-5 \\ x+3\,\overline{)\,2x^2+x-14} \\ \underline{2x^2+6x} \\ -5x-14 \\ \underline{-5x-15} \\ 1 \end{array}$$

The solution is $2x-5+\dfrac{1}{x+3}$.

28. First, subtract the smaller number from the larger to obtain the difference between them: $1200-900=300$.

Then, determine what percentage 300 is of 1200: $\dfrac{300}{1200}=0.25$. Thus the percent change from \$1200 to \$900 is -25%.

29. Let x represent the amount of 3% acid. Then, the 5% acid solution is the sum of the 3% acid and 6% acid solutions. Therefore

$0.03x+0.06(400)=0.05(x+400)$

$\Rightarrow 0.03x+24=0.05x+20$

$\Rightarrow 0.03x-0.03x+24=0.05x-0.03x+20$

$\Rightarrow 24=0.02x+20 \Rightarrow 24-20=0.02x+20-20$

$\Rightarrow 4=0.02x \Rightarrow \dfrac{4}{0.02}=\dfrac{0.02x}{0.02} \Rightarrow 200=x$

200 ml of 3% acid should be added to the 6% acid solution to dilute it to a 5% acid solution.

30. (a) $(x+5)(2x)=x\cdot 2x+5\cdot 2x=2x^{1+1}+10x=2x^2+10x$

(b) $(x+2)(x+5)=x\cdot x+x\cdot 5+2\cdot x+2\cdot 5\ =x^{1+1}+5x+2x+10=x^2+7x+10$

(c) $(x+2)(2x)=x\cdot 2x+2\cdot 2x=2x^{1+1}+4x=2x^2+4x$

(d) 3 sides $=2x^2+10x+x^2+7x+10+2x^2+4x\ =2x^2+x^2+2x^2+10x+7x+4x+10=5x^2+21x+10$;

6 sides $=2(5x^2+21x+10)=2\cdot 5x^2+2\cdot 21x+2\cdot 10\ =10x^2+42x+20$

Critical Thinking Solutions for Chapter 5

Section 5.2

- The result is zero.
- Volume of a cube is side cubed $\Rightarrow L^3$, then multiply by 6 of them $\Rightarrow 6L^3$.

Section 5.4

- Let $x=2$ and $y=3$, then $(x+y)^3=(2+3)^3=5^3=125$ and $x^3+y^3=2^3+3^3=8+27=35$, $125\neq 35$, therefore $(x+y)^3\neq x^3+y^3$.

Chapter 6: Factoring Polynomials and Solving Equations

6.1: Introduction to Factoring

Concepts

1. factoring

2. $ab+ac=a(b+c)\Rightarrow a$ is a common factor.

3. multiplying

4. prime; multiplication

5. greatest common factor (GCF)

6. grouping

7. Factors of $2x^2$ are 1, 2, $2x$, x^2 and $2x^2$; factors of $4x$ are 1, 2, 4, $2x$ and $4x$. Therefore, common factors are 1, 2, x and $2x$.

8. Using common factors from #7 the GCF is $2x$.

Common Factors

9. $2x=2\cdot x$ and $4=2\cdot 2\Rightarrow 2(x+2)$. See Figure 9.

11. $z^2=z\cdot z$ and $4z=4\cdot z\Rightarrow z(z+4)$. See Figure 11.

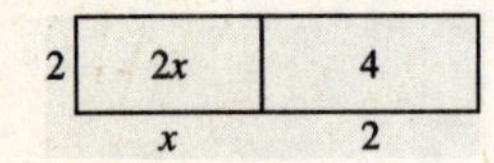

Figure 9

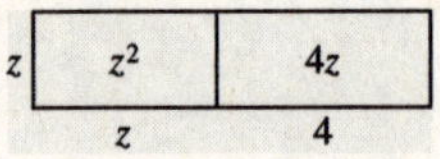

Figure 11

13. In the expression $3x^2+9x$, the terms $3x^2$ and $9x$ both contain a common factor of $3x$ because $3x^2=3x\cdot x$ and $9x=3x\cdot 3$. Therefore $3x^2+9x=3x(x+3)$.

15. In the expression $4y^3-2y^2$, the terms $4y^3$ and $-2y^2$ both contain a common factor of $2y^2$ because $4y^3=2y^2\cdot 2y$ and $-2y^2=2y^2(-1)$. Therefore $4y^3-2y^2=2y^2(2y-1)$.

17. In the expression $2z^3+8z^2-4z$, the terms $2z^3, 8z^2$, and $-4z$ all contain a common factor of $2z$ because $2z^3=2z\cdot z^2$, $8z^2=2z\cdot 4z$, and $-4z=2z(-2)$. Therefore $2z^3+8z^2-4z=2z(z^2+4z-2)$.

19. In the expression $6x^2y-3xy^2$, the terms $6x^2y$ and $-3xy^2$ both contain a common factor of $3xy$ because $6x^2y=3xy\cdot 2x$ and $-3xy^2=3xy(-y)$. Therefore $6x^2y-3xy^2=3xy(2x-y)$.

21. $6x-18x^2$; because $6x=2\cdot 3\cdot x$ and $18x^2=3\cdot 3\cdot 2\cdot x\cdot x$, common factors are $2\cdot 3\cdot x\Rightarrow$ GCF $=6x$ and $\Rightarrow 6x(1-3x)$.

23. $8y^3-12y^2$; because $8y^3=2\cdot 2\cdot 2\cdot y\cdot y\cdot y$ and $12y^2=2\cdot 2\cdot 3\cdot y\cdot y$, common factors are $2\cdot 2\cdot y\cdot y\Rightarrow$ GCF $=4y^2$ and $\Rightarrow 4y^2(2y-3)$.

25. $6z^3+3z^2+9z$; because $6z^3=2\cdot 3\cdot z\cdot z\cdot z$ and $3z^2=3\cdot z\cdot z$ and $9z=3\cdot 3\cdot z$, common factors are $3\cdot z\Rightarrow$ GCF $=3z$ and $\Rightarrow 3z\left(2z^2+z+3\right)$.

27. $x^4-5x^3-4x^2$; because $x^4=x\cdot x\cdot x\cdot x$ and $5x^3=5\cdot x\cdot x\cdot x$ and $4x^2=2\cdot 2\cdot x\cdot x$, common factors are $x\cdot x\Rightarrow$ GCF $=x^2$ and $\Rightarrow x^2\left(x^2-5x-4\right)$.

29. $5y^5+10y^4-15y^3+10y^2$; because $5y^5=5\cdot y\cdot y\cdot y\cdot y\cdot y$ and $10y^4=2\cdot 5\cdot y\cdot y\cdot y\cdot y$ and $15y^3=3\cdot 5\cdot y\cdot y\cdot y$ and $10y^2=2\cdot 5\cdot y\cdot y$, common factors are $5\cdot y\cdot y\Rightarrow$ GCF $=5y^2$ and $\Rightarrow 5y^2\left(y^3+2y^2-3y+2\right)$.

31. $xy+xz$; because $xy=x\cdot y$ and $xz=x\cdot z$, common factors are $x\Rightarrow$ GCF $=x$ and $\Rightarrow x(y+z)$.

33. ab^2-a^2b; because $ab^2=a\cdot b\cdot b$ and $a^2b=a\cdot a\cdot b$, common factors are $a\cdot b\Rightarrow$ GCF $=ab$ and $\Rightarrow ab(b-a)$.

35. $5x^2y^4+10x^3y^3$; because $5x^2y^4=5\cdot x\cdot x\cdot y\cdot y\cdot y\cdot y$ and $10x^3y^3=2\cdot 5\cdot x\cdot x\cdot x\cdot y\cdot y\cdot y$, common factors are $5\cdot x\cdot x\cdot y\cdot y\cdot y\Rightarrow$ GCF $=5x^2y^3$ and $\Rightarrow 5x^2y^3(y+2x)$.

37. a^2b+ab^2+ab; because $a^2b=a\cdot a\cdot b$ and $ab^2=a\cdot b\cdot b$ and $ab=a\cdot b$, common factors $=a\cdot b\Rightarrow$ GCF $=ab$ and $\Rightarrow ab(a+b+1)$.

Factoring by Grouping

39. $x(x+1)-2(x+1)$ has common binomial $(x+1)\Rightarrow(x-2)(x+1)$

41. $(z+5)z+(z+5)4$ has common binomial $(z+5)\Rightarrow(z+4)(z+5)$

43. $4x^3(x-5)-2x(x-5)$ has common binomial $(x-5)\Rightarrow\left(4x^3-2x\right)(x-5)\Rightarrow 2x\left(2x^2-1\right)(x-5)$

45. x^3+2x^2+3x+6 by associative property $=\left(x^3+2x^2\right)+(3x+6)\Rightarrow$ common factors $=x^2(x+2)+3(x+2)\Rightarrow\left(x^2+3\right)(x+2)$

47. $2y^3+y^2+2y+1$ by associative property $=\left(2y^3+y^2\right)+(2y+1)\Rightarrow$ common factors $=y^2(2y+1)+(2y+1)\Rightarrow\left(y^2+1\right)(2y+1)$

49. $2z^3-6z^2+5z-15$ by associative property $=\left(2z^3-6z^2\right)+(5z-15)\Rightarrow$ common factors $=2z^2(z-3)+5(z-3)\Rightarrow\left(2z^2+5\right)(z-3)$

51. $4t^3-20t^2+3t-15$ by associative property $=\left(4t^3-20t^2\right)+(3t-15)\Rightarrow$ common factors $=4t^2(t-5)+3(t-5)\Rightarrow\left(4t^2+3\right)(t-5)$

53. $9r^3+6r^2-6r-4$ by associative property $=\left(9r^3+6r^2\right)-(6r+4)\Rightarrow$

common factors $=3r^2(3r+2)-2(3r+2)\Rightarrow\left(3r^2-2\right)(3r+2)$

55. $7x^3+21x^2-2x-6$ by associative property $=\left(7x^3+21x^2\right)-(2x+6)\Rightarrow$

common factors $=7x^2(x+3)-2(x+3)\Rightarrow\left(7x^2-2\right)(x+3)$

57. $2y^3-7y^2-4y+14$ by associative property $=\left(2y^3-7y^2\right)-(4y-14)\Rightarrow$

common factors $=y^2(2y-7)-2(2y-7)\Rightarrow\left(y^2-2\right)(2y-7)$

59. $z^3-4z^2-7z+28$ by associative property $=\left(z^3-4z^2\right)-(7z-28)\Rightarrow$

common factors $=z^2(z-4)-7(z-4)\Rightarrow\left(z^2-7\right)(z-4)$

61. $2x^4-3x^3+4x-6$ by associative property $=\left(2x^4-3x^3\right)+(4x-6)\Rightarrow$

common factors $=x^3(2x-3)+2(2x-3)\Rightarrow\left(x^3+2\right)(2x-3)$

63. $ax+bx+ay+by$ by associative property $=(ax+bx)+(ay+by)\Rightarrow$

common factors $=x(a+b)+y(a+b)\Rightarrow(x+y)(a+b)$

65. $3x^3+6x^2+3x+6$ has GCF $3\Rightarrow3(x^3+2x^2+x+2)\Rightarrow$ by associative

property $=3[(x^3+2x^2)+(x+2)]\Rightarrow$ common factors $=3[x^2(x+2)+1(x+2)]\Rightarrow3(x^2+1)(x+2)$

67. $6y^4-24y^3-2y^2+8y$ has GCF $2y\Rightarrow2y(3y^3-12y^2-y+4)\Rightarrow$ by associative

property $=2y[(3y^3-12y^2)-(y-4)]\Rightarrow$ common factors $=2y[3y^2(y-4)-1(y-4)]\Rightarrow2y(3y^2-1)(y-4)$

69. $x^5+2x^4-3x^3-6x^2$ has GCF $x^2\Rightarrow x^2(x^3+2x^2-3x-6)\Rightarrow$ by associative

property $=x^2[(x^3+2x^2)-(3x+6)]\Rightarrow$ common factors $=x^2[x^2(x+2)-3(x+2)]\Rightarrow x^2(x^2-3)(x+2)$

71. $4x^5+2x^4-12x^3-6x^2$ has GCF $2x^2\Rightarrow2x^2(2x^3+x^2-6x-3)\Rightarrow$ by associative

property $=2x^2[(2x^3+x^2)-(6x+3)]\Rightarrow$ common factors $=2x^2[x^2(2x+1)-3(2x+1)]\Rightarrow2x^2(x^2-3)(2x+1)$

73. $x^3y+x^2y^2-2x^2y-2xy^2$ has GCF $xy\Rightarrow xy(x^2+xy-2x-2y)\Rightarrow$ by associative

property $=xy[(x^2+xy)-(2x+2y)]\Rightarrow$ common factors $=xy[x(x+y)-2(x+y)]\Rightarrow xy(x-2)(x+y)$

75. $ax^2+bx+c=a\left(x^2+\frac{b}{a}x+\frac{c}{a}\right)$ because $a\cdot x^2=ax^2$, $a\cdot\frac{b}{a}x=bx$, and $a\cdot\frac{c}{a}=c$.

Applications

77. (a) $80t=2\cdot2\cdot2\cdot2\cdot5\cdot t$ and $16t^2=2\cdot2\cdot2\cdot2\cdot t\cdot t$, common factors are

$2\cdot2\cdot2\cdot2\cdot t\Rightarrow\text{GCF}=2\cdot2\cdot2\cdot2\cdot t=16t$.

(b) $80t - 16t^2 = 16t(5 - t)$

79. (a) $x = 3,\ 4x^3 - 60x^2 + 200x \Rightarrow 4(3)^3 - 60(3)^2 + 200(3) = 108 - 540 + 600 = 168 \Rightarrow V = 168 \text{ in}^3.$

(b) $4x^3 = 2 \cdot 2 \cdot x \cdot x \cdot x$ and $60x^2 = 2 \cdot 2 \cdot 3 \cdot 5 \cdot x \cdot x$ and $200x = 2 \cdot 2 \cdot 2 \cdot 5 \cdot 5 \cdot x,$ common factors are $2 \cdot 2 \cdot x \Rightarrow$ GCF $= 4x \Rightarrow 4x\left(x^2 - 15x + 50\right).$

6.2: Factoring Trinomials I ($x^2 + bx + c$)

Concepts

1. 1

2. Product is $x^2 + (m+n)x + mn,$ coefficient of x-term is $m+n$ and the constant term is mn.

3. $x^2 + bx + c,$ so c is the third term $\Rightarrow mn = c$ and b is the second term $\Rightarrow m + n = b.$

4. prime

5. Product of 12 by factors: 1, 12; 2, 6; and 3, 4.

6. Negative integer pairs with a product of 30 by factors: $-1,\ -30;\ \ -2,\ -15;\ \ -3,\ -10;$ and $-5,\ -6.$

7. Factors of 28 are 1, 28; 2, 14; 4, 7 and only $4 + 7 = 11 \Rightarrow 4,\ 7.$

8. Factors of 35 are 1, 35; 5, 7 and only $5 + 7 = 12 \Rightarrow 5,\ 7.$

9. Factors of -30 are $-1, 30; -2, 15; -3, 10; -5, 6;\ 1, -30;\ 2, -15;\ 3, -10;\ 5, -6$ and only $3 + (-10) = -7 \Rightarrow 3,\ -10.$

10. Factors of -100 are $-1, 100; -2, 50; -4, 25; -5, 20;\ 1, -100;\ 2, -50;\ 4, -25;\ 5, -20$ and only $-4 + 25 = 21 \Rightarrow -4,\ 25.$

11. Factors of -50 are $-1,\ 50;\ 1,\ -50;\ -2,\ 25;\ 2,\ -25;\ -5,\ 10;\ 5,\ -10$ and only $-5 + 10 = 5 \Rightarrow -5,\ 10.$

12. Factors of -15 are $-1,\ 15;\ 1,\ -15;\ -3,\ 5;\ 3,\ -5$ and only $3 + (-5) = -2 \Rightarrow 3,\ -5.$

13. Factors of 28 are $1,\ 28;\ -1,\ -28;\ 2,\ 14;\ -2,\ -14;\ 4,\ 7;\ -4,\ -7$ and only $(-4) + (-7) = -11 \Rightarrow -4,\ -7.$

14. Factors of 80 are $1,\ 80;\ -1,\ -80;\ 2,\ 40;\ -2,\ -40;\ 4,\ 20;\ -4,\ -20;\ 5,\ 16;\ -5,\ -16;\ 8,\ 10;\ -8,\ -10$ and only $(-2) + (-40) = -42 \Rightarrow -2,\ -40.$

Factoring Trinomials

15. Factors of 2 with sum of 3 are 1 and 2 $\Rightarrow x^2 + 3x + 2 = (x+1)(x+2).$

17. Factors of 4 with sum of 4 are 2 and 2 $\Rightarrow y^2 + 4y + 4 = (y+2)(y+2).$

19. The only factors of 7 are 1 and 7, whose sum is 8, not 3. The polynomial $z^2 + 3z + 7$ is prime.

21. Factors of 15 with sum of 8 are 3 and 5 $\Rightarrow x^2 + 8x + 15 = (x+3)(x+5).$

23. Factors of 36 with sum of 13 are 4 and 9 $\Rightarrow m^2 + 13m + 36 = (m+4)(m+9).$

25. Factors of 100 with sum of 20 are 10 and 10 $\Rightarrow n^2+20n+100=(n+10)(n+10)$.

27. Factors of 5 with sum of –6 are -1 and $-5 \Rightarrow x^2-6x+5=(x-1)(x-5)$.

29. Factors of 12 with sum of –7 are -3 and $-4 \Rightarrow y^2-7y+12=(y-3)(y-4)$.

31. Factors of 40 with sum of –13 are -5 and $-8 \Rightarrow z^2-13z+40=(z-5)(z-8)$.

33. Factors of 63 with sum of –16 are -7 and $-9 \Rightarrow a^2-16a+63=(a-7)(a-9)$.

35. The factors of 10 are 1, 10 and 2, 5, but neither pair has sum –6. The polynomial $y^2-6y+10$ is prime.

37. Factors of 125 with sum of –30 are -5 and $-25 \Rightarrow b^2-30b+125=(b-5)(b-25)$.

39. Factors of -90 with sum of 13 are -5 and $18 \Rightarrow x^2+13x-90=(x-5)(x+18)$.

41. Factors of -45 with sum of 4 are -5 and $9 \Rightarrow m^2+4m-45=(m-5)(m+9)$.

43. The factors of –63 are –1, 63; 1, –63; 3, –21; –3, 21; 7, –9; and –7, 9 but none of these pairs of factors has sum 16. The polynomial $a^2+16a-63$ is prime.

45. Factors of -200 with sum of 10 are -10 and $20 \Rightarrow n^2+10n-200=(n-10)(n+20)$.

47. Factors of -23 with sum of 22 are -1 and $23 \Rightarrow x^2+22x-23=(x-1)(x+23)$.

49. Factors of -32 with sum of 4 are -4 and $8 \Rightarrow a^2+4a-32=(a-4)(a+8)$.

51. Factors of -20 with sum of -1 are 4 and $-5 \Rightarrow b^2-b-20 \Rightarrow (b+4)(b-5)$.

53. The factors of –22 are 1, –22; –1, 22; –2, 11; and 2, –11 but none of these pairs of factors has sum –14. The polynomial $m^2-14m-22$ is prime.

55. Factors of -72 with sum of -1 are 8 and $-9 \Rightarrow x^2-x-72 \Rightarrow (x+8)(x-9)$.

57. Factors of -34 with sum of -15 are 2 and $-17 \Rightarrow y^2-15y-34 \Rightarrow (y+2)(y-17)$.

59. Factors of -66 with sum of -5 are 6 and $-11 \Rightarrow z^2-5z-66 \Rightarrow (z+6)(z-11)$.

61. The GCF of $5x^2-10x-40$ is $5 \Rightarrow 5(x^2-2x-8)$. Factors of -8 with sum of -2 are -4 and $2 \Rightarrow 5(x^2-2x-8) \Rightarrow 5(x-4)(x+2)$.

63. The GCF of y^3-7y^2+10y is $y \Rightarrow y(y^2-7y+10)$. Factors of 10 with sum of -7 are -5 and $-2 \Rightarrow y(y^2-7y+10) \Rightarrow y(y-5)(y-2)$.

65. The GCF of $3a^3+21a^2+18a$ is $3a \Rightarrow 3a(a^2+7a+6)$. Factors of 6 with sum of 7 are 6 and $1 \Rightarrow 3a(a^2+7a+6) \Rightarrow 3a(a+6)(a+1)$.

67. The GCF of $2x^3-6x^2+8x$ is $2x \Rightarrow 2x(x^2-3x+4)$. Factors of 4 are 1, 4 and 2, 2 but neither of these pairs has sum -3. The polynomial x^2-3x+4 is prime, so factored form of the original polynomial is $2x(x^2-3x+4)$.

69. The GCF of $2m^4 - 10m^3 - 28m^2$ is $2m^2 \Rightarrow 2m^2(m^2 - 5m - 14)$. Factors of -14 with sum of -5 are -7 and $2 \Rightarrow 2m^2(m^2 - 5m - 14) \Rightarrow 2m^2(m-7)(m+2)$.

71. $5 + 6x + x^2$ in standard form is $x^2 + 6x + 5$ and factors of 5 with a sum of $6 \Rightarrow 1,\ 5 \Rightarrow (x+5)(x+1)$.

73. $3 - 4x + x^2$ in standard form is $x^2 - 4x + 3$ and factors of 3 with a sum of $-4 \Rightarrow -1,\ -3 \Rightarrow (x-1)(x-3)$.

75. Using the hint, the answer will be in the form $(m-x)(n+x)$. We need to find m and n. Factors of 12 with a difference of 4 are 6 and $2 \Rightarrow 12 + 4x - x^2 = (6-x)(2+x)$.

77. Factors of 32 with a difference of 4 are 8 and $4 \Rightarrow 32 - 4x - x^2 = (8+x)(4-x)$.

79. Factors of k with a sum of $k+1$ are k and $1 \Rightarrow x^2 + (k+1)x + k \Rightarrow (x+1)(x+k)$.

Geometry

81. $x^2 + 2x + 1 \Rightarrow (x+1)(x+1) \Rightarrow L \cdot W = x^2 + 2x + 1 \Rightarrow L = x + 1$. See Figure 81.

83. $x^2 + 3x + 2 = (x+2)(x+1) = L \cdot W \Rightarrow L = x + 2$ or $x + 1$. See Figure 83.

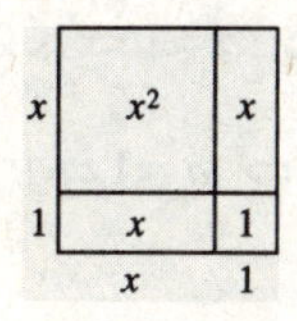

Figure 81

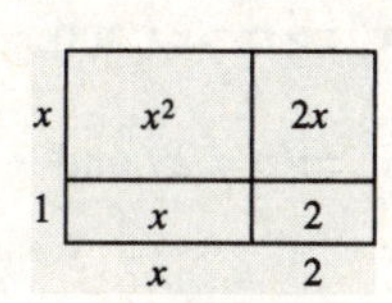

Figure 83

85. $6x^2 + 12x + 6$ divided by 6 (for 6 surfaces) $= x^2 + 2x + 1 = (x+1)(x+1) \Rightarrow$ each side is $x + 1$.

87. Add the four regions: $x^2 + 2x + 6x + 12 = x^2 + 8x + 12 = (x+2)(x+6)$.

Checking Basic Concepts for Sections 6.1 & 6.2

1. $8x^3 = 2 \cdot 2 \cdot 2 \cdot x \cdot x \cdot x$ and $12x^2 = 2 \cdot 2 \cdot 3 \cdot x \cdot x$ and $24x = 2 \cdot 2 \cdot 2 \cdot 3 \cdot x \Rightarrow \text{GCF} = 2 \cdot 2 \cdot x = 4x$.

2. GCF of $12z^3$ and $18z^2$ is $6z^2 \Rightarrow 6z^2(2z-3)$.

3. (a) $6y(y-2) + 5(y-2) = (6y+5)(y-2)$

   (b) $(2x^3 + x^2) + (10x + 5) = x^2(2x+1) + 5(2x+1) = (x^2+5)(2x+1)$

   (c) GCF is $4 \Rightarrow 4z^3 - 12z^2 + 4z - 12 \Rightarrow 4(z^3 - 3z^2 + z - 3) \Rightarrow 4[(z^3 - 3z^2) + (z-3)] \Rightarrow$

   $4[z^2(z-3) + 1(z-3)] \Rightarrow 4(z^2+1)(z-3)$

4. (a) Factors of 8 with a sum of 6 are 2 and 4 $\Rightarrow (x+2)(x+4)$.

   (b) Factors of -42 with a sum of -1 are -7 and 6 $\Rightarrow (x-7)(x+6)$.

   (c) The factors of -5 are $1, -5$ and $-1, 5$ but neither of these pairs has sum 3. The polynomial $a^2 + 3a - 5$ is prime.

   (d) The GCF of $4a^3 + 20a^2 + 24a$ is $4a \Rightarrow 4a(a^2 + 5a + 6) \Rightarrow 4a(a+2)(a+3)$

5. $x^2+5x+5x+25 \Rightarrow x^2+10x+25 \Rightarrow (x+5)(x+5)$

6.3: Factoring Trinomials II ($ax^2 + bx + c$)

Concepts

1. ac; b

2. ax^2; c

3. By ax^2+bx+c, if $a>0$, $b>0$ and $c>0$ then: $+;+$

4. By ax^2+bx+c, if $c<0$ then: $+;-$ or $-;+$ with $+;-$ working by FOIL.

5. By ax^2+bx+c, if $a>0$, $b<0$ and $c>0$ then: $-;-$

6. By ax^2+bx+c, if $c<0$ then: $+;-$ or $-;+$ with $-;+$ working by FOIL.

7. If $(4x+a)(b+2)=4x^2+11x+6$, then by FOIL: $4x\cdot b=4x^2 \Rightarrow b=x$ and $a\cdot 2=6 \Rightarrow a=3 \Rightarrow 3;\ x$.

8. If $(x-a)(b+3)=4x^2-5x-6$, then by FOIL: $x\cdot b=4x^2 \Rightarrow b=4x$ and $-a\cdot 3=-6 \Rightarrow\ a=2 \Rightarrow 2;\ 4x$.

9. If $(2x-a)(b+3)=4x^2+4x-3$, then by FOIL: $2x\cdot b=4x^2 \Rightarrow b=2x$ and $-a\cdot 3=-3 \Rightarrow\ a=1 \Rightarrow 1;\ 2x$.

10. If $(2x-a)(b-3)=4x^2-8x+3$, then by FOIL: $2x\cdot b=4x^2 \Rightarrow b=2x$ and $-a\cdot(-3)=3 \Rightarrow\ a=1 \Rightarrow 1;\ 2x$.

Factoring Trinomials

11. Using factoring by grouping: For $2x^2+7x+3$, $m\cdot n=a\cdot c=2\cdot 3=6$ and $m+n=b=7 \Rightarrow$

    $m=6,\ n=1 \Rightarrow 2x^2+6x+x+3 \Rightarrow 2x(x+3)+(x+3) \Rightarrow (2x+1)(x+3)$.

13. To factor $3y^2+2y+4$, find numbers m and n such that $mn=3\cdot 4=12$ and $m+n=2$. Because no such numbers exist the polynomial is prime.

15. Using factoring by grouping: For $3x^2+4x+1$, $m\cdot n=a\cdot c=3\cdot 1=3$ and $m+n=b=4 \Rightarrow$

    $m=3,\ n=1 \Rightarrow 3x^2+3x+x+1 \Rightarrow 3x(x+1)+(x+1) \Rightarrow (3x+1)(x+1)$.

17. Using factoring by grouping: For $6x^2+11x+3$, $m\cdot n=a\cdot c=6\cdot 3=18$ and $m+n=b=11 \Rightarrow$

    $m=9,\ n=2 \Rightarrow 6x^2+9x+2x+3 \Rightarrow 3x(2x+3)+(2x+3) \Rightarrow (3x+1)(2x+3)$.

19. Using factoring by grouping: For $5x^2-11x+2$, $m\cdot n=a\cdot c=5\cdot 2=10$ and $m+n=b=-11 \Rightarrow$

    $m=-10,\ n=-1 \Rightarrow 5x^2-10x-x+2 \Rightarrow 5x(x-2)-(x-2) \Rightarrow (5x-1)(x-2)$.

21. Using factoring by grouping: For $2y^2-7y+5$, $m\cdot n=a\cdot c=2\cdot 5=10$ and $m+n=b=-7 \Rightarrow$

    $m=-5,\ n=-2 \Rightarrow 2y^2-5y-2y+5 \Rightarrow y(2y-5)-(2y-5) \Rightarrow (y-1)(2y-5)$.

23. To factor $3m^2-11m-6$, find numbers m and n such that $mn=3(-6)=-18$ and $m+n=-11$. Because no such numbers exist the polynomial is prime.

25. Using factoring by grouping: For $7z^2 - 37z + 10$, $m \cdot n = a \cdot c = 7 \cdot 10 = 70$ and $m + n = b = -37 \Rightarrow$

$m = -35,\ n = -2 \Rightarrow 7z^2 - 35z - 2z + 10 \Rightarrow 7z(z-5) - 2(z-5) \Rightarrow (7z-2)(z-5)$.

27. Using factoring by grouping: For $3t^2 - 7t - 6$, $m \cdot n = a \cdot c = 3 \cdot (-6) = -18$ and $m + n = b = -7 \Rightarrow$

$m = -9,\ n = 2 \Rightarrow 3t^2 - 9t + 2t - 6 \Rightarrow 3t(t-3) + 2(t-3) \Rightarrow (3t+2)(t-3)$.

29. Using factoring by grouping: For $15r^2 + r - 6$, $m \cdot n = a \cdot c = 15 \cdot (-6) = -90$ and $m + n = b = 1 \Rightarrow$

$m = 10,\ n = -9 \Rightarrow 15r^2 + 10r - 9r - 6 \Rightarrow 5r(3r+2) - 3(3r+2) \Rightarrow (5r-3)(3r+2)$.

31. Using factoring by grouping:

For $24m^2 - 23m - 12$, $m \cdot n = a \cdot c = 24 \cdot (-12) = -288$ and $m + n = b = -23 \Rightarrow$

$m = -32,\ n = 9 \Rightarrow 24m^2 - 32m + 9m - 12 \Rightarrow 8m(3m-4) + 3(3m-4) \Rightarrow (8m+3)(3m-4)$.

33. Using factoring by grouping: For $25x^2 + 5x - 2$, $m \cdot n = a \cdot c = 25 \cdot (-2) = -50$ and $m + n = b = 5 \Rightarrow$

$m = 10,\ n = -5 \Rightarrow 25x^2 + 10x - 5x - 2 \Rightarrow 5x(5x+2) - (5x+2) \Rightarrow (5x-1)(5x+2)$.

35. Using factoring by grouping: For $6x^2 + 11x - 2$, $m \cdot n = a \cdot c = 6 \cdot (-2) = -12$ and $m + n = b = 11 \Rightarrow$

$m = 12,\ n = -1 \Rightarrow 6x^2 + 12x - x - 2 \Rightarrow 6x(x+2) - (x+2) \Rightarrow (6x-1)(x+2)$.

37. To factor $15y^2 - 7y + 2$, find numbers m and n such that $mn = 15 \cdot 2 = 30$ and $m + n = -7$. Because no such numbers exist the polynomial is prime.

39. Using factoring by grouping: For $21n^2 + 4n - 1$, $m \cdot n = a \cdot c = 21 \cdot (-1) = -21$ and $m + n = b = 4 \Rightarrow$

$m = 7,\ n = -3 \Rightarrow 21n^2 + 7n - 3n - 1 \Rightarrow 7n(3n+1) - (3n+1) \Rightarrow (7n-1)(3n+1)$.

41. Using factoring by grouping: For $14y^2 + 23y + 3$, $m \cdot n = a \cdot c = 14 \cdot 3 = 42$ and $m + n = b = 23 \Rightarrow$

$m = 21,\ n = 2 \Rightarrow 14y^2 + 21y + 2y + 3 \Rightarrow 7y(2y+3) + (2y+3) \Rightarrow (7y+1)(2y+3)$.

43. Using factoring by grouping: For $28z^2 - 25z + 3$, $m \cdot n = a \cdot c = 28 \cdot 3 = 84$ and $m + n = b = -25 \Rightarrow$

$m = -21,\ n = -4 \Rightarrow 28z^2 - 21z - 4z + 3 \Rightarrow 7z(4z-3) - (4z-3) \Rightarrow (7z-1)(4z-3)$.

45. Using factoring by grouping: For $30x^2 - 29x + 6$, $m \cdot n = a \cdot c = 30 \cdot 6 = 180$ and $m + n = b = -29 \Rightarrow$

$m = -20,\ n = -9 \Rightarrow 30x^2 - 20x - 9x + 6 \Rightarrow 10x(3x-2) - 3(3x-2) \Rightarrow (10x-3)(3x-2)$.

47. To factor $20a^2 + 18a - 5$, find numbers m and n such that $mn = 20(-5) = -100$ and $m + n = 18$. Because no such numbers exist the polynomial is prime.

49. Using factoring by grouping: For $18t^2 + 23t - 6$, $m \cdot n = a \cdot c = 18 \cdot (-6) = -108$ and $m + n = b = 23 \Rightarrow$

$m = 27,\ n = -4 \Rightarrow 18t^2 + 27t - 4t - 6 \Rightarrow 9t(2t+3) - 2(2t+3) \Rightarrow (9t-2)(2t+3)$.

51. The GCF of $12a^2+12a-9$ is $3 \Rightarrow 3(4a^2+4a-3)$. Now use factoring by grouping: For $4a^2+4a-3$, $m \cdot n = a \cdot c = 4(-3) = -12$ and $m+n=b=4 \Rightarrow m=6, n=-2 \Rightarrow 3(4a^2+4a-3) \Rightarrow 3[(4a^2+6a)+(-2a-3)]$ $\Rightarrow 3[2a(2a+3)-1(2a+3)] \Rightarrow 3(2a-1)(2a+3)$.

53. The GCF of $12y^3-11y^2+2y$ is $y \Rightarrow y(12y^2-11y+2)$. Now use factoring by grouping: For $12y^2-11y+2$, $m \cdot n = a \cdot c = 12 \cdot 2 = 24$ and $m+n=b=-11 \Rightarrow m=-3, n=-8 \Rightarrow y(12y^2-11y+2) \Rightarrow y[(12y^2-3y)+(-8y+2)]$ $\Rightarrow y[3y(4y-1)-2(4y-1)] \Rightarrow y(3y-2)(4y-1)$.

55. The GCF of $24x^3-30x^2+9x$ is $3x \Rightarrow 3x(8x^2-10x+3)$. Now use factoring by grouping: For $8x^2-10x+3$, $m \cdot n = a \cdot c = 8 \cdot 3 = 24$ and $m+n=b=-10 \Rightarrow m=-6, n=-4 \Rightarrow 3x(8x^2-10x+3) \Rightarrow 3x[(8x^2-6x)+(-4x+3)]$ $\Rightarrow 3x[2x(4x-3)-1(4x-3)] \Rightarrow 3x(2x-1)(4x-3)$.

57. The GCF of $8x^4-6x^3+2x^2$ is $2x^2 \Rightarrow 2x^2(4x^2-3x+1)$. Now use factoring by grouping: For $4x^2-3x+1$, $m \cdot n = a \cdot c = 4 \cdot 1 = 4$ and $m+n=b=-3$, but no such m and n exist. Thus $4x^2-3x+1$ is prime, and the factored form of the original polynomial is $2x^2(4x^2-3x+1)$.

59. The GCF of $28x^4+56x^3+21x^2$ is $7x^2 \Rightarrow 7x^2(4x^2+8x+3)$. Now use factoring by grouping: For $4x^2+8x+3$, $m \cdot n = a \cdot c = 4 \cdot 3 = 12$ and $m+n=b=8 \Rightarrow m=2, n=6$ $\Rightarrow 7x^2(4x^2+8x+3) \Rightarrow 7x^2[(4x^2+2x)+(6x+3)] \Rightarrow 7x^2[2x(2x+1)+3(2x+1)] \Rightarrow 7x^2(2x+3)(2x+1)$.

61. Find numbers m and n such that $m \cdot n = 3k$ and $m+n=3k+1 \Rightarrow m=3k, n=1 \Rightarrow 3x^2+(3k+1)x+k$ $\Rightarrow (3x^2+3kx)+(x+k) \Rightarrow 3x(x+k)+1(x+k) \Rightarrow (3x+1)(x+k)$.

63. Put $2+15x+7x^2$ in standard form $7x^2+15x+2$, $m \cdot n = a \cdot c = 7 \cdot 2 = 14$ and $m+n=b=15 \Rightarrow$ $m=14, n=1 \Rightarrow 7x^2+14x+x+2 \Rightarrow 7x(x+2)+(x+2) \Rightarrow (7x+1)(x+2)$.

65. Put $2-5x+2x^2$ in standard form $2x^2-5x+2$, $m \cdot n = a \cdot c = 2 \cdot 2 = 4$ and $m+n=b=-5 \Rightarrow$ $m=-4, n=-1 \Rightarrow 2x^2-4x-x+2 \Rightarrow 2x(x-2)-(x-2) \Rightarrow (2x-1)(x-2)$.

67. Put $3-2x-8x^2$ in standard form $-8x^2-2x+3 = -(8x^2+2x-3)$, $m \cdot n = a \cdot c = 8 \cdot (-3) = -24$ and $m+n=b=2 \Rightarrow m=6, n=-4 \Rightarrow -(8x^2+6x-4x-3) \Rightarrow -[2x(4x+3)-(4x+3)] \Rightarrow -(2x-1)(4x+3)$.

69. $-2x^2-7x+15 = -\left(2x^2+7x-15\right)$, $m \cdot n = a \cdot c = 2 \cdot (-15) = -30$ and $m+n=b=7 \Rightarrow$ $m=10, n=-3 \Rightarrow -\left(2x^2+10x-3x-15\right) \Rightarrow -\left[2x(x+5)-3(x+5)\right] \Rightarrow -(2x-3)(x+5)$.

71. $-5x^2+14x+3=-(5x^2-14x-3)$, $m\cdot n=a\cdot c=5\cdot(-3)=-15$ and $m+n=b=-14 \Rightarrow$

$m=-15,\ n=1 \Rightarrow -(5x^2-15x+x-3) \Rightarrow -[5x(x-3)+(x-3)] \Rightarrow -(5x+1)(x-3).$

73. $6x^2+7x+2$, $m\cdot n=a\cdot c=6\cdot 2=12$ and $m+n=b=7 \Rightarrow m=4,\ n=3 \Rightarrow$

$(6x^2+4x+3x+2) \Rightarrow 2x(3x+2)+(3x+2) \Rightarrow (2x+1)(3x+2)$. See Figure 73.

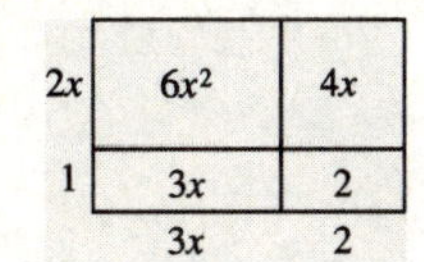

Figure 73

75. $(2x^2+6x+x+3) \Rightarrow 2x(x+3)+(x+3) \Rightarrow (2x+1)(x+3).$

6.4: Special Types of Factoring

Concepts

1. $(a-b)(a+b)$

2. cannot

3. $36x^2-49y^2=(6x)^2-(7y)^2 \Rightarrow a=6x$ and $b=7y$.

4. $a^2+2ab+b^2=(a+b)^2$

5. $a^2-2ab+b^2=(a-b)^2$

6. $(x+3)^2=x^2+6x+9 \Rightarrow 6x$

7. $a^2-2ab+b^2=(a-b)^2 \Rightarrow 4r^2=(2r)^2$ and $25t^2=(5t)^2$ and so $2ab=2(2r)(5t)=20rt$.

8. $a^3+b^3=(a+b)(a^2-ab+b^2)$

9. $a^3-b^3=(a-b)(a^2+ab+b^2)$

10. $8x^3+27y^3=(2x)^3+(3y)^3 \Rightarrow a=2x$ and $b=3y$.

11. $y^3-8=(y-2)(y+2y+4) \Rightarrow -;+$

12. $64z^3+27=(4z)^3+3^3$ so $(4z+3)(16z^2-12z+9) \Rightarrow +;-$

Factoring the Difference of Two Squares

13. $x^2-1=(x)^2-(1)^2 \Rightarrow (x-1)(x+1)$

15. $z^2-100=(z)^2-(10)^2 \Rightarrow (z-10)(z+10)$

17. $4y^2-1=(2y)^2-(1)^2 \Rightarrow (2y-1)(2y+1)$

19. $36z^2-25=(6z)^2-(5)^2 \Rightarrow (6z-5)(6z+5)$

21. $9-x^2=(3)^2-(x)^2 \Rightarrow (3-x)(3+x)$

23. $1-9y^2=(1)^2-(3y)^2 \Rightarrow (1-3y)(1+3y)$

25. $4a^2-9b^2=(2a)^2-(3b)^2 \Rightarrow (2a-3b)(2a+3b)$

27. $36m^2-25n^2=(6m)^2-(5n)^2 \Rightarrow (6m-5n)(6m+5n)$

29. $81r^2-49t^2=(9r)^2-(7t)^2 \Rightarrow (9r-7t)(9r+7t)$

Factoring Perfect Square Trinomials

31. $x^2+8x+16$ is a perfect square trinomial, $a^2=x^2$ so $a=x$, $b^2=4^2$ so $b=4$, $2ab=8x$, the middle term $\Rightarrow$

$(a+b)^2 \Rightarrow (x+4)^2$.

33. $z^2+12z+25$ is not a perfect square trinomial because $a^2=z^2$ so $a=z$, $b^2=5^2$ so $b=5$, $2ab=10z \neq 12z$, the middle term and also, FOIL does not work, therefore: Not possible.

35. x^2-6x+9 is a perfect square trinomial, $a^2=x^2$ so $a=x$, $b^2=(\pm 3)^2$ so $b=\pm 3$, using $b=-3$,

$2ab=-6x$, the middle term $\Rightarrow (a+b)^2 \Rightarrow (x-3)^2$.

37. $9y^2+6y+1$ is a perfect square trinomial, $a^2=(3y)^2$ so $a=3y$, $b^2=1^2$ so $b=1$, $2ab=6y$,

the middle term $\Rightarrow (a+b)^2 \Rightarrow (3y+1)^2$.

39. $4z^2-4z+1$ is a perfect square trinomial, $a^2=(2z)^2$ so $a=2z$, $b^2=(\pm 1)^2$ so $b=\pm 1$,

using $b=-1$, $2ab=-4z$, the middle term $\Rightarrow (a+b)^2 \Rightarrow (2z-1)^2$.

41. $9t^2+16t+4$ is not a perfect square trinomial because $a^2=(3t)^2$ so $a=3t$, $b^2=(\pm 2)^2$ so $b=\pm 2$.

$2ab=\pm 12t \neq 16t$, the middle term and also, FOIL does not work, therefore: Not possible.

43. $9x^2+30x+25$ is a perfect square trinomial, $a^2=(3x)^2$ so $a=3x$, $b^2=(5)^2$ so $b=5$,

$2ab=30x$, the middle term $\Rightarrow (a+b)^2 \Rightarrow (3x+5)^2$.

45. $4a^2-36a+81$ is a perfect square trinomial, $a^2=(2a)^2$ so $a=2a$, $b^2=(\pm 9)^2$ so $b=\pm 9$, using $b=-9$,

$2ab=-36a$, the middle term $\Rightarrow (a+b)^2 \Rightarrow (2a-9)^2$.

47. $x^2+2xy+y^2$ is a perfect square trinomial, $a^2=x^2$ so $a=x$, $b^2=y^2$ so $b=y$,

$2ab=2xy$, the middle term $\Rightarrow (a+b)^2 \Rightarrow (x+y)^2$.

49. $r^2-10rt+25t^2$ is a perfect square trinomial, $a^2=r^2$ so $a=r$, $b^2=(\pm 5t)^2$ so $b=\pm 5t$, using $b=-5t$,

$2ab=-10rt$, the middle term $\Rightarrow (a+b)^2 \Rightarrow (r-5t)^2$.

51. $4y^2 - 10yz + 9z^2$ is not a perfect square trinomial because $a^2 = (2y)^2$ so $a = 2y$, $b^2 = (\pm 3z)^2$ so $b = \pm 3z$, $2ab = \pm 12yz \neq -10yz$, the middle term and also, FOIL does not work, therefore: Not possible.

Factoring Sums and Differences of Two Cubes

53. Using the sum of cubes for $z^3 + 1$, $a^3 = z^3$ so $a = z$, $b^3 = 1^3$ so $b = 1 \Rightarrow$

$(a+b)(a^2 - ab + b^2) \Rightarrow (z+1)(z^2 - z + 1)$.

55. Using the sum of cubes for $x^3 + 64$, $a^3 = x^3$ so $a = x$, $b^3 = 4^3$ so $b = 4 \Rightarrow$

$(a+b)(a^2 - ab + b^2) \Rightarrow (x+4)(x^2 - 4x + 16)$.

57. Using the difference of cubes for $y^3 - 8$, $a^3 = y^3$ so $a = y$, $b^3 = 2^3$ so $b = 2 \Rightarrow$

$(a-b)(a^2 + ab + b^2) \Rightarrow (y-2)(y^2 + 2y + 4)$.

59. Using the difference of cubes for $n^3 - 1$, $a^3 = n^3$ so $a = n$, $b^3 = 1^3$ so $b = 1 \Rightarrow$

$(a-b)(a^2 + ab + b^2) \Rightarrow (n-1)(n^2 + n + 1)$.

61. Using the sum of cubes for $8x^3 + 1$, $a^3 = (2x)^3$ so $a = 2x$, $b^3 = 1^3$ so $b = 1 \Rightarrow$

$(a+b)(a^2 - ab + b^2) \Rightarrow (2x+1)(4x^2 - 2x + 1)$.

63. Using the difference of cubes for $m^3 - 64n^3$, $a^3 = m^3$ so $a = m$, $b^3 = (4n)^3$ so $b = 4n \Rightarrow$

$(a-b)(a^2 + ab + b^2) \Rightarrow (m - 4n)(m^2 + 4mn + 16n^2)$.

65. Using the sum of cubes for $8x^3 + 125y^3$, $a^3 = (2x)^3$ so $a = 2x$, $b^3 = (5y)^3$ so $b = 5y \Rightarrow$

$(a+b)(a^2 - ab + b^2) \Rightarrow (2x + 5y)(4x^2 - 10xy + 25y^2)$.

General Factoring Using Special Methods

67. $4x^2 - 16 = 4(x^2 - 4) = 4((x)^2 - (2)^2) \Rightarrow 4(x-2)(x+2)$

69. $2y^2 - 28y + 98 = 2(y^2 - 14y + 49)$; $y^2 - 14y + 49$ is a perfect square trinomial, $a^2 = y^2$ so $a = y$, $b^2 = 49$ so $b = \pm 7$, using $b = -7$, $2ab = -14y$, the middle term $\Rightarrow (a+b)^2 \Rightarrow (y-7)^2$. Therefore

$2(y^2 - 14y + 49) = 2(y-7)^2$.

71. $5z^3 + 40 = 5\,(z^3 + 8)$; $z^3 + 8$ is a sum of cubes with $a = z$ and

$b = 2 \Rightarrow (a+b)(a^2 - ab + b^2) \Rightarrow (z+2)(z^2 - 2z + 4)$. Therefore $5z^3 + 40 = 5\,(z+2)(z^2 - 2z + 4)$.

73. $x^3y - xy^3 = xy(x^2 - y^2) = xy\,(x - y)(x + y)$

75. $2m^3-10m^2+18m=2m(m^2-5m+9)$; m^2-5m+9 is not a perfect square trinomial because $a^2=m^2$ so $a=m, b^2=9$ so $b=\pm3, 2ab=\pm6m\neq-5m,$ the middle term, and FOIL does not work, therefore $2m^3-10m^2+18m=2m\,(m^2-5m+9).$

77. $700x^4-63x^2y^2=7x^2(100x^2-9y^2)=7x^2((10x)^2-(3y)^2)\;=7x^2\,(10x-3y)(10x+3y)$

79. Using the sum of cubes for $16a^3+2b^3=2\,(8a^3+b^3), a^3=(2a)^3$ so $a=2a, b^3=b^3$ so $b=b\Rightarrow(a+b)(a^2-ab+b^2)\Rightarrow2\,(2a+b)(4a^2-2ab+b^2).$

81. $4b^4+24b^3+36b^2=4b^2(b^2+6b+9)$; b^2+6b+9 is a perfect square trinomial, $a^2=b^2$ so $a=b, b^2=9$ so $b=3, 2ab=6b,$ the middle term $\Rightarrow(a+b)^2\Rightarrow(b+3)^2$. Therefore $4b^4+24b^3+36b^2=4b^2(b+3)^2$.

83. Using the difference of cubes for $500r^3-32t^3=4(125r^3-8t^3), a^3=(5r)^3$ so $a=5r, b^3=(2t)^3$ so $b=2t\Rightarrow(a-b)(a^2+ab+b^2)\Rightarrow4(5r-2t)(25r^2+10rt+4t^2)$

Geometry

85. Sides must be the same $\Rightarrow$ perfect square trinomial, $4x^2+12x+9$, $a^2=(2x)^2$ so $a=2x$, $b^2=3^2$ so $b=3$, $2ab=12x$ the middle term $\Rightarrow$ $(2x+3)^2$. See Figure 85. Length = $2x + 3$.

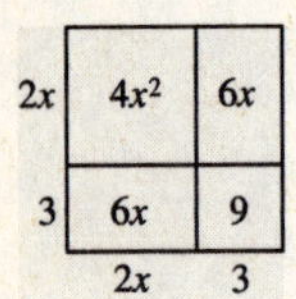

Figure 85

Checking Basic Concepts for Sections 6.3 & 6.4

1. (a) $2x^2-5x-12$ (factor by grouping), $m\cdot n=-24,\ m+n=-5\Rightarrow m=-8,\ n=3\Rightarrow$ $2x^2-8x+3x-12\Rightarrow2x(x-4)+3(x-4)\Rightarrow(2x+3)(x-4).$

   (b) $6x^2+17x-14$ (factor by grouping), $m\cdot n=-84,\ m+n=17\Rightarrow m=21,\ n=-4\Rightarrow$ $6x^2+21x-4x-14\Rightarrow3x(2x+7)-2(2x+7)\Rightarrow(3x-2)(2x+7).$

2. (a) To factor $3y^2+4y-2$, find numbers m and n such that $mn=3(-2)=-6$ and $m+n=4$. Because no such numbers exist the polynomial is prime.

   (b) $6y^3-10y^2-4y=2y\,(3y^2-5y-2)$; $m\cdot n=a\cdot c=3(-2)=-6$ and $m+n=b=-5\Rightarrow m=-6, n=1\Rightarrow$ $2y\,(3y^2-5y-2)\Rightarrow2y\,[(3y^2-6y)+(y-2)]\;\Rightarrow2y\,[3y\,(y-2)+1(y-2)]\Rightarrow2y\,(3y+1)(y-2)$

3. $3x^2+2x+9x+6\Rightarrow x(3x+2)+3(3x+2)\Rightarrow(x+3)(3x+2)$

4. (a) z^2-64 (difference of squares), $a^2=z^2$ so $a=z, b^2=8^2$ so $b=8\Rightarrow(z-8)(z+8).$

   (b) $9r^2-4t^2$ (difference of squares), $a^2=(3r)^2$ so $a=3r, b^2=(2t)^2$ so $b=2t\Rightarrow(3r+2t)(3r-2t).$

5. (a) $x^2+12x+36$ (FOIL in reverse) $\Rightarrow (x+6)(x+6) \Rightarrow (x+6)^2$.

   (b) $9a^2-12ab+4b^2$ (perfect square trinomial), $a^2=(3a)^2$ so $a=3a, b^2=(\pm 2b)^2$ so $b=\pm 2b$, using $b=-2b, 2ab=-12ab$ the second term $\Rightarrow (3a-2b)^2$.

6. (a) m^3-27 (difference of cubes), $a^3=m^3$ so $a=m, b^3=3^3$ so $b=3 \Rightarrow (m-3)(m^2+3m+9)$.

   (b) $125n^3+27$ (sum of cubes), $a^3=(5n)^3$ so $a=5n$, $b^3=3^3$ so $b=3 \Rightarrow (5n+3)(25n^2-15n+9)$.

7. (a) $16x^2-4=4(4x^2-1)=4((2x)^2-(1)^2)=4(2x-1)(2x+1)$

   (b) $3y^4+24y=3y(y^3+8)$ (sum of cubes), $a^3=y^3$ so $a=y, b^3=2^3$ so $b=2 \Rightarrow 3y(y+2)(y^2-2y+4)$.

6.5: Summary of Factoring

Concepts

1. greatest common factor
2. GCF
3. grouping
4. No; a sum of squares cannot be factored.
5. Yes; a sum of cubes can be factored.
6. completely

Warm Up

7. $4x-2=2(2x-1)$

9. $2y^2-4y+4=2(y^2-2y+2)$

11. $z^2-4=(z-2)(z+2)$

13. $a^3+8=(a+2)(a^2-2a+4)$

15. $4b^2-12b+9=(2b-3)^2$

17. m^2+9 is a sum of squares and cannot be factored.

19. $x^3-x^2+5x-5=x^2(x-1)+5(x-1)=(x^2+5)(x-1)$

21. $y^2-5y+4=(y-4)(y-1)$

General Factoring

23. $x^3+4x^2-9x-36=x^2(x+4)-9(x+4)=(x^2-9)(x+4)$

    $=(x-3)(x+3)(x+4)$

25. $8a^3-64=8(a^3-8)=8(a-2)(a^2+2a+4)$

27. $12x^4-18x^3+4x^2-6x=2x(6x^3-9x^2+2x-3)$

    $=2x[3x^2(2x-3)+1(2x-3)]=2x(3x^2+1)(2x-3)$

29. $54t^4 + 16t = 2t(27t^3 + 8) = 2t(3t + 2)(9t^2 - 6t + 4)$

31. $2r^3 + 6r^2 - 2r - 6 = 2r^2(r+3) - 2(r+3) = (2r^2 - 2)(r+3)$

$= 2(r^2 - 1)(r+3) = 2(r-1)(r+1)(r+3)$

33. $6z^4 - 21z^3 - 45z^2 = 3z^2(2z^2 - 7z - 15) = 3z^2(2z+3)(z-5)$

35. $12b^4 - 10b^3 + 2b^2 = 2b^2(6b^2 - 5b + 1) = 2b^2(3b-1)(2b-1)$

37. $6y^2z - 24z^3 = 6z(y^2 - 4z^2) = 6z(y-2z)(y+2z)$

39. $3x^2y - 30xy + 75y = 3y(x^2 - 10x + 25) = 3y(x-5)^2$

41. $27m^3 - 8n^3 = (3m - 2n)(9m^2 + 6mn + 4n^2)$

43. $3x^5 - 12x^3 - 3x^2 + 12 = 3(x^5 - 4x^3 - x^2 + 4)$

$= 3[x^3(x^2 - 4) - 1(x^2 - 4)] = 3(x^3 - 1)(x^2 - 4)$

$= 3(x-1)(x^2 + x + 1)(x-2)(x+2)$

45. $5a^2 - 27a - 18 = 5a^2 - 30a + 3a - 18 = 5a(a-6) + 3(a-6)$

$= (5a+3)(a-6)$

47. $3rt^2 + 33rt + 90r = 3r(t^2 + 11t + 30)$

$= 3r(t+5)(t+6)$

49. $9b^3 + 6b^2 + 12b + 8 = 3b^2(3b+2) + 4(3b+2)$

$= (3b^2 + 4)(3b+2)$

51. $6n^3 + 2n^2 - 10n = 2n(3n^2 + n - 5)$

53. $4x^2 - 36y^2 = 4(x^2 - 9y^2) = 4(x-3y)(x+3y)$

55. $2a^3 - 16a^2 + 32a = 2a(a^2 - 8a + 16) = 2a(a-4)^2$

57. $32xy^3 + 4x = 4x(8y^3 + 1) = 4x(2y+1)(4y^2 - 2y + 1)$

59. $8b^4 + 24b^3 - 2b^2 - 6b = 2b(4b^3 + 12b^2 - b - 3)$

$= 2b[4b^2(b+3) - 1(b+3)] = 2b(4b^2 - 1)(b+3) = 2b(2b-1)(2b+1)(b+3)$

61. Area of one square $= \frac{1}{3}(27x^2 + 18x + 3) = 9x^2 + 6x + 1$; sides must be the same $\Rightarrow$ perfect square trinomial,

$9x^2 + 6x + 1 = (3x+1)^2 \Rightarrow$ one side length is $3x + 1$.

6.6: Solving Equations by Factoring I (Quadratics)

Concepts

1. 0, 0

2. No. In order to apply the zero-product property, one side of the equation must be zero.

3. $2x = 0,\ x + 6 = 0$

4. Set the problem equal to zero by subtracting $4x$ from both sides.

5. Apply the zero-product property by setting $x + 5 = 0$ and $x - 4 = 0$.

6. solving

7. $ax^2 + bx + c = 0$ with $a \neq 0$.

8. Subtract $6x$ from both sides to get $x^2 - 6x + 1 = 0$.

9. zero

10. $3x - 6 = 0 \Rightarrow 3x = 6 \Rightarrow x = 2$; 2 is the zero of the polynomial.

Zero-Product Property

11. $x = 0$ or $y = 0$.

13. $2x = 0$ or $x + 8 = 0$, so $x = -8,\ 0$.

15. $y - 1 = 0$ or $y - 2 = 0$, so $y = 1,\ 2$.

17. $2z - 1 = 0$ or $4z - 3 = 0$, then $2z - 1 = 0 \Rightarrow 2z = 1 \Rightarrow z = \frac{1}{2}$ or $4z - 3 = 0 \Rightarrow 4z = 3 \Rightarrow z = \frac{3}{4}$, so $z = \frac{1}{2}, \frac{3}{4}$.

19. $1 - 3n = 0$ or $3 - 7n = 0$, then $1 - 3n = 0 \Rightarrow -3n = -1 \Rightarrow n = \frac{-1}{-3} = \frac{1}{3}$ or $3 - 7n = 0 \Rightarrow$

    $-7n = -3 \Rightarrow n = \frac{-3}{-7} = \frac{3}{7}$, so $n = \frac{1}{3}, \frac{3}{7}$.

21. $x = 0$ or $x - 5 = 0$ or $x - 8 = 0$, then $x - 5 = 0 \Rightarrow x = 5$ or $x - 8 = 0 \Rightarrow x = 8$, so $x = 0,\ 5,\ 8$.

Solving Quadratic Equations

23. $x^2 - x = 0 \Rightarrow x(x-1) = 0$, then $x = 0$ or $x - 1 = 0 \Rightarrow x = 1$, so $x = 0, 1$.

25. $z^2 - 5z = 0 \Rightarrow z(z-5) = 0$, then $z = 0$ or $z - 5 = 0 \Rightarrow z = 5$, so $z = 0,\ 5$.

27. $10y^2 + 15y = 0 \Rightarrow 5y(2y+3) = 0$, then $5y = 0 \Rightarrow y = 0$ or $2y + 3 = 0 \Rightarrow 2y = -3 \Rightarrow y = \frac{-3}{2}$, so $y = 0, \frac{-3}{2}$.

29. $x^2 - 1 = 0 \Rightarrow (x+1)(x-1)$ then $x + 1 = 0 \Rightarrow x = -1$ or $x - 1 = 0 \Rightarrow x = 1$ so $x = -1, 1$.

31. $4n^2 - 1 = 0 \Rightarrow (2n+1)(2n-1) = 0$, then $2n + 1 = 0 \Rightarrow 2n = -1 \Rightarrow n = \frac{-1}{2}$ or $2n - 1 = 0 \Rightarrow$

    $2n = 1 \Rightarrow n = \frac{1}{2}$, so $n = -\frac{1}{2}, \frac{1}{2}$.

33. $z^2 + 3z + 2 = 0 \Rightarrow (z+2)(z+1) = 0$, then $z + 2 = 0 \Rightarrow z = -2$ or $z + 1 = 0 \Rightarrow z = -1$, so $z = -2, -1$.

35. $x^2-12x+35=0 \Rightarrow (x-7)(x-5)=0$, then $x-7=0 \Rightarrow x=7$ or $x-5=0 \Rightarrow x=5$, so $x=5,\ 7$.

37. $2b^2+3b-2=0 \Rightarrow (2b-1)(b+2)=0$, then $2b-1=0 \Rightarrow 2b=1 \Rightarrow$

$b=\frac{1}{2}$ or $b+2=0 \Rightarrow b=-2$, so $b=-2,\ \frac{1}{2}$.

39. $6y^2+19y+10=0$, (factor by grouping), $m\cdot n=60$, $m+n=19 \Rightarrow m=15$, $n=4 \Rightarrow$

$6y^2+15y+4y+10=0 \Rightarrow 3y(2y+5)+2(2y+5)=0 \Rightarrow (3y+2)(2y+5)=0$, then $3y+2=0 \Rightarrow$

$3y=-2 \Rightarrow y=\frac{-2}{3}$ or $2y+5=0 \Rightarrow 2y=-5 \Rightarrow y=\frac{-5}{2}$, so $y=-\frac{5}{2},-\frac{2}{3}$.

41. $x^2=25 \Rightarrow x^2-25=0 \Rightarrow (x+5)(x-5)=0$, then $x+5=0 \Rightarrow x=-5$ or $x-5=0 \Rightarrow x=5$, so $x=-5,\ 5$.

43. $t^2=5t \Rightarrow t^2-5t=0 \Rightarrow t(t-5)=0$, then $t=0$ or $t-5=0 \Rightarrow t=5$, so $t=0,\ 5$.

45. $3m^2=-9m \Rightarrow 3m^2+9m=0 \Rightarrow 3m(m+3)=0$, then $3m=0 \Rightarrow m=0$ or $m+3=0 \Rightarrow$

$m=-3$, so $m=-3,\ 0$.

47. $x^2=5x+6 \Rightarrow x^2-5x-6=0 \Rightarrow (x+1)(x-6)=0$, then $x+1=0 \Rightarrow x=-1$ or $x-6=0 \Rightarrow$

$x=6$, so $x=-1,\ 6$.

49. $12z^2+11z=15 \Rightarrow 12z^2+11z-15=0$, (factor by grouping), $m\cdot n=-180$, $m+n=11 \Rightarrow$

$m=20$, $n=-9 \Rightarrow 12z^2+20z-9z-15=0 \Rightarrow 4z(3z+5)-3(3z+5)=0 \Rightarrow (4z-3)(3z+5)=0$,

then $4z-3=0 \Rightarrow 4z=3 \Rightarrow z=\frac{3}{4}$ or $3z+5=0 \Rightarrow 3z=-5 \Rightarrow z=\frac{-5}{3}$, so $z=-\frac{5}{3},\frac{3}{4}$.

51. $t(t+1)=2 \Rightarrow t^2+t=2 \Rightarrow t^2+t-2=0 \Rightarrow (t+2)(t-1)=0$, then $t+2=0 \Rightarrow$

$t=-2$ or $t-1=0 \Rightarrow t=1$, so $t=-2,\ 1$.

53. $x(2x+5)=3 \Rightarrow 2x^2+5x=3 \Rightarrow 2x^2+5x-3=0 \Rightarrow (2x-1)(x+3)=0$, then $2x-1=0 \Rightarrow$

$2x=1 \Rightarrow x=\frac{1}{2}$ or $x+3=0 \Rightarrow x=-3$, so $x=-3,\ \frac{1}{2}$.

55. $12x^2+12x=-3 \Rightarrow 12x^2+12x+3=0 \Rightarrow 3(4x^2+4x+1)=0 \Rightarrow 3(2x+1)^2=0$, then

$2x+1=0 \Rightarrow 2x=-1 \Rightarrow x=-\frac{1}{2}$.

57. $30y^2+50y+20=0 \Rightarrow 10(3y^2+5y+2)=0 \Rightarrow 10(3y+2)(y+1)=0$, then $3y+2=0 \Rightarrow 3y=-2 \Rightarrow y=-\frac{2}{3}$

or $y+1=0 \Rightarrow y=-1$, so $y=-1,-\frac{2}{3}$.

Geometry

59. Both sides are the same so $x^2 = 144 \Rightarrow x^2 - 144 = 0 \Rightarrow (x+12)(x-12) = 0$, then $x+12 = 0 \Rightarrow$ $x = -12$ or $x - 12 = 0 \Rightarrow x = 12$, so $x = -12,\ 12$, but length is not negative so $x = 12$ feet.

61. Area of circle πr^2 – circumference of circle $2\pi r = 8\pi \Rightarrow \pi r^2 - 2\pi r = 8\pi \Rightarrow r^2 - 2r = 8 \Rightarrow$ $r^2 - 2r - 8 = 0 \Rightarrow (r-4)(r+2) = 0$, then $r - 4 = 0 \Rightarrow r = 4$ or $r + 2 = 0 \Rightarrow r = -2$, so $r = -2,\ 4$, but radius cannot be -2 so $r = 4$.

63. Using Pythagorean Theorem $a^2 + b^2 = c^2$, $(x-1)^2 + x^2 = (x+1)^2 \Rightarrow x^2 - 2x + 1 + x^2 = x^2 + 2x + 1 \Rightarrow$ $2x^2 - 2x + 1 = x^2 + 2x + 1 \Rightarrow x^2 - 4x = 0 \Rightarrow x(x-4) = 0$, then $x = 0$ or $x - 4 = 0 \Rightarrow x = 4$, so $x = 0,\ 4$, but length cannot be 0 so $x = 4$.

Applications

65. (a) Height is zero at ground, $0 = 96t - 16t^2 \Rightarrow 0 = 16t(6-t)$, then $16t = 0 \Rightarrow t = 0$ or $6 - t = 0 \Rightarrow$ $-t = -6 \Rightarrow t = 6$, so $t = 0,\ 6$, but $t = 0$ is when it was first hit, so after it was hit, it took 6 seconds to hit the ground.

(b) 3 seconds. See Figure 65.

Time (t)	0	1	2	3	4	5	6
Height (h)	0	80	128	144	128	80	0

Figure 65

67. (a) For $x = 30$, $D = \frac{1}{11}(30)^2 \Rightarrow D = \frac{1}{11}(900) \Rightarrow d \approx 81.8$ feet.

For $x = 60$, $D = \frac{1}{11}(60)^2 \Rightarrow D = \frac{1}{11}(3600) \Rightarrow d = 327.3$ feet.

When the speed doubles, the braking distance quadruples.

(b) $33 = \frac{1}{11}x^2 \Rightarrow 363 = x^2 \Rightarrow x \approx 19$ mph.

(c) See Figure 67. About 19 miles per hour: yes.

69. (a) 1930, $x = 30 \Rightarrow W = \frac{19}{3125}(30)^2 + \frac{11}{2} \Rightarrow W = \frac{19}{3125}(900) + \frac{11}{2} \Rightarrow W = 10.972$ million ≈ 11 million.

2000, $x = 100 \Rightarrow W = \frac{19}{3125}(100)^2 + \frac{11}{2} \Rightarrow W = \frac{19}{3125}(10{,}000) + \frac{11}{2} \Rightarrow W = 66.3$ million ≈ 66 million.

(b) See Figure 69. About 1981.

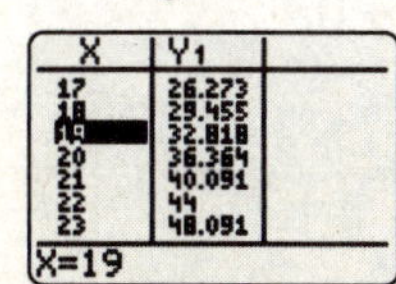

Figure 67

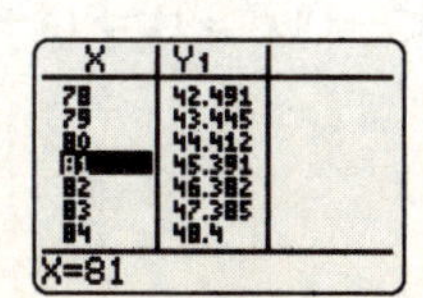

Figure 69

71. Width = x, Length = $x+10$, $x(x+10)=2000 \Rightarrow x^2+10x=2000 \Rightarrow x^2+10x-2000=0 \Rightarrow$ $(x+50)(x-40)=0$, then $x+50=0 \Rightarrow x=-50$ or $x-40=0 \Rightarrow x=40$, so $x=-50,\ 40$, but we can not have -50 pixels, so $x=40 \Rightarrow$ width 40 and length $40+10=50 \Rightarrow 40\times 50$ pixels.

Checking Basic Concepts for Sections 6.5 and 6.6

1. (a) $9a^2-18a+27=9(a^2-2a+3)$

   (b) $7xy^2+28x=7x(y^2+4)$

2. (a) $6z^4-28z^3+16z^2=2z^2(3z^2-14z+8)\ =2z^2(3z-2)(z-4)$

   (b) $2r^2t^2-18r^2=2r^2(t^2-9)=2r^2(t-3)(t+3)$

3. (a) $36x^3-48x^2+16x=4x(9x^2-12x+4)\ =4x(3x-2)^2$

   (b) $24b^3-81=3(8b^3-27)=3(2b-3)(4b^2+6b+9)$

4. (a) $4y^2-6y=0 \Rightarrow 2y(2y-3)=0$, then $2y=0$ or $2y-3=0 \Rightarrow y=0, \frac{3}{2}$

   (b) $5z^2+2z=3 \Rightarrow 5z^2+2z-3=0 \Rightarrow (5z-3)(z+1)=0$, then $5z-3=0 \Rightarrow 5z=3 \Rightarrow z=\frac{3}{5}$ or $z+1=0$, so $z=-1, \frac{3}{5}$

5. $x^2+2x-3=0 \Rightarrow (x+3)(x-1)=0$, then $x+3=0$ or $x-1=0$, so $x=-3,\ 1$

6. $88t-16t^2=0 \Rightarrow -8t(2t-11)=0$, then $-8t=0 \Rightarrow t=0$ or $2t-11=0 \Rightarrow 2t=11 \Rightarrow t=\frac{11}{2}$, so $t=0,\ \frac{11}{2}$; $t=0$ is when the ball was hit so the golf ball will hit the ground 5.5 seconds after it was hit.

6.7: Solving Equations by Factoring II (Higher Degree)

Concepts

1. GCF
2. zero-product
3. factors
4. $(x^2-y^2)^2$
5. $(z^2+1)(z^2+2)$
6. No, x^2-1 can be further factored so $x^4-1 \Rightarrow (x^2-1)(x^2+1) \Rightarrow (x+1)(x-1)(x^2+1)$.
7. Subtract x from both sides.
8. Factor out x^2.

Factoring Polynomials

9. $5x^2-5x-30 \Rightarrow 5\left(x^2-x-6\right) \Rightarrow 5(x+2)(x-3)$

11. $-4y^2-32y-48 \Rightarrow -4(y^2+8y+12) \Rightarrow -4(y+6)(y+2)$

13. $-20z^2-110z-50 \Rightarrow -10(2z^2+11z+5) \Rightarrow -10(2z+1)(z+5)$

15. $60-64t-28t^2 \Rightarrow -4(7t^2+16t-15) \Rightarrow -4(7t-5)(t+3)$

17. $r^3-r \Rightarrow r(r^2-1) \Rightarrow r(r+1)(r-1)$

19. $3x^3+3x^2-18x \Rightarrow 3x(x^2+x-6) \Rightarrow 3x(x+3)(x-2)$

21. $72z^3+12z^2-24z \Rightarrow 12z\ (6z^2+z-2) \Rightarrow 12z(3z+2)(2z-1)$

23. $x^4-4x^2 \Rightarrow x^2(x^2-4) \Rightarrow x^2(x+2)(x-2)$

25. $t^4+t^3-2t^2 \Rightarrow t^2(t^2+t-2) \Rightarrow t^2(t+2)(t-1)$

27. $x^4-5x^2+6 \Rightarrow (x^2-2)(x^2-3)$

29. $2x^4+7x^2+3 \Rightarrow (2x^2+1)(x^2+3)$

31. $y^4+6y^2+9 \Rightarrow (y^2+3)(y^2+3) \Rightarrow (y^2+3)^2$

33. $x^4-9 \Rightarrow (x^2+3)(x^2-3)$

35. $x^4-81 \Rightarrow (x^2+9)(x^2-9) \Rightarrow (x^2+9)(x+3)(x-3)$

37. $z^5+2z^4+z^3 \Rightarrow z^3(z^2+2z+1) \Rightarrow z^3(z+1)(z+1) \Rightarrow z^3(z+1)^2$

39. $2x^2+xy-y^2 \Rightarrow (2x-y)(x+y)$

41. $a^4-2a^2b^2+b^4 \Rightarrow (a^2-b^2)(a^2-b^2) \Rightarrow (a+b)(a-b)(a+b)(a-b) \Rightarrow (a+b)^2(a-b)^2$

43. $x^3-xy^2 \Rightarrow x(x^2-y^2) \Rightarrow x(x+y)(x-y)$

45. $4x^3+4x^2y+xy^2 \Rightarrow x(4x^2+4xy+y^2) \Rightarrow x(2x+y)(2x+y) \Rightarrow x(2x+y)^2$

Solving Equations

47. (a) $x^3-4x \Rightarrow x(x^2-4) \Rightarrow x(x+2)(x-2)$

    (b) $x(x+2)(x-2)=0$, then $x=0$ or $x+2=0$ or $x-2=0 \Rightarrow x=-2,\ 0,\ 2.$

49. (a) $2y^3-6y^2-36y \Rightarrow 2y(y^2-3y-18) \Rightarrow 2y(y-6)(y+3)$

    (b) $2y(y-6)(y+3)=0$, then $2y=0$ or $y-6=0$ or $y+3=0 \Rightarrow y=-3,\ 0,\ 6.$

51. $3x^2+33x+72=0 \Rightarrow 3(x^2+11x+24)=0 \Rightarrow 3(x+8)(x+3)=0$, then $x+8=0$ or $x+3=0 \Rightarrow x=-8,-3.$

53. $25x^2 = 50x + 75 \Rightarrow 25x^2 - 50x - 75 = 0 \Rightarrow 25\left(x^2 - 2x - 3\right) = 0 \Rightarrow 25(x-3)(x+1) = 0,$

then $x - 3 = 0$ or $x + 1 = 0 \Rightarrow x = -1,\ 3.$

55. $y^3 - 3y^2 - 4y = 0 \Rightarrow y\left(y^2 - 3y - 4\right) = 0 \Rightarrow y(y-4)(y+1) = 0,$ then $y = 0$ or $y - 4 = 0$ or

$y + 1 = 0 \Rightarrow y = -1,\ 0,\ 4.$

57. $3z^3 + 6z^2 = 72z \Rightarrow 3z^3 + 6z^2 - 72z = 0 \Rightarrow 3z\left(z^2 + 2z - 24\right) = 0 \Rightarrow 3z(z+6)(z-4) = 0,$ then

$3z = 0$ or $z + 6 = 0$ or $z - 4 = 0 \Rightarrow z = -6,\ 0,\ 4.$

59. $x^4 - 36x^2 = 0 \Rightarrow x^2\left(x^2 - 36\right) = 0 \Rightarrow x^2(x+6)(x-6) = 0,$ then $x^2 = 0$ or $x + 6 = 0$ or $x - 6 = 0 \Rightarrow$

$x = -6, 0, 6.$

61. $r^4 + 6r^3 = 7r^2 \Rightarrow r^4 + 6r^3 - 7r^2 = 0 \Rightarrow r^2\left(r^2 + 6r - 7\right) = 0 \Rightarrow r^2(r+7)(r-1) = 0,$ then $r^2 = 0$ or

$r + 7 = 0$ or $r - 1 = 0 \Rightarrow r = -7, 0, 1.$

63. $x^4 - 13x^2 = -36 \Rightarrow x^4 - 13x^2 + 36 = 0 \Rightarrow \left(x^2 - 9\right)\left(x^2 - 4\right) = 0 \Rightarrow (x+3)(x-3)(x+2)(x-2) = 0,$ then

$x + 3 = 0$ or $x - 3 = 0$ or $x + 2 = 0$ or $x - 2 = 0 \Rightarrow x = -3, -2, 2, 3.$

65. $x^4 + 1 = 2x^2 \Rightarrow x^4 - 2x^2 + 1 = 0 \Rightarrow \left(x^2 - 1\right)\left(x^2 - 1\right) = 0 \Rightarrow (x+1)(x-1)(x+1)(x-1) = 0,$ then

$x + 1 = 0$ or $x - 1 = 0 \Rightarrow x = -1, 1.$

67. $a^4 = 81 \Rightarrow a^4 - 81 = 0 \Rightarrow \left(a^2 + 9\right)\left(a^2 - 9\right) = 0 \Rightarrow (a^2 + 9)(a+3)(a-3) = 0,$ then $a^2 + 9 = 0$ or

$a + 3 = 0$ or $a - 3 = 0,$ but $a^2 + 9 = 0$ does not produce real solutions so $a = -3,\ 3.$

69. $x^3 - 2x^2 - x + 2 = 0 \Rightarrow x^2(x-2) - (x-2) = 0 \Rightarrow \left(x^2 - 1\right)(x-2) = 0 \Rightarrow$

$(x+1)(x-1)(x-2) = 0,$ then $x + 1 = 0$ or $x - 1 = 0$ or $x - 2 = 0 \Rightarrow x = -1, 1, 2.$

71. $x^3 - 5x^2 + x - 5 = 0 \Rightarrow x^2(x-5) + (x-5) = 0 \Rightarrow \left(x^2 + 1\right)(x-5) = 0 \Rightarrow$ then $x^2 + 1 = 0$ or

$x - 5 = 0$ but $x^2 + 1 = 0$ does not produce real solutions so $x = 5.$

Applications

73. (a) $x < 7.5\ in.$ because the width is 15 inches.

(b) Surface area of rectangle with 4 cut-out pieces $15 \times 20 = 300.$ Area of cut-out piece =

$x^2 \cdot 4$ pieces $= 4x^2 \Rightarrow$ surface area $300 - 4x^2.$

(c) $300 - 4x^2 = 275 \Rightarrow 0 = 4x^2 - 25 \Rightarrow (2x+5)(2x-5) = 0,$ then $2x + 5 = 0$ or $2x - 5 = 0,$ but

$2x + 5 = 0$ produces a negative length which we can not have so $2x - 5 = 0 \Rightarrow 2x = 5 \Rightarrow$

$x = \dfrac{5}{2} \Rightarrow x = 2.5$ inches.

75. $1990 = 30 \Rightarrow 0.0013(30)^3 - 0.085(30)^2 + 1.6(30) + 12 = 35.1 - 76.5 + 48 + 12 = 18.6$ trillion ft^3

77. Factoring is very difficult.

Checking Basic Concepts for Section 6.7

1. (a) $3x^2 - 6x - 24 = 3(x^2 - 2x - 8) = 3(x-4)(x+2)$

   (b) $-10y^2 + 5y + 5 = -5(2y^2 - y - 1) = -5(2y+1)(y-1)$

2. (a) $z^4 - 25 = (z^2 - 5)(z^2 + 5)$

   (b) $7t^4 - 7 = 7(t^4 - 1) = 7(t^2 - 1)(t^2 + 1) = 7(t-1)(t+1)(t^2+1)$

3. (a) $x^4 - 8x^2 + 16 \Rightarrow (x^2 - 4)(x^2 - 4) \Rightarrow (x+2)(x-2)(x+2)(x-2) \Rightarrow (x+2)^2(x-2)^2$

   (b) $2y^3 + 17y^2 - 30y \Rightarrow y(2y^2 + 17y - 30) \Rightarrow y(2y-3)(y+10)$

4. $t^4 + t^3 = 12t^2 \Rightarrow t^4 + t^3 - 12t^2 = 0 \Rightarrow t^2(t^2 + t - 12) = 0 \Rightarrow t^2(t+4)(t-3) = 0$, then $t^2 = 0$ or $t + 4 = 0$ or $t - 3 = 0 \Rightarrow t = -4, 0, 3$.

5. $x^3 - 3x^2 + 2x - 6 = 0 \Rightarrow x^2(x-3) + 2(x-3) = 0 \Rightarrow (x^2+2)(x-3) = 0$, then $x^2 + 2 = 0 \Rightarrow x^2 = -2$ (not possible for real numbers) or $x - 3 = 0 \Rightarrow x = 3$.

Chapter 6 Review Exercises

Section 6.1

1. Factors of $8z^3 = 2 \cdot 2 \cdot 2 \cdot z \cdot z \cdot z \Rightarrow$ factors of $4z^2 = 2 \cdot 2 \cdot z \cdot z \Rightarrow \text{GCF} = 2 \cdot 2 \cdot z \cdot z = 4z^2$; $4z^2(2z-1)$

2. Factors of $6x^4 = 2 \cdot 3 \cdot x \cdot x \cdot x \cdot x \Rightarrow$ factors of $3x^3 = 3 \cdot x \cdot x \cdot x \Rightarrow$ factors of $12x^2 = 2 \cdot 2 \cdot 3 \cdot x \cdot x \Rightarrow$ $\text{GCF} = 3 \cdot x \cdot x = 3x^2$; $3x^2(2x^2 + x - 4)$.

3. Factors of $9xy = 3 \cdot 3 \cdot x \cdot y \Rightarrow$ factors of $15yz^2 = 3 \cdot 5 \cdot y \cdot z \cdot z \Rightarrow \text{GCF}=3 \cdot y = 3y$; $3y(3x + 5z^2)$.

4. Factors of $a^2b^3 = a \cdot a \cdot b \cdot b \cdot b \Rightarrow$ factors of $a^3b^2 = a \cdot a \cdot a \cdot b \cdot b \Rightarrow \text{GCF} = a \cdot a \cdot b \cdot b = a^2b^2$; $a^2b^2(b+a)$.

5. $x(x+2) - 3(x+2) \Rightarrow (x-3)(x+2)$

6. $y^2(x-5) + 3y(x-5) \Rightarrow (y^2 + 3y)(x-5) \Rightarrow y(y+3)(x-5)$

7. $z^3 - 2z^2 + 5z - 10 \Rightarrow z^2(z-2) + 5(z-2) \Rightarrow (z^2+5)(z-2)$

8. $t^3 + t^2 + 8t + 8 \Rightarrow t^2(t+1) + 8(t+1) \Rightarrow (t^2+8)(t+1)$

9. $x^3 - 3x^2 + 6x - 18 \Rightarrow x^2(x-3) + 6(x-3) \Rightarrow (x^2+6)(x-3)$

10. $ax+bx-ay-by \Rightarrow x(a+b)-y(a+b) \Rightarrow (x-y)(a+b)$

11. $x^5+3x^4-2x^3-6x^2=x^2\left(x^3+3x^2-2x-6\right) = x^2\left[x^2(x+3)-2(x+3)\right]=x^2\left(x^2-2\right)(x+3)$

12. $2y^4+6y^3+2y^2+6y=2y\left(y^3+3y^2+y+3\right) = 2y\left[y^2(y+3)+1(y+3)\right]=2y\left(y^2+1\right)(y+3)$

Section 6.2

13. 4, 5

14. –3, 7

15. –9, – 4

16. –25, 4

17. Product -12, sum -1 is $-4, 3 \Rightarrow (x-4)(x+3)$.

18. Product 24, sum 10 is 4, 6 $\Rightarrow (x+4)(x+6)$.

19. Product -16, sum 6 is $-2, 8 \Rightarrow (x-2)(x+8)$.

20. Product -42, sum -1 is $-7, 6 \Rightarrow (x-7)(x+6)$.

21. Product -3, sum 2 is $-1, 3 \Rightarrow (x-1)(x+3)$.

22. Product 120, sum 22 is 10, 12 $\Rightarrow (x+10)(x+12)$.

23. $2x^3+6x^2-20x=2x\left(x^2+3x-10\right)$; product -10, sum 3 is $5, -2 \Rightarrow 2x(x+5)(x-2)$.

24. $x^4-3x^3-28x^2=x^2\left(x^2-3x-28\right)$; product -28, sum -3 is $-7, 4 \Rightarrow x^2(x-7)(x+4)$.

25. Product 10, sum -7 is $-5, -2 \Rightarrow (2-x)(5-x)$.

26. Product 24, sum 2 is $6, -4 \Rightarrow (6-x)(4+x)$.

Section 6.3

27. Using FOIL: $9x^2+3x-2=(3x+2)(3x-1)$.

28. Using FOIL: $2x^2+3x-5=(2x+5)(x-1)$.

29. Using FOIL: $3x^2+14x+15=(3x+5)(x+3)$.

30. Using FOIL: $35x^2-2x-1=(5x-1)(7x+1)$.

31. Using FOIL: $24x^2-7x-5=(8x-5)(3x+1)$.

32. Using FOIL: $4x^2+33x-27=(x+9)(4x-3)$.

33. $12x^3+48x^2+21x=3x\left(4x^2+16x+7\right)$; using FOIL: $3x(2x+7)(2x+1)$.

34. $8x^4+14x^3-30x^2=2x^2\left(4x^2+7x-15\right)$; using FOIL: $2x^2(x+3)(4x-5)$.

35. Using FOIL: $12-5x-2x^2=(3-2x)(4+x)$.

36. Using FOIL: $1+3x-10x^2=(1+5x)(1-2x)$.

Section 6.4

37. Difference of squares, $z^2-4=(z+2)(z-2)$.

38. Difference of squares, $9z^2-64=(3z+8)(3z-8)$.

39. Difference of squares, $36-y^2=(6+y)(6-y)$.

40. Difference of squares, $100a^2-81b^2=(10a+9b)(10a-9b)$.

41. Perfect square trinomial, $a^2=x^2$ so $a=x$, $b^2=7^2$ so $b=7$, then $2ab=2\cdot 7\cdot x=14x$ the middle term $\Rightarrow$ $(x+7)^2$.

42. Perfect square trinomial, $a^2=x^2$ so $a=x$, $b^2=(\pm 5)^2$ so $b=\pm 5$, using $b=-5$, then $2ab=2\cdot(-5)x=$ $-10x$ the middle term $\Rightarrow$ $(x-5)^2$.

43. Perfect square trinomial, $a^2=(2x)^2$ so $a=2x$, $b^2=(\pm 3)^2$ so $b=\pm 3$, using $b=-3$, then $2ab=2\cdot(-3)2x=$ $-12x$ the middle term $\Rightarrow$ $(2x-3)^2$.

44. Perfect square trinomial, $a^2=(3x)^2$ so $a=3x$, $b^2=8^2$ so $b=8$,

then $2ab=2\cdot 8\cdot 3x=48x$ the middle term $\Rightarrow$ $(3x+8)^2$.

45. Difference of cubes, $8t^3-1$, $a^3=(2t)^3$ so $a=2t$, $b^3=1^3$ so $b=1\Rightarrow(2t-1)(4t^2+2t+1)$.

46. Sum of cubes, $27r^3+8t^3$, $a^3=(3r)^3$ so $a=3r$, $b^3=(2t)^3$ so $b=2t\Rightarrow(3r+2t)(9r^2-6rt+4t^2)$.

47. $2x^3-50x=2x(x^2-25)=2x(x-5)(x+5)$ (difference of squares).

48. $24x^3+81=3(8x^3+27)=3(2x+3)(4x^2-6x+9)$ (sum of cubes).

49. $2x^3+28x^2+98x=2x(x^2+14x+49)=2x(x+7)^2$ (perfect square trinomial).

50. $2x^4-128x=2x(x^3-64)=2x(x-4)(x^2+4x+16)$ (difference of cubes).

Section 6.5

51. $9y^2-6y+6=3(3y^2-2y+2)$

52. $yz^2-9y=y(z^2-9)=y(z-3)(z+3)$

53. $x^4+7x^3-4x^2-28x=x(x^3+7x^2-4x-28)$

$=x[x^2(x+7)-4(x+7)]=x(x^2-4)(x+7)$

$=x(x-2)(x+2)(x+7)$

54. $12x^3+36x^2+27x=3x\left(4x^2+12x+9\right)=3x(2x+3)^2$

55. $3ab^3-24a=3a\left(b^3-8\right)=3a(b-2)\left(b^2+2b+4\right)$

56. $5x^3+20x=5x\left(x^2+4\right)$

57. $24x^3-6xy^2=6x\left(4x^2-y^2\right)=6x(2x-y)(2x+y)$

58. $x^3y+27y=y\left(x^3+27\right)=y(x+3)\left(x^2-3x+9\right)$

Section 6.6

59. $m=0$ or $n=0$

60. $y=0$

61. $(4x-3)(x+9)=0$, then $4x-3=0 \Rightarrow 4x=3 \Rightarrow x=\frac{3}{4}$ or $x+9=0 \Rightarrow x=-9$, so $x=-9, \frac{3}{4}$.

62. $(1-4x)(6+5x)=0$, then $1-4x=0 \Rightarrow -4x=-1 \Rightarrow x=\frac{-1}{-4} \Rightarrow x=\frac{1}{4}$ or $6+5x=0 \Rightarrow 5x=-6 \Rightarrow$

$x=\frac{-6}{5}$, so $x=-\frac{6}{5}, \frac{1}{4}$.

63. $z(z-1)(z-2)=0$, then $z=0$ or $z-1=0 \Rightarrow z=1$ or $z-2=0 \Rightarrow z=2$, so $z=0, 1, 2$.

64. $z^2-7z=0 \Rightarrow z(z-7)=0$, then $z=0$ or $z-7=0 \Rightarrow z=7$, so $z=0, 7$.

65. $y^2-64=0$, difference of squares $(y+8)(y-8)=0$, then $y+8=0 \Rightarrow y=-8$ or $y-8=0 \Rightarrow$

$y=8$, so $y=-8, 8$.

66. $y^2+9y+14=0 \Rightarrow (y+2)(y+7)=0$, then $y+2=0 \Rightarrow y=-2$ or $y+7=0 \Rightarrow y=-7$, so $y=-2, -7$.

67. $x^2=x+6 \Rightarrow x^2-x-6=0 \Rightarrow (x-3)(x+2)=0$, then $x-3=0 \Rightarrow x=3$ or $x+2=0 \Rightarrow$

$x=-2$, so $x=-2, 3$.

68. $10x^2+11x=6 \Rightarrow 10x^2+11x-6=0 \Rightarrow (2x+3)(5x-2)=0$, then $2x+3=0 \Rightarrow 2x=-3 \Rightarrow$

$x=\frac{-3}{2}$ or $5x-2=0 \Rightarrow 5x=2 \Rightarrow x=\frac{2}{5}$, so $x=-\frac{3}{2}, \frac{2}{5}$.

69. $t(t-14)=72 \Rightarrow t^2-14t=72 \Rightarrow t^2-14t-72=0 \Rightarrow (t-18)(t+4)=0$, then $t-18=0 \Rightarrow$

$t=18$ or $t+4=0 \Rightarrow t=-4$, so $t=-4, 18$.

70. $t(2t-1)=10 \Rightarrow 2t^2-t=10 \Rightarrow 2t^2-t-10=0 \Rightarrow (2t-5)(t+2)=0$, then $2t-5=0 \Rightarrow$

$2t=5 \Rightarrow t=\frac{5}{2}$ or $t+2=0 \Rightarrow t=-2$, so $t=-2, \frac{5}{2}$.

Section 6.7

71. $5x^2-15x-50 \Rightarrow 5\left(x^2-3x-10\right) \Rightarrow 5(x-5)(x+2)$

72. $-3x^2-6x+45 \Rightarrow -3\left(x^2+2x-15\right) \Rightarrow -3(x+5)(x-3)$

73. $y^3-4y \Rightarrow y\left(y^2-4\right) \Rightarrow y(y+2)(y-2)$

74. $3y^3+6y^2-9y \Rightarrow 3y\left(y^2+2y-3\right) \Rightarrow 3y(y+3)(y-1)$

75. $2z^4+14z^3+20z^2 \Rightarrow 2z^2\left(z^2+7z+10\right) \Rightarrow 2z^2(z+2)(z+5)$

76. $8z^4-32z^2 \Rightarrow 8z^2\left(z^2-4\right) \Rightarrow 8z^2(z+2)(z-2)$

77. $x^4-6x^2+9 \Rightarrow \left(x^2-3\right)\left(x^2-3\right) \Rightarrow \left(x^2-3\right)^2$

78. $2x^4-15x^2-27 \Rightarrow \left(2x^2+3\right)\left(x^2-9\right) \Rightarrow \left(2x^2+3\right)(x+3)(x-3)$

79. $a^2+10ab+25b^2 \Rightarrow (a+5b)(a+5b) \Rightarrow (a+5b)^2$

80. $x^3-xy^2 \Rightarrow x\left(x^2-y^2\right) \Rightarrow x(x+y)(x-y)$

81. $16x^2-72x-40=0 \Rightarrow 8\left(2x^2-9x-5\right)=0 \Rightarrow 8(2x+1)(x-5)=0$, then $2x+1=0 \Rightarrow 2x=-1 \Rightarrow$

$x=-\frac{1}{2}$ or $x-5=0 \Rightarrow x=5$, so $x=-\frac{1}{2}, 5$.

82. $2x^3-11x^2+15x=0 \Rightarrow x\left(2x^2-11x+15\right)=0 \Rightarrow x(2x-5)(x-3)=0$, then $x=0$ or $2x-5=0 \Rightarrow$

$2x=5 \Rightarrow x=\frac{5}{2}$ or $x-3=0 \Rightarrow x=3$, so $x=0, \frac{5}{2}, 3$.

83. $t^3-25t=0 \Rightarrow t\left(t^2-25\right)=0 \Rightarrow t(t+5)(t-5)=0$, then $t=0$ or $t+5=0 \Rightarrow$

$t=-5$ or $t-5=0 \Rightarrow t=5$, so $t=-5, 0, 5$.

84. $t^4-7t^3+12t^2=0 \Rightarrow t^2\left(t^2-7t+12\right)=0 \Rightarrow t^2(t-3)(t-4)=0$, then $t^2=0$ or $t-3=0 \Rightarrow$

$t=3$ or $t-4=0 \Rightarrow t=4$, so $t=0, 3, 4$.

85. $z^4+16=8z^2 \Rightarrow z^4-8z^2+16=0 \Rightarrow \left(z^2-4\right)\left(z^2-4\right)=0 \Rightarrow (z+2)(z-2)(z+2)(z-2)=0$, then

$z+2=0 \Rightarrow z=-2$ or $z-2=0 \Rightarrow z=2$, so $z=-2, 2$.

86. $z^4-256=0 \Rightarrow \left(z^2+16\right)\left(z^2-16\right)=0 \Rightarrow \left(z^2+16\right)(z+4)(z-4)=0$, then $z^2+16=0$ which

produces no real solutions, or $z+4=0 \Rightarrow z=-4$ or $z-4=0 \Rightarrow z=4$, so $z=-4, 4$.

87. $y^3=-64 \Rightarrow y^3+64=0 \Rightarrow (y+4)\left(y^2-4y+16\right)=0$, then $y+4=0 \Rightarrow$

$y=-4$ or $y^2-4y+16=0$ which produces no real solutions, so $y=-4$.

88. $y^3-y^2-y+1=0 \Rightarrow y^2(y-1)-1(y-1)=0 \Rightarrow \left(y^2-1\right)(y-1)=0 \Rightarrow$

$(y+1)(y-1)(y-1)=0$, then $y+1=0 \Rightarrow y=-1$ or $y-1=0 \Rightarrow y=1$, so $y=-1, 1$.

Applications

89. The sides of a square are equal so the trinomial must be a perfect square trinomial,

$a^2 = (3x)^2$ so $a = 3x$, $b^2 = 7^2$ so

$b = 7$, then $2ab = 2 \cdot 3x \cdot 7 = 42x$ the middle term, so $(3x+7)^2$, so each side is $3x+7$. See Figure 89.

90. $x^2 + 6x + 5 = 0 \Rightarrow (x+5)(x+1) = 0$ so the sides are $(x+5)$ by $(x+1)$. See Figure 90.

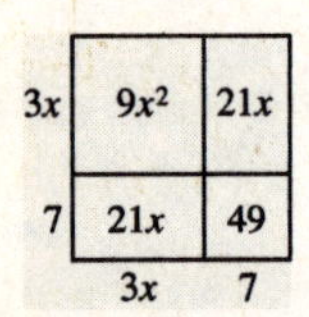

Figure 89

	x	5
x	x^2	$5x$
1	x	5

Figure 90

91. A cube has six sides, so $(6x^2 + 12x + 6) \div 6 = x^2 + 2x + 1$, the area of each side $\Rightarrow (x+1)(x+1) \Rightarrow$

each side is $(x+1)$.

92. $x^2 + 3x + x + 3 = x^2 + 4x + 3 \Rightarrow (x+3)(x+1)$

93. $2x^2 + 3x + 12x + 18 = 2x^2 + 15x + 18 \Rightarrow (2x+3)(x+6)$

94. $(\text{area})\, \pi r^2 = 2r\pi(\text{circumference}) \Rightarrow r^2 = 2r \Rightarrow r = 2$

95. $x(x+7) = 120 \Rightarrow x^2 + 7x - 120 = 0 \Rightarrow (x+15)(x-8) = 0$, $x = -15$ (not possible), 8, so rectangle 8×15 ft.

96. $100 = -16t^2 + 80t + 4 \Rightarrow 16t^2 - 80t + 96 = 0 \Rightarrow 16(t^2 - 5t + 6) = 0 \Rightarrow 16(t-2)(t-3) = 0$, so

$t = 2, 3$ so at 2 seconds and 3 seconds.

97. (a) $D = \frac{1}{9}(45)^2 + \frac{11}{3}(45) \Rightarrow D = \frac{1}{9}(2025) + \frac{11}{3}(45) \Rightarrow D = 225 + 165 \Rightarrow D = 390$ feet.

(b) $80 = \frac{1}{9}x^2 + \frac{11}{3}x \Rightarrow \frac{1}{9}x^2 + \frac{11}{3}x - 80 = 0 \Rightarrow$ multiply by 9 $\Rightarrow x^2 + 33x - 720 = 0 \Rightarrow$

$(x+48)(x-15) = 0$, so $D = -48$ (not possible) or 15 mph.

(c) See Figure 97. 15 mph, yes.

98. (a) $R = 100(200 - 100) \Rightarrow R = 100(100) \Rightarrow R = \$10{,}000$

(b) $7500 = p(200 - p) \Rightarrow 7500 = 200p - p^2 \Rightarrow p^2 - 200p + 7500 = 0 \Rightarrow (p-50)(p-150) = 0 \Rightarrow$

$p = 50, 150 \Rightarrow$ \$50 or \$150.

(c) See Figure 98. \$50 or \$150, yes.

99. (a) $Y = 20$ for 1970, $N = 0.68(20)^2 + 3.8(20) + 24 \Rightarrow N = 272 + 76 + 24 \Rightarrow N = 372$ million.

(b) See Figure 99. 1975.

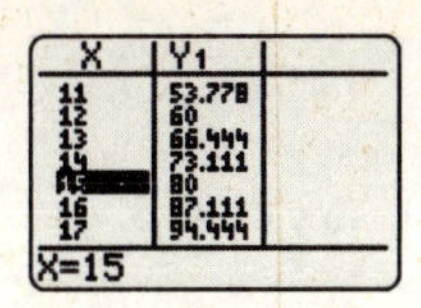

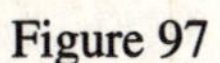
Figure 97

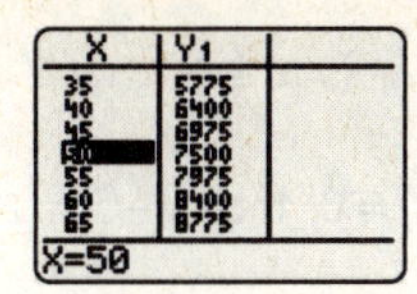

Figure 98

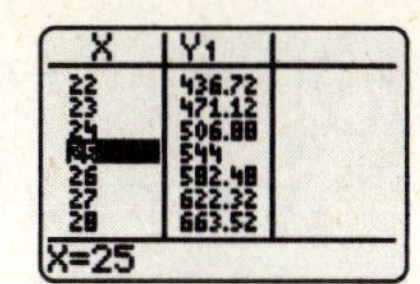

Figure 99

100. $x(x+30)=4000 \Rightarrow x^2+30x-4000=0 \Rightarrow (x+80)(x-50)=0$, so $x=-80$ (not possible) or 50, so 50×80 pixels.

101. (a) Bottom $(50-2x)(40-2x)=2000-80x-100x+4x^2=2000-180x+4x^2$; 2 sides $x(50-2x)\Rightarrow(50x-2x^2)(2)=100x-4x^2$; 2 sides $x(40-2x)\Rightarrow(40x-2x^2)(2)=80x-4x^2$. Add all sides $2000-180x+4x^2+100x-4x^2+80x-4x^2=2000-4x^2$.

(b) $1900=2000-4x^2 \Rightarrow -100=-4x^2 \Rightarrow 25=x^2 \Rightarrow x=5$ inches.

Chapter 6 Test

1. $4x^2y=2\cdot2\cdot x\cdot x\cdot y$; $20xy^2=2\cdot2\cdot5\cdot x\cdot y\cdot y$; $12xy=2\cdot2\cdot3\cdot x\cdot y \Rightarrow$ $\text{GCF}=2\cdot2\cdot x\cdot y=4xy \Rightarrow 4x^2y-20xy^2+12xy=4xy(x-5y+3)$

2. $9a^3b^2=3\cdot3\cdot a\cdot a\cdot a\cdot b\cdot b$; $3a^2b^2=3\cdot a\cdot a\cdot b\cdot b \Rightarrow \text{GCF}=3\cdot a\cdot a\cdot b\cdot b=3a^2b^2 \Rightarrow$ $9a^3b^2+3a^2b^2=3a^2b^2(3a+1)$

3. $ay+by+az+bz \Rightarrow y(a+b)+z(a+b) \Rightarrow (y+z)(a+b)$

4. $3x^3+x^2-15x-5 \Rightarrow x^2(3x+1)-5(3x+1) \Rightarrow (x^2-5)(3x+1)$

5. $y^2+4y-12=(y+6)(y-2)$

6. $4x^2+20x+25=(2x+5)(2x+5)=(2x+5)^2$

7. $4z^2-19z+12=(4z-3)(z-4)$

8. $21-17t+2t^2=2t^2-17t+21=(2t-3)(t-7)$

9. $6x^3+3x^2-3x \Rightarrow 3x(2x^2+x-1) \Rightarrow 3x(2x-1)(x+1)$

10. $2z^4-12z^2-54 \Rightarrow 2(z^4-6z^2-27) \Rightarrow 2(z^2-9)(z^2+3) \Rightarrow 2(z+3)(z-3)(z^2+3)$

11. $36y^3-100y=4y(9y^2-25)=4y(3y-5)(3y+5)$

12. $7x^4+56x=7x(x^3+8)=7x(x+2)(x^2-2x+4)$

13. $16a^4+24a^3+9a^2=a^2(16a^2+24a+9)=a^2(4a+3)^2$

14. $2b^4-32=2(b^4-16)=2(b^2-4)(b^2+4)=2(b-2)(b+2)(b^2+4)$

15. $x^2-16=0 \Rightarrow (x+4)(x-4)=0$, then $x+4=0$ or $x-4=0$, so $x=-4, 4$.

16. $y^2=y+20 \Rightarrow y^2-y-20=0 \Rightarrow (y-5)(y+4)=0$, then $y-5=0$ or $y+4=0$, so $y=-4, 5$.

17. $9z^2+16=24z \Rightarrow 9z^2-24z+16=0$ is a perfect square trinomial $a^2=(3z)^2$ so $a=3z$, $b^2=(\pm 4)^2$ so $b=\pm 4$, using $b=-4$, $2ab=2\cdot 3z\cdot -4=-24z$ the middle term $\Rightarrow (3z-4)^2$, then $3z-4=0 \Rightarrow 3z=4 \Rightarrow z=\frac{4}{3}$.

18. $x(x-5)=66 \Rightarrow x^2-5x-66=0 \Rightarrow (x-11)(x+6)=0$, then $x-11=0$ or $x+6=0$, so $x=-6, 11$

19. $y^3=9y \Rightarrow y^3-9y=0 \Rightarrow y(y^2-9)=0 \Rightarrow y(y+3)(y-3)=0$, then $y=0$ or $y+3=0$ or $y-3=0$, so $y=-3, 0, 3$.

20. $x^4-5x^2+4=0 \Rightarrow (x^2-4)(x^2-1)=0 \Rightarrow (x+2)(x-2)(x+1)(x-1)=0$, then $x+2=0$ or $x-2=0$ or $x+1=0$ or $x-1=0$, so $x=-2, -1, 1, 2$.

21. In a square the sides must be equal so $9x^2+30x+25$ must be a perfect square trinomial $\Rightarrow a^2=(3x)^2$ so $a=3x$,
$b^2=5^2$ so $b=5$, $2ab=2(3x)5=30x$ the middle term $\Rightarrow (3x+5)^2 \Rightarrow$ each side is $3x+5$.

22. $x^2+3x+2x+6=x^2+5x+6=(x+2)(x+3)$

23. (a) $D=\frac{1}{11}(55)^2=\frac{1}{11}(3025)=275$ ft.

(b) $99=\frac{1}{11}x^2 \Rightarrow \frac{1}{11}x^2-99=0 \Rightarrow x^2-1089=0 \Rightarrow (x+33)(x-33)=0$, then $x+33=0$ or $x-33=0$, so $x=-33$ (not possible) or 33, so 33 mph.

24. $36=-16t^2+48t+4 \Rightarrow 16t^2-48t+32=0 \Rightarrow 16(t^2-3t+2)=0 \Rightarrow 16(t-2)(t-1)=0$, then $t-2=0$ or $t-1=0$, so $t=1, 2$ so 1 sec. or 2 sec.

Chapter 6 Extended and Discovery Exercises

1. $x^2-5=(x+\sqrt{5})(x-\sqrt{5})$

2. $y^2-7=(y+\sqrt{7})(y-\sqrt{7})$

3. $3z^2-25=(\sqrt{3}z+5)(\sqrt{3}z-5)$

4. $7t^2-2=(\sqrt{7}t+\sqrt{2})(\sqrt{7}t-\sqrt{2})$

5. $x-4=(\sqrt{x}+2)(\sqrt{x}-2)$

6. $x-7=(\sqrt{x}+\sqrt{7})(\sqrt{x}-\sqrt{7})$

7. $x^2-3=0 \Rightarrow (x+\sqrt{3})(x-\sqrt{3})=0$, then $x+\sqrt{3}=0 \Rightarrow x=-\sqrt{3}$ or $x-\sqrt{3}=0 \Rightarrow x=\sqrt{3}$, so $x=-\sqrt{3}, \sqrt{3}$.

8. $y^2-7=0 \Rightarrow (y+\sqrt{7})(y-\sqrt{7})=0$, then $y+\sqrt{7}=0 \Rightarrow y=-\sqrt{7}$ or $y-\sqrt{7}=0 \Rightarrow y=\sqrt{7}$,

so $x=-\sqrt{7}, \sqrt{7}$.

9. $3x^2-25=0 \Rightarrow (\sqrt{3}x+5)(\sqrt{3}x-5)=0$, then $\sqrt{3}x+5=0 \Rightarrow \sqrt{3}x=-5 \Rightarrow x=\frac{-5}{\sqrt{3}}$ or

$\sqrt{3}x-5=0 \Rightarrow \sqrt{3}x=5 \Rightarrow x=\frac{5}{\sqrt{3}}$, so $x=\frac{-5}{\sqrt{3}}, \frac{5}{\sqrt{3}}$.

10. $7x^2-11=0 \Rightarrow (\sqrt{7}x+\sqrt{11})(\sqrt{7}x-\sqrt{11})=0$, then $\sqrt{7}x+\sqrt{11}=0 \Rightarrow \sqrt{7}x=-\sqrt{11} \Rightarrow$

$x=\frac{-\sqrt{11}}{\sqrt{7}}$ or $\sqrt{7}x-\sqrt{11}=0 \Rightarrow \sqrt{7}x=\sqrt{11} \Rightarrow x=\frac{\sqrt{11}}{\sqrt{7}}$, so $x=\frac{-\sqrt{11}}{\sqrt{7}}, \frac{\sqrt{11}}{\sqrt{7}}$.

11. $x^4-9=0 \Rightarrow (x^2+3)(x^2-3)=0 \Rightarrow (x^2+3)(x+\sqrt{3})(x-\sqrt{3})=0$, then $x^2+3=0 \Rightarrow x^2=-3$, which has no real solutions, or $x+\sqrt{3}=0 \Rightarrow x=-\sqrt{3}$ or $x-\sqrt{3}=0 \Rightarrow x=\sqrt{3}$, so $x=-\sqrt{3}, \sqrt{3}$.

12. $x^4-25=0 \Rightarrow (x^2+5)(x^2-5)=0 \Rightarrow (x^2+5)(x+\sqrt{5})(x-\sqrt{5})=0$, then $x^2+5=0 \Rightarrow x^2=-5$, which has no real solutions, or $x+\sqrt{5}=0 \Rightarrow x=-\sqrt{5}$ or $x-\sqrt{5}=0 \Rightarrow x=\sqrt{5}$, so $x=-\sqrt{5}, \sqrt{5}$.

Chapters 1-6 Cumulative Review Exercises

1. $144=2\times2\times2\times2\times3\times3$

2. $-2(-2)+3(4)=4+12=16$

3. $\frac{3}{5}\cdot\frac{15}{21}=\frac{45}{105}=\frac{3}{7}$

4. $\frac{4}{5}-\frac{1}{10}=\frac{8}{10}-\frac{1}{10}=\frac{7}{10}$

5. $26-3\cdot6\div2=26-18\div2=26-9=17$

6. $-2^2+\frac{3+2}{8+2}=-(2^2)+\frac{5}{10}=-4+\frac{1}{2}=-\frac{8}{2}+\frac{1}{2}=-\frac{7}{2}$

7. $4t-7=25 \Rightarrow 4t=32 \Rightarrow t=\frac{32}{4}$, so $t=8$.

8. See Figure 8. $2x+3=5$ when $x=1$.

x	-2	-1	0	1	2
$2x+3$	-1	1	3	5	7

Figure 8

9. $3n-5=n-7$; $3n-5=n-7 \Rightarrow 3n-n=-7+5 \Rightarrow 2n=-2 \Rightarrow n=-1$

10. $\frac{5.7}{100}=\frac{57}{1000}$; 0.057

11. $0.123 = 12.3\%$

12. $P = 2W + 2L \Rightarrow P - 2L = 2W \Rightarrow \dfrac{P-2L}{2} = W \Rightarrow W = \dfrac{P-2L}{2}$

13. $5 - 3z < -1 \Rightarrow -3z < -6 \Rightarrow z > 2$

14. See Figure 14.

15. See Figure 15.

x-intercept: $0 = 3x - 2 \Rightarrow 2 = 3x \Rightarrow x = \dfrac{2}{3}$; y-intercept: $y = 3(0) - 2 \Rightarrow y = 0 - 2 \Rightarrow y = -2$

16. See Figure 16. x-intercept: None; y-intercept: $y = -2$.

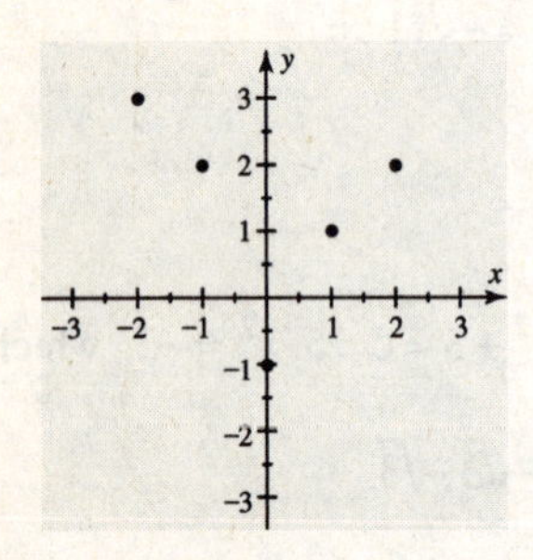
Figure 14

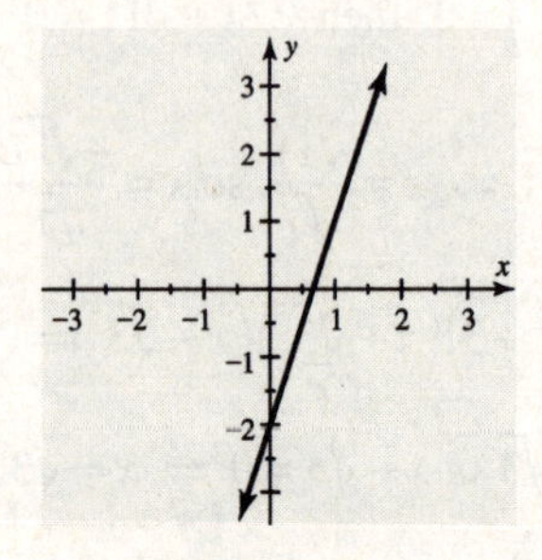
Figure 15

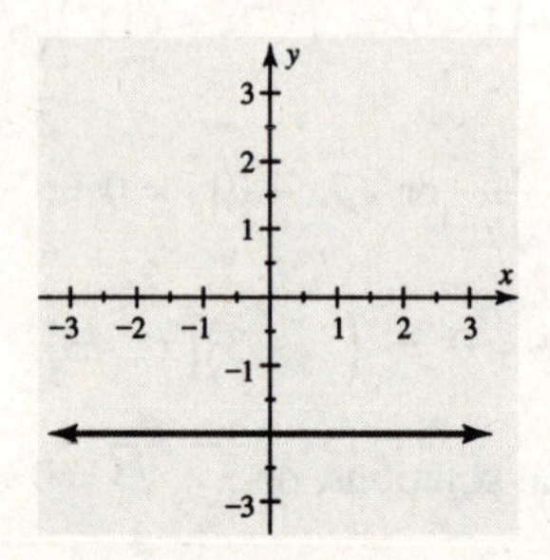
Figure 16

17. x-intercept: $x = -1 \Rightarrow (-1, 0)$; y-intercept: $y = -2 \Rightarrow (0, -2)$; $m = \dfrac{-2-0}{0-(-1)} = \dfrac{-2}{1} = -2$;

$y = mx + b \Rightarrow y = -2x - 2$.

18. $2x - 3y = -6 \Rightarrow -3y = -2x - 6 \Rightarrow y = \dfrac{2}{3}x + 2$, since perpendicular lines have slopes that are negative reciprocals, $m = -\dfrac{3}{2}$, now using $(1, 2)$ and $y - y_1 = m(x - x_1) \Rightarrow y - 2 = -\dfrac{3}{2}(x-1) \Rightarrow$

$y - 2 = -\dfrac{3}{2}x + \dfrac{3}{2} \Rightarrow y = -\dfrac{3}{2}x + \dfrac{7}{2}$.

19. $m = \dfrac{5-1}{1-(-2)} = \dfrac{4}{3}$, using $(-2, 1)$ and $y - y_1 = m(x - x_1) \Rightarrow y - 1 = \dfrac{4}{3}[x - (-2)] \Rightarrow$

$y - 1 = \dfrac{4}{3}x + \dfrac{8}{3} \Rightarrow y = \dfrac{4}{3}x + \dfrac{11}{3}$.

20. $(1, 2)$; $x + y = 3 \Rightarrow 1 + 2 = 3$, Yes; $-2x + y = 0 \Rightarrow -2(1) + 2 = 0 \Rightarrow -2 + 2 = 0$, Yes

21. Using substitution $y = -1$ into $2x + y = 1 \Rightarrow 2x + (-1) = 1 \Rightarrow 2x = 2 \Rightarrow x = 1 \Rightarrow (1, -1)$.

22. Using substitution $5x + y = -5 \Rightarrow y = -5x - 5$, substituting into $-x + 2y = 12 \Rightarrow$

$-x + 2(-5x - 5) = 12 \Rightarrow -x - 10x - 10 = 12 \Rightarrow -11x = 22 \Rightarrow x = -2$ and $y = -5(-2) - 5 \Rightarrow$

$y = 10 - 5 \Rightarrow y = 5 \Rightarrow (-2, 5)$.

23. (a) One solution, the intersection of the lines.

(b) Consistent, independent

24. (a) No solutions, no intersection of the lines.

(b) Inconsistent

25. See Figure 25. $x \le 2$

26. See Figure 26. $2x+3y \ge 6 \Rightarrow 3y \ge -2x+6 \Rightarrow y \ge -\frac{2}{3}x+2$

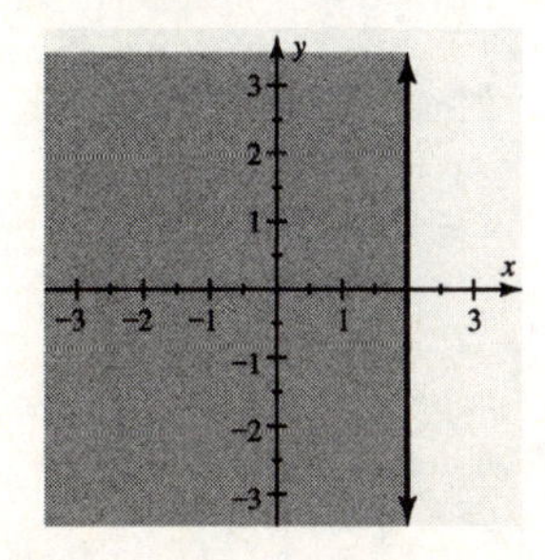

Figure 25

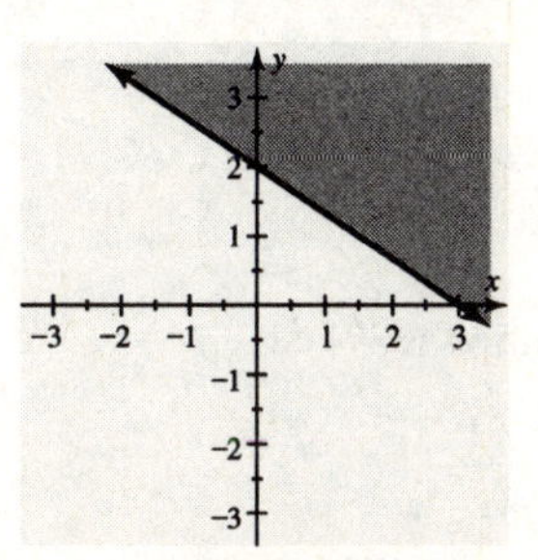

Figure 26

27. See Figure 27.

28. See Figure 28. $2x-y \le 4 \Rightarrow -y \le -2x+4 \Rightarrow y \ge 2x-4$ and $x+2y \ge 2 \Rightarrow 2y \ge -x+2 \Rightarrow y \ge -\frac{1}{2}x+1.$

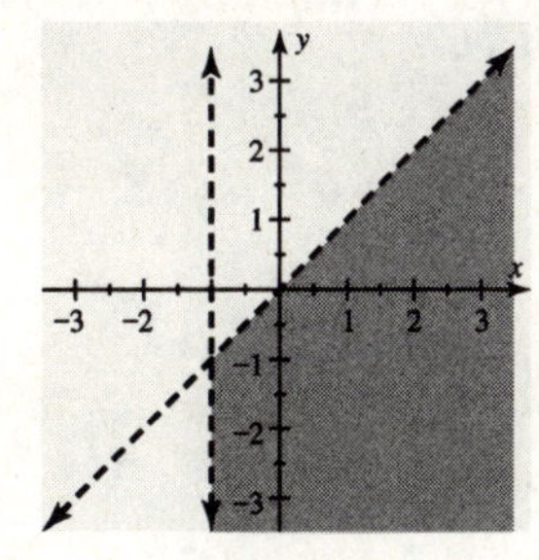

Figure 27

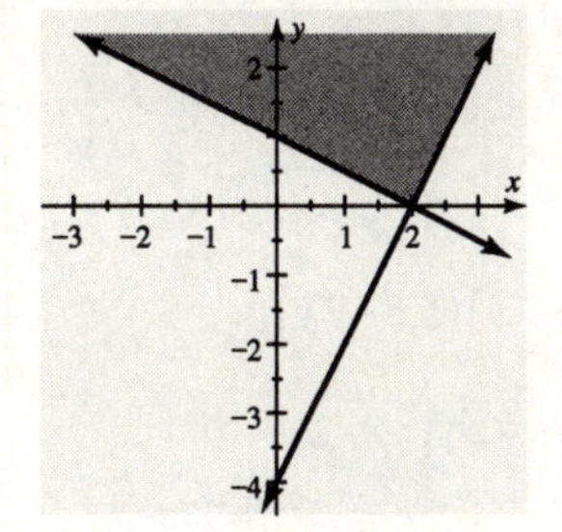

Figure 28

29. $-2^4 = -\left(2^4\right) = -16$

30. $(xy)^0 = 1$

31. $(xy)^4\left(x^3y^{-4}\right)^2 = x^4 \cdot y^4 \cdot x^6 \cdot y^{-8} = x^{10}y^{-4} = \dfrac{x^{10}}{y^4}$

32. $7x^3\left(-2x^2+3x\right) = -14x^5+21x^4$

33. By FOIL, $(7x-2)(3x+5) = 21x^2+35x-6x-10 = 21x^2+29x-10.$

34. By FOIL, $(x-7)^2 = (x-7)(x-7) = x^2-7x-7x+49 = x^2-14x+49.$

35. $a^{-4} \cdot a^2 = a^{-2} = \dfrac{1}{a^2}$

36. $\left(2t^3\right)^{-2} = 2^{-2} \cdot t^{-6} = \dfrac{1}{2^2} \cdot \dfrac{1}{t^6} = \dfrac{1}{4t^6}$

37. $(xy)^{-3}\left(x^{-1}y^{2}\right)^{-1} = x^{-3}y^{-3}\cdot x^{1}\cdot y^{-2} = x^{-2}\cdot y^{-5} = \dfrac{1}{x^{2}y^{5}}$

38. $\left(\dfrac{2x}{y^{-2}}\right)^{5} = \dfrac{2^{5}x^{5}}{y^{-10}} = \dfrac{32x^{5}}{y^{-10}} = 32x^{5}y^{10}$

39. $6.23\times10^{-3} = 0.00623$

40. $543{,}000 = 5.43\times10^{5}$

41. $\dfrac{6x^{3}+12x^{2}}{3x} = \dfrac{6x^{3}}{3x}+\dfrac{12x^{2}}{3x} = 2x^{2}+4x$

42.
$$\begin{array}{r} 3x \\ x^{2}+1\overline{)3x^{3}+0x^{2}-\ x+1} \\ \underline{3x^{3}+0x^{2}+3x} \\ -4x+1 \end{array}$$

The solution is

$3x+\dfrac{-4x+1}{x^{2}+1}$

43. $2y^{2}(x+2)-5(x+2) = \left(2y^{2}-5\right)(x+2)$

44. $t^{3}+6t^{2}+t+6 = \left(t^{3}+6t^{2}\right)+(t+6) = t^{2}(t+6)+(t+6) = \left(t^{2}+1\right)(t+6)$

45. FOIL, $x^{2}+3x-28 = (x-4)(x+7)$

46. FOIL, $6y^{2}+y-12 = (2y+3)(3y-4)$

47. Difference of squares, $25x^{2}-4y^{2} = (5x+2y)(5x-2y)$

48. FOIL, $64x^{2}-16x+1 = (8x-1)(8x-1) = (8x-1)^{2}$

49. Difference of cubes, $27t^{3}-8 = (3t-2)\left(9t^{2}+6t+4\right)$

50. $-4x^{2}+4x+24 = -4\left(x^{2}-x-6\right) \Rightarrow$ FOIL, $-4(x-3)(x+2)$

51. FOIL, $x^{4}-12x^{2}+27 = \left(x^{2}-3\right)\left(x^{2}-9\right) = \left(x^{2}-3\right)(x+3)(x-3)$

52. $x^{3}y-x^{2}y^{2} = x^{2}y(x-y)$

53. $y^{4} = 25y^{2} \Rightarrow y^{4}-25y^{2} = 0 \Rightarrow y^{2}\left(y^{2}-25\right) = 0 \Rightarrow y^{2}(y+5)(y-5) = 0$, then $y^{2} = 0 \Rightarrow y = 0$ or

$y+5 = 0 \Rightarrow y = -5$ or $y-5 = 0 \Rightarrow y = 5$, so $y = -5, 0, 5$.

54. $8z^{2}+8z-16 = 0 \Rightarrow 8\left(z^{2}+z-2\right) = 0 \Rightarrow 8(z+2)(z-1) = 0$, then $z+2 = 0 \Rightarrow z = -2$ or

$z-1 = 0 \Rightarrow z = 1$, so $z = -2, 1$.

55. $4z^{3} = 49z \Rightarrow 4z^{3}-49z = 0 \Rightarrow z\left(4z^{2}-49\right) = 0 \Rightarrow z(2z+7)(2z-7) = 0$, then $z = 0$ or

$2z+7 = 0 \Rightarrow 2z = -7 \Rightarrow z = \dfrac{-7}{2}$ or $2z-7 = 0 \Rightarrow 2z = 7 \Rightarrow z = \dfrac{7}{2}$, so $z = \dfrac{-7}{2}, 0, \dfrac{7}{2}$.

56. $x^4-18x^2+81=0 \Rightarrow (x^2-9)(x^2-9)=0 \Rightarrow (x+3)(x-3)(x+3)(x-3)=0$, then

$x+3=0 \Rightarrow x=-3$ or $x-3=0 \Rightarrow x=3$, so $x=-3, 3$.

Applications

57. (a) $10x+8x=18x$

(b) $18x=900 \Rightarrow x=50$ min.

58. $1\frac{3}{4}+2\frac{1}{2}+2\frac{2}{3}=1\frac{9}{12}+2\frac{6}{12}+2\frac{8}{12}=5\frac{23}{12}=6\frac{11}{12}$ miles.

59. $0.07(C)=1470 \Rightarrow C=\$21,000$

60. Let x represent the minutes cross-country skiing and $60-x$ represent minutes running.

$12(x)+9(60-x)=615 \Rightarrow 12x+540-9x=615 \Rightarrow 3x=75 \Rightarrow x=25$, so 25 minutes skiing and 35 minutes running.

61. (a) $C=0.25x+20$

(b) $100=0.25x+20 \Rightarrow 80=0.25x \Rightarrow x=320$ or 320 miles.

62. (a) $x=$ one angle $\Rightarrow x+x+y=180 \Rightarrow 2x+y=180$ and $2x=y+20 \Rightarrow 2x-y=20$

(b) Using elimination,

$$\begin{array}{r} 2x+y=180 \\ \underline{2x-y=20} \\ 4x=200 \end{array}$$

$4x=200 \Rightarrow x=50 \Rightarrow 50, 50, 80$ degrees.

63. See Figure 63.

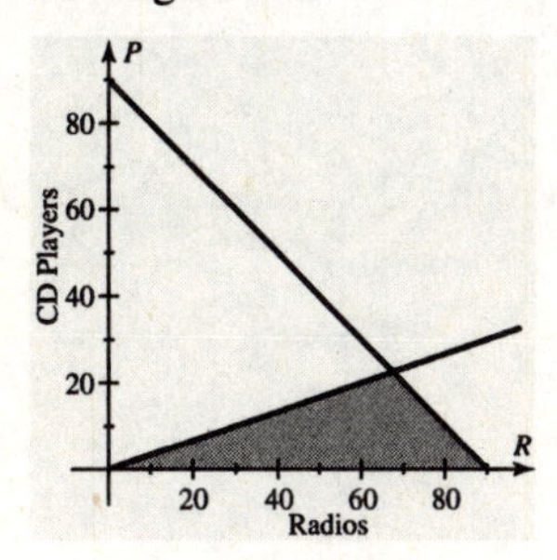

Figure 63

64. $3(3x)(2y)=18xy; 18(2)(3)=108$ yd^2.

65. $V=s^3 \Rightarrow V=(2xy^2)^3 \Rightarrow V=2^3\cdot x^3\cdot(y^2)^3 \Rightarrow V=8x^3y^6$

66. (a) $(x+2)(x+2)=x^2+4x+4$

(b) $2(x)+x\cdot x+2\cdot 2+2\cdot x=2x+x^2+4+2x=x^2+4x+4$

67. $x^2+12x+36=(x+6)(x+6)=(x+6)^2 \Rightarrow$ each side is $x+6$

68. (a) $0=64t-16t^2 \Rightarrow -16t(t-4)=0$, then $-16t=0 \Rightarrow t=0$ (when hit) or $t-4=0 \Rightarrow t=4$ sec.

(b) $48 = 64t - 16t^2 \Rightarrow 16t^2 - 64t + 48 = 0 \Rightarrow 16(t^2 - 4t + 3) = 0 \Rightarrow 16(t-3)(t-1) = 0$, then

$t - 3 = 0 \Rightarrow t = 3$ or $t - 1 = 0 \Rightarrow t = 1$, so at 1 and 3 sec.

Critical Thinking Solutions for Chapter 6

Section 6.2

- A cube has 6 sides so $(6x^2 + 24x + 24) \div 6 = x^2 + 4x + 4$, the area of each side. Factoring gives $(x+2)(x+2)$, so each side is $(x+2)$.

Section 6.6

- No, the zero-product property does not work for 1. We solve instead for zero, $x(2x+1) = 1 \Rightarrow$

  $2x^2 + x - 1 = 0 \Rightarrow (2x-1)(x+1) = 0;\ 2x - 1 = 0 \Rightarrow x = \frac{1}{2}$ and $x + 1 = 0 \Rightarrow x = -1$.

  The solutions are $\frac{1}{2}$ and -1.

Section 6.7

- The area of the metal square is 100 in^2 and the area of the four cutout squares is $4x^2$, so the surface area is $100 - 4x^2$. The sum of the areas of the four sides and the bottom is $4x(10-2x) + (10-2x)^2 = 100 - 4x^2$.

Chapter 7: Rational Expressions

7.1: Introduction to Rational Expressions

Concepts

1. $\frac{P}{Q}$; polynomials

2. Yes, both x and $2x^2+1$ are polynomials.

3. the denominator is zero

4. a

5. $\frac{a}{b}$

6. rational

7. $\frac{18}{60}=\frac{3\cdot 6}{10\cdot 6}=\frac{3}{10}\cdot\frac{6}{6}=\frac{3}{10}\cdot 1=\frac{3}{10}$

8. $\frac{3(x+1)}{2x(x+1)}=\frac{3}{2x}\cdot\frac{x+1}{x+1}=\frac{3}{2x}\cdot 1=\frac{3}{2x}$

9. $-1, \frac{x-a}{a-x}=\frac{1(x-a)}{a-x}=\frac{-1(-x+a)}{a-x}=\frac{-1(a-x)}{a-x}=-1$

10. $1-x, -\frac{x-1}{x+1}=\frac{-(x-1)}{x+1}=\frac{-x+1}{x+1}=\frac{1-x}{x+1}$

Evaluating Rational Expressions

11. $x=-7, \frac{3}{x}=\frac{3}{-7}=-\frac{3}{7}$

13. $x=-4, -\frac{x}{x-5}=-\frac{-4}{-4-5}=\frac{-(-4)}{-9}=\frac{4}{-9}=-\frac{4}{9}$

15. $y=-2, \frac{y+1}{y^2}=\frac{-2+1}{(-2)^2}=\frac{-1}{4}=-\frac{1}{4}$

17. $z=-2, \frac{7z}{z^2-4}=\frac{7(-2)}{(-2)^2-4}=\frac{-14}{4-4}=\frac{-14}{0}=$ undefined

19. $t=-2, \frac{5}{3t+6}=\frac{5}{3(-2)+6}=\frac{5}{-6+6}=\frac{5}{0}=$ undefined

21. $x=-2, \frac{4-x}{x-4}=\frac{4-(-2)}{-2-4}=\frac{6}{-6}=-1$

23. $x=0, -\frac{6-x}{x-6}=-\frac{6-0}{0-6}=-\frac{6}{-6}=\frac{-(6)}{-6}=\frac{-6}{-6}=1$

25. $x=-2, \frac{x}{x+1}=\frac{-2}{-2+1}=\frac{-2}{-1}=2$; $x=-1, \frac{x}{x+1}=\frac{-1}{-1+1}=\frac{-1}{0}=$ undefined;

$x=0, \frac{x}{x+1}=\frac{0}{0+1}=\frac{0}{1}=0$; $x=1, \frac{x}{x+1}=\frac{1}{1+1}=\frac{1}{2}$; $x=2, \frac{x}{x+1}=\frac{2}{2+1}=\frac{2}{3}$. See Figure 25.

x	−2	−1	0	1	2
$\frac{x}{x+1}$	2	—	0	$\frac{1}{2}$	$\frac{2}{3}$

Figure 25

x	−2	−1	0	1	2
$\frac{3x}{2x^2+1}$	$-\frac{2}{3}$	−1	0	1	$\frac{2}{3}$

Figure 27

27. $x=-2, \frac{3x}{2x^2+1}=\frac{3(-2)}{2(-2)^2+1}=\frac{-6}{2(4)+1}=\frac{-6}{8+1}=\frac{-6}{9}=\frac{-2}{3}=-\frac{2}{3}$;

$x=-1, \frac{3x}{2x^2+1}=\frac{3(-1)}{2(-1)^2+1}=\frac{-3}{2(1)+1}=\frac{-3}{2+1}=\frac{-3}{3}=-1$;

$x=0, \frac{3x}{2x^2+1}=\frac{3(0)}{2(0)^2+1}=\frac{0}{2(0)+1}=\frac{0}{0+1}=\frac{0}{1}=0$;

$x=1, \frac{3x}{2x^2+1}=\frac{3(1)}{2(1)^2+1}=\frac{3}{2(1)+1}=\frac{3}{2+1}=\frac{3}{3}=1$;

$x=2, \frac{3x}{2x^2+1}=\frac{3(2)}{2(2)^2+1}=\frac{6}{2(4)+1}=\frac{6}{8+1}=\frac{6}{9}=\frac{2}{3}$. See Figure 27.

29. A rational expression is undefined when the denominator = 0. $x=0$

31. A rational expression is undefined when the denominator = 0. $z-3=0 \Rightarrow z=3$

33. A rational expression is undefined when the denominator = 0. $5y+4=0 \Rightarrow 5y=-4 \Rightarrow y=\frac{-4}{5}$

35. A rational expression is undefined when the denominator = 0. $t^2+1=0 \Rightarrow t^2=-1$ which is impossible, so none.

37. A rational expression is undefined when the denominator = 0. $x^2-25=0 \Rightarrow x^2=25 \Rightarrow x=-5, 5$

39. A rational expression is undefined when the denominator = 0. $x^2+5x+6=0 \Rightarrow (x+3)(x+2)=0 \Rightarrow$ $x+3=0 \Rightarrow x=-3$ or $x+2=0 \Rightarrow x=-2$, so $x=-3, -2$

41. A rational expression is undefined when the denominator = 0. $2z^2-7z+5=0 \Rightarrow (2z-5)(z-1)=0 \Rightarrow$ $2z-5=0 \Rightarrow 2z=5 \Rightarrow z=\frac{5}{2}$ or $z-1=0 \Rightarrow z=1$, so $z=1, \frac{5}{2}$

Simplifying Rational Expressions

43. $\frac{12}{18}=\frac{2}{3}\cdot\frac{6}{6}=\frac{2}{3}\cdot 1=\frac{2}{3}$

45. $\frac{24}{48}=\frac{1}{2}\cdot\frac{24}{24}=\frac{1}{2}\cdot 1=\frac{1}{2}$

47. $-\frac{6}{15}=-\frac{2}{5}\cdot\frac{3}{3}=-\frac{2}{5}\cdot 1=-\frac{2}{5}$

49. $-\frac{25}{75}=-\frac{1}{3}\cdot\frac{25}{25}=-\frac{1}{3}\cdot 1=-\frac{1}{3}$

51. (a) $\frac{8}{16}=\frac{1}{2}\cdot\frac{8}{8}=\frac{1}{2}\cdot 1=\frac{1}{2}$

(b) $\frac{x+2}{2x+4}=\frac{x+2}{2(x+2)}=\frac{1}{2}\cdot\frac{x+2}{x+2}=\frac{1}{2}\cdot 1=\frac{1}{2}$

53. (a) $\frac{7-3}{3-7}=\frac{4}{-4}=-1$

(b) $\frac{7-x}{x-7}=\frac{7-x}{-1(7-x)}=-1\cdot\frac{7-x}{7-x}=-1\cdot 1=-1$

55. $\frac{5x^4}{10x^6}=\frac{1}{2}x^{4-6}=\frac{1}{2}x^{-2}=\frac{1}{2x^2}$

57. $\frac{8xy^3}{6x^2y^2}=\frac{4}{3}x^{1-2}y^{3-2}=\frac{4}{3}x^{-1}y=\frac{4y}{3x}$

59. $\frac{x+4}{2x+8}=\frac{x+4}{2(x+4)}=\frac{1}{2}\cdot\frac{x+4}{x+4}=\frac{1}{2}\cdot 1=\frac{1}{2}$

61. $\frac{3z-9}{5z-15}=\frac{3(z-3)}{5(z-3)}=\frac{3}{5}\cdot\frac{z-3}{z-3}=\frac{3}{5}\cdot 1=\frac{3}{5}$

63. $\frac{(x+1)(x-1)}{(x+6)(x-1)}=\frac{x+1}{x+6}\cdot\frac{x-1}{x-1}=\frac{x+1}{x+6}\cdot 1=\frac{x+1}{x+6}$

65. $\frac{(5y+3)(2y-1)}{(2y-1)(y+2)}=\frac{(5y+3)(2y-1)}{(y+2)(2y-1)}=\frac{5y+3}{y+2}\cdot\frac{2y-1}{2y-1}=\frac{5y+3}{y+2}\cdot 1=\frac{5y+3}{y+2}$

67. $\frac{x-7}{7-x}=\frac{x-7}{-1(-7+x)}=\frac{x-7}{-1(x-7)}=-1$

69. $\frac{a-b}{b-a}=\frac{a-b}{-1(-b+a)}=\frac{a-b}{-1(a-b)}=-1$

71. $\frac{-6-x}{18+3x}=\frac{-1(6+x)}{3(6+x)}=-\frac{1}{3}\cdot\frac{6+x}{6+x}=-\frac{1}{3}\cdot 1=-\frac{1}{3}$

73. $\frac{x+1}{-2x-2}=\frac{1(x+1)}{-2(x+1)}=-\frac{1}{2}\cdot\frac{x+1}{x+1}=-\frac{1}{2}\cdot 1=-\frac{1}{2}$

75. $-\frac{9-x}{x-9}=\frac{x-9}{x-9}=1$

77. $\frac{(3x+5)(x-1)}{(3x-5)(1-x)}=\frac{3x+5}{3x-5}\cdot\frac{(x-1)}{-1(-1+x)}=\frac{3x+5}{3x-5}\cdot\frac{x-1}{-1(x-1)}=\frac{3x+5}{3x-5}\cdot -1=-\frac{3x+5}{3x-5}$

79. $\frac{n^2-n}{n^2-5n}=\frac{n(n-1)}{n(n-5)}=\frac{n}{n}\cdot\frac{n-1}{n-5}=1\cdot\frac{n-1}{n-5}=\frac{n-1}{n-5}$

81. $\frac{x^2-3x}{6x-18}=\frac{x(x-3)}{6(x-3)}=\frac{x}{6}\cdot\frac{x-3}{x-3}=\frac{x}{6}\cdot 1=\frac{x}{6}$

83. $\frac{z^2-3z+2}{z^2-4z+3}=\frac{(z-2)(z-1)}{(z-3)(z-1)}=\frac{z-2}{z-3}\cdot\frac{z-1}{z-1}=\frac{z-2}{z-3}\cdot 1=\frac{z-2}{z-3}$

85. $\frac{2x^2+7x-4}{6x^2+x-2}=\frac{(x+4)(2x-1)}{(3x+2)(2x-1)}=\frac{x+4}{3x+2}\cdot\frac{2x-1}{2x-1}=\frac{x+4}{3x+2}\cdot 1=\frac{x+4}{3x+2}$

87. $\frac{x-3}{3x^2-11x+6}=\frac{x-3}{(x-3)(3x-2)}=\frac{x-3}{x-3}\cdot\frac{1}{3x-2}=1\cdot\frac{1}{3x-2}=\frac{1}{3x-2}$

89. $-\frac{a-9}{9-a}=\frac{-(a-9)}{9-a}=\frac{-a+9}{9-a}=\frac{9-a}{9-a}=1$

91. $\frac{-2x-1}{4x+2}=\frac{-2x-1}{2(2x+1)}=\frac{-1(2x+1)}{2(2x+1)}=-\frac{1}{2}\cdot\frac{2x+1}{2x+1}=-\frac{1}{2}\cdot 1=-\frac{1}{2}$

93. (a) Equation

(b) $x+1=7 \Rightarrow x+1-1=7-1 \Rightarrow x=6$

95. (a) Expression

(b) $\frac{x}{x(x+1)}=\frac{x}{x}\cdot\frac{1}{x+1}=1\cdot\frac{1}{x+1}=\frac{1}{x+1}$

97. (a) Expression

(b) $\frac{x^2-4}{x+2}=\frac{(x-2)(x+2)}{x+2}=\frac{x-2}{1}\cdot\frac{x+2}{x+2}=(x-2)\cdot 1=x-2$

99. (a) Equation

(b) $\frac{x}{2(1+3)}=1\Rightarrow\frac{x}{2(4)}=1\Rightarrow\frac{x}{8}=1\Rightarrow 8\cdot\frac{x}{8}=8\cdot 1\Rightarrow x=8$

Applications

101. (a) For $x=3$, $T=\frac{1}{5-x}\Rightarrow T=\frac{1}{5-3}=\frac{1}{2}$; When traffic arrives at an average rate of 3 vehicles/min., the average wait is one-half minute.

(b) For $x=2$, $T=\frac{1}{5-x}\Rightarrow T=\frac{1}{5-2}=\frac{1}{3}$;

For $x=4$, $T=\frac{1}{5-x}\Rightarrow T=\frac{1}{5-4}=\frac{1}{1}=1$;

For $x=4.5$, $T=\frac{1}{5-x}\Rightarrow T=\frac{1}{5-4.5}=\frac{1}{0.5}=2$;

For $x = 4.9$, $T = \frac{1}{5-x} \Rightarrow T = \frac{1}{5-4.9} = \frac{1}{0.1} = 10$;

For $x = 4.99$, $T = \frac{1}{5-x} \Rightarrow T = \frac{1}{5-4.99} = \frac{1}{0.01} = 100$; See Figure 101.

As x nears 5 vehicles/min., a small increase in x increases the wait dramatically.

x	2	4	4.5	4.9	4.99
T	$\frac{1}{3}$	1	2	10	100

Figure 101

103. $\frac{1}{2}$

105. (a) $\frac{3}{n}$

(b) $\frac{n-3}{n}$; $n = 100$, $\frac{n-3}{n} = \frac{100-3}{100} = \frac{97}{100}$; There is a 97% chance that a winning ball will not be drawn.

107. (a) $\frac{360}{60} = 6$ hours

(b) $\frac{M}{60}$

109. (a) $T = \frac{1}{5-x}$, $5 - x = 0 \Rightarrow -x = -5 \Rightarrow x = 5$

(b) As the average arrival rate nears 5 cars/min., a small increase in x increases the waiting time dramatically.

7.2: Multiplication and Division of Rational Expressions

Concepts

1. $\frac{10}{21}$

2. $\frac{2}{3} \div \frac{5}{7} = \frac{2}{3} \cdot \frac{7}{5} = \frac{14}{15}$

3. $\frac{AC}{BD}$

4. $\frac{AD}{BC}$

5. $\frac{(x+7)(x+2)}{(x+1)(x+2)} = \frac{x+7}{x+1} \cdot \frac{x+2}{x+2} = \frac{x+7}{x+1} \cdot 1 = \frac{x+7}{x+1}$

6. $\frac{AC}{BC} = \frac{A}{B} \cdot \frac{C}{C} = \frac{A}{B} \cdot 1 = \frac{A}{B}$

7. No, it equals $1 + \frac{2}{x}$.

8. $\frac{1}{x}; \frac{z}{y}$

Review of Fractions

9. $\frac{1}{2} \cdot \frac{4}{5} = \frac{4}{10} = \frac{2}{5}$

11. $\frac{3}{7} \cdot 4 = \frac{3}{7} \cdot \frac{4}{1} = \frac{12}{7}$

13. $\frac{5}{4} \cdot \frac{8}{15} = \frac{5 \cdot 2 \cdot 4}{4 \cdot 3 \cdot 5} = \frac{5}{5} \cdot \frac{4}{4} \cdot \frac{2}{3} = 1 \cdot 1 \cdot \frac{2}{3} = \frac{2}{3}$

15. $\frac{1}{3} \cdot \frac{2}{3} \cdot \frac{9}{11} = \frac{2}{9} \cdot \frac{9}{11} = \frac{18}{99} = \frac{2}{11}$

17. $\frac{2}{3} \div \frac{1}{6} = \frac{2}{3} \cdot \frac{6}{1} = \frac{12}{3} = 4$

19. $\frac{8}{9} \div \frac{5}{3} = \frac{8}{9} \cdot \frac{3}{5} = \frac{24}{45} = \frac{8}{15}$

21. $8 \div \frac{4}{5} = \frac{8}{1} \cdot \frac{5}{4} = \frac{40}{4} = 10$

23. $\frac{4}{5} \div \frac{2}{3} \div \frac{1}{2} = \frac{4}{5} \cdot \frac{3}{2} \div \frac{1}{2} = \frac{12}{10} \div \frac{1}{2} = \frac{6}{5} \cdot \frac{2}{1} = \frac{12}{5}$

Multiplying Rational Expressions

25. $\frac{x+5}{x+5} = 1$

27. $\frac{(z+1)(z+2)}{(z+4)(z+2)} = \frac{z+1}{z+4} \cdot \frac{z+2}{z+2} = \frac{z+1}{z+4} \cdot 1 = \frac{z+1}{z+4}$

29. $\frac{8y(y+7)}{12y(y+7)} = \frac{8}{12} \cdot \frac{y}{y} \cdot \frac{y+7}{y+7} = \frac{8}{12} \cdot 1 \cdot 1 = \frac{8}{12} = \frac{2}{3}$

31. $\frac{x(x+2)(x+3)}{x(x-2)(x+3)} = \frac{x}{x} \cdot \frac{x+2}{x-2} \cdot \frac{x+3}{x+3} = 1 \cdot \frac{x+2}{x-2} \cdot 1 = \frac{x+2}{x-2}$

33. $\frac{8}{x} \cdot \frac{x+1}{x} = \frac{8(x+1)}{x^2}$

35. $\frac{8+x}{x} \cdot \frac{x-3}{x+8} = \frac{8+x}{8+x} \cdot \frac{x-3}{x} = 1 \cdot \frac{x-3}{x} = \frac{x-3}{x}$

37. $\frac{z+3}{z+4} \cdot \frac{z+4}{z-7} = \frac{z+4}{z+4} \cdot \frac{z+3}{z-7} = 1 \cdot \frac{z+3}{z-7} = \frac{z+3}{z-7}$

39. $\frac{5x+1}{3x+2} \cdot \frac{3x+2}{5x+1} = \frac{5x+1}{5x+1} \cdot \frac{3x+2}{3x+2} = 1 \cdot 1 = 1$

41. $\frac{(t+1)^2}{t+2}\cdot\frac{(t+2)^2}{t+1}=\frac{(t+1)(t+1)}{t+2}\cdot\frac{(t+2)(t+2)}{t+1}=\frac{t+1}{t+1}\cdot\frac{t+1}{1}\cdot\frac{t+2}{t+2}\cdot\frac{t+2}{1}=1\cdot(t+1)\cdot 1\cdot(t+2)=(t+1)(t+2)$

43. $\frac{x^2}{x^2+4}\cdot\frac{x+4}{x}=\frac{x^2}{x}\cdot\frac{x+4}{x^2+4}=\frac{x}{x}\cdot\frac{x(x+4)}{x^2+4}=1\cdot\frac{x(x+4)}{x^2+4}=\frac{x(x+4)}{x^2+4}$

45. $\frac{(z^2-1)}{(z^2-4)}\cdot\frac{z-2}{z+1}=\frac{(z-1)(z+1)}{(z-2)(z+2)}\cdot\frac{z-2}{z+1}=\frac{z-1}{z+2}\cdot\frac{z+1}{z+1}\cdot\frac{z-2}{z-2}=\frac{z-1}{z+2}\cdot 1\cdot 1=\frac{z-1}{z+2}$

47. $\frac{y^2-2y}{y^2-1}\cdot\frac{y+1}{y-2}=\frac{y(y-2)}{(y-1)(y+1)}\cdot\frac{y+1}{y-2}=\frac{y}{y-1}\cdot\frac{y+1}{y+1}\cdot\frac{y-2}{y-2}=\frac{y}{y-1}\cdot 1\cdot 1=\frac{y}{y-1}$

49. $\frac{2x^2-x-3}{3x^2-8x-3}\cdot\frac{3x+1}{2x-3}=\frac{(2x-3)(x+1)}{(3x+1)(x-3)}\cdot\frac{3x+1}{2x-3}=\frac{2x-3}{2x-3}\cdot\frac{3x+1}{3x+1}\cdot\frac{x+1}{x-3}=1\cdot 1\cdot\frac{x+1}{x-3}=\frac{x+1}{x-3}$

51. $\frac{(x-3)^3}{x^2-2x+1}\cdot\frac{x-1}{(x-3)^2}=\frac{(x-3)(x-3)^2}{(x-1)(x-1)}\cdot\frac{x-1}{(x-3)^2}=\frac{x-3}{x-1}\cdot\frac{(x-3)^2}{(x-3)^2}\cdot\frac{x-1}{x-1}=\frac{x-3}{x-1}\cdot 1\cdot 1=\frac{x-3}{x-1}$

53. $\frac{2}{x}\div\frac{2x+3}{x}=\frac{2}{x}\cdot\frac{x}{2x+3}=\frac{x}{x}\cdot\frac{2}{2x+3}=1\cdot\frac{2}{2x+3}=\frac{2}{2x+3}$

55. $\frac{x-2}{3x}\div\frac{2-x}{6x}=\frac{x-2}{3x}\cdot\frac{6x}{2-x}=\frac{x-2}{-1(-2+x)}\cdot\frac{6x}{3x}=\frac{x-2}{-1(-2+x)}\cdot 2=-1\cdot 2=-2$

57. $\frac{z+2}{z+1}\div\frac{z+2}{z-1}=\frac{z+2}{z+1}\cdot\frac{z-1}{z+2}=\frac{z+2}{z+2}\cdot\frac{z-1}{z+1}=1\cdot\frac{z-1}{z+1}=\frac{z-1}{z+1}$

59. $\frac{3y+4}{2y+1}\div\frac{3y+4}{y+2}=\frac{3y+4}{2y+1}\cdot\frac{y+2}{3y+4}=\frac{3y+4}{3y+4}\cdot\frac{y+2}{2y+1}=1\cdot\frac{y+2}{2y+1}=\frac{y+2}{2y+1}$

61. $\frac{t^2-1}{t^2+1}\div\frac{t+1}{4}=\frac{t^2-1}{t^2+1}\cdot\frac{4}{t+1}=\frac{(t+1)(t-1)}{t^2+1}\cdot\frac{4}{t+1}=\frac{t+1}{t+1}\cdot\frac{4(t-1)}{t^2+1}=1\cdot\frac{4(t-1)}{t^2+1}=\frac{4(t-1)}{t^2+1}$

63. $\frac{y^2-9}{y^2-25}\div\frac{y+3}{y+5}=\frac{y^2-9}{y^2-25}\cdot\frac{y+5}{y+3}=\frac{(y-3)(y+3)}{(y-5)(y+5)}\cdot\frac{y+5}{y+3}=\frac{y-3}{y-5}\cdot\frac{y+3}{y+3}\cdot\frac{y+5}{y+5}=\frac{y-3}{y-5}\cdot 1\cdot 1=\frac{y-3}{y-5}$

65. $\frac{2x^2-4x}{2x-1}\div\frac{x-2}{2x-1}=\frac{2x^2-4x}{2x-1}\cdot\frac{2x-1}{x-2}=\frac{2x(x-2)}{x-2}\cdot\frac{2x-1}{2x-1}=2x\cdot 1\cdot 1=2x$

67. $\frac{2z^2-5z-3}{z^2+z-20}\div\frac{z-3}{z-4}=\frac{(2z+1)(z-3)}{(z+5)(z-4)}\cdot\frac{z-4}{z-3}=\frac{2z+1}{z+5}\cdot\frac{z-3}{z-3}\cdot\frac{z-4}{z-4}=\frac{2z+1}{z+5}\cdot 1\cdot 1=\frac{2z+1}{z+5}$

69. $\frac{t^2-1}{t^2+5t-6}\div(t+1)=\frac{(t+1)(t-1)}{(t-1)(t+6)}\cdot\frac{1}{(t+1)}=\frac{t+1}{t+1}\cdot\frac{t-1}{t-1}\cdot\frac{1}{t+6}=1\cdot 1\cdot\frac{1}{t+6}=\frac{1}{t+6}$

71. $\frac{a-b}{a+b}\div\frac{a-b}{2a+3b}=\frac{a-b}{a+b}\cdot\frac{2a+3b}{a-b}=\frac{a-b}{a-b}\cdot\frac{2a+3b}{a+b}=1\cdot\frac{2a+3b}{a+b}=\frac{2a+3b}{a+b}$

73. $\frac{x-y}{x^2+2xy+y^2}\div\frac{1}{(x+y)^2}=\frac{x-y}{(x+y)^2}\cdot\frac{(x+y)^2}{1}=\frac{x-y}{1}\cdot\frac{(x+y)^2}{(x+y)^2}=(x-y)\cdot 1=x-y$

75. $\frac{a-b}{b-c}\cdot\frac{c-b}{b-a}=\frac{a-b}{b-c}\cdot\frac{-1(b-c)}{-1(a-b)}=\frac{-1}{-1}\cdot\frac{a-b}{a-b}\cdot\frac{b-c}{b-c}=1\cdot 1\cdot 1=1$

Applications

77. (a) $D=\frac{900}{30}\cdot\frac{1}{x}=30\cdot\frac{1}{x}=\frac{30}{x}\Rightarrow D=\frac{30}{x}$

(b) $x=0.1, D=\frac{30}{0.1}=300$ ft; $x=0.4, D=\frac{30}{0.4}=75$ ft; stopping distance on dry pavement is 75 ft, one-fourth as far as the stopping distance on the icy road.

79. (a) $\frac{1}{n}\cdot\frac{n}{n+1}=\frac{n}{n}\cdot\frac{1}{n+1}=1\cdot\frac{1}{n+1}=\frac{1}{n+1}$

(b) $\frac{1}{n+1}\Rightarrow\frac{1}{99+1}=\frac{1}{100}$

Checking Basic Concepts for Sections 7.1 & 7.2

1. For $x=-1, \frac{3}{x^2-1}=\frac{3}{(-1)^2-1}=\frac{3}{1-1}=\frac{3}{0}=$ undefined. For $x=3, \frac{3}{x^2-1}=\frac{3}{(3)^2-1}=\frac{3}{9-1}=\frac{3}{8}$

2. (a) $\frac{6x^3y^2}{15x^2y^3}=\frac{2}{5}xy^{-1}=\frac{2x}{5y}$

(b) $\frac{5x-15}{x-3}=\frac{5(x-3)}{x-3}=\frac{5}{1}\cdot\frac{x-3}{x-3}=5\cdot 1=5$

(c) $\frac{x^2-x-6}{x^2+x-12}=\frac{(x+2)(x-3)}{(x+4)(x-3)}=\frac{x+2}{x+4}\cdot\frac{x-3}{x-3}=\frac{x+2}{x+4}\cdot 1=\frac{x+2}{x+4}$

3. (a) $\frac{4}{3x}\cdot\frac{2x}{6}=\frac{8x}{18x}=\frac{8}{18}\cdot\frac{x}{x}=\frac{8}{18}\cdot 1=\frac{8}{18}=\frac{4}{9}$

(b) $\frac{2x+4}{x^2-1}\cdot\frac{x+1}{x+2}=\frac{2(x+2)}{(x-1)(x+1)}\cdot\frac{x+1}{x+2}=\frac{2}{x-1}\cdot\frac{x+2}{x+2}\cdot\frac{x+1}{x+1}=\frac{2}{x-1}\cdot 1\cdot 1=\frac{2}{x-1}$

4. (a) $\frac{7}{3z^2}\div\frac{14}{5z^3}=\frac{7}{3z^2}\cdot\frac{5z^3}{14}=\frac{35z^3}{42z^2}=\frac{35z}{42}\cdot\frac{z^2}{z^2}=\frac{35z}{42}\cdot 1=\frac{35z}{42}=\frac{5z}{6}$

(b) $\frac{x^2+x}{x-3}\div\frac{x}{x-3}=\frac{x(x+1)}{x-3}\cdot\frac{x-3}{x}=\frac{x}{x}\cdot\frac{x-3}{x-3}\cdot\frac{x+1}{1}=1\cdot 1\cdot(x+1)=x+1$

5. (a) For $x=0.5, T=\frac{1}{2-x}\Rightarrow T=\frac{1}{2-0.5}=\frac{1}{1.5}=\frac{2}{3}$;

For $x=1, T=\frac{1}{2-x}\Rightarrow T=\frac{1}{2-1}=\frac{1}{1}=1$;

For $x=1.5, T=\frac{1}{2-x}\Rightarrow T=\frac{1}{2-1.5}=\frac{1}{0.5}=2$;

For $x = 1.9, T = \frac{1}{2-x} \Rightarrow T = \frac{1}{2-1.9} = \frac{1}{0.1} = 10$; See Figure 5.

(b) As x nears 2 customers/min., a small increase in x increases the wait dramatically.

x	0.5	1.0	1.5	1.9
T	$\frac{2}{3}$	1	2	10

Figure 5

7.3: Addition and Subtraction with Like Denominators

Concepts

1. add, numerators; denominators

2. subtract, numerators; denominators

3. $\frac{2}{5} + \frac{1}{5} = \frac{2+1}{5} = \frac{3}{5}$

4. $\frac{6}{7} - \frac{2}{7} = \frac{6-2}{7} = \frac{4}{7}$

5. $\frac{A+B}{C}$

6. $\frac{A-B}{C}$

7. $5-(x+1) = 5-x-1 = 4-x$

8. $\frac{5}{x-1} - \frac{x+1}{x-1} = \frac{5-(x+1)}{x-1} = \frac{4-x}{x-1}$

9. $3x-(2x-5) = 3x-2x+5 = x+5$

10. $\frac{3x}{7} - \frac{2x-5}{7} = \frac{3x-(2x-5)}{7} = \frac{3x-2x+5}{7} = \frac{x+5}{7}$

Addition and Subtraction of Fractions

11. $\frac{1}{2} + \frac{1}{2} = \frac{1+1}{2} = \frac{2}{2} = 1$

13. $\frac{4}{5} + \frac{2}{5} = \frac{4+2}{5} = \frac{6}{5}$

15. $\frac{1}{6} + \frac{5}{6} = \frac{1+5}{6} = \frac{6}{6} = 1$

17. $\frac{4}{7} - \frac{1}{7} = \frac{4-1}{7} = \frac{3}{7}$

19. $\frac{7}{8} - \frac{3}{8} = \frac{7-3}{8} = \frac{4}{8} = \frac{1}{2}$

21. $\frac{11}{12} - \frac{5}{12} = \frac{11-5}{12} = \frac{6}{12} = \frac{1}{2}$

23. $\frac{7}{15}+\frac{4}{15}-\frac{1}{15}=\frac{7+4-1}{15}=\frac{11-1}{15}=\frac{10}{15}=\frac{2}{3}$

Addition and Subtraction of Rational Expressions

25. $\frac{2}{x}+\frac{1}{x}=\frac{2+1}{x}=\frac{3}{x}$

27. $\frac{7+2x}{4x}-\frac{7}{4x}=\frac{(7+2x)-7}{4x}=\frac{2x}{4x}=\frac{1}{2}\cdot\frac{x}{x}=\frac{1}{2}\cdot 1=\frac{1}{2}$

29. $\frac{y+3}{y-3}+\frac{2y-12}{y-3}=\frac{(y+3)+(2y-12)}{y-3}=\frac{3y-9}{y-3}=\frac{3(y-3)}{y-3}=3\cdot\frac{y-3}{y-3}=3\cdot 1=3$

31. $\frac{x}{x-3}-\frac{3}{x-3}=\frac{x-3}{x-3}=1$

33. $\frac{5z}{4z+3}-\frac{z}{4z+3}=\frac{5z-z}{4z+3}=\frac{4z}{4z+3}$

35. $\frac{t+5}{t+6}+\frac{t+7}{t+6}=\frac{(t+5)+(t+7)}{t+6}=\frac{2t+12}{t+6}=\frac{2(t+6)}{t+6}=2\cdot\frac{t+6}{t+6}=2\cdot 1=2$

37. $\frac{5x}{2x+3}-\frac{3x-3}{2x+3}=\frac{5x-(3x-3)}{2x+3}=\frac{2x+3}{2x+3}=1$

39. $\frac{x-4}{x^2-x}+\frac{4}{x^2-x}=\frac{x-4+4}{x^2-x}=\frac{x}{x(x-1)}=\frac{x}{x}\cdot\frac{1}{x-1}=1\cdot\frac{1}{x-1}=\frac{1}{x-1}$

41. $\frac{z^2-1}{z-2}+\frac{3-3z}{z-2}=\frac{z^2-1+3-3z}{z-2}=\frac{z^2-3z+2}{z-2}=\frac{(z-1)(z-2)}{z-2}$

$=\frac{z-1}{1}\cdot\frac{z-2}{z-2}=\frac{z-1}{1}\cdot 1=z\cdot 1$

43. $\frac{x^2+4x-1}{4x+2}-\frac{x^2-4x-5}{4x+2}=\frac{(x^2+4x-1)-(x^2-4x-5)}{4x+2}=\frac{8x+4}{4x+2}=\frac{2(4x+2)}{4x+2}=2\cdot\frac{4x+2}{4x+2}=2\cdot 1=2$

45. $\frac{3y}{5}+\frac{2y-5}{5}=\frac{3y+(2y-5)}{5}=\frac{5y-5}{5}=\frac{5(y-1)}{5}=\frac{5}{5}\cdot(y-1)=1\cdot(y-1)=y-1$

47. $\frac{x+y}{4}+\frac{x-y}{4}=\frac{(x+y)+(x-y)}{4}=\frac{2x}{4}=\frac{1x}{2}=\frac{x}{2}$

49. $\frac{z^2+4}{z-2}-\frac{4z}{z-2}=\frac{(z^2+4)-4z}{z-2}=\frac{z^2-4z+4}{z-2}=\frac{(z-2)(z-2)}{z-2}=\frac{z-2}{z-2}\cdot\frac{z-2}{1}=1\cdot(z-2)=z-2$

51. $\frac{2x^2-5x}{2x+1}-\frac{3}{2x+1}=\frac{(2x^2-5x)-3}{2x+1}=\frac{2x^2-5x-3}{2x+1}=\frac{(2x+1)(x-3)}{2x+1}=\frac{2x+1}{2x+1}\cdot\frac{x-3}{1}=1\cdot(x-3)=x-3$

53. $\frac{3n}{2n^2-n+5}+\frac{4n}{2n^2-n+5}=\frac{3n+4n}{2n^2-n+5}=\frac{7n}{2n^2-n+5}$

55. $\frac{1}{x+3}+\frac{2}{x+3}+\frac{3}{x+3}=\frac{1+2+3}{x+3}=\frac{6}{x+3}$

57. $\frac{8}{ab}+\frac{1}{ab}=\frac{8+1}{ab}=\frac{9}{ab}$

59. $\frac{x}{(x+y)^2}+\frac{y}{(x+y)^2}=\frac{x+y}{(x+y)^2}=\frac{x+y}{x+y}\cdot\frac{1}{x+y}=1\cdot\frac{1}{x+y}=\frac{1}{x+y}$

61. $\frac{8}{a-b}+\frac{-8}{b-a}=\frac{8}{a-b}+\frac{-8}{-1(a-b)}=\frac{8}{a-b}+\frac{8}{a-b}=\frac{8+8}{a-b}=\frac{16}{a-b}$

63. $\frac{a+b}{4a}-\frac{a-b}{4a}=\frac{a+b-a+b}{4a}=\frac{2b}{4a}=\frac{2}{2}\cdot\frac{b}{2a}=1\cdot\frac{b}{2a}=\frac{b}{2a}$

65. $\frac{x}{x+y}+\frac{y}{x+y}=\frac{x+y}{x+y}=1$

67. $\frac{a^2}{a+b}-\frac{b^2}{a+b}=\frac{a^2-b^2}{a+b}=\frac{(a+b)(a-b)}{a+b}=\frac{a+b}{a+b}\cdot\frac{a-b}{1}=1\cdot(a-b)=a-b$

69. $\frac{4x^2}{2x+3y}-\frac{9y^2}{2x+3y}=\frac{4x^2-9y^2}{2x+3y}=\frac{(2x+3y)(2x-3y)}{2x+3y}=\frac{2x+3y}{2x+3y}\cdot\frac{2x-3y}{1}=1\cdot(2x-3y)=2x-3y$

71. $\frac{2}{3+x}+\frac{3}{3+x}=\frac{5}{10}\Rightarrow\frac{2+3}{3+x}=\frac{5}{10}\Rightarrow\frac{5}{3+x}=\frac{5}{10}\Rightarrow 3+x=10\Rightarrow x=7$

Applications

73. (a) $\frac{6}{n+1}+\frac{5}{n+1}+\frac{3}{n+1}=\frac{6+5+3}{n+1}=\frac{14}{n+1}$

(b) For $n=99$, $\frac{14}{n+1}\Rightarrow\frac{14}{99+1}=\frac{14}{100}=\frac{7}{50}$; there are 7 chances in 50 that a defective battery is chosen.

7.4: Addition and Subtraction with Unlike Denominators

Concepts

1. Examples include 36 and 54. *Answers may vary.*

2. xy

3. $\frac{3}{3}$

4. $\frac{x+1}{x+1}$

5. Factoring 4 and 6 completely yields: 2^2 and $2\cdot 3$, a list of factors is $2^2\cdot 3=12$.

6. Factoring $2x$ and $6x^2$ completely yields: $2\cdot x$ and $2\cdot 3\cdot x^2$, a list of factors is $2\cdot 3\cdot x^2=6x^2$.

7. Factoring 4 and 6 completely yields: 2^2 and $2\cdot 3$, a list of factors is $2^2\cdot 3=12$.

8. Factoring 6 and 9 completely yields: $2\cdot 3$ and 3^2, a list of factors is $2\cdot 3^2=18$.

9. 2 and 3 are both prime $\Rightarrow$ a list of factors is $2\cdot 3=6$.

10. Factoring 5 and 4 completely yields: 5 and 2^2, a list of factors is $5\cdot 2^2=20$.

11. Factoring 10 and 15 completely yields: $2 \cdot 5$ and $3 \cdot 5$, a list of factors is $2 \cdot 3 \cdot 5 = 30$.

12. Factoring 8 and 12 completely yields: 2^3 and $3 \cdot 2^2$, a list of factors is $2^3 \cdot 3 = 24$.

13. Factoring 24 and 36 completely yields: $2^3 \cdot 3$ and $2^2 \cdot 3^2$, a list of factors is $2^3 \cdot 3^2 = 72$.

14. Factoring 32 and 40 completely yields: 2^5 and $2^3 \cdot 5$, a list of factors is $2^5 \cdot 5 = 160$.

15. Factoring $4x$ and $6x$ completely yields: $2^2 \cdot x$ and $2 \cdot 3 \cdot x$, a list of factors is $2^2 \cdot 3 \cdot x = 12x$.

16. Factoring $6x$ and $9x$ completely yields: $2 \cdot 3 \cdot x$ and $3^2 \cdot x$, a list of factors is $2 \cdot 3^2 \cdot x = 18x$.

17. Factoring $5x$ and $10x^2$ completely yields: $5 \cdot x$ and $2 \cdot 5 \cdot x^2$, a list of factors is $2 \cdot 5 \cdot x^2 = 10x^2$.

18. Factoring $4x^2$ and $12x$ completely yields: $2^2 \cdot x^2$ and $2^2 \cdot 3 \cdot x$, a list of factors is $2^2 \cdot 3 \cdot x^2 = 12x^2$.

19. Both x and $x+1$ are prime $\Rightarrow$ a list of factors is $x(x+1)$.

20. Factoring $4x$ and $x-1$ completely yields: $2^2 \cdot x$ and $x-1$, a list of factors is $2^2 \cdot x \cdot (x-1) = 4x(x-1)$.

21. Both $2x+1$ and $x+3$ are prime $\Rightarrow$ a list of factors is $(2x+1)(x+3)$.

22. Both $5x+3$ and $x+9$ are prime $\Rightarrow$ a list of factors is $(5x+3)(x+9)$.

23. Factoring $4x^2$ and $9x^3$ completely yields: $2^2 \cdot x^2$ and $3^2 \cdot x^3$, a list of factors is $3^2 \cdot 2^2 \cdot x^3 = 36x^3$.

24. Factoring $12x^3$ and $15x^5$ completely yields: $2^2 \cdot 3 \cdot x^3$ and $3 \cdot 5 \cdot x^5$, a list of factors is $2^2 \cdot 3 \cdot 5 \cdot x^5 = 60x^5$.

25. Factoring $x^2 - x$ and $x^2 + x$ completely yields: $x(x-1)$ and $x(x+1)$, a list of factors is $x(x-1)(x+1)$.

26. Factoring $x^2 + 2x$ and x^2 completely yields: $x(x+2)$ and x^2, a list of factors is $x^2(x+2)$.

27. Both $(x+1)^2$ and $x+1$ are factored completely $\Rightarrow$ a list of factors is $(x+1)^2$.

28. Both $(x-8)^2$ and $(x-8)(x+1)$ are factored completely $\Rightarrow$ a list of factors is $(x-8)^2(x+1)$.

29. Both $(2x-1)^3$ and $(2x-1)(x+3)$ are factored completely $\Rightarrow$ a list of factors is $(2x-1)^3(x+3)$.

30. Both $(x-4)(x+4)$ and $(x-4)(x+3)$ are factored completely $\Rightarrow$ a list of factors is $(x-4)(x+4)(x+3)$.

31. Factoring $4x^2 - 1$ and $2x+1$ completely yields: $(2x+1)(2x-1)$ and $(2x+1)$, a list of factors is $(2x+1)(2x-1)$.

32. Factoring $x^2 + 4x + 3$ and $x+3$ completely yields: $(x+3)(x+1)$ and $(x+3)$, a list of factors is $(x+3)(x+1)$.

33. Factoring $x^2 - 1$ and $x+1$ completely yields: $(x+1)(x-1)$ and $(x+1)$, a list of factors is $(x+1)(x-1)$.

34. Factoring $x^2 - 4$ and $x-2$ completely yields: $(x+2)(x-2)$ and $(x-2)$, a list of factors is $(x+2)(x-2)$.

35. Both $x^2 + 4$ and $4x$ are factored completely $\Rightarrow$ a list of factors is $4x(x^2 + 4)$.

36. Factoring $4x^2+x$ and x completely yields: $x(4x+1)$ and x, a list of factors is $x(4x+1)$.

37. Factoring $2x^2+7x+6$ and x^2+5x+6 completely yields: $(2x+3)(x+2)$ and $(x+2)(x+3)$, a list of factors is $(2x+3)(x+2)(x+3)$.

38. Factoring x^2-3x+2 and x^2+2x-3 completely yields: $(x-2)(x-1)$ and $(x+3)(x-1)$, a list of factors is $(x-2)(x-1)(x+3)$.

Addition and Subtraction of Rational Expressions

39. $\frac{1}{3},\ D=9 \Rightarrow \frac{1}{3}\cdot\frac{3}{3}=\frac{3}{9}$

41. $\frac{5}{7},\ D=21 \Rightarrow \frac{5}{7}\cdot\frac{3}{3}=\frac{15}{21}$

43. $\frac{1}{4x},\ D=8x^3 \Rightarrow \frac{1}{4x}\cdot\frac{2x^2}{2x^2}=\frac{2x^2}{8x^3}$

45. $\frac{1}{x+2},\ D=x^2-4 \Rightarrow \frac{1}{x+2}\cdot\frac{x-2}{x-2}=\frac{x-2}{x^2-4}$

47. $\frac{1}{x+1},\ D=x^2+x \Rightarrow \frac{1}{x+1}\cdot\frac{x}{x}=\frac{x}{x^2+x}$

49. $\frac{2x}{x+1},\ D=x^2+2x+1 \Rightarrow \frac{2x}{x+1}\cdot\frac{x+1}{x+1}=\frac{2x^2+2x}{x^2+2x+1}$

51. $\frac{4}{5}+\frac{1}{2}=\frac{8}{10}+\frac{5}{10}=\frac{13}{10}$

53. $\frac{5}{9}-\frac{1}{3}=\frac{5}{9}-\frac{3}{9}=\frac{2}{9}$

55. $\frac{4}{25}+\frac{2}{5}=\frac{4}{25}+\frac{10}{25}=\frac{14}{25}$

57. $\frac{1}{5}+\frac{3}{4}-\frac{1}{2}=\frac{4}{20}+\frac{15}{20}-\frac{10}{20}=\frac{9}{20}$

59. $\frac{1}{3x}+\frac{3}{4x}=\frac{4}{12x}+\frac{9}{12x}=\frac{13}{12x}$

61. $\frac{5}{z^2}-\frac{7}{z^3}=\frac{5z}{z^3}-\frac{7}{z^3}=\frac{5z-7}{z^3}$

63. $\frac{1}{x}-\frac{1}{y}=\frac{y}{xy}-\frac{x}{xy}=\frac{y-x}{xy}$

65. $\frac{a}{b}+\frac{b}{a}=\frac{a^2}{ab}+\frac{b^2}{ab}=\frac{a^2+b^2}{ab}$

67. $\frac{1}{2x+4}+\frac{3}{x+2}=\frac{1}{2(x+2)}+\frac{3}{x+2}\cdot\frac{2}{2}=\frac{1}{2(x+2)}+\frac{6}{2(x+2)}=\frac{7}{2(x+2)}$

69. $\dfrac{2}{t-2}-\dfrac{1}{t}=\dfrac{2}{t-2}\cdot\dfrac{t}{t}-\dfrac{1}{t}\cdot\dfrac{t-2}{t-2}=\dfrac{2t}{t(t-2)}-\dfrac{t-2}{t(t-2)}=\dfrac{t+2}{t(t-2)}$

71. $\dfrac{5}{n-1}+\dfrac{n}{n+1}=\dfrac{5}{n-1}\cdot\dfrac{n+1}{n+1}+\dfrac{n}{n+1}\cdot\dfrac{n-1}{n-1}=\dfrac{5n+5}{(n-1)(n+1)}+\dfrac{n^2-n}{(n-1)(n+1)}=\dfrac{n^2+4n+5}{(n-1)(n+1)}$

73. $\dfrac{3}{x-3}+\dfrac{6}{3-x}=\dfrac{3}{x-3}+\dfrac{6}{-(x-3)}=\dfrac{3}{x-3}-\dfrac{6}{x-3}=-\dfrac{3}{x-3}$

75. $\dfrac{1}{5k-1}+\dfrac{1}{1-5k}=\dfrac{1}{5k-1}+\dfrac{1}{-(5k-1)}=\dfrac{1}{5k-1}-\dfrac{1}{5k-1}=0$

77. $\dfrac{2x}{(x-1)^2}+\dfrac{4}{x-1}=\dfrac{2x}{(x-1)^2}+\dfrac{4}{x-1}\cdot\dfrac{x-1}{x-1}=\dfrac{2x}{(x-1)^2}+\dfrac{4x-4}{(x-1)^2}=\dfrac{6x-4}{(x-1)^2}$

79. $\dfrac{2y}{y(2y-1)}+\dfrac{1}{2y-1}=\dfrac{2y}{y(2y-1)}+\dfrac{1}{2y-1}\cdot\dfrac{y}{y}=\dfrac{2y}{y(2y-1)}+\dfrac{y}{y(2y-1)}=\dfrac{3y}{y(2y-1)}=\dfrac{3}{2y-1}\cdot\dfrac{y}{y}=$

$\dfrac{3}{2y-1}\cdot 1=\dfrac{3}{2y-1}$

81. $\dfrac{1}{x+2}-\dfrac{1}{x^2+2x}=\dfrac{1}{x+2}\cdot\dfrac{x}{x}-\dfrac{1}{x(x+2)}=\dfrac{x}{x(x+2)}-\dfrac{1}{x(x+2)}=\dfrac{x-1}{x(x+2)}$

83. $\dfrac{3}{x-2}-\dfrac{1}{x^2-4}=\dfrac{3}{x-2}\cdot\dfrac{x+2}{x+2}-\dfrac{1}{(x-2)(x+2)}=\dfrac{3(x+2)}{(x-2)(x+2)}-\dfrac{1}{(x-2)(x+2)}=\dfrac{3x+6-1}{(x-2)(x+2)}=\dfrac{3x+5}{(x-2)(x+2)}$

85. $\dfrac{2}{x^2-3x}-\dfrac{1}{x^2+3x}=\dfrac{2}{x(x-3)}\cdot\dfrac{x+3}{x+3}-\dfrac{1}{x(x+3)}\cdot\dfrac{x-3}{x-3}$

$=\dfrac{2(x+3)}{x(x-3)(x+3)}-\dfrac{x-3}{x(x-3)(x+3)}=\dfrac{2x+6-x+3}{x(x-3)(x+3)}=\dfrac{x+9}{x(x-3)(x+3)}$

87. $\dfrac{1}{x-2}-\dfrac{1}{x+2}+\dfrac{1}{x}=\dfrac{1}{x-2}\cdot\dfrac{x(x+2)}{x(x+2)}-\dfrac{1}{x+2}\cdot\dfrac{x(x-2)}{x(x-2)}+\dfrac{1}{x}\cdot\dfrac{(x-2)(x+2)}{(x-2)(x+2)}$

$=\dfrac{x(x+2)-x(x-2)+(x-2)(x+2)}{x(x-2)(x+2)}=\dfrac{x^2+2x-x^2+2x+x^2-4}{x(x-2)(x+2)}=\dfrac{x^2+4x-4}{x(x-2)(x+2)}$

89. $\dfrac{x}{x^2+4x+4}+\dfrac{1}{x+2}=\dfrac{x}{(x+2)^2}+\dfrac{1}{x+2}\cdot\dfrac{x+2}{x+2}=\dfrac{x}{(x+2)^2}+\dfrac{x+2}{(x+2)^2}=\dfrac{2x+2}{(x+2)^2}$

91. $\dfrac{x}{(x+1)(x+2)}-\dfrac{1}{(x+2)(x+3)}=\dfrac{x}{(x+1)(x+2)}\cdot\dfrac{x+3}{x+3}-\dfrac{1}{(x+2)(x+3)}\cdot\dfrac{x+1}{x+1}=$

$\dfrac{x^2+3x}{(x+1)(x+2)(x+3)}-\dfrac{x+1}{(x+1)(x+2)(x+3)}=\dfrac{x^2+2x-1}{(x+1)(x+2)(x+3)}$

93. $\dfrac{1}{a+b}-\dfrac{1}{a-b}=\dfrac{1}{a+b}\cdot\dfrac{a-b}{a-b}-\dfrac{1}{a-b}\cdot\dfrac{a+b}{a+b}=\dfrac{a-b}{(a+b)(a-b)}-\dfrac{a+b}{(a+b)(a-b)}=-\dfrac{2b}{(a+b)(a-b)}$

95. $\frac{r}{r-t}+\frac{t}{t-r}-1=\frac{r}{r-t}+\frac{t}{-(r-t)}-1=\frac{r}{r-t}-\frac{t}{r-t}-1=\frac{r-t}{r-t}-1=1-1=0$

97. $\frac{1}{2a}+\frac{1}{3a}+\frac{1}{4a}=\frac{1}{2a}\cdot\frac{6}{6}+\frac{1}{3a}\cdot\frac{4}{4}+\frac{1}{4a}\cdot\frac{3}{3}=\frac{6}{12a}+\frac{4}{12a}+\frac{3}{12a}=\frac{13}{12a}$

99. $\frac{2}{x-y}+\frac{3}{y-x}+\frac{1}{x-y}=\frac{2}{x-y}+\frac{3}{-(x-y)}+\frac{1}{x-y}=\frac{2}{x-y}-\frac{3}{x-y}+\frac{1}{x-y}=\frac{0}{x-y}=0$

101. $\frac{3}{x-3}-\frac{3}{x^2-3x}-\frac{6}{x(x-3)}=\frac{3}{x-3}\cdot\frac{x}{x}-\frac{3}{x(x-3)}-\frac{6}{x(x-3)}=\frac{3x}{x(x-3)}-\frac{3}{x(x-3)}-\frac{6}{x(x-3)}=$

$\frac{3x-9}{x(x-3)}=\frac{3(x-3)}{x(x-3)}=\frac{3}{x}\cdot\frac{x-3}{x-3}=\frac{3}{x}\cdot 1=\frac{3}{x}$

Applications

103. $R=120,\ S=200;\ \frac{1}{R}+\frac{1}{S}\Rightarrow\frac{1}{120}+\frac{1}{200}=\frac{1}{120}\cdot\frac{5}{5}+\frac{1}{200}\cdot\frac{3}{3}=\frac{5}{600}+\frac{3}{600}=\frac{8}{600}=\frac{1}{75}$ and

$\frac{S+R}{RS}\Rightarrow\frac{120+200}{120(200)}=\frac{320}{24{,}000}=\frac{1}{75}$. Yes, they are the same.

105. $\frac{1}{F}-\frac{1}{D}\Rightarrow F,\ D=F\cdot D=FD\Rightarrow\frac{1}{F}\cdot\frac{D}{D}-\frac{1}{D}\cdot\frac{F}{F}=\frac{D}{FD}-\frac{F}{FD}=\frac{D-F}{FD}$

Checking Basic Concepts for Sections 7.3 & 7.4

1. (a) $\frac{x}{x+2}+\frac{2}{x+2}=\frac{x+2}{x+2}=1$

   (b) $\frac{2}{3x}-\frac{x}{3x}=\frac{2-x}{3x}$

   (c) $\frac{z^2+z}{z+2}+\frac{z}{z+2}=\frac{z^2+2z}{z+2}=\frac{z(z+2)}{z+2}=\frac{z}{1}\cdot\frac{z+2}{z+2}=\frac{z}{1}\cdot 1=z$

2. (a) Factoring $3x$ and $5x$ completely yields: $3\cdot x$ and $5\cdot x$, a list of factors is $3\cdot 5\cdot x=15x$.

   (b) Factoring $4x$ and x^2+x completely yields:

   $2^2\cdot x$ and $x(x+1)$, a list of factors is $2^2\cdot x\cdot(x+1)=4x(x+1)$.

   (c) Both $x+1$ and $x-1$ are factored completely, a list of factors is $(x+1)(x-1)$.

3. (a) $\frac{1}{x+1}+\frac{5}{x}=\frac{1}{x+1}\cdot\frac{x}{x}+\frac{5}{x}\cdot\frac{x+1}{x+1}=\frac{x}{x(x+1)}+\frac{5x+5}{x(x+1)}=\frac{6x+5}{x(x+1)}$

   (b) $\frac{5}{x-3}+\frac{1}{3-x}=\frac{5}{x-3}+\frac{1}{-(x-3)}=\frac{5}{x-3}-\frac{1}{x-3}=\frac{4}{x-3}$

   (c) $\frac{-4}{4x+2}-\frac{x+2}{2x+1}=\frac{-4}{2(2x+1)}-\frac{x+2}{2x+1}=\frac{-4}{2(2x+1)}-\frac{x+2}{2x+1}\cdot\frac{2}{2}=\frac{-4}{2(2x+1)}-\frac{2x+4}{2(2x+1)}=$

$$\frac{-2x-8}{2(2x+1)}=\frac{-2(x+4)}{2(2x+1)}=-\frac{x+4}{2x+1}$$

4. $\frac{a}{a-b}-\frac{b}{a+b}=\frac{a}{a-b}\cdot\frac{a+b}{a+b}-\frac{b}{a+b}\cdot\frac{a-b}{a-b}=\frac{a^2+ab}{(a-b)(a+b)}-\frac{ab-b^2}{(a-b)(a+b)}=\frac{a^2+b^2}{(a-b)(a+b)}$

7.5: Complex Fractions

Concepts

1. $\frac{\frac{1}{2}}{\frac{3}{4}}=\frac{1}{2}\cdot\frac{4}{3}=\frac{4}{6}=\frac{2}{3}$

2. $\frac{\frac{a}{b}}{\frac{c}{d}}=\frac{a}{b}\cdot\frac{d}{c}=\frac{ad}{bc}$

3. fractions

4. $\frac{x+\frac{1}{2}}{x-\frac{1}{2}}$

5. Division

6. $\frac{\frac{x}{2}}{\frac{1}{x-1}}$

7. $\frac{\frac{a}{b}}{\frac{c}{d}}$

8. $\frac{1}{x+2},\ \frac{1}{x};\ x,\ x+2\Rightarrow x(x+2)$

Simplifying Complex Fractions

9. For $\frac{\frac{x}{5}-\frac{1}{6}}{\frac{2}{15}-3x}$, the denominators 5, 6 and 15 have prime factorization of $5,\ 2\cdot 3$ and $3\cdot 5$,

therefore $2\cdot 3\cdot 5=30$.

11. For $\frac{\frac{2}{x+1}-x}{\frac{2}{x-1}+x}$, the denominators $x+1$ and $x-1$ are prime, therefore $(x+1)(x-1)$.

13. For $\frac{\frac{1}{2x-1}-\frac{1}{2x+1}}{\frac{x+1}{x}}$, the denominators $2x-1,\ 2x+1$ and x are prime, therefore $x(2x-1)(2x+1)$.

15. $\frac{\frac{2}{3}}{\frac{5}{6}}=\frac{2}{3}\div\frac{5}{6}=\frac{2}{3}\cdot\frac{6}{5}=\frac{12}{15}=\frac{4}{5}$

17. $\frac{2\frac{1}{2}}{1\frac{3}{4}}=\frac{\frac{5}{2}}{\frac{7}{4}}=\frac{5}{2}\div\frac{7}{4}=\frac{5}{2}\cdot\frac{4}{7}=\frac{20}{14}=\frac{10}{7}$

19. $\dfrac{1\frac{1}{2}}{2\frac{1}{3}}=\dfrac{\frac{3}{2}}{\frac{7}{3}}=\dfrac{3}{2}\div\dfrac{7}{3}=\dfrac{3}{2}\cdot\dfrac{3}{7}=\dfrac{9}{14}$

21. $\dfrac{\frac{r}{t}}{\frac{2r}{t}}=\dfrac{r}{t}\div\dfrac{2r}{t}=\dfrac{r}{t}\cdot\dfrac{t}{2r}=\dfrac{t}{t}\cdot\dfrac{1}{2}\cdot\dfrac{r}{r}=1\cdot\dfrac{1}{2}\cdot1=\dfrac{1}{2}$

23. $\dfrac{\frac{6}{x}}{\frac{2}{y}}=\dfrac{6}{x}\div\dfrac{2}{y}=\dfrac{6}{x}\cdot\dfrac{y}{2}=\dfrac{6y}{2x}=\dfrac{3y}{x}$

25. $\dfrac{\frac{6}{m-2}}{\frac{2}{m-2}}=\dfrac{6}{m-2}\div\dfrac{2}{m-2}=\dfrac{6}{m-2}\cdot\dfrac{m-2}{2}=\dfrac{m-2}{m-2}\cdot\dfrac{6}{2}=1\cdot3=3$

27. $\dfrac{\frac{p+1}{p}}{\frac{p+2}{p}}=\dfrac{p+1}{p}\div\dfrac{p+2}{p}=\dfrac{p+1}{p}\cdot\dfrac{p}{p+2}=\dfrac{p}{p}\cdot\dfrac{p+1}{p+2}=1\cdot\dfrac{p+1}{p+2}=\dfrac{p+1}{p+2}$

29. $\dfrac{\frac{5}{z^2-1}}{\frac{z}{z^2-1}}=\dfrac{5}{z^2-1}\div\dfrac{z}{z^2-1}=\dfrac{5}{z^2-1}\cdot\dfrac{z^2-1}{z}=\dfrac{z^2-1}{z^2-1}\cdot\dfrac{5}{z}=1\cdot\dfrac{5}{z}=\dfrac{5}{z}$

31. $\dfrac{\frac{y}{y^2-9}}{\frac{1}{y+3}}=\dfrac{y}{y^2-9}\div\dfrac{1}{y+3}=\dfrac{y}{y^2-9}\cdot\dfrac{y+3}{1}=\dfrac{y}{(y+3)(y-3)}\cdot\dfrac{y+3}{1}=\dfrac{y}{y-3}\cdot\dfrac{y+3}{y+3}=\dfrac{y}{y-3}\cdot1=\dfrac{y}{y-3}$

33. $\dfrac{x-\frac{1}{x}}{x+\frac{1}{x}}=\dfrac{x-\frac{1}{x}}{x+\frac{1}{x}}\cdot\dfrac{x}{x}=\dfrac{x^2-1}{x^2+1}$

35. $\dfrac{x}{\frac{2}{x}+\frac{1}{x}}=\dfrac{x}{\frac{2}{x}+\frac{1}{x}}\cdot\dfrac{x}{x}=\dfrac{x^2}{2+1}=\dfrac{x^2}{3}$

37. $\dfrac{\frac{3}{x+1}}{\frac{4}{x+1}-\frac{1}{x+1}}=\dfrac{\frac{3}{x+1}}{\frac{3}{x+1}}=\dfrac{3}{x+1}\div\dfrac{3}{x+1}=\dfrac{3}{x+1}\cdot\dfrac{x+1}{3}=\dfrac{x+1}{x+1}\cdot\dfrac{3}{3}=1\cdot1=1$

39. $\dfrac{\frac{1}{m^2n}+\frac{1}{mn^2}}{\frac{1}{m^2n}-\frac{1}{mn^2}}=\dfrac{\frac{1}{m^2n}+\frac{1}{mn^2}}{\frac{1}{m^2n}-\frac{1}{mn^2}}\cdot\dfrac{m^2n^2}{m^2n^2}=\dfrac{\frac{m^2n^2}{m^2n}+\frac{m^2n^2}{mn^2}}{\frac{m^2n^2}{m^2n}-\frac{m^2n^2}{mn^2}}=\dfrac{\left(\frac{m^2}{m^2}\cdot\frac{n}{n}\cdot n\right)+\left(m\cdot\frac{m}{m}\cdot\frac{n^2}{n^2}\right)}{\left(\frac{m^2}{m^2}\cdot\frac{n}{n}\cdot n\right)-\left(m\cdot\frac{m}{m}\cdot\frac{n^2}{n^2}\right)}=\dfrac{(1\cdot1\cdot n)+(m\cdot1\cdot1)}{(1\cdot1\cdot n)-(m\cdot1\cdot1)}=\dfrac{n+m}{n-m}$

41. $\dfrac{\frac{1}{2x}+\frac{1}{y}}{\frac{1}{y}-\frac{1}{2x}}=\dfrac{\frac{1}{2x}+\frac{1}{y}}{\frac{1}{y}-\frac{1}{2x}}\cdot\dfrac{2xy}{2xy}=\dfrac{\frac{2xy}{2x}+\frac{2xy}{y}}{\frac{2xy}{y}-\frac{2xy}{2x}}=\dfrac{\frac{2x}{2x}\cdot y+2x\cdot\frac{y}{y}}{2x\cdot\frac{y}{y}-\frac{2x}{2x}\cdot y}=\dfrac{1\cdot y+2x\cdot1}{2x\cdot1-1\cdot y}=\dfrac{2x+y}{2x-y}$

43. $\dfrac{\frac{1}{ab}+\frac{1}{a}}{\frac{1}{ab}-\frac{1}{b}}=\dfrac{\frac{1}{ab}+\frac{b}{ab}}{\frac{1}{ab}-\frac{a}{ab}}=\dfrac{\frac{1+b}{ab}}{\frac{1-a}{ab}}=\dfrac{1+b}{ab}\div\dfrac{1-a}{ab}=\dfrac{1+b}{ab}\cdot\dfrac{ab}{1-a}=\dfrac{ab}{ab}\cdot\dfrac{1+b}{1-a}=1\cdot\dfrac{1+b}{1-a}=\dfrac{1+b}{1-a}$

45. $\dfrac{\frac{2}{q}-\frac{1}{q+1}}{\frac{1}{q+1}}=\dfrac{\frac{2}{q}-\frac{1}{q+1}}{\frac{1}{q+1}}\cdot\dfrac{q(q+1)}{q(q+1)}=\dfrac{\frac{2q(q+1)}{q}-\frac{q(q+1)}{q+1}}{\frac{q(q+1)}{q+1}}=\dfrac{\frac{q}{q}\cdot2(q+1)-q\cdot\frac{q+1}{q+1}}{q\cdot\frac{q+1}{q+1}}=\dfrac{2q+2-q}{q}=\dfrac{q+2}{q}$

47. $\dfrac{\frac{1}{x+1}+\frac{1}{x+2}}{\frac{1}{x+1}-\frac{1}{x+2}}=\dfrac{\frac{1}{x+1}+\frac{1}{x+2}}{\frac{1}{x+1}-\frac{1}{x+2}}\cdot\dfrac{(x+1)(x+2)}{(x+1)(x+2)}=\dfrac{\frac{(x+1)(x+2)}{x+1}+\frac{(x+1)(x+2)}{x+2}}{\frac{(x+1)(x+2)}{x+1}-\frac{(x+1)(x+2)}{x+2}}=$

$\dfrac{\frac{x+1}{x+1}\cdot(x+2)+\frac{x+2}{x+2}\cdot(x+1)}{\frac{x+1}{x+1}\cdot(x+2)-\frac{x+2}{x+2}\cdot(x+1)}=\dfrac{1\cdot(x+2)+1\cdot(x+1)}{1\cdot(x+2)-1\cdot(x+1)}=\dfrac{x+2+x+1}{x+2-x-1}=\dfrac{2x+3}{1}=2x+3$

49. $\dfrac{\frac{1}{2x-1}-\frac{1}{2x+1}}{\frac{x+1}{x}}=\dfrac{\frac{2x+1}{(2x-1)(2x+1)}-\frac{2x-1}{(2x-1)(2x+1)}}{\frac{x+1}{x}}=\dfrac{\frac{2}{(2x-1)(2x+1)}}{\frac{x+1}{x}}=$

$\dfrac{2}{(2x-1)(2x+1)}\div\dfrac{x+1}{x}=\dfrac{2}{(2x-1)(2x+1)}\cdot\dfrac{x}{x+1}=\dfrac{2x}{(x+1)(2x-1)(2x+1)}$

51. $\dfrac{\frac{1}{ab^2}-\frac{1}{a^2b}}{\frac{1}{b}-\frac{1}{a}}=\dfrac{\frac{a}{a^2b^2}-\frac{b}{a^2b^2}}{\frac{a}{ab}-\frac{b}{ab}}=\dfrac{\frac{a-b}{a^2b^2}}{\frac{a-b}{ab}}=\dfrac{a-b}{a^2b^2}\div\dfrac{a-b}{ab}=\dfrac{a-b}{a^2b^2}\cdot\dfrac{ab}{a-b}=\dfrac{a-b}{a-b}\cdot\dfrac{ab}{a^2b^2}=1\cdot\dfrac{1}{ab}=\dfrac{1}{ab}$

53. $\dfrac{1}{a^{-1}+b^{-1}}=\dfrac{1}{\frac{1}{a}+\frac{1}{b}}=\dfrac{1}{\frac{1}{a}\cdot\frac{b}{b}+\frac{1}{b}\cdot\frac{a}{a}}=\dfrac{1}{\frac{b}{ab}+\frac{a}{ab}}=\dfrac{1}{\frac{a+b}{ab}}=1\div\dfrac{a+b}{ab}=1\cdot\dfrac{ab}{a+b}=\dfrac{ab}{a+b}$

Applications

55. $\dfrac{P\left(1+\frac{r}{26}\right)^{52}-P}{\frac{r}{26}}$

57. $R=\dfrac{1}{\frac{1}{T}+\frac{1}{S}}=\dfrac{1}{\frac{1}{T}+\frac{1}{S}}\cdot\dfrac{ST}{ST}=\dfrac{ST}{\frac{ST}{T}+\frac{ST}{S}}=\dfrac{ST}{S\cdot\frac{T}{T}+\frac{S}{S}\cdot T}=\dfrac{ST}{S\cdot1+1\cdot T}=\dfrac{ST}{S+T}$

7.6: Rational Equations and Formulas

Concepts

1. rational

2. $\dfrac{2x+5}{3x}$; $\dfrac{2x+5}{3x}=9$; *Answers may vary.*

3. $ad=bc$; b; d

4. Yes, $\dfrac{2}{x-1}\cdot\dfrac{x-1}{1}=5\cdot\dfrac{x-1}{1}\Rightarrow 2=5(x-1)$

5. $12x$

6. V; T

Solving Rational Equations

7. $\dfrac{x}{2}=\dfrac{3}{4}\Rightarrow x\cdot4=2\cdot3\Rightarrow 4x=6\Rightarrow\dfrac{4x}{4}=\dfrac{6}{4}\Rightarrow x=\dfrac{6}{4}\Rightarrow x=\dfrac{3}{2}$

Check: $\dfrac{\frac{3}{2}}{2}=\dfrac{3}{4}\Rightarrow\dfrac{3}{2}\div\dfrac{2}{1}=\dfrac{3}{2}\cdot\dfrac{1}{2}=\dfrac{3}{4}$

9. $\dfrac{3}{z}=\dfrac{6}{5}\Rightarrow 3\cdot5=z\cdot6\Rightarrow 15=6z\Rightarrow\dfrac{15}{6}=\dfrac{6z}{6}\Rightarrow z=\dfrac{15}{6}\Rightarrow z=\dfrac{5}{2}$

Check: $\frac{3}{\frac{5}{2}}=\frac{6}{5}\Rightarrow\frac{3}{1}\div\frac{5}{2}=\frac{3}{1}\cdot\frac{2}{5}=\frac{6}{5}$

11. $\frac{12}{7}=\frac{2}{t}\Rightarrow 12\cdot t=7\cdot 2\Rightarrow 12t=14\Rightarrow\frac{12t}{12}=\frac{14}{12}\Rightarrow t=\frac{14}{12}\Rightarrow t=\frac{7}{6}$

Check: $\frac{12}{7}=\frac{2}{\frac{7}{6}}\Rightarrow\frac{2}{1}\div\frac{7}{6}=\frac{2}{1}\cdot\frac{6}{7}=\frac{12}{7}$

13. $\frac{3y}{4}=\frac{7y}{2}\Rightarrow 3y\cdot 2=4\cdot 7y\Rightarrow 6y=28y\Rightarrow -22y=0\Rightarrow\frac{-22y}{-22}=\frac{0}{-22}\Rightarrow y=0$

Check: $\frac{3(0)}{4}=\frac{7(0)}{2}\Rightarrow\frac{0}{4}=\frac{0}{2}\Rightarrow 0=0$

15. $\frac{2}{3}=\frac{1}{2x+1}\Rightarrow 2(2x+1)=3\cdot 1\Rightarrow 4x+2=3\Rightarrow 4x=1\Rightarrow\frac{4x}{4}=\frac{1}{4}\Rightarrow x=\frac{1}{4}$

Check: $\frac{2}{3}=\frac{1}{2\left(\frac{1}{4}\right)+1}\Rightarrow\frac{1}{\frac{2}{4}+1}=\frac{1}{\frac{6}{4}}=\frac{1}{1}\div\frac{6}{4}=\frac{1}{1}\cdot\frac{4}{6}=\frac{4}{6}=\frac{2}{3}$

17. $\frac{5}{2x}=\frac{8}{x+2}\Rightarrow 5(x+2)=2x\cdot 8\Rightarrow 5x+10=16x\Rightarrow -11x=-10\Rightarrow\frac{-11x}{-11}=\frac{-10}{-11}\Rightarrow x=\frac{10}{11}$

Check: $\frac{5}{2\left(\frac{10}{11}\right)}=\frac{8}{\frac{10}{11}+2}\Rightarrow\frac{5}{\frac{20}{11}}=\frac{8}{\frac{32}{11}}\Rightarrow\frac{5}{1}\div\frac{20}{11}=\frac{8}{1}\div\frac{32}{11}\Rightarrow\frac{5}{1}\cdot\frac{11}{20}=\frac{8}{1}\cdot\frac{11}{32}\Rightarrow\frac{55}{20}=\frac{88}{32}\Rightarrow\frac{11}{4}=\frac{11}{4}$

19. $\frac{1}{z-1}=\frac{2}{z+1}\Rightarrow 1\cdot(z+1)=(z-1)\cdot 2\Rightarrow z+1=2z-2\Rightarrow 3=z\Rightarrow z=3$

Check: $\frac{1}{3-1}=\frac{2}{3+1}\Rightarrow\frac{1}{2}=\frac{2}{4}\Rightarrow\frac{1}{2}=\frac{1}{2}$

21. $\frac{3}{n+5}=\frac{2}{n-5}\Rightarrow 3\cdot(n-5)=(n+5)\cdot 2\Rightarrow 3n-15=2n+10\Rightarrow n=25$

Check: $\frac{3}{25+5}=\frac{2}{25-5}\Rightarrow\frac{3}{30}=\frac{2}{20}\Rightarrow\frac{1}{10}=\frac{1}{10}$

23. $\frac{m}{m-1}=\frac{5}{4}\Rightarrow m\cdot 4=(m-1)\cdot 5\Rightarrow 4m=5m-5\Rightarrow -m=-5\Rightarrow\frac{-m}{-1}=\frac{-5}{-1}\Rightarrow m=5$

Check: $\frac{5}{5-1}=\frac{5}{4}\Rightarrow\frac{5}{4}=\frac{5}{4}$

25. $\frac{5x}{5-x}=\frac{1}{3}\Rightarrow 5x\cdot 3=(5-x)\cdot 1\Rightarrow 15x=5-x\Rightarrow 16x=5\Rightarrow\frac{16x}{16}=\frac{5}{16}\Rightarrow x=\frac{5}{16}$

Check: $\frac{5\left(\frac{5}{16}\right)}{5-\frac{5}{16}}=\frac{1}{3}\Rightarrow\frac{\frac{25}{16}}{\frac{75}{16}}=\frac{25}{16}\div\frac{75}{16}=\frac{25}{16}\cdot\frac{16}{75}=\frac{16}{16}\cdot\frac{25}{75}=1\cdot\frac{1}{3}=\frac{1}{3}$

27. $\frac{6}{5-2x}=\frac{2}{1}\Rightarrow 6\cdot 1=(5-2x)\cdot 2\Rightarrow 6=10-4x\Rightarrow -4=-4x\Rightarrow\frac{-4}{-4}=\frac{-4x}{-4}\Rightarrow 1=x\Rightarrow x=1$

Check: $\dfrac{6}{5-2(1)}=2\Rightarrow\dfrac{6}{5-2}=\dfrac{6}{3}\Rightarrow 2=2$

29. $\dfrac{2x}{2x+1}=\dfrac{-1}{2x+1}\Rightarrow 2x(2x+1)=-1(2x+1)\Rightarrow 4x^2+2x=-2x-1$

$\Rightarrow 4x^2+4x+1=0\Rightarrow(2x+1)^2=0\Rightarrow 2x+1=0\Rightarrow 2x=-1\Rightarrow x=-\dfrac{1}{2}$

Check: $\dfrac{2\left(-\frac{1}{2}\right)}{2\left(-\frac{1}{2}\right)+1}\overset{?}{=}\dfrac{-1}{2\left(-\frac{1}{2}\right)+1}\Rightarrow\dfrac{-1}{-1+1}\overset{?}{=}\dfrac{-1}{-1+1}$, false because the denominators are 0 so both expressions are undefined.

There are no solutions because $-\dfrac{1}{2}$ is extraneous.

31. $\dfrac{1}{1-x}=\dfrac{3}{1+x}\Rightarrow 1\cdot(1+x)=(1-x)\cdot 3\Rightarrow 1+x=3-3x\Rightarrow 4x=2\Rightarrow\dfrac{4x}{4}=\dfrac{2}{4}\Rightarrow x=\dfrac{1}{2}$

Check: $\dfrac{1}{1-\frac{1}{2}}=\dfrac{3}{1+\frac{1}{2}}\Rightarrow\dfrac{1}{\frac{1}{2}}=\dfrac{3}{\frac{3}{2}}\Rightarrow\dfrac{1}{1}\div\dfrac{1}{2}=\dfrac{3}{1}\div\dfrac{3}{2}\Rightarrow\dfrac{1}{1}\cdot\dfrac{2}{1}=\dfrac{3}{1}\cdot\dfrac{2}{3}\Rightarrow 2=\dfrac{6}{3}\Rightarrow 2=2$

33. $\dfrac{1}{z+2}=\dfrac{-z}{1}\Rightarrow 1\cdot 1=-z(z+2)\Rightarrow 1=-z^2-2z\Rightarrow z^2+2z+1=0\Rightarrow(z+1)^2=0\Rightarrow\ z=-1$

Check: $\dfrac{1}{-1+2}=\dfrac{-(-1)}{1}\Rightarrow\dfrac{1}{1}=\dfrac{1}{1}\Rightarrow 1=1$

35. $\dfrac{-1}{2x+5}=\dfrac{x}{3}\Rightarrow -1\cdot 3=(2x+5)\cdot x\Rightarrow -3=2x^2+5x\Rightarrow 0=2x^2+5x+3\Rightarrow$

$0=(2x+3)(x+1)\Rightarrow 0=2x+3\Rightarrow -\dfrac{3}{2}=\dfrac{2}{2}x\Rightarrow -\dfrac{3}{2}=x,\ 0=x+1\Rightarrow -1=x\Rightarrow x=-\dfrac{3}{2},\ -1$

Check: $\dfrac{-1}{2\left(\frac{-3}{2}\right)+5}=\dfrac{\frac{-3}{2}}{3}\Rightarrow\dfrac{-1}{\frac{4}{2}}=\dfrac{\frac{-3}{2}}{3}\Rightarrow\dfrac{-1}{1}\div\dfrac{4}{2}=\dfrac{-3}{2}\div\dfrac{3}{1}\Rightarrow\dfrac{-1}{1}\cdot\dfrac{2}{4}=\dfrac{-3}{2}\cdot\dfrac{1}{3}\Rightarrow\dfrac{-2}{4}=\dfrac{-3}{6}\Rightarrow$

$\dfrac{-1}{2}=\dfrac{-1}{2};\ \dfrac{-1}{2(-1)+5}=\dfrac{-1}{3}\Rightarrow\dfrac{-1}{-2+5}\Rightarrow\dfrac{-1}{3}$

37. $\dfrac{x}{2}+\dfrac{x}{4}=3\Rightarrow\dfrac{x(4)}{2}+\dfrac{x(4)}{4}=3(4)\Rightarrow\dfrac{4x}{2}+\dfrac{4x}{4}=12\Rightarrow 2x+x=12\Rightarrow 3x=12\Rightarrow\dfrac{3x}{3}=\dfrac{12}{3}\Rightarrow x=4$

Check: $\dfrac{4}{2}+\dfrac{4}{4}=3\Rightarrow 2+1=3\Rightarrow 3=3$

39. $\dfrac{3x}{4}-\dfrac{x}{2}=1\Rightarrow\dfrac{3x(4)}{4}-\dfrac{x(4)}{2}=1(4)\Rightarrow\dfrac{12x}{4}-\dfrac{4x}{2}=4\Rightarrow 3x-2x=4\Rightarrow x=4$

Check: $\dfrac{3(4)}{4}-\dfrac{4}{2}=1\Rightarrow\dfrac{12}{4}-\dfrac{4}{2}=1\Rightarrow 3-2=1=1$

41. $\dfrac{4}{t+1}+\dfrac{1}{t+1}=-1\Rightarrow\dfrac{5}{t+1}=-1\Rightarrow(t+1)\cdot\dfrac{5}{t+1}=-1\cdot(t+1)\Rightarrow 5=-t-1\Rightarrow 6=-t\Rightarrow\ \dfrac{6}{-1}=\dfrac{-t}{-1}\Rightarrow t=-6$

Check: $\frac{4}{t+1}+\frac{1}{t+1}=-1 \Rightarrow \frac{5}{t+1}=-1 \Rightarrow \frac{5}{-6+1}=-1 \Rightarrow \frac{5}{-5}=-1 \Rightarrow -1=-1$

43. $\frac{1}{x}+\frac{2}{x}=\frac{1}{2} \Rightarrow \frac{3}{x}=\frac{1}{2} \Rightarrow x\cdot\frac{3}{x}=\frac{1}{2}\cdot x \Rightarrow 3=\frac{1}{2}x \Rightarrow 3\cdot 2=\frac{1}{2}x\cdot 2 \Rightarrow 6=x \Rightarrow x=6$

Check: $\frac{1}{6}+\frac{2}{6}=\frac{1}{2} \Rightarrow \frac{3}{6}=\frac{1}{2} \Rightarrow \frac{1}{2}=\frac{1}{2}$

45. $\frac{2}{x-1}+1=\frac{4}{x^2-1} \Rightarrow \frac{2+x-1}{x-1}=\frac{4}{x^2-1} \Rightarrow \frac{x+1}{x-1}=\frac{4}{(x-1)(x+1)}$

$\Rightarrow (x+1)(x-1)(x+1)=4(x-1) \Rightarrow \frac{(x+1)(x-1)(x+1)}{(x-1)}=\frac{4(x-1)}{x-1}$

$\Rightarrow (x+1)(x+1)=4 \Rightarrow x^2+2x+1=4 \Rightarrow x^2+2x-3=0$

$\Rightarrow (x+3)(x-1)=0 \Rightarrow x+3=0,\ x=-3$ and $x-1=0,\ x=1$

Check: $\frac{2}{-3-1}+1=\frac{4}{(-3)^2-1} \Rightarrow \frac{2}{-4}+\frac{-4}{-4}=\frac{4}{9-1} \Rightarrow \frac{-2}{-4}=\frac{4}{8} \Rightarrow \frac{1}{2}=\frac{1}{2}$

$\frac{2}{1-1}+1 \stackrel{?}{=} \frac{4}{1-1}$, false because the denominators are 0 so both expressions are undefined.

The only solution is -3 because 1 is extraneous.

47. $\frac{1}{x+2}=\frac{4}{4-x^2}-1 \Rightarrow \frac{1}{x+2}=\frac{4-(4-x^2)}{4-x^2} \Rightarrow (4-x^2)\cdot 1$

$=(x+2)(4-4+x^2) \Rightarrow 4-x^2=(x+2)(x^2)$

$\Rightarrow (2-x)(2+x)=(x+2)(x^2) \Rightarrow \frac{(2-x)(2+x)}{x+2}=\frac{(x+2)(x^2)}{x+2}$

$\Rightarrow 2-x=x^2 \Rightarrow x^2+x-2=0 \Rightarrow (x-1)(x+2)=0$

$\Rightarrow x-1=0,\ x=1$ and $x+2=0,\ x=-2$

Check: -2 is extraneous because $\frac{1}{-2+2}$ is undefined. $\frac{1}{1+2}=\frac{4}{4-(1)^2}-1 \Rightarrow \frac{1}{3}=\frac{4}{3}-\frac{3}{3} \Rightarrow \frac{1}{3}=\frac{1}{3}$

49. $\frac{5}{4z}-\frac{2}{3z}=1 \Rightarrow \frac{5(12z)}{4z}-\frac{2(12z)}{3z}=1(12z) \Rightarrow \frac{60z}{4z}-\frac{24z}{3z}=12z \Rightarrow 15-8=12z \Rightarrow \frac{7}{12}=\frac{12z}{12} \Rightarrow$

$\frac{7}{12}=z \Rightarrow z=\frac{7}{12}$ Check: $\frac{5}{4\left(\frac{7}{12}\right)}-\frac{2}{3\left(\frac{7}{12}\right)}=1 \Rightarrow \frac{5}{\frac{7}{3}}-\frac{2}{\frac{7}{4}}=1 \Rightarrow \frac{5}{1}\div\frac{7}{3}-\frac{2}{1}\div\frac{7}{4}=1 \Rightarrow$

$\frac{5}{1}\cdot\frac{3}{7}-\frac{2}{1}\cdot\frac{4}{7}=1 \Rightarrow \frac{15}{7}-\frac{8}{7}=1 \Rightarrow \frac{7}{7}=1 \Rightarrow 1=1$

51. $\frac{4}{y-1}+\frac{1}{y}=\frac{6}{5} \Rightarrow \frac{4(5)(y)(y-1)}{y-1}+\frac{1(5)(y)(y-1)}{y}=\frac{6(5)(y)(y-1)}{5} \Rightarrow$

$\frac{20y(y-1)}{y-1}+\frac{5y(y-1)}{y}=\frac{(6y^2-6y)(5)}{5} \Rightarrow 20y+5y-5=6y^2-6y \Rightarrow 25y-5=6y^2-6y \Rightarrow$

$6y^2-31y+5=0 \Rightarrow (6y-1)(y-5)=0 \Rightarrow 6y-1=0 \Rightarrow 6y=1 \Rightarrow y=\frac{1}{6},\ y-5=0 \Rightarrow\ y=5 \Rightarrow y=\frac{1}{6},\ 5$

Check: $\frac{4}{y-1}+\frac{1}{y}=\frac{6}{5} \Rightarrow \frac{4}{\frac{1}{6}-1}+\frac{1}{\frac{1}{6}}=\frac{6}{5} \Rightarrow \frac{4}{-\frac{5}{6}}+\frac{1}{\frac{1}{6}}=\frac{6}{5} \Rightarrow \frac{4}{1}\cdot-\frac{6}{5}+\frac{1}{1}\cdot\frac{6}{1}=\frac{6}{5} \Rightarrow -\frac{24}{5}+\frac{6}{1}=\frac{6}{5} \Rightarrow$

$-\frac{24}{5}+\frac{30}{5}=\frac{6}{5} \Rightarrow \frac{6}{5}=\frac{6}{5},\ \frac{4}{5-1}+\frac{1}{5}=\frac{6}{5} \Rightarrow \frac{4}{4}+\frac{1}{5}=\frac{6}{5} \Rightarrow \frac{20}{20}+\frac{4}{20}=\frac{24}{20} \Rightarrow \frac{24}{20}=\frac{24}{20}=\frac{6}{5}=\frac{6}{5}$

53. $\frac{1}{2x}-\frac{1}{x+3}=0 \Rightarrow \frac{1(2x)(x+3)}{2x}-\frac{1(2x)(x+3)}{x+3}=0(2x)(x+3) \Rightarrow \frac{(x+3)(2x)}{2x}-\frac{2x(x+3)}{x+3}=0 \Rightarrow$

$x+3-2x=0 \Rightarrow -x+3=0 \Rightarrow -x=-3 \Rightarrow x=3$

Check: $\frac{1}{2(3)}-\frac{1}{3+3}=0 \Rightarrow \frac{1}{6}-\frac{1}{6}=0 \Rightarrow 0=0$

55. $\frac{1}{x-1}+\frac{1}{x+1}=\frac{2}{x^2-1} \Rightarrow \frac{1}{x-1}\cdot\frac{x+1}{x+1}+\frac{1}{x+1}\cdot\frac{x-1}{x-1}=\frac{2}{x^2-1}$

$\Rightarrow \frac{x+1+x-1}{x^2-1}=\frac{2}{x^2-1} \Rightarrow \frac{2x}{x^2-1}=\frac{2}{x^2-1} \Rightarrow 2x(x^2-1)=2(x^2-1)$

$\Rightarrow \frac{2x(x^2-1)}{2(x^2-1)}=\frac{2(x^2-1)}{2(x^2-1)} \Rightarrow x=1$, extraneous because $\frac{1}{1-1}$ is undefined. There are no solutions.

57. $\frac{1}{x-2}+\frac{1}{x+2}=\frac{6}{x^2-4} \Rightarrow \frac{1(x-2)(x+2)}{x-2}+\frac{1(x-2)(x+2)}{x+2}=\frac{6(x-2)(x+2)}{(x-2)(x+2)} \Rightarrow$

$\frac{(x+2)(x-2)}{x-2}+\frac{(x-2)(x+2)}{x+2}=\frac{6(x-2)(x+2)}{(x-2)(x+2)} \Rightarrow x+2+x-2=6 \Rightarrow 2x=6 \Rightarrow x=\frac{6}{2} \Rightarrow x=3$

Check: $\frac{1}{3-2}+\frac{1}{3+2}=\frac{6}{(3)^2-4} \Rightarrow \frac{1}{1}+\frac{1}{5}=\frac{6}{5} \Rightarrow \frac{5}{5}+\frac{1}{5}=\frac{6}{5} \Rightarrow \frac{6}{5}=\frac{6}{5}$

59. $\frac{1}{p+1}+\frac{1}{p+2}=\frac{1}{p^2+3p+2} \Rightarrow \frac{1}{p+1}\cdot\frac{p+2}{p+2}+\frac{1}{p+2}\cdot\frac{p+1}{p+1}=\frac{1}{(p+1)(p+2)} \Rightarrow$

$\frac{p+2}{(p+1)(p+2)}+\frac{p+1}{(p+1)(p+2)}=\frac{1}{(p+1)(p+2)} \Rightarrow 2p+3=1 \Rightarrow 2p=-2 \Rightarrow$

$p=-1$, but $p=-1$ makes $\frac{1}{p+1}=\frac{1}{0}$ which is undefined $\Rightarrow$ no solution

61. $\frac{1}{x-2}+\frac{3}{2x-4}=\frac{6}{3x-6} \Rightarrow \frac{1}{x-2}+\frac{3}{2(x-2)}=\frac{6}{3(x-2)} \Rightarrow$

$\frac{1}{x-2}\cdot\frac{6}{6}+\frac{3}{2(x-2)}\cdot\frac{3}{3}=\frac{6}{3(x-2)}\cdot\frac{2}{2} \Rightarrow \frac{6}{6(x-2)}+\frac{9}{6(x-2)}=\frac{12}{6(x-2)} \Rightarrow 6+9=12 \Rightarrow$

$15=12$, which is false $\Rightarrow$ no solution

63. $\frac{1}{r^2-r-2}+\frac{2}{r^2-2r}=\frac{1}{r^2+r} \Rightarrow \frac{1(r)(r-2)(r+1)}{(r-2)(r+1)}+\frac{2(r)(r-2)(r+1)}{r(r-2)}=\frac{1(r)(r-2)(r+1)}{r(r+1)} \Rightarrow$

$\frac{r(r-2)(r+1)}{(r-2)(r+1)}+\frac{(2r+2)(r)(r-2)}{r(r-2)}=\frac{(r-2)(r)(r+1)}{r(r+1)} \Rightarrow r+2r+2=r-2 \Rightarrow 3r+2=r-2 \Rightarrow$

$3r=r-4 \Rightarrow 2r=-4 \Rightarrow r=\frac{-4}{2} \Rightarrow r=-2$

Check: $\frac{1}{(-2)^2-(-2)-2}+\frac{2}{(-2)^2-2(-2)}=\frac{1}{(-2)^2+(-2)} \Rightarrow \frac{1}{4}+\frac{2}{8}=\frac{1}{2} \Rightarrow \frac{1}{4}+\frac{1}{4}=\frac{1}{2} \Rightarrow \frac{2}{4}=\frac{1}{2} \Rightarrow \frac{1}{2}=\frac{1}{2}$

65. Expression; $\frac{1}{x}-\frac{1-x}{x}=\frac{1-1+x}{x}=\frac{x}{x}=1$; when $x=2$, the expression has the value 1.

67. Equation; $\frac{1}{2x}-\frac{1}{4x}=\frac{1}{8} \Rightarrow \frac{2}{4x}-\frac{1}{4x}=\frac{1}{8} \Rightarrow \frac{1}{4x}=\frac{1}{8} \Rightarrow 4x=8 \Rightarrow x=2$

69. Equation; $\frac{x+1}{x-1}=\frac{2x-3}{2x-5} \Rightarrow (2x-5)(x+1)=(2x-3)(x-1)$

$\Rightarrow 2x^2-3x-5=2x^2-5x+3 \Rightarrow 2x-8=0 \Rightarrow 2x=8 \Rightarrow x=4$

71. Expression; $\frac{4x+4}{x+2}+\frac{x^2}{x+2}=\frac{x^2+4x+4}{x+2}=\frac{(x+2)^2}{x+2}=x+2$; $x=2, 2+2=4$

Graphical and Numerical Solutions

73. The graphs intersect at $\left(-\frac{1}{2},-2\right)$ and $\left(\frac{1}{2},2\right)$. Check: $\frac{1}{-\frac{1}{2}}=4\left(-\frac{1}{2}\right) \Rightarrow -2=-2$; $\frac{1}{\frac{1}{2}}=4\left(\frac{1}{2}\right) \Rightarrow 2=2$ The solutions are $-\frac{1}{2}, \frac{1}{2}$.

75. The graphs intersect at $(-1, 2)$ and $\left(\frac{1}{2},-1\right)$. Check:

$\frac{-1-1}{-1}=-2(-1) \Rightarrow \frac{-2}{-1}=2 \Rightarrow 2=2$; $\frac{\frac{1}{2}-1}{\frac{1}{2}}=-2\left(\frac{1}{2}\right) \Rightarrow \frac{-\frac{1}{2}}{\frac{1}{2}}=-1 \Rightarrow -1=-1$ The solutions are $-1, \frac{1}{2}$.

77. (a) $-3, 1$; $y_1=\frac{3}{x}$, $y_2=x+2$. See Figure 77a.

(b) See Figure 77b. $-3, 1$

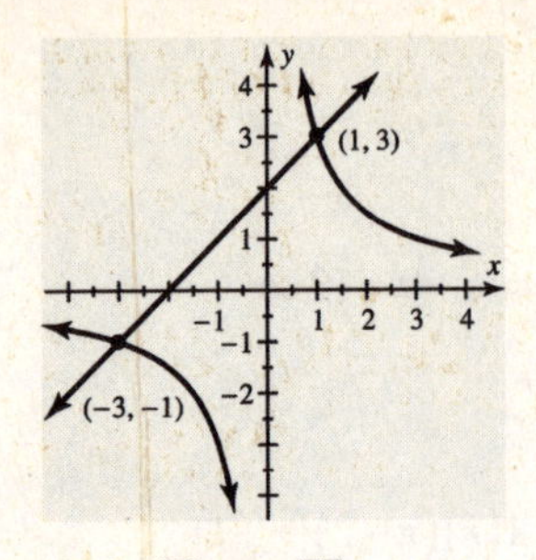

Figure 77a

x	−3	−1	1	3
$y = 3/x$	−1	−3	3	1
$y = x + 2$	−1	1	3	5

Figure 77b

79. (a) −1; $y_1 = \dfrac{3x}{2}$, $y_2 = \dfrac{1}{2}x - 1$. See Figure 79a.

(b) See Figure 79b. −1

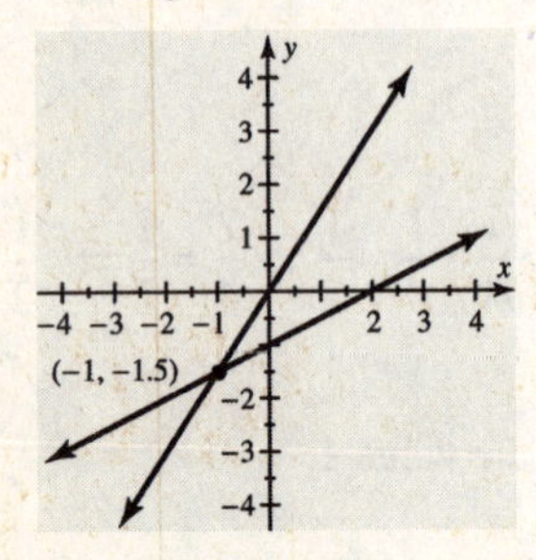

Figure 79a

x	−2	−1	1	2
$y = 3x/2$	−3	−1.5	1.5	3
$y = x/2 - 1$	−2	−1.5	−0.5	0

Figure 79b

81. (a) 2; $y_1 = \dfrac{3}{x-1}$, $y_2 = 3$. See Figure 81a.

(b) See Figure 81b. 2

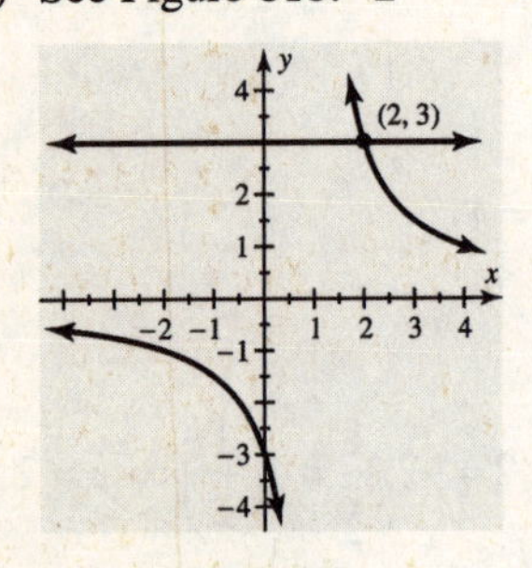

Figure 81a

x	−2	0	2	4
$y = 3/(x-1)$	−1	−3	3	1
$y = 3$	3	3	3	3

Figure 81b

83. (a) −2, 2; $y_1 = \dfrac{4}{x^2}$, $y_2 = 1$. See Figure 83a.

(b) See Figure 83b. −2, 2

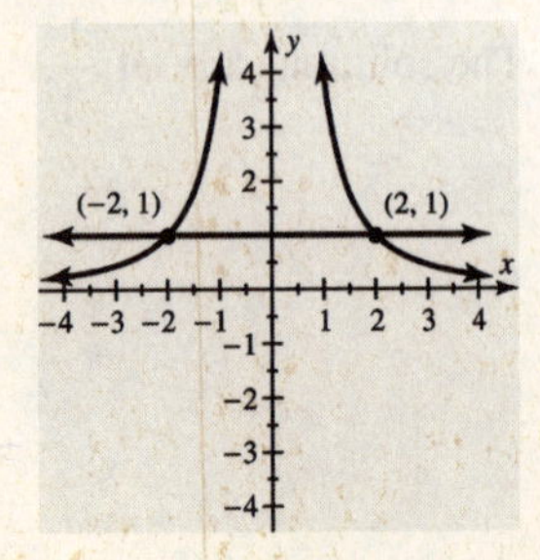

Figure 83a

x	−2	−1	1	2
$y = 4/x^2$	1	4	4	1
$y = 1$	1	1	1	1

Figure 83b

85.

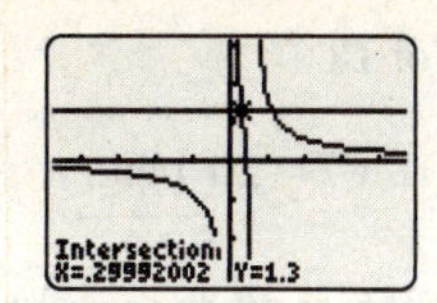

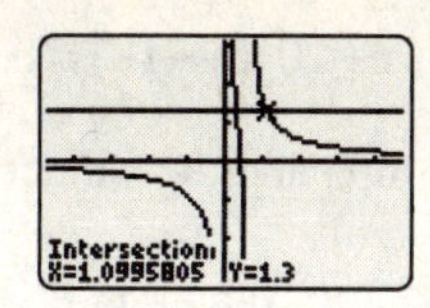

The solutions are 0.300 and 1.100.

87.

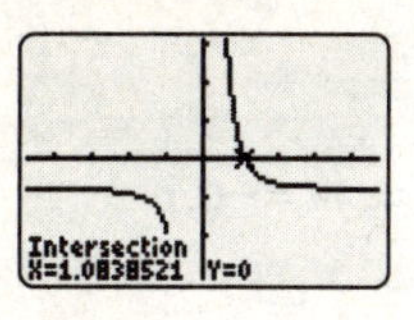

The solution is 1.084.

Solving an Equation for a Variable

89. $m=\frac{F}{a}$ for $a \Rightarrow a \cdot m=\frac{F}{a} \cdot a \Rightarrow am=F \Rightarrow \frac{am}{m}=\frac{F}{m} \Rightarrow a=\frac{F}{m}$

91. $I=\frac{V}{R+r}$ for $r \Rightarrow I(R+r)=\frac{V}{R+r} \cdot (R+r) \Rightarrow I(R+r)=V \Rightarrow \frac{I(R+r)}{I}=\frac{V}{I} \Rightarrow R+r=\frac{V}{I} \Rightarrow r=\frac{V}{I}-R$

93. $h=\frac{2A}{b}$ for $b \Rightarrow b \cdot h=\frac{2A}{b} \cdot b \Rightarrow bh=2A \Rightarrow \frac{bh}{h}=\frac{2A}{h} \Rightarrow b=\frac{2A}{h}$

95. $\frac{3}{k}=\frac{z}{z+5}$ for $z \Rightarrow 3(z+5)=kz \Rightarrow 3z+15=kz \Rightarrow 15=kz-3z \Rightarrow 15=z(k-3) \Rightarrow z=\frac{15}{k-3}$

97. $T=\frac{ab}{a+b}$ for $b \Rightarrow T(a+b)=ab \cdot 1 \Rightarrow aT+bT=ab \Rightarrow aT=ab-bT \Rightarrow aT=b(a-T) \Rightarrow b=\frac{aT}{a-T}$

99. $\frac{3}{k}=\frac{1}{x}-\frac{2}{y}$ for $x \Rightarrow \frac{3}{k} \cdot \frac{xy}{xy}=\frac{1}{x} \cdot \frac{ky}{ky}-\frac{2}{y} \cdot \frac{xk}{xk} \Rightarrow \frac{3xy}{kxy}=\frac{ky}{kxy}-\frac{2xk}{kxy} \Rightarrow 3xy=ky-2xk \Rightarrow$

$3xy+2xk=ky \Rightarrow x(3y+2k)=ky \Rightarrow x=\frac{ky}{3y+2k}$

Applications

101. $\frac{1}{10-x}=1 \Rightarrow (10-x) \cdot \frac{1}{10-x}=1(10-x) \Rightarrow 1=10-x \Rightarrow x=9$ cars/minute.

103. $\frac{t}{4}+\frac{t}{3}=1 \Rightarrow \frac{t(12)}{4}+\frac{t(12)}{3}=12 \Rightarrow \frac{12t}{4}+\frac{12t}{3}=12 \Rightarrow 3t+4t=12 \Rightarrow 7t=12 \Rightarrow t=\frac{12}{7} \Rightarrow t=1.7$ hr

105. $\frac{d}{8}+\frac{d}{4}=1 \Rightarrow \frac{(8)d}{8}+\frac{(8)d}{4}=8 \Rightarrow \frac{8d}{8}+\frac{8d}{4}=8 \Rightarrow d+2d=8 \Rightarrow 3d=8 \Rightarrow d=\frac{8}{3} \Rightarrow d \approx 2.7$ days

107. Let x represent the speed of the teammate. Then $x+2$ represents the speed of the winner. The finishing time for the teammate is $\frac{6}{x}$ and the finishing time for the winner is $\frac{6}{x+2}$. Since the winner's time is 2 minutes, or $\frac{1}{30}$ hour, faster than the teammate's time, the needed equation is $\frac{6}{x}-\frac{6}{x+2}=\frac{1}{30}$. Multiply each term by the LCD, $30x(x+2)$. $\frac{6}{x}-\frac{6}{x+2}=\frac{1}{30} \Rightarrow 30x(x+2) \cdot \frac{6}{x}-30x(x+2) \cdot \frac{6}{x+2}=30x(x+2) \cdot \frac{1}{30} \Rightarrow$

$180x+360-180x=x^2+2x \Rightarrow x^2+2x-360=0 \Rightarrow (x+20)(x-18)=0 \Rightarrow x=-20$ or 18

Since x must be positive, the teammate's speed is 18 mph and the winner's speed is $18+2=20$ mph.

109\. (a) $m=-0.05,\ B=\dfrac{30}{0.3+(-0.05)} \Rightarrow B=\dfrac{30}{0.25} \Rightarrow B=120$, the braking distance is 120 feet when the slope of the road is -0.05.

(b) $B=150,\ 150=\dfrac{30}{0.3+m} \Rightarrow (0.3+m)(150)=30 \Rightarrow 0.3+m=\dfrac{30}{150} \Rightarrow 0.3+m=.2 \Rightarrow m=-0.1$

111\. $\dfrac{36}{x-3}=\dfrac{54}{x+3} \Rightarrow 36x+108=54x-162 \Rightarrow 270=18x \Rightarrow 15=x \Rightarrow x=15$ mph

113\. $\dfrac{450}{x-50}=\dfrac{750}{x+50} \Rightarrow 450x+22{,}500=750x-37{,}500 \Rightarrow 60{,}000=300x \Rightarrow 200=x \Rightarrow x=200$ mph

115\. Let $x=$ walking in mph and $x+5=$ running in mph, then $x\cdot(t+1)=10$ and $(x+5)t=10$, so

$t+1=\dfrac{10}{x} \Rightarrow t=\dfrac{10}{x}-1$ and $t=\dfrac{10}{x+5}$.

Now, $\dfrac{10}{x}-1=\dfrac{10}{x+5} \Rightarrow \dfrac{10}{x}\cdot\dfrac{x+5}{x+5}-\dfrac{1}{1}\cdot\dfrac{x(x+5)}{x(x+5)}=\dfrac{10}{x+5}\cdot\dfrac{x}{x} \Rightarrow 10(x+5)-x(x+5)=10x \Rightarrow$

$10x+50-x^2-5x=10x \Rightarrow x^2+5x-50=0 \Rightarrow (x+10)(x-5)=0$, so $x=-10,\ 5$ but -10 cannot be the rate so $x=5$. So 10 mph running and 5 mph walking.

117\. $\dfrac{2+n}{8-n}=1 \Rightarrow 2+n=8-n \Rightarrow 2n=6 \Rightarrow n=3$

119\. $\dfrac{4+6n}{4-4n}=-\dfrac{49}{31} \Rightarrow 124+186n=-196+196n \Rightarrow 320=10n \Rightarrow 32=n \Rightarrow n=32$

Checking Basic Concepts for Sections 7.5 & 7.6

1\. (a) $\dfrac{\frac{x}{3}}{\frac{2x}{5}}=\dfrac{x}{3}\div\dfrac{2x}{5}=\dfrac{x}{3}\cdot\dfrac{5}{2x}=\dfrac{5x}{6x}=\dfrac{5}{6}\cdot\dfrac{x}{x}=\dfrac{5}{6}\cdot 1=\dfrac{5}{6}$

(b) $\dfrac{\frac{2}{2x}-\frac{1}{3x}}{6x}=\dfrac{\frac{6}{6x}-\frac{2}{6x}}{6x}=\dfrac{\frac{4}{6x}}{6x}=\dfrac{4}{6x}\div\dfrac{6x}{1}=\dfrac{4}{6x}\cdot\dfrac{1}{6x}=\dfrac{4}{36x^2}=\dfrac{1}{9x^2}$

(c) $\dfrac{\frac{1}{a}-\frac{1}{b}}{\frac{1}{a}+\frac{1}{b}}=\dfrac{\frac{b}{ab}-\frac{a}{ab}}{\frac{b}{ab}+\frac{a}{ab}}=\dfrac{\frac{b-a}{ab}}{\frac{b+a}{ab}}=\dfrac{b-a}{ab}\div\dfrac{b+a}{ab}=\dfrac{b-a}{ab}\cdot\dfrac{ab}{b+a}=\dfrac{ab}{ab}\cdot\dfrac{b-a}{b+a}=1\cdot\dfrac{b-a}{b+a}=\dfrac{b-a}{b+a}$

(d) $\dfrac{\frac{1}{r^2}-\frac{1}{t^2}}{\frac{2}{r}-\frac{2}{t}}=\dfrac{\frac{r^2}{r^2t^2}-\frac{r^2}{r^2t^2}}{\frac{2t}{rt}-\frac{2r}{rt}}=\dfrac{\frac{t^2-r^2}{r^2t^2}}{\frac{2t-2r}{rt}}=\dfrac{t^2-r^2}{r^2t^2}\div\dfrac{2t-2r}{rt}=\dfrac{t^2-r^2}{r^2t^2}\cdot\dfrac{rt}{2t-2r}=\dfrac{rt}{r^2t^2}\cdot\dfrac{t^2-r^2}{2t-2r}=$

$\dfrac{1}{2rt}\cdot\dfrac{(t+r)(t-r)}{t-r}=\dfrac{t+r}{2rt}=\dfrac{r+t}{2rt}$

2\. (a) $\dfrac{1}{2x}=\dfrac{3}{x+1} \Rightarrow 6x=x+1 \Rightarrow 5x=1 \Rightarrow x=\dfrac{1}{5}$

Check: $\frac{1}{2\left(\frac{1}{5}\right)}=\frac{3}{\left(\frac{1}{5}\right)+1} \Rightarrow \frac{1}{\frac{2}{5}}=\frac{3}{\frac{6}{5}} \Rightarrow \frac{1}{1}\cdot\frac{5}{2}=\frac{3}{1}\cdot\frac{5}{6} \Rightarrow \frac{5}{2}=\frac{15}{6} \Rightarrow \frac{5}{2}=\frac{5}{2}$

(b) $\frac{x}{2x+3}=\frac{4}{5} \Rightarrow 5x=8x+12 \Rightarrow -3x=12 \Rightarrow x=\frac{12}{-3} \Rightarrow x=-4$

Check: $\frac{-4}{2(-4)+3}=\frac{4}{5} \Rightarrow \frac{-4}{-5}=\frac{4}{5} \Rightarrow \frac{4}{5}=\frac{4}{5}$

3. (a) $\frac{1}{2x}+\frac{3}{2x}=1 \Rightarrow \frac{4}{2x}=1 \Rightarrow 2x=4 \Rightarrow x=\frac{4}{2} \Rightarrow x=2$

Check: $\frac{1}{2(2)}+\frac{3}{2(2)}=1 \Rightarrow \frac{1}{4}+\frac{3}{4}=1 \Rightarrow \frac{4}{4}=1 \Rightarrow 1=1$

(b) $\frac{3}{x+1}-\frac{2}{x}=-2 \Rightarrow \frac{3}{x+1}\cdot\frac{x}{x}-\frac{2}{x}\cdot\frac{x+1}{x+1}=\frac{-2}{1}\cdot\frac{x(x+1)}{x(x+1)} \Rightarrow 3x-2x-2=-2x^2-2x \Rightarrow$

$x-2=-2x^2-2x \Rightarrow 2x^2+3x-2=0 \Rightarrow (2x-1)(x+2)=0$, so $x=\frac{1}{2}$, -2

Check: $\frac{3}{\frac{1}{2}+1}-\frac{2}{\frac{1}{2}}=-2 \Rightarrow \frac{3}{1\frac{1}{2}}-\frac{2}{\frac{1}{2}}=-2 \Rightarrow \frac{6}{3}-\frac{12}{3}=-2 \Rightarrow \frac{-6}{3}=-2 \Rightarrow -2=-2$

Check: $\frac{3}{-2+1}-\frac{2}{-2}=-2 \Rightarrow \frac{3}{-1}-\frac{2}{-2}=-2 \Rightarrow -3-(-1)=-2 \Rightarrow -2=-2$

4. $\frac{1}{x-1}=\frac{2}{x^2-1}-\frac{1}{2} \Rightarrow \frac{1}{x-1}=\frac{2}{x^2-1}\cdot\frac{2}{2}-\frac{1}{2}\cdot\frac{x^2-1}{x^2-1}$

$\Rightarrow \frac{1}{x-1}=\frac{4-x^2+1}{2\left(x^2-1\right)} \Rightarrow 2\left(x^2-1\right)=(x-1)\left(5-x^2\right)$

$\Rightarrow \frac{2(x-1)(x+1)}{x-1}=\frac{(x-1)\left(5-x^2\right)}{x-1} \Rightarrow 2(x+1)=\left(5-x^2\right)$

$\Rightarrow 2x+2=5-x^2 \Rightarrow x^2+2x-3=0 \Rightarrow (x+3)(x-1)=0$

$\Rightarrow x+3=0,\ x=-3$ and $x-1=0,\ x=1$

Check: $\frac{1}{-3-1}=\frac{2}{(-3)^2-1}-\frac{1}{2} \Rightarrow \frac{1}{-4}=\frac{2}{8}-\frac{1}{2} \Rightarrow -\frac{1}{4}=\frac{1}{4}-\frac{2}{4}$

$\Rightarrow -\frac{1}{4}=-\frac{1}{4}$; 1 is an extraneous solution because $\frac{1}{1-1}$ is undefined.

The solution is -3.

5. (a) $\frac{ax}{2}-3y=b$ for $x \Rightarrow \frac{ax}{2}=b+3y \Rightarrow ax=2(b+3y) \Rightarrow x=\frac{2(b+3y)}{a}$

(b) $\frac{1}{2m-1}=\frac{k}{m}$ for $m \Rightarrow m\cdot 1=k(2m-1) \Rightarrow m=2km-k \Rightarrow m-2km=-k \Rightarrow$

$m(1-2k)=-k \Rightarrow m=\frac{-k}{1-2k} \Rightarrow m=\frac{-k}{-(2k-1)} \Rightarrow m=\frac{k}{2k-1}$

6. (a) $m=0.1,\ D=\frac{120}{0.3+0.1} \Rightarrow D=\frac{120}{0.4} \Rightarrow D=300;$ when the slope of the hill is 0.1, the braking distance is 300 ft.

(b) $D=200,\ 200=\frac{120}{0.3+m} \Rightarrow 200(0.3+m)=120 \Rightarrow 0.3+m=\frac{120}{200} \Rightarrow 0.3+m=0.6 \Rightarrow\ m=0.3;$ the braking distance is 200 ft when the slope of the road is 0.3.

7.7: Proportions and Variation

Concepts

1. A statement that two ratios are equal
2. $\frac{5}{6}=\frac{x}{7}$
3. It doubles
4. It is halved
5. constant
6. constant
7. Directly; doubling the number being fed doubles the bill.
8. Inversely; doubling the number of painters will halve the time.

Proportions

9. $\frac{x}{24}=\frac{5}{8} \Rightarrow 120=8x \Rightarrow \frac{120}{8}=x \Rightarrow x=15$

11. $\frac{14}{x}=\frac{2}{3} \Rightarrow 42=2x \Rightarrow \frac{42}{2}=x \Rightarrow x=21$

13. $\frac{3}{16}=\frac{h}{256} \Rightarrow 768=16h \Rightarrow \frac{768}{16}=h \Rightarrow h=48$

15. $\frac{3}{4}=\frac{2x}{7} \Rightarrow 21=8x \Rightarrow \frac{21}{8}=x \Rightarrow x=\frac{21}{8}$

17. $\frac{x}{6}=\frac{8}{3x} \Rightarrow 3x^2=48 \Rightarrow x^2=16 \Rightarrow x^2-16=0 \Rightarrow (x-4)(x+4)=0$

$\Rightarrow x-4=0, x=4$ and $x+4=0,\ x=-4$

19. $\frac{x}{7}=\frac{7}{4x} \Rightarrow 4x^2=49 \Rightarrow x^2=\frac{49}{4} \Rightarrow x^2-\frac{49}{4}=0$

$\Rightarrow \left(x-\frac{7}{2}\right)\left(x+\frac{7}{2}\right)=0 \Rightarrow x-\frac{7}{2}=0,\ x=\frac{7}{2}$ and $x+\frac{7}{2}=0,\ x=-\frac{7}{2}$

21. $\frac{a}{b} = \frac{c}{d} \Rightarrow ad = bc \Rightarrow \frac{ad}{c} = b$ or $b = \frac{ad}{c}$

23. (a) $\frac{5}{8} = \frac{9}{x}$

(b) $\frac{5}{8} = \frac{9}{x} \Rightarrow 72 = 5x \Rightarrow \frac{72}{5} = x \Rightarrow x = \frac{72}{5}$

25. (a) $\frac{4}{8} = \frac{10}{x}$

(b) $\frac{4}{8} = \frac{10}{x} \Rightarrow 4x = 80 \Rightarrow x = \frac{80}{4} \Rightarrow x = 20$

27. (a) $\frac{98}{7} = \frac{x}{11}$

(b) $\frac{98}{7} = \frac{x}{11} \Rightarrow 1078 = 7x \Rightarrow \frac{1078}{7} = x \Rightarrow x = \$154$

29. (a) $\frac{3}{180} = \frac{7}{x}$

(b) $\frac{3}{180} = \frac{7}{x} \Rightarrow 3x = 1260 \Rightarrow x = \frac{1260}{3} \Rightarrow x = 420$ min.

Variation

31. (a) $k = \frac{y}{x},\ y = 4,\ x = 2 \Rightarrow k = \frac{4}{2} = 2$

(b) $y = kx,\ k = 2,\ x = 6 \Rightarrow y = (2)(6) = 12$

33. (a) $k = \frac{y}{x},\ y = 3,\ x = 2 \Rightarrow k = \frac{3}{2}$

(b) $y = kx,\ k = \frac{3}{2},\ x = 6 \Rightarrow y = \left(\frac{3}{2}\right)(6) = 9$

35. (a) $k = \frac{y}{x},\ y = -60,\ x = 8 \Rightarrow k = \frac{-60}{8} = -\frac{15}{2}$

(b) $y = kx,\ k = -\frac{15}{2},\ x = 6 \Rightarrow y = \left(-\frac{15}{2}\right)(6) = -45$

37. (a) $k = yx,\ y = 6,\ x = 4 \Rightarrow k = 6 \cdot 4 = 24$

(b) $y = \frac{k}{x},\ k = 24,\ x = 8 \Rightarrow y = \frac{24}{8} = 3$

39. (a) $k = yx,\ y = 80\ x = \frac{1}{2} \Rightarrow k = 80 \cdot \frac{1}{2} = \frac{80}{2} = 40$

(b) $y = \frac{k}{x},\ k = 40,\ x = 8 \Rightarrow y = \frac{40}{8} = 5$

41. (a) $k = yx,\ y = 20,\ x = 20 \Rightarrow k = 20 \cdot 20 = 400$

(b) $y = \dfrac{k}{x},\ k = 400,\ x = 8 \Rightarrow y = \dfrac{400}{8} = 50$

43. (a) Direct, because as x increases, y increases and $k = \dfrac{y}{x} \Rightarrow k = \dfrac{3}{2} = \dfrac{4.5}{3} = \dfrac{6}{4}$, etc.

(b) $k = \dfrac{y}{x} \Rightarrow k = \dfrac{3}{2} \Rightarrow y = kx \Rightarrow y = \dfrac{3}{2}x$

(c) See Figure 43.

45. (a) Inverse, because as x increases, y decreases and $k = xy \Rightarrow k = 3 \cdot 12 = 6 \cdot 6 = 9 \cdot 4$, etc.

(b) $k = xy \Rightarrow k = 3 \cdot 12 = 36 \Rightarrow y = \dfrac{k}{x} \Rightarrow y = \dfrac{36}{x}$

(c) See Figure 45.

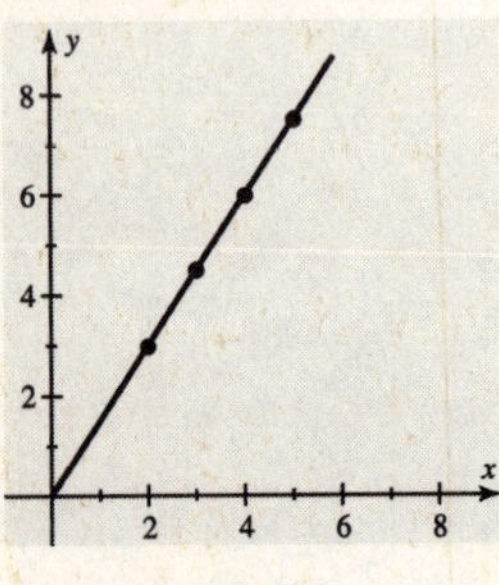

Figure 43

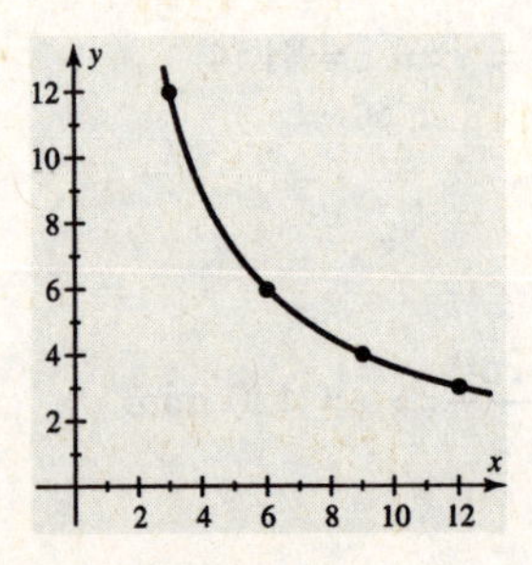

Figure 45

47. (a) Neither, because as x increases, y increases, but $k = \dfrac{y}{x} \Rightarrow \dfrac{10}{4} \neq \dfrac{40}{20}$

(b) NA

(c) NA

49. Direct, $k = \dfrac{y}{x} \Rightarrow k = \dfrac{6}{3} \Rightarrow k = 2$

51. Neither

53. Inverse, $k = xy \Rightarrow k = 4 \cdot 2 \Rightarrow k = 8$

Applications

55. $\dfrac{85}{750} = \dfrac{x}{420} \Rightarrow 750x = 35{,}700 \Rightarrow x = \dfrac{35{,}700}{750} \Rightarrow x = 47.6$ minutes

57. $\dfrac{8}{1} = \dfrac{13}{x} \Rightarrow 8x = 13 \Rightarrow x = \dfrac{13}{8} \Rightarrow x = 1.625$ inches

59. $y = kx,\ x = 6.2,\ y = 2800 \Rightarrow 2800 = k \cdot 6.2 \Rightarrow k \approx 451.613;$

$y = 451.613x,\ x = 4.7 \Rightarrow y = 451.613(4.7) \Rightarrow y \approx 2123$ lb

61. $\dfrac{2\frac{2}{3}}{14} = \dfrac{x}{49} \Rightarrow \dfrac{8}{3} \cdot 49 = 14x \Rightarrow \dfrac{392}{3} = 14x \Rightarrow \dfrac{1}{14} \cdot \dfrac{392}{3} = \dfrac{1}{14} \cdot 14x$

$\Rightarrow x = \dfrac{28}{3}$ or $9\dfrac{1}{3}$ c

63. (a) Direct, the ratios $\dfrac{R}{W}$ always equal 0.012

(b) $\dfrac{R}{W} = 0.012 \Rightarrow R = 0.012W$. See Figure 63.

(c) $W = 3200,\ R = 0.012(3200) \Rightarrow R = 38.4$ pounds.

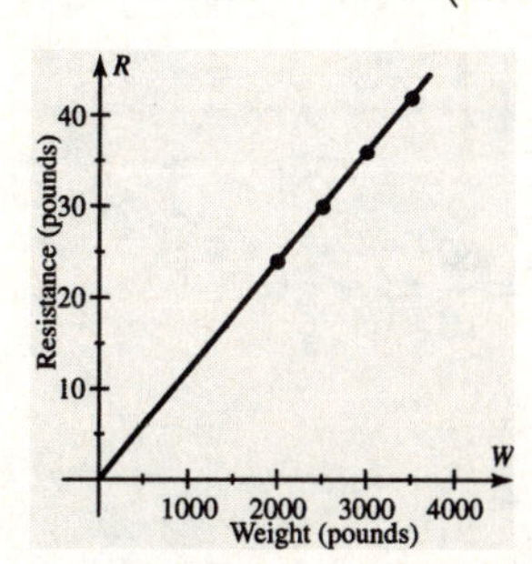

Figure 63

65. (a) Direct, the ratios $\dfrac{G}{A}$ always equal 27.

(b) $\dfrac{G}{A} = 27 \Rightarrow G = 27A$. See Figure 65.

(c) For each square-inch increase in the cross-sectional area of the hose, the flow increases by 27 gal/min.

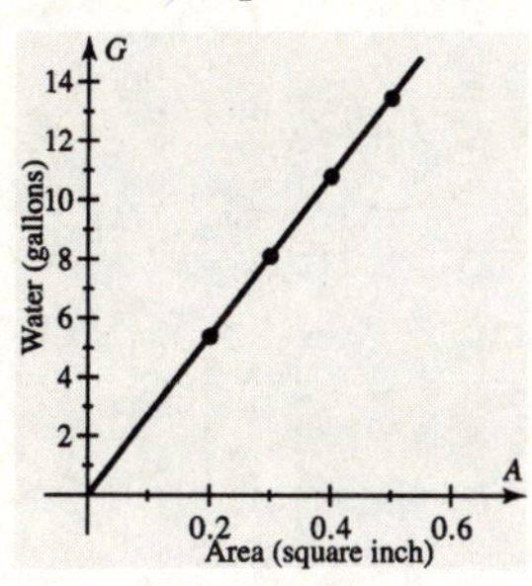

Figure 65

67. (a) $k = FL \Rightarrow k = 120 \cdot 10 \Rightarrow k = 1200 \Rightarrow F = \dfrac{k}{L} \Rightarrow F = \dfrac{1200}{L}$

(b) $L = 15 \Rightarrow F = \dfrac{1200}{15} \Rightarrow F = 80$ pounds

69. (a) Direct

(b) $k = \dfrac{y}{x} \Rightarrow k = \dfrac{-95}{5} \Rightarrow k = -19 \Rightarrow y = kx \Rightarrow y = -19x$

(c) Negative, for each 1-mile increase in altitude the temperature decreases by 19°F.

(d) $y = kx,\ k = -19,\ x = 2.5,\ y = -19 \cdot 2.5 \Rightarrow y = -47.5 \Rightarrow 47.5$°F decrease

71. $\dfrac{30}{3} = \dfrac{18}{x} \Rightarrow 30x = 54 \Rightarrow x = \dfrac{54}{30} \Rightarrow x = 1.8$ ohms

Checking Basic Concepts for Section 7.7

1. (a) $\frac{x}{9}=\frac{2}{5} \Rightarrow 5x=18 \Rightarrow x=\frac{18}{5}$

   (b) $\frac{4}{3}\cdot\frac{5}{b} \Rightarrow 4b=15 \Rightarrow b=\frac{15}{4}$

2. (a) $\frac{4}{6}=\frac{8}{x} \Rightarrow 4x=48 \Rightarrow x=\frac{48}{4} \Rightarrow x=12$

   (b) $\frac{2}{148}=\frac{5}{x} \Rightarrow 2x=740 \Rightarrow x=\frac{740}{2} \Rightarrow x=370$ minutes

3. $k=xy,\ x=15,\ y=4 \Rightarrow k=15\cdot 4 \Rightarrow k=60;\ y=\frac{k}{x},\ k=60,\ x=10 \Rightarrow y=\frac{60}{10} \Rightarrow y=6$

4. (a) Direct, the ratios $\frac{y}{x}$ always equal $\frac{3}{2}, \frac{3}{2}; \frac{x}{y}=\frac{3}{2}, \frac{6}{4}=\frac{3}{2}, \frac{9}{6}=\frac{3}{2}, \frac{12}{8}=\frac{3}{2}$

   (b) Inverse, the products xy always equal 24, 24; $xy=24,\ 2\cdot 12=24,\ 4\cdot 6=24,\ 6\cdot 4=24,\ 8\cdot 3=24$

5. $\frac{272}{17}=\frac{x}{10} \Rightarrow 2720=17x \Rightarrow \frac{2720}{17}=x \Rightarrow x=160 \Rightarrow x=\$160$

Chapter 7 Review Exercises

Section 7.1

1. $x=-2,\ \frac{3}{x-3}=\frac{3}{-2-3}=\frac{3}{-5}=-\frac{3}{5}$

2. $x=3,\ \frac{4x}{5-x^2}=\frac{4(3)}{5-(3)^2}=\frac{12}{5-9}=\frac{12}{-4}=-3$

3. $x=7,\ \frac{-x}{7-x}=\frac{-(7)}{7-(7)}=\frac{-7}{0}=$ undefined

4. $x=2,\ \frac{4x}{x^2-3x+2}=\frac{4(2)}{(2)^2-3(2)+2}=\frac{8}{4-6+2}=\frac{8}{0}=$ undefined

5. $x=-2,\ \frac{3x}{x-1}=\frac{3(-2)}{-2-1}=\frac{-6}{-3}=2;\ x=-1,\ \frac{3x}{x-1}=\frac{3(-1)}{-1-1}=\frac{-3}{-2}=\frac{3}{2};$

   $x=0,\ \frac{3x}{x-1}=\frac{3(0)}{0-1}=\frac{0}{-1}=0;\ x=1,\ \frac{3x}{x-1}=\frac{3(1)}{1-1}=\frac{3}{0}=$ undefined;

   $x=2,\ \frac{3x}{x-1}=\frac{3(2)}{2-1}=\frac{6}{1}=6$. See Figure 5.

x	−2	−1	0	1	2
$\frac{3x}{x-1}$	2	$\frac{3}{2}$	0	—	6

Figure 5

6. $\frac{8}{x^2-4} \Rightarrow x^2-4 \neq 0 \Rightarrow (x-2)(x+2) \neq 0 \Rightarrow x-2=0 \Rightarrow x \neq 2 \text{ or } x+2=0 \Rightarrow x \neq -2$

Therefore $x=2,\ -2$ make the expression undefined.

7. $\frac{25x^3y^4}{15x^5y} = \frac{5}{3}x^{3-5}y^{4-1} = \frac{5}{3}x^{-2}y^3 = \frac{5y^3}{3x^2}$

8. $\frac{x^2-36}{x+6} = \frac{(x+6)(x-6)}{x+6} = \frac{x+6}{x+6} \cdot \frac{x-6}{1} = 1 \cdot (x-6) = x-6$

9. $\frac{x-9}{9-x} = \frac{x-9}{-1(-9+x)} = \frac{x-9}{-1(x-9)} = -1$

10. $\frac{x^2-5x}{5x} = \frac{x(x-5)}{x(5)} = \frac{x}{x} \cdot \frac{x-5}{5} = 1 \cdot \frac{x-5}{5} = \frac{x-5}{5}$

11. $\frac{2x^2+5x-3}{2x^2+x-1} = \frac{(x+3)(2x-1)}{(x+1)(2x-1)} = \frac{x+3}{x+1} \cdot \frac{2x-1}{2x-1} = \frac{x+3}{x+1} \cdot 1 = \frac{x+3}{x+1}$

12. $\frac{3x^2+10x-8}{3x^2+x-2} = \frac{(x+4)(3x-2)}{(x+1)(3x-2)} = \frac{x+4}{x+1} \cdot \frac{3x-2}{3x-2} = \frac{x+4}{x+1} \cdot 1 = \frac{x+4}{x+1}$

Section 7.2

13. $\frac{x-3}{x+1} \cdot \frac{2x+2}{x-3} = \frac{x-3}{x+1} \cdot \frac{2(x+1)}{x-3} = \frac{x-3}{x-3} \cdot \frac{x+1}{x+1} \cdot 2 = 1 \cdot 1 \cdot 2 = 2$

14. $\frac{2x+5}{(x+5)(x-1)} \cdot \frac{x-1}{2x+5} = \frac{2x+5}{2x+5} \cdot \frac{x-1}{x-1} \cdot \frac{1}{x+5} = 1 \cdot 1 \cdot \frac{1}{x+5} = \frac{1}{x+5}$

15. $\frac{z+3}{z-4} \cdot \frac{z-4}{(z+3)^2} = \frac{z+3}{z+3} \cdot \frac{1}{z+3} \cdot \frac{z-4}{z-4} = 1 \cdot \frac{1}{z+3} \cdot 1 = \frac{1}{z+3}$

16. $\frac{x^2}{x^2-4} \cdot \frac{x+2}{x} = \frac{x(x)}{(x-2)(x+2)} \cdot \frac{x+2}{x} = \frac{x}{x} \cdot \frac{x+2}{x+2} \cdot \frac{x}{x-2} = 1 \cdot 1 \cdot \frac{x}{x-2} = \frac{x}{x-2}$

17. $\frac{x+1}{2x} \div \frac{3x+3}{5x} = \frac{x+1}{2x} \cdot \frac{5x}{3(x+1)} = \frac{5}{6} \cdot \frac{x+1}{x+1} \cdot \frac{x}{x} = \frac{5}{6} \cdot 1 \cdot 1 = \frac{5}{6}$

18. $\frac{4}{x^3} \div \frac{x+1}{2x^2} = \frac{4}{x^3} \cdot \frac{2x^2}{x+1} = \frac{8}{x} \cdot \frac{x^2}{x^2} \cdot \frac{1}{x+1} = \frac{8}{x} \cdot 1 \cdot \frac{1}{x+1} = \frac{8}{x(x+1)}$

19. $\frac{x-5}{x+2} \div \frac{2x-10}{x+2} = \frac{x-5}{x+2} \cdot \frac{x+2}{2(x-5)} = \frac{1}{2} \cdot \frac{x-5}{x-5} \cdot \frac{x+2}{x+2} = \frac{1}{2} \cdot 1 \cdot 1 = \frac{1}{2}$

20. $\frac{x^2-6x+5}{x^2-25} \div \frac{x-1}{x+5} = \frac{(x-1)(x-5)}{(x-5)(x+5)} \cdot \frac{x+5}{x-1} = \frac{x-1}{x-1} \cdot \frac{x+5}{x+5} \cdot \frac{x-5}{x-5} = 1 \cdot 1 \cdot 1 = 1$

21. $\frac{x^2-y^2}{x+y} \div \frac{x-y}{x+y} = \frac{(x-y)(x+y)}{x+y} \cdot \frac{x+y}{x-y} = \frac{x-y}{x-y} \cdot \frac{x+y}{x+y} \cdot (x+y) = 1 \cdot 1 \cdot (x+y) = x+y$

22. $\frac{a^3-b^3}{a+b} \div \frac{a-b}{2a+2b} = \frac{(a-b)(a^2+ab+b^2)}{a+b} \cdot \frac{2(a+b)}{a-b} = \frac{a-b}{a-b} \cdot \frac{a+b}{a+b} \cdot \frac{2(a^2+ab+b^2)}{1} =$

$1 \cdot 1 \cdot 2(a^2+ab+b^2) = 2(a^2+ab+b^2)$

Section 7.3

23. $\frac{2}{x+10} + \frac{8}{x+10} = \frac{2+8}{x+10} = \frac{10}{x+10}$

24. $\frac{9}{x-1} - \frac{8}{x-1} = \frac{9-8}{x-1} = \frac{1}{x-1}$

25. $\frac{x+2y}{2x} + \frac{x-2y}{2x} = \frac{(x+2y)+(x-2y)}{2x} = \frac{2x}{2x} = 1$

26. $\frac{x}{x+3} + \frac{3}{x+3} = \frac{x+3}{x+3} = 1$

27. $\frac{x}{x^2-1} - \frac{1}{x^2-1} = \frac{x-1}{x^2-1} = \frac{x-1}{(x-1)(x+1)} = \frac{x-1}{x-1} \cdot \frac{1}{x+1} = 1 \cdot \frac{1}{x+1} = \frac{1}{x+1}$

28. $\frac{2x}{x^2-25} + \frac{10}{x^2-25} = \frac{2x+10}{x^2-25} = \frac{2(x+5)}{(x-5)(x+5)} = \frac{2}{x-5} \cdot \frac{x+5}{x+5} = \frac{2}{x-5} \cdot 1 = \frac{2}{x-5}$

29. $\frac{3}{xy} - \frac{1}{xy} = \frac{3-1}{xy} = \frac{2}{xy}$

30. $\frac{x+y}{2y} + \frac{x-y}{2y} = \frac{x+y+x-y}{2y} = \frac{2x}{2y} = \frac{2}{2} \cdot \frac{x}{y} = 1 \cdot \frac{x}{y} = \frac{x}{y}$

Section 7.4

31. $3x = 3 \cdot x$ and $5x = 5 \cdot x \Rightarrow 3 \cdot 5 \cdot x = 15x$

32. $5x^2 = 5 \cdot x^2$ and $10x = 2 \cdot 5 \cdot x \Rightarrow 2 \cdot 5 \cdot x^2 = 10x^2$

33. x and $x-5$ are both prime $\Rightarrow x(x-5)$

34. $10x^2 = 2 \cdot 5 \cdot x^2$ and $x^2 - x = x(x-1) \Rightarrow 2 \cdot 5 \cdot x^2 \cdot (x-1) = 10x^2(x-1)$

35. $x^2 - 1 = (x+1)(x-1)$ and $(x+1)^2$ is prime $\Rightarrow (x+1)^2(x-1)$

36. $x^2 - 4x = x(x-4)$ and $x^2 - 16 = (x+4)(x-4) \Rightarrow x(x-4)(x+4)$

37. $\frac{3}{8}$, $D = 24 \Rightarrow \frac{3}{8} \cdot \frac{3}{3} = \frac{9}{24}$

38. $\frac{4}{3x}$, $D = 12x \Rightarrow \frac{4}{3x} \cdot \frac{4}{4} = \frac{16}{12x}$

39. $\frac{3x}{x-2}$, $D = x^2 - 4 \Rightarrow \frac{3x}{x-2} \cdot \frac{x+2}{x+2} = \frac{3x^2+6}{x^2-4}$

40. $\frac{2}{x+1}$, $D=x^2+x \Rightarrow \frac{2}{x+1}\cdot\frac{x}{x}=\frac{2x}{x^2+x}$

41. $\frac{3}{5x}$, $D=5x^2-5x \Rightarrow \frac{3}{5x}\cdot\frac{x-1}{x-1}=\frac{3x-3}{5x^2-5x}$

42. $\frac{2x}{2x-3}$, $D=2x^2+x-6 \Rightarrow \frac{2x}{2x-3}\cdot\frac{x+2}{x+2}=\frac{2x^2+4x}{2x^2+x-6}$

43. $\frac{5}{8}+\frac{1}{6}=\frac{5}{8}\cdot\frac{3}{3}+\frac{1}{6}\cdot\frac{4}{4}=\frac{15}{24}+\frac{4}{24}=\frac{19}{24}$

44. $\frac{3}{4x}+\frac{1}{x}=\frac{3}{4x}+\frac{1}{x}\cdot\frac{4}{4}=\frac{3}{4x}+\frac{4}{4x}=\frac{7}{4x}$

45. $\frac{5}{9x}-\frac{2}{3x}=\frac{5}{9x}-\frac{2}{3x}\cdot\frac{3}{3}=\frac{5}{9x}-\frac{6}{9x}=\frac{-1}{9x}=-\frac{1}{9x}$

46. $\frac{7}{x-1}-\frac{3}{x}=\frac{7}{x-1}\cdot\frac{x}{x}-\frac{3}{x}\cdot\frac{x-1}{x-1}=\frac{7x}{x(x-1)}-\frac{3x-3}{x(x-1)}=\frac{4x+3}{x(x-1)}$

47. $\frac{1}{x+1}+\frac{1}{x-1}=\frac{1}{x+1}\cdot\frac{x-1}{x-1}+\frac{1}{x-1}\cdot\frac{x+1}{x+1}=\frac{x-1}{(x+1)(x-1)}+\frac{x+1}{(x+1)(x-1)}=\frac{2x}{(x-1)(x+1)}$

48. $\frac{4}{3x^2}-\frac{3}{2x}=\frac{4}{3x^2}\cdot\frac{2}{2}-\frac{3}{2x}\cdot\frac{3x}{3x}=\frac{8}{6x^2}-\frac{9x}{6x^2}=\frac{8-9x}{6x^2}$

49. $\frac{1+x}{3x}-\frac{3}{2x}=\frac{1+x}{3x}\cdot\frac{2}{2}-\frac{3}{2x}\cdot\frac{3}{3}=\frac{2+2x}{6x}-\frac{9}{6x}=\frac{2x-7}{6x}$

50. $\frac{x}{x^2-1}-\frac{1}{x-1}=\frac{x}{(x+1)(x-1)}-\frac{1}{x-1}\cdot\frac{x+1}{x+1}=\frac{x}{(x+1)(x-1)}-\frac{x+1}{(x+1)(x-1)}=-\frac{1}{(x-1)(x+1)}$

51. $\frac{2}{x-y}-\frac{3}{x+y}=\frac{2}{x-y}\cdot\frac{x+y}{x+y}-\frac{3}{x+y}\cdot\frac{x-y}{x-y}=\frac{2x+2y}{(x-y)(x+y)}-\frac{3x-3y}{(x-y)(x+y)}=\frac{5y-x}{(x-y)(x+y)}$

52. $\frac{2}{x}-\frac{1}{2x}+\frac{2}{3x}=\frac{2}{x}\cdot\frac{6}{6}-\frac{1}{2x}\cdot\frac{3}{3}+\frac{2}{3x}\cdot\frac{2}{2}=\frac{12}{6x}-\frac{3}{6x}+\frac{4}{6x}=\frac{12-3+4}{6x}=\frac{12+1}{6x}=\frac{13}{6x}$

53. $\frac{3}{2y}+\frac{1}{2x}=\frac{3}{2y}\cdot\frac{x}{x}+\frac{1}{2x}\cdot\frac{y}{y}=\frac{3x}{2xy}+\frac{y}{2xy}=\frac{3x+y}{2xy}$

54. $\frac{x}{y-x}+\frac{y}{x-y}=\frac{x}{y-x}+\frac{y}{-1(y-x)}=\frac{x}{y-x}+\frac{-y}{y-x}=\frac{x-y}{y-x}=\frac{x-y}{-1(x-y)}=-1\cdot 1=-1$

Section 7.5

55. $\frac{\frac{3}{4}}{\frac{7}{11}}=\frac{3}{4}\div\frac{7}{11}=\frac{3}{4}\cdot\frac{11}{7}=\frac{33}{28}$

56. $\frac{\frac{x}{5}}{\frac{2x}{7}}=\frac{x}{5}\div\frac{2x}{7}=\frac{x}{5}\cdot\frac{7}{2x}=\frac{7x}{10x}=\frac{7}{10}\cdot\frac{x}{x}=\frac{7}{10}\cdot 1=\frac{7}{10}$

57. $\dfrac{\frac{m}{n}}{\frac{2m}{n^2}}=\dfrac{m}{n}\div\dfrac{2m}{n^2}=\dfrac{m}{n}\cdot\dfrac{n^2}{2m}=\dfrac{m}{2m}\cdot\dfrac{n^2}{n}=\left(\dfrac{1}{2}\cdot\dfrac{m}{m}\right)\cdot\left(\dfrac{n}{n}\cdot n\right)=\left(\dfrac{1}{2}\cdot 1\right)\cdot 1\cdot n=\dfrac{n}{2}$

58. $\dfrac{\frac{3}{p-1}}{\frac{1}{p+1}}=\dfrac{3}{p-1}\div\dfrac{1}{p+1}=\dfrac{3}{p-1}\cdot\dfrac{p+1}{1}=\dfrac{3(p+1)}{p-1}$

59. $\dfrac{\frac{3}{m-1}}{\frac{2m-2}{m+1}}=\dfrac{3}{m-1}\div\dfrac{2m-2}{m+1}=\dfrac{3}{m-1}\cdot\dfrac{m+1}{2m-2}=\dfrac{3}{m-1}\cdot\dfrac{m+1}{2(m-1)}=\dfrac{3(m+1)}{2(m-1)^2}$

60. $\dfrac{\frac{2}{2n+1}}{\frac{8}{2n-1}}=\dfrac{2}{2n+1}\div\dfrac{8}{2n-1}=\dfrac{2}{2n+1}\cdot\dfrac{2n-1}{8}=\dfrac{2}{8}\cdot\dfrac{2n-1}{2n+1}=\dfrac{2n-1}{4(2n+1)}$

61. $\dfrac{\frac{1}{2x}-\frac{1}{3x}}{\frac{2}{3x}-\frac{1}{6x}}=\dfrac{\frac{3}{6x}-\frac{2}{6x}}{\frac{4}{6x}-\frac{1}{6x}}=\dfrac{\frac{1}{6x}}{\frac{3}{6x}}=\dfrac{1}{6x}\div\dfrac{3}{6x}=\dfrac{1}{6x}\cdot\dfrac{6x}{3}=\dfrac{1}{3}\cdot\dfrac{6x}{6x}=\dfrac{1}{3}\cdot 1=\dfrac{1}{3}$

62. $\dfrac{\frac{2}{xy}-\frac{1}{y}}{\frac{2}{xy}+\frac{1}{y}}=\dfrac{\frac{2}{xy}-\frac{1}{y}}{\frac{2}{xy}+\frac{1}{y}}\cdot\dfrac{xy}{xy}=\dfrac{\frac{2xy}{xy}-\frac{xy}{y}}{\frac{2xy}{xy}+\frac{xy}{y}}=\dfrac{2\cdot\frac{xy}{xy}-x\cdot\frac{y}{y}}{2\cdot\frac{xy}{xy}+x\cdot\frac{y}{y}}=\dfrac{2\cdot 1-x\cdot 1}{2\cdot 1+x\cdot 1}=\dfrac{2-x}{2+x}$

63. $\dfrac{\frac{1}{x}-\frac{1}{x+1}}{\frac{x}{x+1}}=\dfrac{\frac{x+1}{x(x+1)}-\frac{x}{x(x+1)}}{\frac{x}{x+1}}=\dfrac{\frac{1}{x(x+1)}}{\frac{x}{x+1}}=\dfrac{1}{x(x+1)}\div\dfrac{x}{x+1}=\dfrac{1}{x(x+1)}\cdot\dfrac{x+1}{x}=\dfrac{1}{x\cdot x}\cdot\dfrac{x+1}{x+1}=\dfrac{1}{x^2}\cdot 1=\dfrac{1}{x^2}$

64. $\dfrac{\frac{2}{x-1}-\frac{1}{x+1}}{\frac{1}{x^2-1}}=\dfrac{\frac{2x+2}{(x-1)(x+1)}-\frac{x-1}{(x-1)(x+1)}}{\frac{1}{x^2-1}}=\dfrac{\frac{x+3}{x^2-1}}{\frac{1}{x^2-1}}=\dfrac{x+3}{x^2-1}\div\dfrac{1}{x^2-1}=\dfrac{x+3}{x^2-1}\cdot\dfrac{x^2-1}{1}=$

$\dfrac{x+3}{1}\cdot\dfrac{x^2-1}{x^2-1}=(x+3)\cdot 1=x+3$

Section 7.6

65. $\dfrac{x}{5}=\dfrac{4}{7}\Rightarrow 7x=20\Rightarrow x=\dfrac{20}{7}$ Check: $\dfrac{\frac{20}{7}}{5}=\dfrac{4}{7}\Rightarrow 7\left(\dfrac{20}{7}\right)=20\Rightarrow 20=20$

66. $\dfrac{4}{x}=\dfrac{3}{2}\Rightarrow 3x=8\Rightarrow x=\dfrac{8}{3}$ Check: $\dfrac{4}{\frac{8}{3}}=\dfrac{3}{2}\Rightarrow 8=3\left(\dfrac{8}{3}\right)\Rightarrow 8=8$

67. $\dfrac{3}{z+1}=\dfrac{1}{2z}\Rightarrow 6z=z+1\Rightarrow 5z=1\Rightarrow z=\dfrac{1}{5}$

Check: $\dfrac{3}{\frac{1}{5}+1}=\dfrac{1}{2\left(\frac{1}{5}\right)}\Rightarrow\dfrac{3}{\frac{6}{5}}=\dfrac{1}{\frac{2}{5}}\Rightarrow\dfrac{3}{1}\cdot\dfrac{5}{6}=\dfrac{1}{1}\cdot\dfrac{5}{2}\Rightarrow\dfrac{15}{6}=\dfrac{5}{2}\Rightarrow\dfrac{5}{2}=\dfrac{5}{2}$

68. $\dfrac{x+2}{x}=\dfrac{3}{5}\Rightarrow 3x=5x+10\Rightarrow -2x=10\Rightarrow x=\dfrac{10}{-2}\Rightarrow x=-5$

Check: $\dfrac{-5+2}{-5}=\dfrac{3}{5}\Rightarrow\dfrac{-3}{-5}=\dfrac{3}{5}\Rightarrow\dfrac{3}{5}=\dfrac{3}{5}$

69. $\frac{1}{x+1}=\frac{2}{x-2}\Rightarrow x-2=2x+2\Rightarrow -x=4\Rightarrow x=-4$

Check: $\frac{1}{-4+1}=\frac{2}{-4-2}\Rightarrow\frac{1}{-3}=\frac{2}{-6}\Rightarrow-\frac{1}{3}=-\frac{1}{3}$

70. $\frac{x}{3}=\frac{-1}{x+4}\Rightarrow x^2+4x=-3\Rightarrow x^2+4x+3=0\Rightarrow(x+1)(x+3)=0\Rightarrow x+1=0\Rightarrow x=-1,$

$x+3=0\Rightarrow x=-3\Rightarrow x=-1,\ -3$

Check: $\frac{-1}{3}=\frac{-1}{-1+4}\Rightarrow-\frac{1}{3}=-\frac{1}{3},\ \frac{-3}{3}=\frac{-1}{-3+4}\Rightarrow-1=\frac{-1}{1}\Rightarrow-1=-1$

71. $\frac{1}{5x}+\frac{3}{5x}=\frac{1}{5}\Rightarrow\frac{4}{5x}=\frac{1}{5}\Rightarrow 5x=20\Rightarrow x=4$ Check: $\frac{1}{5(4)}+\frac{3}{5(4)}=\frac{1}{5}\Rightarrow\frac{4}{20}=\frac{1}{5}\Rightarrow\frac{1}{5}=\frac{1}{5}$

72. $\frac{1}{x-1}+\frac{2x}{x-1}=1\Rightarrow\frac{2x+1}{x-1}=1\Rightarrow\frac{(2x+1)(x-1)}{x-1}=\frac{x-1}{1}\Rightarrow 2x+1=x-1\Rightarrow x=-2$

Check: $\frac{1}{-2-1}+\frac{2(-2)}{-2-1}=1\Rightarrow\frac{-3}{-3}=1\Rightarrow 1=1$

73. $\frac{1}{x}+\frac{2}{3x}=\frac{1}{3}\Rightarrow\frac{1(3)}{3(x)}+\frac{2}{3x}=\frac{1(x)}{3(x)}\Rightarrow\frac{3}{3x}+\frac{2}{3x}=\frac{x}{3x}\Rightarrow\frac{5}{3x}=\frac{x}{3x}\Rightarrow 15x=3x^2\Rightarrow\frac{15}{3}=\frac{x^2}{x}\Rightarrow$

$5=x\Rightarrow x=5$ Check: $\frac{1}{5}+\frac{2}{3(5)}=\frac{1}{3}\Rightarrow\frac{1}{5}+\frac{2}{15}=\frac{1}{3}\Rightarrow\frac{3}{15}+\frac{2}{15}=\frac{5}{15}\Rightarrow\frac{5}{15}=\frac{5}{15}\Rightarrow\frac{1}{3}=\frac{1}{3}$

74. $\frac{1}{x+3}+\frac{2x}{x+3}=\frac{3}{2}\Rightarrow\frac{2x+1}{x+3}=\frac{3}{2}\Rightarrow\frac{(2x+1)(x+3)(2)}{x+3}=\frac{3(x+3)(2)}{2}\Rightarrow$

$\frac{(4x+2)(x+3)}{x+3}=\frac{(3x+9)(2)}{2}\Rightarrow 4x+2=3x+9\Rightarrow x=7$

Check: $\frac{1}{7+3}+\frac{2(7)}{7+3}=\frac{3}{2}\Rightarrow\frac{15}{10}=\frac{3}{2}\Rightarrow\frac{3}{2}=\frac{3}{2}$

75. $\frac{5}{x}-\frac{3}{x+1}=\frac{1}{2}\Rightarrow\frac{5(x)(x+1)(2)}{x}-\frac{3(x)(x+1)(2)}{x+1}=\frac{1(x)(x+1)(2)}{2}\Rightarrow$

$\frac{(10x+10)(x)}{x}-\frac{6x(x+1)}{x+1}=\frac{(x^2+x)(2)}{2}\Rightarrow 10x+10-6x=x^2+x\Rightarrow 4x+10=x^2+x\Rightarrow$

$x^2-3x-10=0\Rightarrow(x-5)(x+2)=0,\ x-5=0\Rightarrow x=5,\ x+2=0\Rightarrow x=-2\Rightarrow x=5,\ -2$

Check: $\frac{5}{-2}-\frac{3}{-2+1}=\frac{1}{2}\Rightarrow\frac{5}{-2}-\frac{3}{-1}=\frac{1}{2}\Rightarrow\frac{5}{-2}-\frac{6}{-2}=\frac{1}{2}\Rightarrow\frac{-1}{-2}=\frac{1}{2}\Rightarrow\frac{1}{2}=\frac{1}{2}$

Check: $\frac{5}{5}-\frac{3}{5+1}=\frac{1}{2}\Rightarrow\frac{5}{5}-\frac{3}{6}=\frac{1}{2}\Rightarrow 1-\frac{1}{2}=\frac{1}{2}\Rightarrow\frac{1}{2}=\frac{1}{2}$

76. $\frac{1}{x-1}-\frac{1}{x+1}=\frac{1}{4}\Rightarrow\frac{1(x-1)(x+1)(4)}{x-1}-\frac{1(x-1)(x+1)(4)}{x+1}=\frac{1(x-1)(x+1)(4)}{4}\Rightarrow$

$\frac{(4x+4)(x-1)}{x-1}-\frac{(4x-4)(x+1)}{x+1}=\frac{(x^2-1)(4)}{4}\Rightarrow 4x+4-(4x-4)=x^2-1\Rightarrow 8=x^2-1\Rightarrow$

$9=x^2\Rightarrow\pm3=x\Rightarrow x=-3,\ 3$

Check: $\frac{1}{-3-1}-\frac{1}{-3+1}=\frac{1}{4}\Rightarrow\frac{1}{-4}-\frac{1}{-2}=\frac{1}{4}\Rightarrow\frac{1}{-4}-\frac{2}{-4}=\frac{1}{4}\Rightarrow\frac{-1}{-4}=\frac{1}{4}\Rightarrow\frac{1}{4}=\frac{1}{4}$

Check: $\frac{1}{3-1}-\frac{1}{3+1}=\frac{1}{4}\Rightarrow\frac{1}{2}-\frac{1}{4}=\frac{1}{4}\Rightarrow\frac{2}{4}-\frac{1}{4}=\frac{1}{4}\Rightarrow\frac{1}{4}=\frac{1}{4}$

77. $\frac{4}{p}-\frac{5}{p+2}=0\Rightarrow\frac{4(p)(p+2)}{p}-\frac{5(p)(p+2)}{p+2}=0(p)(p+2)\Rightarrow\frac{(4p+8)(p)}{p}-\frac{5p(p+2)}{p+2}=0\Rightarrow$

$4p+8-5p=0\Rightarrow -p+8=0\Rightarrow -p=-8\Rightarrow p=8$

Check: $\frac{4}{8}-\frac{5}{8+2}=0\Rightarrow\frac{4}{8}-\frac{5}{10}=0\Rightarrow\frac{1}{2}-\frac{1}{2}=0\Rightarrow 0=0$

78. $\frac{1}{x-3}-\frac{1}{x+3}=\frac{1}{x^2-9}\Rightarrow\frac{1}{x-3}\cdot\frac{x+3}{x+3}-\frac{1}{x+3}\cdot\frac{x-3}{x-3}=\frac{1}{(x+3)(x-3)}\Rightarrow$

$\frac{x+3}{(x-3)(x+3)}-\frac{x-3}{(x-3)(x+3)}=\frac{1}{(x-3)(x+3)}\Rightarrow 6=1,$ which is false, so no solution.

79. $\frac{1}{x+1}=\frac{-x}{x+1}\Rightarrow x+1=-x(x+1)\Rightarrow x+1=-x^2-x$

$\Rightarrow x^2+2x+1=0\Rightarrow(x+1)^2=0\Rightarrow x+1=0,\ x=-1$

Check: -1 is extraneous because $\frac{1}{-1+1}$ is undefined. There are no solutions.

80. $\frac{2}{x}=\frac{2}{x^2+x}-4\Rightarrow\frac{2}{x}=\frac{2-4(x^2+x)}{x^2+x}\Rightarrow\frac{2}{x}=\frac{2-4x^2-4x}{x^2+x}$

$\Rightarrow 2(x^2+x)=x(2-4x^2-4x)\Rightarrow 2x^2+2x=2x-4x^3-4x^2$

$\Rightarrow 4x^3+6x^2=0\Rightarrow 2x^2(2x+3)=0\Rightarrow x=0$ and $2x+3=0,\ x=-\frac{3}{2}$

Check: 0 is extraneous because $\frac{2}{0}$ is undefined.

$\frac{2}{-\frac{3}{2}}=\frac{2}{(-\frac{3}{2})^2+(-\frac{3}{2})}-4\Rightarrow 2\div\left(-\frac{3}{2}\right)=\frac{2}{\frac{9}{4}-\frac{3}{2}}-4\Rightarrow 2\cdot\left(-\frac{2}{3}\right)=\frac{2}{\frac{3}{4}}-4$

$\Rightarrow -\frac{4}{3}=2\div\frac{3}{4}-4\Rightarrow -\frac{4}{3}=2\cdot\frac{4}{3}-4\Rightarrow -\frac{4}{3}=\frac{8}{3}-\frac{12}{3}\Rightarrow -\frac{4}{3}=-\frac{4}{3}$

The solution is $-\frac{3}{2}$.

81. $\frac{2}{x^2-2x}+\frac{1}{x^2-4}=\frac{1}{x^2+2x}\Rightarrow\frac{2(x)(x-2)(x+2)}{x(x-2)}+\frac{1(x)(x-2)(x+2)}{(x-2)(x+2)}=\frac{1(x)(x-2)(x+2)}{x(x+2)}\Rightarrow$

$\frac{(2x+4)(x)(x-2)}{x(x-2)}+\frac{x(x^2-4)}{x^2-4}=\frac{(x-2)(x)(x+2)}{x(x+2)}\Rightarrow 2x+4+x=x-2\Rightarrow 3x+4=x-2\Rightarrow$

$2x=-6\Rightarrow x=-3$ Check: $\frac{2}{(-3)^2-2(-3)}+\frac{1}{(-3)^2-4}=\frac{1}{(-3)^2+2(-3)}\Rightarrow\frac{2}{15}+\frac{1}{5}=\frac{1}{3}\Rightarrow$

$\frac{2}{15}+\frac{3}{15}=\frac{5}{15}\Rightarrow\frac{5}{15}=\frac{5}{15}\Rightarrow\frac{1}{3}=\frac{1}{3}$

82. $\frac{3}{x^2-3x}-\frac{1}{x^2-9}=\frac{1}{x^2+3x}\Rightarrow\frac{3(x)(x-3)(x+3)}{x(x-3)}-\frac{1(x)(x-3)(x+3)}{(x-3)(x+3)}=\frac{1(x)(x-3)(x+3)}{x(x+3)}\Rightarrow$

$\frac{(3x+9)(x)(x-3)}{x(x-3)}-\frac{x(x^2-9)}{x^2-9}=\frac{(x-3)(x)(x+3)}{x(x+3)}\Rightarrow 3x+9-x=x-3\Rightarrow 2x+9=x-3\Rightarrow x=-12$ Check:

$\frac{3}{(-12)^2-3(-12)}-\frac{1}{(-12)^2-9}=\frac{1}{(-12)^2+3(-12)}\Rightarrow\frac{3}{180}-\frac{1}{135}=\frac{1}{108}\Rightarrow$

$\frac{9}{540}-\frac{4}{540}=\frac{5}{540}\Rightarrow\frac{5}{540}=\frac{5}{540}\Rightarrow\frac{1}{108}=\frac{1}{108}$

83. $\frac{1}{x^2}-\frac{5}{x^2+4x}=\frac{1}{x^2+4x}\Rightarrow\frac{1(x^2)(x^2+4x)}{x^2}-\frac{5(x^2)(x^2+4x)}{(x^2+4x)}=\frac{1(x^2)(x^2+4x)}{x^2+4x}\Rightarrow$

$\frac{(x^2+4x)(x^2)}{x^2}-\frac{5x^2(x^2+4x)}{x^2+4x}=\frac{x^2(x^2+4x)}{x^2+4x}\Rightarrow x^2+4x-5x^2=x^2\Rightarrow -4x^2+4x=x^2\Rightarrow$

$-5x^2=-4x\Rightarrow\frac{x^2}{x}=\frac{-4}{-5}\Rightarrow x=\frac{4}{5}$

Check: $\frac{1}{\left(\frac{4}{5}\right)^2}-\frac{5}{\left(\frac{4}{5}\right)^2+4\left(\frac{4}{5}\right)}=\frac{1}{\left(\frac{4}{5}\right)^2+4\left(\frac{4}{5}\right)}\Rightarrow\frac{1}{\frac{16}{25}}-\frac{5}{\frac{96}{25}}=\frac{1}{\frac{96}{25}}\Rightarrow\frac{1}{1}\cdot\frac{25}{16}-\frac{5}{1}\cdot\frac{25}{96}=\frac{1}{1}\cdot\frac{25}{96}\Rightarrow$

$\frac{25}{16}-\frac{125}{96}=\frac{25}{96}\Rightarrow\frac{150}{96}-\frac{125}{96}=\frac{25}{96}\Rightarrow\frac{25}{96}=\frac{25}{96}$

84. $\frac{5}{x^2-1}-\frac{1}{x^2+2x+1}=\frac{3}{x^2-1}\Rightarrow$

$\frac{5(x+1)(x+1)(x-1)}{(x+1)(x-1)}-\frac{1(x+1)(x+1)(x-1)}{(x+1)(x+1)}=\frac{3(x+1)(x+1)(x-1)}{(x+1)(x-1)}\Rightarrow$

$$\frac{(5x+5)(x^2-1)}{(x^2-1)} - \frac{(x-1)(x^2+2x+1)}{x^2+2x+1} = \frac{(3x+3)(x^2-1)}{x^2-1} \Rightarrow 5x+5-(x-1)=3x+3 \Rightarrow$$

$4x+6=3x+3 \Rightarrow x=-3$

Check: $\frac{5}{(-3)^2-1} - \frac{1}{(-3)^2+2(-3)+1} = \frac{3}{(-3)^2-1} \Rightarrow \frac{5}{8}-\frac{1}{4}=\frac{3}{8} \Rightarrow \frac{5}{8}-\frac{2}{8}=\frac{3}{8} \Rightarrow \frac{3}{8}=\frac{3}{8}$

85. $\frac{1}{a}+\frac{2}{b}=\frac{3}{c}$ for $b \Rightarrow \frac{1}{a}\cdot\frac{bc}{bc}+\frac{2}{b}\cdot\frac{ac}{ac}=\frac{3}{c}\cdot\frac{ab}{ab} \Rightarrow bc+2ac=3ab \Rightarrow 2ac=3ab-bc \Rightarrow$

$2ac=b(3a-c) \Rightarrow b=\frac{2ac}{3a-c}$

86. $y=\frac{x}{x-1}$ for $x \Rightarrow y(x-1)=x\cdot 1 \Rightarrow xy-y=x \Rightarrow xy-x=y \Rightarrow x(y-1)=y \Rightarrow x=\frac{y}{y-1}$

Section 7.7

87. $\frac{x}{6}=\frac{1}{3} \Rightarrow 3x=6 \Rightarrow x=\frac{6}{3}=2$

88. $\frac{5}{x}=\frac{7}{3} \Rightarrow 15=7x \Rightarrow \frac{15}{7}=x \Rightarrow x=\frac{15}{7}$

89. (a) $\frac{6}{x}=\frac{13}{20}$

(b) $\frac{6}{x}=\frac{13}{20} \Rightarrow 120=13x \Rightarrow \frac{120}{13}=x \Rightarrow x=\frac{120}{13}$

90. (a) $\frac{341}{11}=\frac{x}{8}$

(b) $\frac{341}{11}=\frac{x}{8} \Rightarrow 11x=2728 \Rightarrow x=\frac{2728}{11} \Rightarrow x=\$248$

91. (a) $k=\frac{y}{x},\ y=8,\ x=2 \Rightarrow k=\frac{8}{2} \Rightarrow k=4$

(b) $y=kx,\ k=4,\ x=5 \Rightarrow y=4\cdot 5 \Rightarrow y=20$

92. (a) $k=\frac{y}{x},\ y=21,\ x=7 \Rightarrow k=\frac{21}{7} \Rightarrow k=3$

(b) $y=kx,\ k=3,\ x=5 \Rightarrow y=3\cdot 5 \Rightarrow y=15$

93. (a) $k=xy,\ x=4,\ y=2.5 \Rightarrow k=4\cdot 2.5 \Rightarrow k=10$

(b) $y=\frac{k}{x},\ k=10,\ x=5 \Rightarrow y=\frac{10}{5} \Rightarrow y=2$

94. (a) $k=xy,\ x=3,\ y=7 \Rightarrow k=3\cdot 7 \Rightarrow k=21$

(b) $y=\frac{k}{x},\ k=21,\ x=5 \Rightarrow y=\frac{21}{5}$

95. (a) Inverse, because as x increases y decreases and $k = xy \Rightarrow 2 \cdot 30 = 3 \cdot 20 = 4 \cdot 15$, etc.

(b) $k = xy,\ x = 2,\ y = 30 \Rightarrow k = 2 \cdot 30 \Rightarrow k = 60;\ y = \frac{k}{x} \Rightarrow y = \frac{60}{x}$

(c) See Figure 95.

96. (a) Direct, because as x increases y also increases and $k = \frac{y}{x} \Rightarrow \frac{6}{2} = \frac{12}{4} = \frac{18}{6}$, etc.

(b) $k = \frac{y}{x},\ x = 2,\ y = 6 \Rightarrow k = \frac{6}{2} \Rightarrow k = 3;\ y = kx \Rightarrow y = 3x$

(c) See Figure 96.

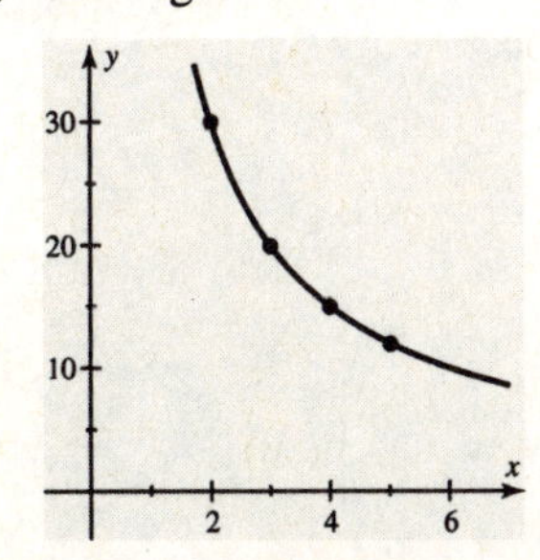

Figure 95

Figure 96

97. Direct, $k = \frac{y}{x},\ y = 2,\ x = 4 \Rightarrow k = \frac{2}{4} \Rightarrow k = \frac{1}{2}$

98. Inverse, $k = xy,\ x = 2,\ y = 6 \Rightarrow k = 2 \cdot 6 \Rightarrow k = 12$

Applications

99. (a) $x = 10,\ T = \frac{1}{15 - x} \Rightarrow T = \frac{1}{15 - 10} = \frac{1}{5} = 0.2$, when the rate of arrival is 10 cars/min., the wait is 0.2 minutes or 12 seconds.

(b) $x = 5,\ T = \frac{1}{15 - x} \Rightarrow T = \frac{1}{15 - 5} = \frac{1}{10};\ x = 10,\ T = \frac{1}{15 - x} \Rightarrow T = \frac{1}{15 - 10} = \frac{1}{5};$

$x = 13,\ T = \frac{1}{15 - x} \Rightarrow T = \frac{1}{15 - 13} = \frac{1}{2};\ x = 14,\ T = \frac{1}{15 - x} \Rightarrow T = \frac{1}{15 - 14} = \frac{1}{1} = 1;$

$x = 14.9,\ T = \frac{1}{15 - x} \Rightarrow T = \frac{1}{15 - 14.9} = \frac{1}{0.1} = 10$; See Figure 99.

(c) It increases dramatically.

x	5	10	13	14	14.9
T	$\frac{1}{10}$	$\frac{1}{5}$	$\frac{1}{2}$	1	10

Figure 99

100. If $r \cdot t = d$, then $50 \cdot t = 150 \Rightarrow t = 3$ and $75t = 150 \Rightarrow t = 2$. The combined $d = 300$ and the combined $t = 5$. Therefore, $r \cdot 5 = 300 \Rightarrow r = 60 \Rightarrow 60$ mph.

101. $\frac{t}{100} + \frac{t}{160} = 1 \Rightarrow \frac{800t}{100} + \frac{800t}{160} = 800 \Rightarrow 8t + 5t = 800 \Rightarrow 13t = 800 \Rightarrow t = \frac{800}{13} \Rightarrow t = 61.5$ hrs.

102. Let x = slower jogger speed and $x+2$ = faster jogger speed, then $\frac{10}{x}-\frac{10}{x+2}=1\Rightarrow$

$10x+20-10x=x^2+2x\Rightarrow 20=x^2+2x\Rightarrow x^2+2x-20=0\Rightarrow(x-10)(x+12)=0\Rightarrow$

$x-10=0\Rightarrow x=10,\ x+12=0\Rightarrow x=-12\Rightarrow x=10,-12.$ Therefore 10 mph and 12 mph.

103. $\frac{16}{x-4}=\frac{48}{x+4}\Rightarrow 16x+64=48x-192\Rightarrow 256=32x\Rightarrow\frac{256}{32}=x\Rightarrow x=8$ mph

104. $\frac{5}{6}=\frac{x}{40}\Rightarrow 6x=200\Rightarrow x=\frac{200}{6}\Rightarrow x\approx 33.3$ ft.

105. $k=xy,\ x=0.25,\ y=400\Rightarrow k=0.25\cdot 400\Rightarrow k=100$

$y=\frac{k}{x},\ x=0.50,\ k=100\Rightarrow y=\frac{100}{0.50}\Rightarrow y=200$ vehicles

106. $\frac{20}{1}=\frac{32}{x}\Rightarrow 20x=32\Rightarrow x=\frac{32}{20}\Rightarrow x=1.6$ in.

107. $k=xy,\ x=12,\ y=30\Rightarrow k=12\cdot 30\Rightarrow k=360$

$y=\frac{k}{x},\ x=10,\ k=360\Rightarrow y=\frac{360}{10}\Rightarrow y=36$ lbs.

108. $k=\frac{y}{x},\ x=17,\ y=612\Rightarrow k=\frac{612}{17}\Rightarrow k=36$

$y=kx,\ k=36,\ x=13\Rightarrow y=36\cdot 13\Rightarrow y=\$468$

109. $30\cdot 25=750$ and $90\cdot 25=2250$, therefore 750 to 2250 sec. or 12.5 to 37.5 min.

110. $y=\frac{k}{x},\ x=18,\ y=900\Rightarrow 900=\frac{k}{18}\Rightarrow 16{,}200=k;$

$y=\frac{16{,}200}{x},\ x=21\Rightarrow y=\frac{16{,}200}{21}\Rightarrow y\approx 771$ lb

Chapter 7 Test

1. $x=3,\ \frac{3x}{2x-1}\Rightarrow\frac{3(3)}{2(3)-1}=\frac{9}{6-1}=\frac{9}{5}$

2. $\frac{x-1}{x+2}$, the equation is undefined when $x+2=0\Rightarrow x=-2$.

3. $\frac{x^2-25}{x-5}=\frac{(x-5)(x+5)}{x-5}=\frac{x-5}{x-5}\cdot(x+5)=1\cdot(x+5)=x+5$

4. $\frac{3x^2-15x}{3x}=\frac{3x(x-5)}{3x}=\frac{3x}{3x}\cdot(x-5)=1\cdot(x-5)=x-5$

5. $\frac{x-2}{x+4}\cdot\frac{3x+12}{x-2}=\frac{x-2}{x-2}\cdot\frac{3(x+4)}{x+4}=1\cdot 1\cdot 3=3$

6. $\frac{z+1}{z+3}\cdot\frac{2z+6}{z+1}=\frac{z+1}{z+1}\cdot\frac{2(z+3)}{z+3}=1\cdot1\cdot2=2$

7. $\frac{x+1}{5x}\div\frac{2x+2}{x-1}=\frac{x+1}{5x}\cdot\frac{x-1}{2x+2}=\frac{x+1}{2(x+1)}\cdot\frac{x-1}{5x}=\frac{x-1}{2(5x)}\cdot1=\frac{x-1}{10x}$

8. $\frac{2}{x^2}\div\frac{x+3}{3x}=\frac{2}{x^2}\cdot\frac{3x}{x+3}=\frac{x}{x}\cdot\frac{2\cdot3}{x(x+3)}=1\cdot\frac{6}{x(x+3)}=\frac{6}{x(x+3)}$

9. $\frac{x}{x+4}+\frac{3x+1}{x+4}=\frac{4x+1}{x+4}$

10. $\frac{4t+1}{2t-3}-\frac{3t-6}{2t-3}=\frac{4t+1-(3t-6)}{2t-3}=\frac{t+7}{2t-3}$

11. $\frac{1}{y^2+y}-\frac{y-1}{y^2-y}=\frac{1}{y(y+1)}\cdot\frac{y-1}{y-1}-\frac{y-1}{y(y-1)}\cdot\frac{y+1}{y+1}=\frac{y-1}{y(y+1)(y-1)}-\frac{y^2-1}{y(y+1)(y-1)}=$

$\frac{y-y^2}{y(y+1)(y-1)}=\frac{-y(y-1)}{y(y+1)(y-1)}=\frac{-1}{y+1}=-\frac{1}{y+1}$

12. $\frac{1}{xy}+\frac{x}{y}-\frac{1}{y^2}=\frac{y}{xy^2}+\frac{x^2y}{xy^2}-\frac{x}{xy^2}=\frac{x^2y-x+y}{xy^2}$

13. $\frac{\frac{a}{3b}}{\frac{5a}{b^2}}=\frac{a}{3b}\div\frac{5a}{b^2}=\frac{a}{3b}\cdot\frac{b^2}{5a}=\frac{a}{a}\cdot\frac{b}{b}\cdot\frac{b}{15}=1\cdot1\cdot\frac{b}{15}=\frac{b}{15}$

14. $\frac{1+\frac{1}{p-1}}{1-\frac{1}{p-1}}=\frac{\frac{p-1}{p-1}+\frac{1}{p-1}}{\frac{p-1}{p-1}-\frac{1}{p-1}}=\frac{\frac{p}{p-1}}{\frac{p-2}{p-1}}=\frac{p}{p-1}\div\frac{p-2}{p-1}=\frac{p}{p-1}\cdot\frac{p-1}{p-2}=\frac{p-1}{p-1}\cdot\frac{p}{p-2}=1\cdot\frac{p}{p-2}=\frac{p}{p-2}$

15. $\frac{2}{7}=\frac{5}{x}\Rightarrow2x=35\Rightarrow x=\frac{35}{2}$

16. $\frac{x+3}{2x}=1\Rightarrow x+3=2x\Rightarrow3=x\Rightarrow x=3$

17. $\frac{1}{2x}+\frac{2}{5x}=\frac{9}{10}=\frac{1\cdot10x}{2x}+\frac{2\cdot10x}{5x}=\frac{9\cdot10x}{10}\Rightarrow5+4=9x\Rightarrow9=9x\Rightarrow x=1$

18. $\frac{1}{x-1}+\frac{2}{x+2}=\frac{3}{2}\Rightarrow\frac{2(x-1)(x+2)}{x-1}+\frac{2(2)(x-1)(x+2)}{x+2}=\frac{3(2)(x-1)(x+2)}{2}\Rightarrow$

$2x+4+4x-4=3x^2+3x-6\Rightarrow6x=3x^2+3x-6\Rightarrow3x^2-3x-6=0\Rightarrow$

$3(x+1)(x-2)=0,\ x+1=0\Rightarrow x=-1,\ x-2=0\Rightarrow x=2\Rightarrow\ x=-1,\ 2$

19. $\frac{1}{x^2-1}-\frac{4}{x+1}=\frac{3}{x-1}\Rightarrow\frac{1}{(x+1)(x-1)}-\frac{4}{x+1}=\frac{3}{x-1}\Rightarrow$

$\frac{1(x+1)(x-1)}{(x+1)(x-1)}-\frac{4(x+1)(x-1)}{x+1}=\frac{3(x+1)(x-1)}{x-1}\Rightarrow1-(4x-4)=3x+3\Rightarrow$

$-4x+5=3x+3 \Rightarrow -7x=-2 \Rightarrow x=\frac{-2}{-7} \Rightarrow x=\frac{2}{7}$

20. $\frac{1}{x^2-4x}+\frac{2}{x^2-16}=\frac{2}{x^2+4x} \Rightarrow \frac{1}{x(x-4)}+\frac{2}{(x-4)(x+4)}=\frac{2}{x(x+4)} \Rightarrow$

$\frac{1(x)(x+4)(x-4)}{x(x-4)}+\frac{2(x)(x+4)(x-4)}{(x-4)(x+4)}=\frac{2(x)(x+4)(x-4)}{x(x+4)} \Rightarrow x+4+2x=2x-8 \Rightarrow$

$3x+4=2x-8 \Rightarrow x=-12$

21. $\frac{x}{2x-1}=\frac{1-x}{2x-1} \Rightarrow x=1-x \Rightarrow x+x=1-x+x \Rightarrow 2x=1 \Rightarrow \frac{2x}{2}=\frac{1}{2} \Rightarrow x=\frac{1}{2}$

However, $x=\frac{1}{2}$ is an extraneous solution because $\frac{x}{2x-1}=\frac{\frac{1}{2}}{2\left(\frac{1}{2}\right)-1}=\frac{\frac{1}{2}}{1-1}=\frac{\frac{1}{2}}{0}$ is undefined.

So there are no solutions.

22. $\frac{x}{x-5}+\frac{x}{x+5}=\frac{10x}{x^2-25} \Rightarrow \frac{x(x+5)}{(x-5)(x+5)}+\frac{x(x-5)}{(x-5)(x+5)}=\frac{10x}{(x-5)(x+5)}$

$\Rightarrow \frac{x(x+5)+x(x-5)}{(x-5)(x+5)}=\frac{10x}{(x-5)(x+5)} \Rightarrow x(x+5)+x(x-5)=10x$

$\Rightarrow x^2+5x+x^2-5x=10x \Rightarrow 2x^2=10x \Rightarrow 2x^2-10x=0$

$\Rightarrow 2x(x-5)=0 \Rightarrow 2x=0,\ x-5=0 \Rightarrow x=0,\ 5$

However, $x=5$ is an extraneous solution because $\frac{x}{x-5}=\frac{5}{5-5}=\frac{5}{0}$ is undefined.

So the solution is $x=0$.

23. $y=\frac{2}{3x-5}$ for x, $y(3x-5)=2 \Rightarrow 3x-5=\frac{2}{y} \Rightarrow 3x=\frac{2}{y}+5 \Rightarrow x=\left(\frac{2}{y}+5\right)\frac{1}{3} \Rightarrow$

$x=\frac{2}{3y}+\frac{5}{3} \Rightarrow x=\frac{2}{3y}+\frac{5y}{3y} \Rightarrow x=\frac{2+5y}{3y}$

24. $\frac{a+b}{ab}=1$ for b, $ab=a+b \Rightarrow ab-b=a \Rightarrow b(a-1)=a \Rightarrow b=\frac{a}{a-1}$

25. (a) $k=\frac{y}{x} \Rightarrow k=\frac{14}{4} \Rightarrow k=\frac{7}{2}$

(b) $y=kx \Rightarrow y=\frac{7}{2}(6) \Rightarrow y=\frac{42}{2} \Rightarrow y=21$

26. Inversely, as x increases y decreases and for all $k=xy$, $k=32$.

27. $\frac{t}{40}+\frac{t}{60}=1 \Rightarrow \frac{120t}{40}+\frac{120t}{60}=120 \Rightarrow 3t+2t=120 \Rightarrow 5t=120 \Rightarrow t=24$ hours

28. $\frac{5}{4}=\frac{x}{54} \Rightarrow 4x=270 \Rightarrow x=67.5$ ft.

29. $N = \dfrac{x^2}{900 - 30x}$ for $x = 24$, $N = \dfrac{24^2}{900 - 30(24)} \Rightarrow N = \dfrac{576}{900 - 720} \Rightarrow N = \dfrac{576}{180} \Rightarrow N = \dfrac{16}{5}$ or 3.2; when the arrival rate is 24 people/hr., there are about 3 people in line.

Chapter 7 Extended and Discovery Exercises

1. (a) $x = 3,\ N = \dfrac{3^2}{225 - 15(3)} = \dfrac{9}{225 - 45} = \dfrac{9}{180} = \dfrac{1}{20} \Rightarrow N = 0.05;$

$x = 9,\ N = \dfrac{9^2}{225 - 15(9)} = \dfrac{81}{225 - 135} = \dfrac{81}{90} = \dfrac{9}{10} \Rightarrow N = 0.9;$

$x = 12,\ N = \dfrac{12^2}{225 - 15(12)} = \dfrac{144}{225 - 180} = \dfrac{144}{45} = \dfrac{16}{5} \Rightarrow N = 3.2;$

$x = 13,\ N = \dfrac{13^2}{225 - 15(13)} = \dfrac{169}{225 - 195} = \dfrac{169}{30} \Rightarrow N = 5.6\overline{3};$

$x = 14,\ N = \dfrac{14^2}{225 - 15(14)} = \dfrac{196}{225 - 210} = \dfrac{196}{15} \Rightarrow N = 13.0\overline{6}$ See Figure 1a.

(b) $x = 15$, because $\dfrac{15^2}{225 - 15(15)} = \dfrac{225}{0}$, which is undefined.

(c) See Figure 1c.

(d) See Figure 1d.

(e) As x approaches 15, the wait increases dramatically.

(f) The formula is only valid for arrival rates under 15 cars per hour, but it can be inferred that the line of cars will continue to grow.

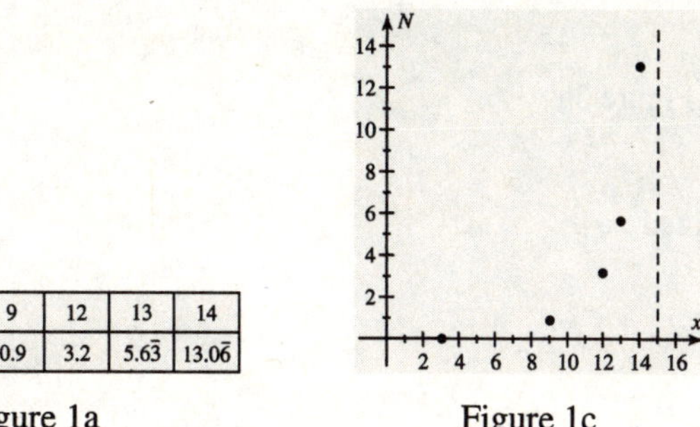

x	3	9	12	13	14
N	0.05	0.9	3.2	$5.6\overline{3}$	$13.0\overline{6}$

Figure 1a

Figure 1c

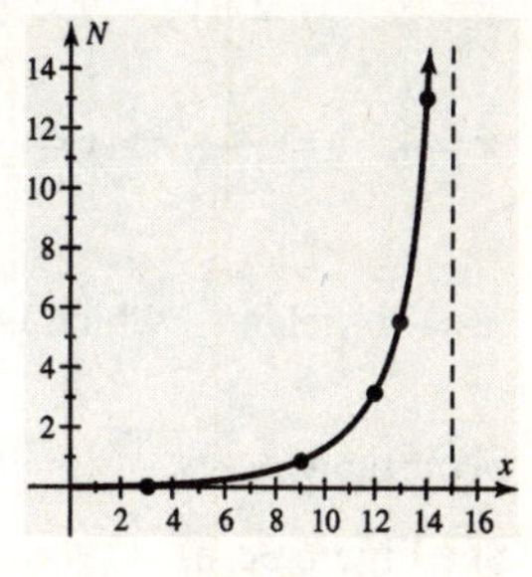

Figure 1d

2. (a) $x = -4,\ y = \dfrac{1}{-4 - 1} \Rightarrow y = -\dfrac{1}{5};\quad x = -3,\ y = \dfrac{1}{-3 - 1} \Rightarrow y = -\dfrac{1}{4};$

$x = -2,\ y = \dfrac{1}{-2 - 1} \Rightarrow y = -\dfrac{1}{3};\quad x = -1,\ y = \dfrac{1}{-1 - 1} \Rightarrow y = -\dfrac{1}{2};$

$x = 0,\ y = \dfrac{1}{0 - 1} \Rightarrow y = -\dfrac{1}{1} = -1;\quad x = 1,\ y = \dfrac{1}{1 - 1} \Rightarrow y = \dfrac{1}{0} \Rightarrow y = \text{undefined};$

$x = 2,\ y = \dfrac{1}{2-1} \Rightarrow y = \dfrac{1}{1} \Rightarrow y = 1;\quad x = 3,\ y = \dfrac{1}{3-1} \Rightarrow y = \dfrac{1}{2};$

$x = 4,\ y = \dfrac{1}{4-1} \Rightarrow y = \dfrac{1}{3};$ See Figure 2a.

(b) $x = 1$ yields $\dfrac{1}{0}$ which is undefined.

(c) See Figure 2c-e.

(d) See Figure 2c-e.

(e) See Figure 2c-e.

x	−4	−3	−2	−1	0	1	2	3	4
y	−0.2	−0.25	$-0.\overline{3}$	−0.5	−1	—	1	0.5	$0.\overline{3}$

Figure 2a

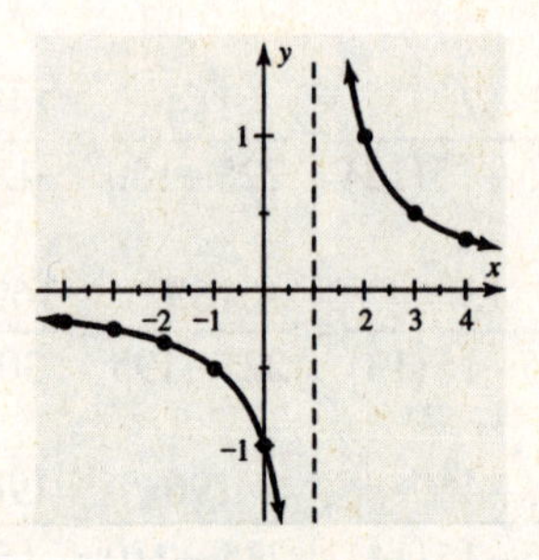

Figure 2c-e

3. (a) $x = -4,\ y = \dfrac{1}{-4+1} \Rightarrow y = -\dfrac{1}{3};\quad x = -3,\ y = \dfrac{1}{-3+1} \Rightarrow y = -\dfrac{1}{2};$

$x = -2,\ y = \dfrac{1}{-2+1} \Rightarrow y = -\dfrac{1}{1} = -1;\quad x = -1,\ y = \dfrac{1}{-1+1} \Rightarrow y = \dfrac{1}{0} \Rightarrow y = \text{undefined};$

$x = 0,\ y = \dfrac{1}{0+1} \Rightarrow y = \dfrac{1}{1} = 1;\quad x = 1,\ y = \dfrac{1}{1+1} \Rightarrow y = \dfrac{1}{2};$

$x = 2,\ y = \dfrac{1}{2+1} \Rightarrow y = \dfrac{1}{3};\quad x = 3,\ y = \dfrac{1}{3+1} \Rightarrow y = \dfrac{1}{4};$

$x = 4,\ y = \dfrac{1}{4+1} \Rightarrow y = \dfrac{1}{5};$ See Figure 3a.

(b) $x = -1$ yields $\dfrac{1}{0}$ which is undefined.

(c) See Figure 3c-e.

(d) See Figure 3c-e.

(e) See Figure 3c-e.

x	−4	−3	−2	−1	0	1	2	3	4
y	$-0.\overline{3}$	−0.5	−1	—	1	0.5	$0.\overline{3}$	0.25	0.2

Figure 3a

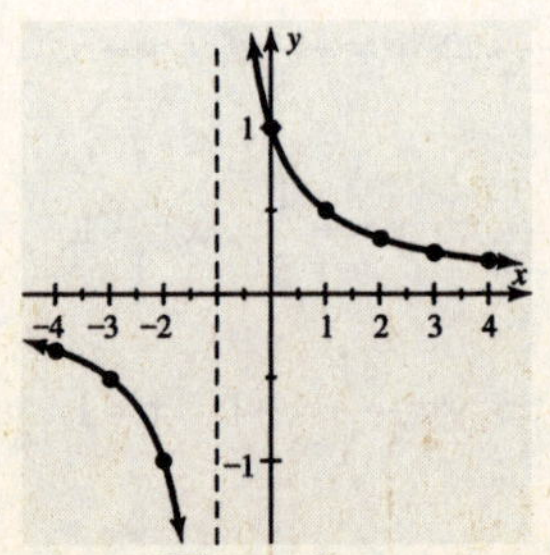

Figure 3c-e

4. (a) $x=-4,\ y=\dfrac{4}{16+1} \Rightarrow y=\dfrac{4}{17};\quad x=-3,\ y=\dfrac{4}{9+1} \Rightarrow y=\dfrac{4}{10};$

$x=-2,\ y=\dfrac{4}{4+1} \Rightarrow y=\dfrac{4}{5};\quad x=-1,\ y=\dfrac{4}{1+1} \Rightarrow y=\dfrac{4}{2} \Rightarrow y=2;$

$x=0,\ y=\dfrac{4}{0+1} \Rightarrow y=\dfrac{4}{1}=4;\quad x=1,\ y=\dfrac{4}{1+1} \Rightarrow y=\dfrac{4}{2}=2;$

$x=2,\ y=\dfrac{4}{4+1} \Rightarrow y=\dfrac{4}{5};\quad x=3,\ y=\dfrac{4}{9+1} \Rightarrow y=\dfrac{4}{10};$

$x=4,\ y=\dfrac{4}{16+1} \Rightarrow y=\dfrac{4}{17};$ See Figure 4a.

(b) No points are undefined.

(c) See Figure 4c-e.

(d) See Figure 4c-e.

(e) See Figure 4c-e.

x	−4	−3	−2	−1	0	1	2	3	4
y	$1.\overline{3}$	1.5	2	—	0	0.5	$0.\overline{6}$	0.75	0.8

Figure 4a

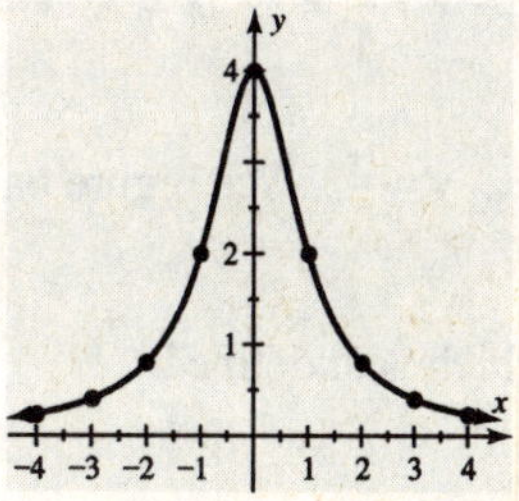

Figure 4c-e

5. (a) $x=-4,\ y=\dfrac{-4}{-4+1} \Rightarrow y=\dfrac{4}{3};\quad x=-3,\ y=\dfrac{-3}{-3+1} \Rightarrow y=\dfrac{3}{2};$

$x=-2,\ y=\dfrac{-2}{-2+1} \Rightarrow y=\dfrac{2}{1}=2;\quad x=-1,\ y=\dfrac{-1}{-1+1} \Rightarrow y=\dfrac{-1}{0} \Rightarrow y=\text{undefined};$

$x=0,\ y=\dfrac{0}{0+1} \Rightarrow y=\dfrac{0}{1}=0;\quad x=1,\ y=\dfrac{1}{1+1} \Rightarrow y=\dfrac{1}{2};$

$x=2,\ y=\dfrac{2}{2+1} \Rightarrow y=\dfrac{2}{3};\quad x=3,\ y=\dfrac{3}{3+1} \Rightarrow y=\dfrac{3}{4};$

$x=4,\ y=\dfrac{4}{4+1} \Rightarrow y=\dfrac{4}{5};$ See Figure 5a.

(b) $x=-1$ yields $\dfrac{-1}{0}$ which is undefined.

(c) See Figure 5c-e.

(d) See Figure 5c-e.

(e) See Figure 5c-e.

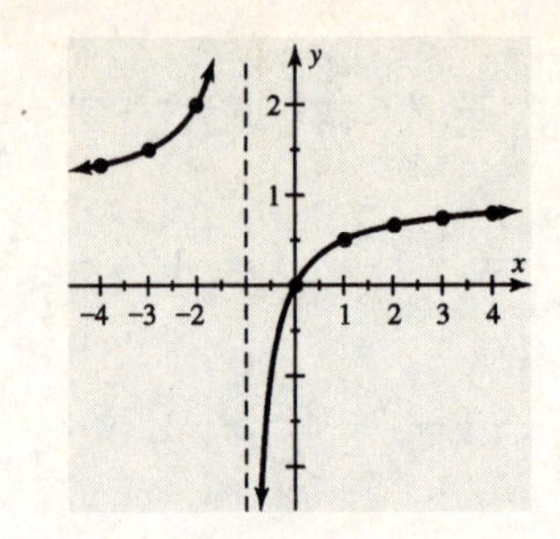

x	−4	−3	−2	−1	0	1	2	3	4
y	$1.\overline{3}$	1.5	2	—	0	0.5	$0.\overline{6}$	0.75	0.8

Figure 5a

Figure 5c-e

6. (a) $x=-4,\ y=\dfrac{-4}{-4-1}\Rightarrow y=\dfrac{4}{5};\quad x=-3,\ y=\dfrac{-3}{-3-1}\Rightarrow y=\dfrac{3}{4};$

$x=-2,\ y=\dfrac{-2}{-2-1}\Rightarrow y=\dfrac{2}{3};\quad x=-1,\ y=\dfrac{-1}{-1-1}\Rightarrow y=\dfrac{1}{2};$

$x=0,\ y=\dfrac{0}{0-1}\Rightarrow y=\dfrac{0}{-1}=0;\quad x=1,\ y=\dfrac{1}{1-1}\Rightarrow y=\dfrac{1}{0}\Rightarrow y=\text{undefined};$

$x=2,\ y=\dfrac{2}{2-1}\Rightarrow y=\dfrac{2}{1}=2;\quad x=3,\ y=\dfrac{3}{3-1}\Rightarrow y=\dfrac{3}{2};$

$x=4,\ y=\dfrac{4}{4-1}\Rightarrow y=\dfrac{4}{3};$ See Figure 6a.

(b) $x=1$ yields $\dfrac{1}{0}$ which is undefined.

(c) See Figure 6c-e.

(d) See Figure 6c-e.

(e) See Figure 6c-e.

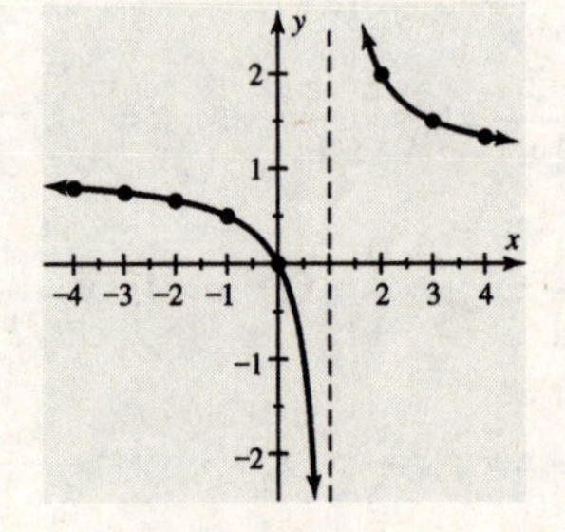

x	−4	−3	−2	−1	0	1	2	3	4
y	0.8	0.75	$0.\overline{6}$	0.5	0	—	2	1.5	$1.\overline{3}$

Figure 6a

Figure 6c-e

Chapters 1–7 Cumulative Review Exercises

1. $\pi r^2 h=\pi(2)^2(6)=\pi(4)(6)=24\pi\approx 75.4$

2. $2x-2$

3. $-\dfrac{5}{8}$

4. $\dfrac{1}{2}\div\dfrac{5}{4}=\dfrac{1}{2}\cdot\dfrac{4}{5}=\dfrac{4}{10}=\dfrac{2}{5}$

5. $\frac{5}{8}+\frac{1}{8}=\frac{6}{8}=\frac{3}{4}$

6. $\frac{4x}{9y}\div\frac{6x}{3y}=\frac{4x}{9y}\cdot\frac{3y}{6x}=\frac{12xy}{54xy}=\frac{2}{9}$

7. $-2+7x+4-5x=2x+2$

8. $-4(4-y)+(5-3y)=-16+4y+5-3y=y-11$

9. $-2x+11=13\Rightarrow -2x=2\Rightarrow x=-1$

   Check: $-2(-1)+11\stackrel{?}{=}13\Rightarrow 2+11\stackrel{?}{=}13\Rightarrow 13=13$, so $x=-1$

10. $0.045=0.045\cdot 100\%=4.5\%$

11. $V=6LW\Rightarrow W=\frac{V}{6L}$

12. $-3x+1\geq x\Rightarrow -4x+1\geq 0\Rightarrow -4x\geq -1\Rightarrow x\leq\frac{1}{4}$

13. See Figure 13.

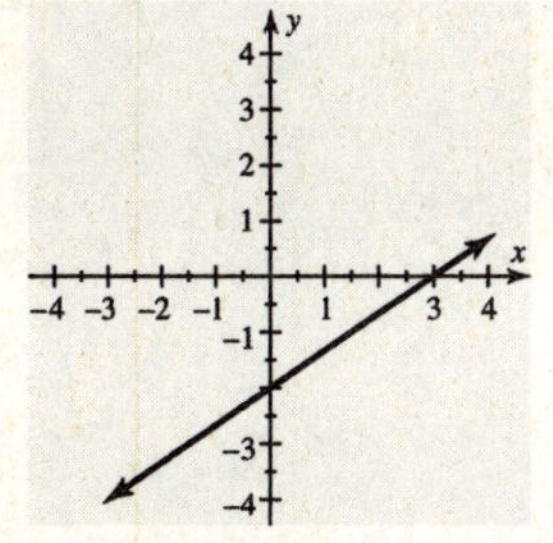

Figure 13

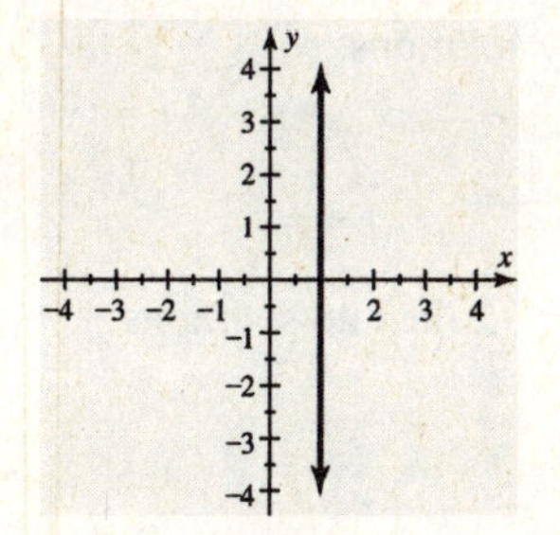

Figure 14

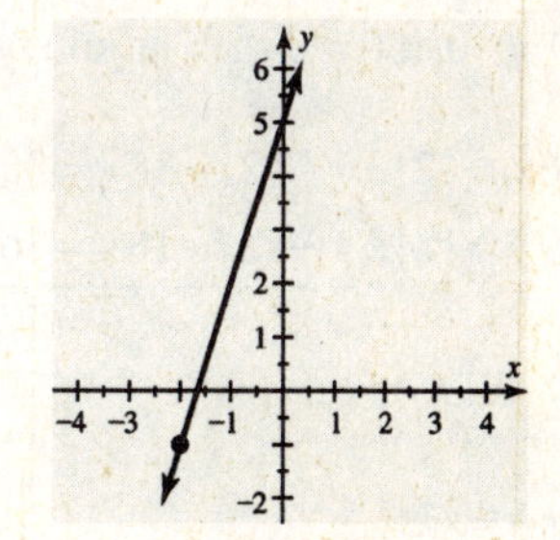

Figure 15

   x-intercept: $2x-3(0)=6\Rightarrow 2x=6\Rightarrow x=3$;

   y-intercept: $2(0)-3y=6\Rightarrow -3y=6\Rightarrow y=-2$

14. See Figure 14.

   x-intercept: $x=1$

   y-intercept: none

15. See Figure 15.

   Using $y=mx+b\Rightarrow y=3x+5$

16. Using $(-2,-5)$ and $(-1,-3)$, $m=\frac{-5-(-3)}{-2-(-1)}=\frac{-2}{-1}=2$, the y-intercept is $(0,-1)$ or $b=-1$, so $y=2x-1$.

17. Parallel lines have equal slopes $\Rightarrow m=-\frac{2}{3}$, now using $(2,-1)$ and

   $y-y_1=m(x-x_1)\Rightarrow y-(-1)=-\frac{2}{3}(x-2)\Rightarrow y+1=-\frac{2}{3}x+\frac{4}{3}\Rightarrow y=-\frac{2}{3}x+\frac{1}{3}$

18. $m=\dfrac{4-2}{2-(-1)}=\dfrac{2}{3}$, using $(2,4)$ and $y-y_1=m(x-x_1)\Rightarrow y-4=\dfrac{2}{3}(x-2)\Rightarrow y-4=\dfrac{2}{3}x-\dfrac{4}{3}\Rightarrow y=\dfrac{2}{3}x+\dfrac{8}{3}$

19. $N=200x+2000$

20. Using $(2,-6)$ for $4x+y=2\Rightarrow 4(2)+(-6)=2\Rightarrow 8+(-6)=2$, Yes; Using $(2,-6)$ for $x-4y=9\Rightarrow 2-4(-6)=9\Rightarrow 2+24=9$, No; Using $(1,-2)$ for $4x+y=2\Rightarrow 4(1)+(-2)=2\Rightarrow 4+(-2)=2$, Yes; Using $(1,-2)$ for $x-4y=9\Rightarrow 1-4(-2)=9\Rightarrow 1+8=9$, Yes; $(1,-2)$ is true for both equations, therefore a solution to the system of equations.

21. When $x=1$ both solutions are 2, therefore $(1,2)$ is a solution to both equations.

22. Using substitution $2r+t=-4\Rightarrow t=-2r-4$, substituting into

$-3r-t=2\Rightarrow -3r-(-2r-4)=2\Rightarrow -r+4=2$

$\Rightarrow -r=-2\Rightarrow r=2$ and $t=-2(2)-4\Rightarrow t=-4-4\Rightarrow t=-8\Rightarrow(2,-8)$ is the solution.

23.
$$\begin{array}{r} 2x-y=\ \ 5 \\ \underline{-2x+y=-5} \\ 0=\ \ 0 \end{array}$$
if $0=0$ then infinitely many solutions.

24.
$$\begin{array}{rr} (4x-6y=12)3= & 12x-18y=36 \\ (-6x+9y=18)2= & \underline{-12x+18y=36} \\ & 0=72 \end{array}$$
if $0=72$ then no solutions.

25.
$$\begin{array}{r} -2x+y=\ \ 0 \\ \underline{-x-y=-3} \\ -3x=-3 \end{array}$$
$\Rightarrow x=1$, then $1+y=3\Rightarrow y=2\Rightarrow(1,2)$ is the solution.

26. See Figure 26.

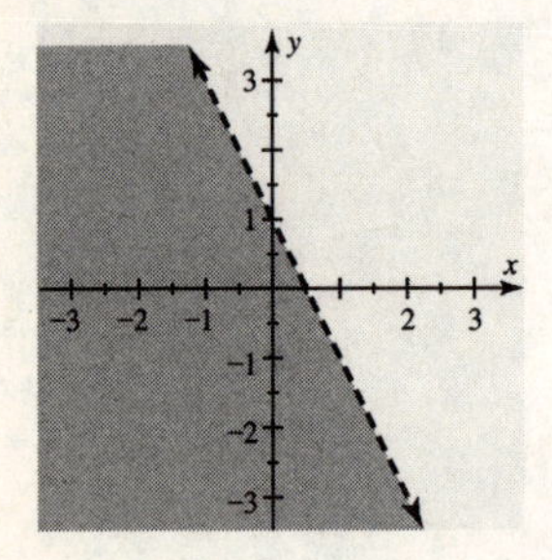

Figure 26

$(0,0)$; *answers may vary.*

27. $(5x^2-3)+(-x^2+4)\Rightarrow 5x^2-3-x^2+4\Rightarrow 4x^2+1$

28. $3z^2\cdot 5z^6=3\cdot 5\cdot z^2\cdot z^6=15z^8$

29. $(ab)^3=a^3b^3$

30. $(2y-3)(5y+2)=10y^2+4y-15y-6=10y^2-11y-6$

31. $\left(x^2-y^2\right)^2=\left(x^2-y^2\right)\left(x^2-y^2\right)=x^4-x^2y^2-x^2y^2+y^4=x^4-2x^2y^2+y^4$

32. $(2t+5)^2=(2t+5)(2t+5)=4t^2+10t+10t+25=4t^2+20t+25$

33. $2^{-4}\cdot 2^5=2^1=2$

34. $\frac{1}{3^{-2}}=\frac{3^2}{1}=9$

35. $\left(3x^2\right)^{-3}=3^{-3}x^{-6}=\frac{1}{3^3}\cdot\frac{1}{x^6}=\frac{1}{27x^6}$

36. $\frac{4x^2}{2x^4}=\frac{4}{2}\cdot\frac{x^2}{x^4}=\frac{2}{1}\cdot\frac{1}{x^2}=\frac{2}{x^2}$

37. $0.00123=1.23\times 10^{-3}$

38.
$$\begin{array}{r|l} & 2x+1+\frac{4}{x-1} \\ x-1 & 2x^2-\ x+3 \\ & \underline{2x^2-2x} \\ & \qquad x+3 \\ & \qquad \underline{x-1} \\ & \qquad\quad 4 \end{array}$$

39. $20z^3=2\cdot2\cdot5\cdot z\cdot z\cdot z$ and $15z^2=3\cdot5\cdot z\cdot z\Rightarrow \text{GCF}=5\cdot z\cdot z=5z^2$; $20z^3-15z^2=5z^2(4z-3)$

40. $12x^2y=2\cdot2\cdot3\cdot x\cdot x\cdot y$ and $15xy^2=3\cdot5\cdot x\cdot y\cdot y\Rightarrow \text{GCF}=3\cdot x\cdot y=3xy; 12x^2y+15xy^2=3xy(4x+5y)$

41. $6+13x-5x^2=(3-x)(2+5x)$ by FOIL

42. $9z^2-4=(3z+2)(3z-2)$, difference of squares

43. $t^2+16t+64=(t+8)(t+8)=(t+8)^2$ by FOIL

44. $x^3-16x=x\left(x^2-16\right)\Rightarrow$ difference of squares, $x(x-4)(x+4)$

45. $x^3+2x^2-99x=x\left(x^2+2x-99\right)\Rightarrow$ FOIL, $x(x+11)(x-9)$

46. $a^2+6ab+9b^2=(a+3b)(a+3b)=(a+3b)^2$ by FOIL

47. $(2x-7)(x+5)=0\Rightarrow 2x-7=0\Rightarrow 2x=7\Rightarrow x=\frac{7}{2}$ or $x+5=0\Rightarrow x=-5$, so $x=-5,\ \frac{7}{2}$

48. $2x^2-4x=0\Rightarrow 2x(x-2)=0$, then $2x=0\Rightarrow x=0$ or $x-2=0\Rightarrow x=2$, so $x=0, 2$

49. $y^2+5y-14=0\Rightarrow(y+7)(y-2)=0$, then $y+7=0\Rightarrow y=-7$ or $y-2=0\Rightarrow y=2$, so $y=-7, 2$

50. $x^3=4x\Rightarrow x^3-4x=0\Rightarrow x\left(x^2-4\right)=0\Rightarrow x(x-2)(x+2)=0$

$\Rightarrow x=0$ or $x-2=0\Rightarrow x=2$ or $x+2=0\Rightarrow x=-2$, so $x=0,\ -2,\ 2$

51. $x=3, \frac{1}{x-2}=\frac{1}{3-2}=\frac{1}{1}=1;\ \frac{1}{x-2}$ is undefined when $x-2=0 \Rightarrow x=2$

52. $\frac{x^2+2x+1}{x^2-1}=\frac{(x+1)(x+1)}{(x-1)(x+1)}=\frac{(x+1)}{(x-1)}\cdot\frac{(x+1)}{(x+1)}=\frac{x+1}{x-1}$

53. $\frac{x}{x+1}+\frac{1}{x+1}=\frac{x+1}{x+1}=1$

54. $\frac{1}{x-1}-\frac{1}{x^2-1}=\frac{1}{x-1}\cdot\frac{x+1}{x+1}-\frac{1}{(x-1)(x+1)}=\frac{x+1}{(x-1)(x+1)}-\frac{1}{(x-1)(x+1)}$

$=\frac{x+1-1}{(x-1)(x+1)}=\frac{x}{x^2-1}$

55. $\frac{3x}{4y}\cdot\frac{y}{9x^2}=\frac{3xy}{36x^2y}=\frac{3}{36}\cdot\frac{x}{x^2}\cdot\frac{y}{y}=\frac{1}{12}\cdot\frac{1}{x}\cdot 1=\frac{1}{12x}$

56. $\frac{x}{x^2-4}\div\frac{2x}{x-2}=\frac{x}{(x-2)(x+2)}\cdot\frac{x-2}{2x}=\frac{x(x-2)}{2x(x-2)(x+2)}$

$=\frac{1}{2}\cdot\frac{x}{x}\cdot\frac{x-2}{x-2}\cdot\frac{1}{x+2}=\frac{1}{2}\cdot 1\cdot 1\cdot\frac{1}{x+2}=\frac{1}{2(x+2)}$

57. $\frac{1+\frac{2}{x}}{1-\frac{2}{x}}\cdot\frac{x}{x}=\frac{x+2}{x-2}$

58. $\frac{5}{x}=\frac{7}{8}\Rightarrow 40=7x\Rightarrow x=\frac{40}{7}$

59. $\frac{4}{3x}-\frac{3}{4x}=1\Rightarrow\frac{4}{3x}\cdot\frac{4}{4}-\frac{3}{4x}\cdot\frac{3}{3}=1\Rightarrow\frac{16}{12x}-\frac{9}{12x}=1$

$\Rightarrow\frac{7}{12x}=1\Rightarrow 7=12x\Rightarrow x=\frac{7}{12}$

60. $\frac{1}{x-1}+\frac{2}{x+2}=\frac{3}{2}\Rightarrow\frac{x+2}{(x-1)(x+2)}+\frac{2(x-1)}{(x-1)(x+2)}=\frac{3}{2}$

$\Rightarrow\frac{x+2+2x-2}{(x-1)(x+2)}=\frac{3}{2}\Rightarrow\frac{3x}{(x-1)(x+2)}=\frac{3}{2}$

$\Rightarrow 6x=3(x-1)(x+2)\Rightarrow 6x=3\left(x^2+x-2\right)\Rightarrow 2x=x^2+x-2$

$\Rightarrow x^2-x-2=0\Rightarrow(x-2)(x+1)=0\Rightarrow x-2=0,\ x=2$ or $x+1=0,\ x=-1,$ so $x=-1,\ 2$

61. $z=3x-2y\Rightarrow z+2y=3x\Rightarrow x=\frac{z+2y}{3}$

62. $y=kx,\ x=14,\ y=7\Rightarrow 7=k\cdot 14\Rightarrow k=\frac{7}{14}=\frac{1}{2};$

$y=\frac{1}{2}x,\ x=11 \Rightarrow y=\frac{1}{2}(11) \Rightarrow y=5.5$

63. $y=\frac{k}{x},\ x=20,\ y=8 \Rightarrow 8=\frac{k}{20} \Rightarrow k=160;$

$y=\frac{160}{x},\ x=2 \Rightarrow y=\frac{160}{2} \Rightarrow y=80$

64. Inverse variation, because as x increases, y decreases and $k=xy \Rightarrow k=1\cdot 20=2\cdot 10=4\cdot 5$, etc.

$k=xy \Rightarrow k=1\cdot 20=20 \Rightarrow y=\frac{k}{x} \Rightarrow y=\frac{20}{x}$

Applications

65. (a) $12x+9x=21x$

(b) $21x=1890 \Rightarrow x=90$ min

66. Let x represent the minutes running and $60-x$ represent the minutes walking.

$10x+4(60-x)=450 \Rightarrow 10x+240-4x=450$

$\Rightarrow 6x=210 \Rightarrow x=35$, so 35 minutes running and 25 minutes walking.

67. (a) $m=\frac{y_2-y_1}{x_2-x_1}$; so $m=\frac{858-376}{2000-1980}=\frac{482}{20}=24.1$

(b) Participation in Head Start increased by 24.1 thousand children per year, on average.

(c) $24.1(10)+858=241+858=1099$ thousand or about 1.1 million

68. (a) $x+x+y=180 \Rightarrow 2x+y=180$ and $2x=y+32 \Rightarrow 2x-y=32$

(b) Using elimination,

$$\begin{array}{r} 2x+y=180 \\ \underline{2x-y=\ \ 32} \\ 4x=212 \Rightarrow x=53 \Rightarrow 53°,\ 53°,\ 74° \end{array}$$

69. (a) $G=-0.036(5)^2+0.76(5)+18.7=-0.036(25)+3.80+18.7$

$=-0.900+3.80+18.7=21.6 \Rightarrow G=21.6$ trillion ft^3 was the natural gas consumption in 1995.

(b) $22.7=-0.036x^2+0.76x+18.7 \Rightarrow 0=-0.036x^2+0.76x-4$

$\Rightarrow 0=-36x^2+760x+4000 \Rightarrow 0=-4\left(9x^2-190x+1000\right)$

factor by grouping: $m\cdot n=9000,\ m+n=-190 \Rightarrow m=-90$ and

$n=-100 \Rightarrow 9x^2-90x-100x+1000=0 \Rightarrow 9x(x-10)-100(x-10)=0 \Rightarrow (9x-100)(x-10)=0$, then

$9x-100=0 \Rightarrow 9x=100 \Rightarrow x=\frac{100}{9} \Rightarrow x=11\frac{1}{9}$ or $x-10=0 \Rightarrow x=10$, so $x=10,\ 11\frac{1}{9}$. Thus,

$x=10 \Rightarrow 2000$ and $x=11\frac{1}{9} \Rightarrow >2001$, therefore 2000.

70. $y=\frac{k}{x},\ x=10,\ y=1100 \Rightarrow 1100=\frac{k}{10} \Rightarrow k=11{,}000;$

$y=\frac{11{,}000}{x},\ x=22 \Rightarrow y=\frac{11{,}000}{22} \Rightarrow y=500$ lb

Critical Thinking Solutions for Chapter 7

Section 7.4

- $x+\frac{1}{x}=\frac{x}{1}+\frac{1}{x}=\frac{x^2}{x}+\frac{1}{x}=\frac{x^2+1}{x}$, then the reciprocal is $\frac{x}{x^2+1}$.

Section 7.5

- No, $\frac{\frac{a}{b}}{\frac{a}{b}+1}=\frac{\frac{a}{b}}{\frac{a}{b}+\frac{b}{b}}=\frac{\frac{a}{b}}{\frac{a+b}{b}}=\frac{a}{b}\div\frac{a+b}{b}=\frac{a}{b}\cdot\frac{b}{a+b}=\frac{a}{a+b}$, which is not equal to $\frac{1}{1+1}$.

  Yes, $\frac{\frac{a}{b}+1}{\frac{a}{b}}=\frac{\frac{a}{b}}{\frac{a}{b}}+\frac{1}{\frac{a}{b}}=1+1\div\frac{a}{b}=1+1\cdot\frac{b}{a}=1+\frac{b}{a}$

Section 7.6

- $\frac{2}{x}=x+1 \Rightarrow 2=x^2+x \Rightarrow x^2+x-2=0 \Rightarrow$

  $(x-1)(x+2)=0 \Rightarrow x-1=0 \Rightarrow x=1$ and $x+2=0 \Rightarrow x=-2$, so $x=1,\ -2$, the same answers from the graph and table.

- $\frac{t}{x}+\frac{t}{y}=1 \Rightarrow \frac{txy}{x}+\frac{txy}{y}=xy \Rightarrow ty+tx=xy \Rightarrow t(x+y)=xy \Rightarrow t=\frac{xy}{x+y}$

Chapter 8: Radical Expressions

8.1: Introduction to Radical Expressions

Concepts

1. square root
2. $3, -3$
3. radicand
4. 2
5. 3
6. 2
7. Irrational
8. cube root
9. $\sqrt[3]{64} = 4$
10. $c^2 = a^2 + b^2$
11. $3^2 + 4^2 = c^2 \Rightarrow 9 + 16 = c^2 \Rightarrow 25 = c^2 \Rightarrow \sqrt{c^2} = \sqrt{25} \Rightarrow c = 5$
12. $\sqrt{(x_2 - x_1)^2 + (y_2 - y_1)^2}$

Square and Cube Roots

13. $\pm\sqrt{4} = \pm 2$
15. $\pm\sqrt{49} = \pm 7$
17. $\pm\sqrt{121} = \pm 11$
19. $\pm\sqrt{400} = \pm 20$
21. $\pm\sqrt{196} = \pm 14$
23. $\pm\sqrt{8} = \pm 2.828$
25. $\pm\sqrt{24} = \pm 4.899$
27. $\sqrt{16} = 4$
29. $\sqrt{4} = 2$
31. $\sqrt{\frac{1}{9}} = \frac{1}{3}$
33. $\sqrt{\frac{9}{4}} = \frac{3}{2}$
35. $\sqrt{0.04} = 0.2$
37. $\sqrt{100} = 10 \Rightarrow$ real, rational

39. $\sqrt{2500} = 50 \Rightarrow$ real, rational

41. $\sqrt{6} = 2.45 \Rightarrow$ real, irrational

43. $\sqrt{150} = 12.25 \Rightarrow$ real, irrational

45. None, no real solution to a square root of a negative number.

47. $\sqrt{144} = 12$

49. $-\sqrt{9} = -(3) = -3$

51. $\sqrt{10} = 3.16$

53. $\sqrt[3]{8} = 2$ because $2^3 = 2 \cdot 2 \cdot 2 = 8$

55. $\sqrt[3]{-125} = -5$ because $(-5)^3 = (-5)(-5)(-5) = -125$

57. $-\sqrt[3]{27} = -(3) = -3$ because $-3^3 = -(3 \cdot 3 \cdot 3) = -(27)$

59. $-\sqrt[3]{-1} = -(-1) = 1$ because $-1^3 = (-1)(-1)(-1) = -1$

61. $\sqrt[3]{1000} = 10$ because $10^3 = 10 \cdot 10 \cdot 10 = 1000$

63. $\sqrt[3]{-343} = -7$ because $(-7)^3 = (-7)(-7)(-7) = -343$

65. $\sqrt[3]{5} = 1.710$ because $1.710^3 = 1.710 \cdot 1.710 \cdot 1.710 = 5$

67. $\sqrt[3]{-16} = -2.520$ because $(-2.520)^3 = (-2.520)(-2.520)(-2.520) = -16$

69. $\sqrt[3]{-9} = -2.080$ because $(-2.080)^3 = (-2.080)(-2.080)(-2.080) = -9$

71. $\sqrt{a^2} = a$ if $a > 0$

Pythagorean Theorem

73. Given $a = 4$ inches, $b = 3$ inches, and using Pythagorean Theorem: $a^2 + b^2 = c^2$, we get $4^2 + 3^2 = c^2 \Rightarrow$

$16 + 9 = c^2 \Rightarrow 25 = c^2 \Rightarrow \sqrt{25} = \sqrt{c^2} \Rightarrow c = 5$ inches.

75. Given $a = 5$ meters, $b = 12$ meters, and using Pythagorean Theorem: $a^2 + b^2 = c^2$, we get $5^2 + 12^2 = c^2 \Rightarrow$

$25 + 144 = c^2 \Rightarrow 169 = c^2 \Rightarrow \sqrt{169} = \sqrt{c^2} \Rightarrow c = 13$ meters.

77. Given $c = 3$ miles, $b = 2$ miles, and using Pythagorean Theorem: $a^2 + b^2 = c^2$, we get $a^2 + 2^2 = 3^2 \Rightarrow$

$a^2 + 4 = 9 \Rightarrow a^2 = 5 \Rightarrow \sqrt{a^2} = \sqrt{5} \Rightarrow a = \sqrt{5}$ miles.

79. Given $a = 6$ feet, $c = 10$ feet, and using Pythagorean Theorem: $a^2 + b^2 = c^2$, we get $6^2 + b^2 = 10^2 \Rightarrow$

$36 + b^2 = 100 \Rightarrow b^2 = 64 \Rightarrow \sqrt{b^2} = \sqrt{64} \Rightarrow b = 8$ feet.

81. Given $a = 5, b = 12$, $c = 13$, and using Pythagorean Theorem: $a^2 + b^2 = c^2$, we get $5^2 + 12^2 = 13^2 \Rightarrow$

$25 + 144 = 169 \Rightarrow 169 = 169$, yes a right triangle.

83. Given $a = 6, b = 4, c = 8$, and using Pythagorean Theorem: $a^2 + b^2 = c^2$, we get $6^2 + 4^2 = 8^2 \Rightarrow$

$36 + 16 = 64 \Rightarrow 52 \neq 64$, no not a right triangle.

85. Given $a = 60, b = 11, c = 61$, and using Pythagorean Theorem: $a^2 + b^2 = c^2$, we get $60^2 + 11^2 = 61^2 \Rightarrow$

$3600 + 121 = 3721 \Rightarrow 3721 = 3721$, yes a right triangle.

The Distance Formula

87. Using $d = \sqrt{(x_2 - x_1)^2 + (y_2 - y_1)^2}$, for $(-3, -1), (1, 2)$, we get $d = \sqrt{(1-(-3))^2 + (2-(-1))^2} \Rightarrow$

$d = \sqrt{(4)^2 + (3)^2} \Rightarrow d = \sqrt{16+9} \Rightarrow d = \sqrt{25} \Rightarrow d = 5$

89. Using $d = \sqrt{(x_2 - x_1)^2 + (y_2 - y_1)^2}$, for $(-20, 30), (20, -20)$, we get

$d = \sqrt{(20-(-20))^2 + (-20-30)^2} \Rightarrow d = \sqrt{(40)^2 + (-50)^2} \Rightarrow d = \sqrt{1600 + 2500} \Rightarrow$

$d = \sqrt{4100} \Rightarrow d = \sqrt{4100} \approx 64.03$

91. Using $d = \sqrt{(x_2 - x_1)^2 + (y_2 - y_1)^2}$, for $(1, 3), (4, 7)$, we get $d = \sqrt{(4-1)^2 + (7-3)^2} \Rightarrow$

$d = \sqrt{(3)^2 + (4)^2} \Rightarrow d = \sqrt{9+16} \Rightarrow d = \sqrt{25} \Rightarrow d = 5.$

93. Using $d = \sqrt{(x_2 - x_1)^2 + (y_2 - y_1)^2}$, for $(4, -4), (-3, 20)$, we get $d = \sqrt{(-3-4)^2 + (20-(-4))^2} \Rightarrow$

$d = \sqrt{(-7)^2 + (24)^2} \Rightarrow d = \sqrt{49+576} \Rightarrow d = \sqrt{625} \Rightarrow d = 25.$

95. Using $d = \sqrt{(x_2 - x_1)^2 + (y_2 - y_1)^2}$, for $(-3, 5), (2, -6)$, we get $d = \sqrt{(2-(-3))^2 + (-6-5)^2} \Rightarrow$

$d = \sqrt{(5)^2 + (-11)^2} \Rightarrow d = \sqrt{25+121} \Rightarrow d = \sqrt{146} \Rightarrow d = \sqrt{146} \approx 12.083.$

97. Using $d = \sqrt{(x_2 - x_1)^2 + (y_2 - y_1)^2}$, for $(-2, -1), (0, -9)$, we get $d = \sqrt{(0-(-2))^2 + (-9-(-1))^2} \Rightarrow$

$d = \sqrt{(2)^2 + (-8)^2} \Rightarrow d = \sqrt{4+64} \Rightarrow d = \sqrt{68} \Rightarrow d = \sqrt{68} \approx 8.246.$

99.

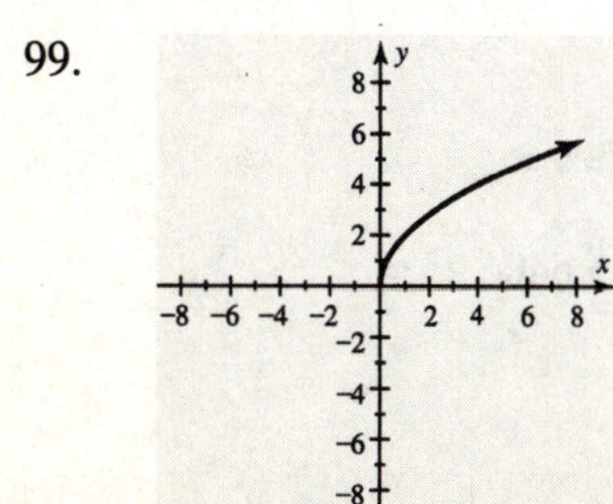

101.

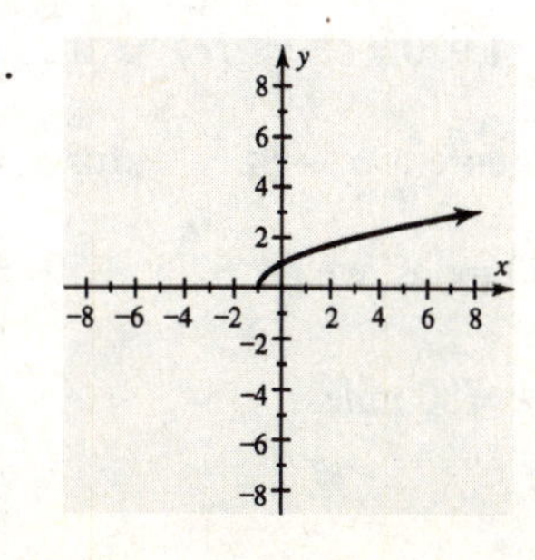

103.

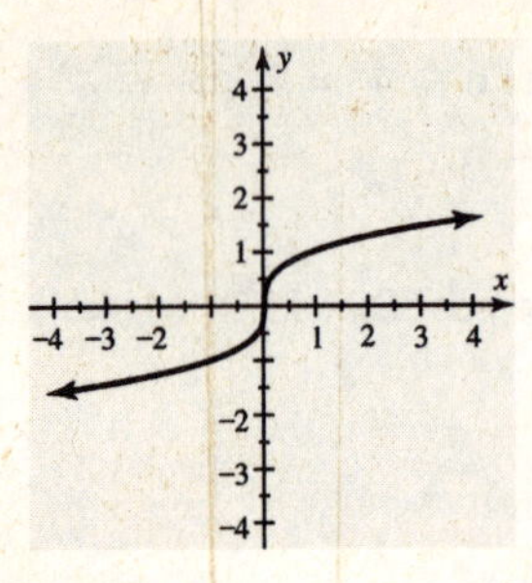

Applications

105. Using Pythagorean Theorem: $a^2+b^2=c^2$, for $b=5$ inches, $c=13$ inches, we get $a^2+5^2=13^2 \Rightarrow$

$a^2+25=169 \Rightarrow a^2=144 \Rightarrow \sqrt{a^2}=\sqrt{144} \Rightarrow a=12$ inches. Therefore the perimeter is $a+b+c \Rightarrow$

$12+5+13=30$ inches.

107. Using Pythagorean Theorem: $a^2+b^2=c^2$, for $a=9$ feet, $b=12$ feet, we get $9^2+12^2=c^2 \Rightarrow$

$81+144=c^2 \Rightarrow 225=c^2 \Rightarrow \sqrt{225}=\sqrt{c^2} \Rightarrow c=15$ feet. Therefore the diagonal is 15 feet.

109. (a) For $W=400$, $R=\dfrac{885}{\sqrt{W}} \Rightarrow R=\dfrac{885}{\sqrt{400}} \Rightarrow R=\dfrac{885}{20} \approx 44$ bpm

(b) For $W=25$, $R=\dfrac{885}{\sqrt{W}} \Rightarrow R=\dfrac{885}{\sqrt{25}} \Rightarrow R=\dfrac{885}{5}=177$ bpm

(c) For $w=169$, $R=\dfrac{885}{\sqrt{w}} \Rightarrow R=\dfrac{885}{\sqrt{169}} \Rightarrow R=\dfrac{885}{13} \approx 68$ bpm

111. Using Pythagorean Theorem: $a^2+b^2=c^2$, for $a=20$ inches, $b=15$ inches, we get $20^2+15^2=c^2 \Rightarrow$

$400+225=c^2 \Rightarrow 625=c^2 \Rightarrow \sqrt{c^2}=\sqrt{625} \Rightarrow c=25$ inches.

113. Using Pythagorean Theorem: $a^2+b^2=c^2$, for $a=15$ feet, $c=40$ feet, we get $15^2+h^2=40^2 \Rightarrow$

$225+h^2=1600 \Rightarrow h^2=1375 \Rightarrow \sqrt{h^2}=\sqrt{1375} \Rightarrow h=\sqrt{1375}=37.1$ feet.

115. Using Pythagorean Theorem: $a^2+b^2=c^2$, for $a=52$ feet, $b=96$ feet, we get $52^2+96^2=c^2 \Rightarrow$

$2704+9216=c^2 \Rightarrow 11{,}920=c^2 \Rightarrow \sqrt{c^2}=\sqrt{11{,}920} \Rightarrow c=109.2$ feet.

117. Using Pythagorean Theorem: $a^2+b^2=c^2$, for $a=25\cdot 2.5$ miles, $b=20\cdot 2.5$ miles $\Rightarrow$

$a=62.5$ miles, $b=50$ miles, we get $62.5^2+50^2=c^2 \Rightarrow 3906.25+2500=c^2 \Rightarrow 6406.25=c^2 \Rightarrow$

$\sqrt{c^2}=\sqrt{6406.25} \Rightarrow c=80$ miles.

119. The height h is one leg of a right triangle with hypotenuse a and base $\frac{a}{2}$. Using Pythagorean Theorem:

$$a^2+b^2=c^2 \text{ we get } h^2+\left(\frac{a}{2}\right)^2=a^2 \Rightarrow h^2=a^2-\frac{a^2}{4} \Rightarrow h^2=\frac{4a^2}{4}-\frac{a^2}{4} \Rightarrow h^2=\frac{3a^2}{4} \Rightarrow$$

$$\sqrt{h^2}=\sqrt{\frac{3a^2}{4}} \Rightarrow h=\frac{a\sqrt{3}}{2}$$

8.2: Multiplication and Division of Radical Expressions

Concepts

1. $\sqrt{2}\cdot\sqrt{32} \Rightarrow \sqrt{2\cdot 32} \Rightarrow \sqrt{64} \Rightarrow 8$

2. $\sqrt{a}\cdot\sqrt{b} \Rightarrow \sqrt{a\cdot b} \Rightarrow \sqrt{ab}$

3. $\sqrt{25}=5$, yes

4. 16

5. $\sqrt{8} \Rightarrow \sqrt{4}\cdot\sqrt{2} \Rightarrow 2\sqrt{2}$

6. x

7. $\frac{\sqrt{32}}{\sqrt{2}} \Rightarrow \frac{\sqrt{16}\cdot\sqrt{2}}{\sqrt{2}} \Rightarrow 4\cdot\frac{\sqrt{2}}{\sqrt{2}} \Rightarrow 4\cdot 1 \Rightarrow 4$

8. $\sqrt{\frac{a}{b}} \Rightarrow \frac{\sqrt{a}}{\sqrt{b}}$

9. $\sqrt{b^4}=\sqrt{\left(b^2\right)^2}=b^2$

10. $\sqrt{x^3}=\sqrt{x^2\cdot x}=\sqrt{x^2}\cdot\sqrt{x}=x\sqrt{x}$

Multiplying Square Roots

11. $\sqrt{5}\cdot\sqrt{5}=\sqrt{5\cdot 5}=\sqrt{25}=5$

13. $\sqrt{3}\cdot\sqrt{7}=\sqrt{3\cdot 7}=\sqrt{21}$

15. $\sqrt{\frac{1}{2}}\cdot\sqrt{\frac{1}{2}}=\sqrt{\frac{1}{2}\cdot\frac{1}{2}}=\sqrt{\frac{1}{4}}=\frac{1}{2}$

17. $\sqrt{\frac{3}{5}}\cdot\sqrt{\frac{2}{7}}=\sqrt{\frac{3}{5}\cdot\frac{2}{7}}=\sqrt{\frac{6}{35}}$

19. $\sqrt{2}\cdot\sqrt{32}=\sqrt{2\cdot 32}=\sqrt{64}=8$

21. $\sqrt{48}\cdot\sqrt{3}=\sqrt{48\cdot 3}=\sqrt{144}=12$

23. $\sqrt{5x}\cdot\sqrt{3}=\sqrt{5x\cdot 3}=\sqrt{15x}$

25. $\sqrt{a}\cdot\sqrt{5b}=\sqrt{a\cdot 5b}=\sqrt{5ab}$

27. $\sqrt{7x} \cdot \sqrt{3z} = \sqrt{7x \cdot 3z} = \sqrt{21xz}$

29. $\sqrt{33} = \sqrt{3 \cdot 11} = \sqrt{3} \cdot \sqrt{11}$

31. $\sqrt{85} = \sqrt{5 \cdot 17} = \sqrt{5} \cdot \sqrt{17}$

33. $\sqrt{7z} = \sqrt{7} \cdot \sqrt{z}$

35. $\sqrt{30} = \sqrt{5 \cdot 6} = \sqrt{5} \cdot \sqrt{6}$ *Answers may vary*

37. 4, because $4 \cdot 5 = 20$

39. 144, because $144 \cdot 1 = 144$

41. 100, because $100 \cdot 2 = 200$

43. 64, because $64 \cdot 2 = 128$

45. $\sqrt{75} = \_\_\sqrt{3} \Rightarrow \sqrt{75} = \underline{\sqrt{25}} \cdot \sqrt{3} \Rightarrow \sqrt{75} = \underline{5}\sqrt{3}$; therefore 5

47. $\sqrt{24} = \_\_\sqrt{6} \Rightarrow \sqrt{24} = \underline{\sqrt{4}} \cdot \sqrt{6} \Rightarrow \sqrt{24} = \underline{2}\sqrt{6}$; therefore 2

49. $\sqrt{28} = \_\_\sqrt{7} \Rightarrow \sqrt{28} = \underline{\sqrt{4}} \cdot \sqrt{7} \Rightarrow \sqrt{28} = \underline{2}\sqrt{7}$; therefore 2

51. $\sqrt{20} = \sqrt{4} \cdot \sqrt{5} = 2\sqrt{5}$

53. $\sqrt{63} = \sqrt{9} \cdot \sqrt{7} = 3\sqrt{7}$

55. $\sqrt{12} = \sqrt{4} \cdot \sqrt{3} = 2\sqrt{3}$

57. $\sqrt{300} = \sqrt{100} \cdot \sqrt{3} = 10\sqrt{3}$

59. $\sqrt{180} = \sqrt{36} \cdot \sqrt{5} = 6\sqrt{5}$

61. $\sqrt{t^3} = \sqrt{t^2} \cdot \sqrt{t} = t\sqrt{t}$

63. $\sqrt{4a^2} = \sqrt{4} \cdot \sqrt{a^2} = 2a$

65. $\sqrt{x^4} = \sqrt{x^2 \cdot x^2} = \sqrt{x^2} \cdot \sqrt{x^2} = x \cdot x = x^2$

67. $\sqrt{8x^2} = \sqrt{4x^2} \cdot \sqrt{2} = 2x\sqrt{2}$

69. $\sqrt{125t^3} = \sqrt{25t^2} \cdot \sqrt{5t} = 5t\sqrt{5t}$

71. $\sqrt{x^2y^4} = \sqrt{x^2 \cdot y^2 \cdot y^2} = \sqrt{x^2} \cdot \sqrt{y^2} \cdot \sqrt{y^2} = x \cdot y \cdot y = xy^2$

73. $\sqrt{16x^2y^3} = \sqrt{16x^2y^2} \cdot \sqrt{y} = 4xy\sqrt{y}$

75. $\sqrt{\frac{4}{49}} = \frac{\sqrt{4}}{\sqrt{49}} = \frac{2}{7}$

77. $\sqrt{\frac{16}{81}} = \frac{\sqrt{16}}{\sqrt{81}} = \frac{4}{9}$

79. $\sqrt{\frac{17}{36}} = \frac{\sqrt{17}}{\sqrt{36}} = \frac{\sqrt{17}}{6}$

81. $\sqrt{\frac{0.01}{0.49}} = \frac{\sqrt{0.01}}{\sqrt{0.49}} = \frac{0.1}{0.7} = \frac{1}{7}$

83. $\frac{\sqrt{72}}{\sqrt{2}} = \frac{\sqrt{36} \cdot \sqrt{2}}{\sqrt{2}} = 6 \cdot \frac{\sqrt{2}}{\sqrt{2}} = 6 \cdot 1 = 6$

85. $\frac{\sqrt{75}}{\sqrt{3}} = \frac{\sqrt{25} \cdot \sqrt{3}}{\sqrt{3}} = 5 \cdot \frac{\sqrt{3}}{\sqrt{3}} = 5 \cdot 1 = 5$

87. $\sqrt{5} \cdot \frac{\sqrt{5}}{\sqrt{9}} = \frac{\sqrt{5 \cdot 5}}{\sqrt{9}} = \frac{\sqrt{25}}{\sqrt{9}} = \frac{5}{3}$

89. $\sqrt{\frac{2}{3}} \cdot \frac{\sqrt{2}}{\sqrt{3}} = \frac{\sqrt{2}}{\sqrt{3}} \cdot \frac{\sqrt{2}}{\sqrt{3}} = \frac{\sqrt{2 \cdot 2}}{\sqrt{3 \cdot 3}} = \frac{\sqrt{4}}{\sqrt{9}} = \frac{2}{3}$

91. $\sqrt{\frac{x^2}{9}} = \frac{\sqrt{x^2}}{\sqrt{9}} = \frac{x}{3}$

93. $\sqrt{\frac{2y^2}{49}} = \frac{\sqrt{2y^2}}{\sqrt{49}} = \frac{y\sqrt{2}}{7}$

95. $\sqrt{\frac{20}{y^4}} = \frac{\sqrt{20}}{\sqrt{y^4}} = \frac{\sqrt{4} \cdot \sqrt{5}}{\sqrt{y^4}} = \frac{2\sqrt{5}}{y^2}$

97. $\sqrt{2xy} \cdot \sqrt{2xy} = \sqrt{2xy \cdot 2xy} = \sqrt{4x^2y^2} = 2xy$

99. $\sqrt{2xy} \cdot \sqrt{8xy^3} = \sqrt{2xy \cdot 8xy^3} = \sqrt{16x^2y^4} = 4xy^2$

101. $\sqrt{\frac{2}{x^2y^2}} = \frac{\sqrt{2}}{\sqrt{x^2y^2}} = \frac{\sqrt{2}}{xy}$

103. $\frac{\sqrt{75x^3}}{\sqrt{3x}} = \frac{\sqrt{25x^2} \cdot \sqrt{3x}}{\sqrt{3x}} = 5x \cdot \frac{\sqrt{3x}}{\sqrt{3x}} = 5x \cdot 1 = 5x$

105. $\frac{\sqrt{50xy}}{\sqrt{2xy}} = \frac{\sqrt{25} \cdot \sqrt{2xy}}{\sqrt{2xy}} = 5 \cdot \frac{\sqrt{2xy}}{\sqrt{2xy}} = 5 \cdot 1 = 5$

107. $\frac{\sqrt{2x}}{\sqrt{3y}} \cdot \frac{\sqrt{32xy^2}}{\sqrt{3x^2y}} = \frac{\sqrt{2x \cdot 32xy^2}}{\sqrt{3y \cdot 3x^2y}} = \frac{\sqrt{64x^2y^2}}{\sqrt{9x^2y^2}} = \frac{8xy}{3xy} = \frac{8}{3} \cdot \frac{xy}{xy} = \frac{8}{3} \cdot 1 = \frac{8}{3}$

109. $\sqrt{\frac{32}{xy}} \cdot \sqrt{\frac{x^3y}{2}} = \frac{\sqrt{32 \cdot x^3y}}{\sqrt{xy \cdot 2}} = \frac{\sqrt{32x^3y}}{\sqrt{2xy}} = \frac{\sqrt{16x^2} \cdot \sqrt{2xy}}{\sqrt{2xy}} = 4x \cdot \frac{\sqrt{2xy}}{\sqrt{2xy}} = 4x \cdot 1 = 4x$

111. $\sqrt{x^{2n}} = \sqrt{\left(x^n\right)^2} = x^n$

Applications

113. (a) If $h = 81$ and $T = \sqrt{\frac{h}{4}}$, then $T = \sqrt{\frac{81}{4}} \Rightarrow T = \frac{\sqrt{81}}{\sqrt{4}} \Rightarrow T = \frac{9}{2}$ seconds

(b) $T = \sqrt{\frac{h}{4}} \Rightarrow T = \frac{\sqrt{h}}{\sqrt{4}} \Rightarrow T = \frac{\sqrt{h}}{2}$

115. (a) $r = \sqrt{\frac{A}{\pi}}, A = 10 \Rightarrow r = \sqrt{\frac{10}{\pi}} \approx 1.78$ in.

(b) $r = \sqrt{\frac{A}{\pi}} \Rightarrow r = \frac{\sqrt{A}}{\sqrt{\pi}}$

117. (a) If $v = 40$mph, $D = 225$ feet, $d = 100$ feet, and $V = \sqrt{\frac{v^2 D}{d}}$, then $V = \sqrt{\frac{40^2 \cdot 225}{100}} \Rightarrow$

$V = \frac{\sqrt{1600 \cdot 225}}{\sqrt{100}} \Rightarrow V = \frac{40 \cdot 15}{10} \Rightarrow V = \frac{600}{10} \Rightarrow V = 60$mph.

(b) $V = \sqrt{\frac{v^2 D}{d}} \Rightarrow V = \sqrt{v^2} \cdot \sqrt{\frac{D}{d}} \Rightarrow V = v\sqrt{\frac{D}{d}}$

Checking Basic Concepts for Sections 8.1 & 8.2

1. (a) $\sqrt{81} = 9$

(b) $\pm\sqrt{625} = \pm 25$

2. $\sqrt{7} = 2.646$

3. (a) $\sqrt[3]{8} = 2$ because $2^3 = 2 \cdot 2 \cdot 2 = 8$

(b) $-\sqrt[3]{-64} = -(-4) = 4$ because $(-4)^3 = (-4) \cdot (-4) \cdot (-4) = -64$

(c) $\sqrt[3]{-27} = -3$ because $(-3)^3 = (-3) \cdot (-3) \cdot (-3) = -27$

4. Using the distance formula: $d = \sqrt{(x_2 - x_1)^2 + (y_2 - y_1)^2}$, for $(-3, 4), (5, -2)$, we get

$d = \sqrt{(5-(-3))^2 + (-2-4)^2} \Rightarrow d = \sqrt{(8)^2 + (-6)^2} \Rightarrow d = \sqrt{64+36} \Rightarrow d = \sqrt{100} \Rightarrow d = 10.$

5. (a) $\sqrt{80} = \sqrt{16} \cdot \sqrt{5} = 4\sqrt{5}$

(b) $\sqrt{16x^3} = \sqrt{16x^2} \cdot \sqrt{x} = 4x\sqrt{x}$

(c) $\sqrt{20x^2y^2} = \sqrt{4x^2y^2} \cdot \sqrt{5} = 2xy\sqrt{5}$

6. (a) $\sqrt{8} \cdot \sqrt{2} = \sqrt{8 \cdot 2} = \sqrt{16} = 4$

(b) $\sqrt{\frac{a^3}{100}} = \frac{\sqrt{a^3}}{\sqrt{100}} = \frac{\sqrt{a^2} \cdot \sqrt{a}}{\sqrt{100}} = \frac{a\sqrt{a}}{10}$

(c) $\sqrt{\frac{8}{x^3}} \cdot \sqrt{\frac{x}{2}} = \frac{\sqrt{8 \cdot x}}{\sqrt{x^3 \cdot 2}} = \frac{\sqrt{8x}}{\sqrt{2x^3}} = \frac{\sqrt{4} \cdot \sqrt{2x}}{\sqrt{x^2} \cdot \sqrt{2x}} = \frac{2}{x} \cdot 1 = \frac{2}{x}$

7. Using Pythagorean Theorem: $a^2 + b^2 = c^2$, for $b = 7$ feet, $c = 25$ feet, we get $a^2 + 7^2 = 25^2 \Rightarrow$

$a^2 + 49 = 625 \Rightarrow a^2 = 576 \Rightarrow \sqrt{a^2} = \sqrt{576} \Rightarrow a = 24$ feet.

8. Using Pythagorean Theorem: $a^2 + b^2 = c^2$, for $a = 20 \cdot 3.5$ miles, $b = 25 \cdot 3.5$ miles $\Rightarrow$

a = 70 miles, b = 87.5 miles, we get $70^2 + 87.5^2 = c^2 \Rightarrow 4900 + 7656.25 = c^2 \Rightarrow 12556.25 = c^2 \Rightarrow$

$\sqrt{c^2} = \sqrt{12556.25} \Rightarrow c \approx 112$ miles.

8.3: Addition and Subtraction of Radical Expressions

Concepts

1. $\sqrt{2} + \sqrt{2} = (1+1)\sqrt{2} = 2\sqrt{2}$

2. $3\sqrt{b} - 2\sqrt{b} = (3-2)\sqrt{b} = \sqrt{b}$

3. $3\sqrt[3]{b} - 2\sqrt[3]{b} = (3-2)\sqrt[3]{b} = \sqrt[3]{b}$

4. $3\sqrt[3]{b} + 2\sqrt[3]{b} = (3+2)\sqrt[3]{b} = 5\sqrt[3]{b}$

5. Yes, it simplifies to $3\sqrt{3}$, $\sqrt{3} + \sqrt{12} = \sqrt{3} + \left(\sqrt{4} \cdot \sqrt{3}\right) = \sqrt{3} + 2\sqrt{3} = 3\sqrt{3}$.

6. No, they are unlike radicals and can not simplify.

Like Radicals

7. $\sqrt{2}, \sqrt{8} \Rightarrow \sqrt{2}, \sqrt{4} \cdot \sqrt{2} \Rightarrow \sqrt{2}, 2\sqrt{2}$

9. Not possible

11. $\sqrt{63}, \sqrt{28} \Rightarrow \sqrt{63} = \sqrt{9} \cdot \sqrt{7}, \sqrt{28} = \sqrt{4} \cdot \sqrt{7} \Rightarrow 3\sqrt{7}, 2\sqrt{7}$

13. $\sqrt{5}, \sqrt{125} \Rightarrow \sqrt{5}, \sqrt{125} = \sqrt{25} \cdot \sqrt{5} = \sqrt{5}, 5\sqrt{5}$

Operations on Radical Expressions

15. $2\sqrt{5} + 3\sqrt{5} = (2+3)\sqrt{5} = 5\sqrt{5}$

17. $\sqrt{13} + 8\sqrt{13} = (1+8)\sqrt{13} = 9\sqrt{13}$

19. $2\sqrt{8} + 3\sqrt{8} = 2\left(\sqrt{4} \cdot \sqrt{2}\right) + 3\left(\sqrt{4} \cdot \sqrt{2}\right) = 2(2)\sqrt{2} + 3(2)\sqrt{2} = 4\sqrt{2} + 6\sqrt{2} = (4+6)\sqrt{2} = 10\sqrt{2}$

21. $4\sqrt{3} + \sqrt{27} + \sqrt{18} = 4\sqrt{3} + \left(\sqrt{9} \cdot \sqrt{3}\right) + \left(\sqrt{9} \cdot \sqrt{2}\right) = 4\sqrt{3} + 3\sqrt{3} + 3\sqrt{2} = (4+3)\sqrt{3} + 3\sqrt{2} = 7\sqrt{3} + 3\sqrt{2}$

23. $9\sqrt{72} + \sqrt{32} = 9\left(\sqrt{36} \cdot \sqrt{2}\right) + \left(\sqrt{16} \cdot \sqrt{2}\right) = 9(6)\sqrt{2} + 4\sqrt{2} = 54\sqrt{2} + 4\sqrt{2} = (54+4)\sqrt{2} = 58\sqrt{2}$

25. $6\sqrt{15} - 2\sqrt{15} = (6-2)\sqrt{15} = 4\sqrt{15}$

27. $5\sqrt{75} - 7\sqrt{75} = 5\left(\sqrt{25} \cdot \sqrt{3}\right) - 7\left(\sqrt{25} \cdot \sqrt{3}\right) = 5(5)\sqrt{3} - 7(5)\sqrt{3} = 25\sqrt{3} - 35\sqrt{3} = (25-35)\sqrt{3} = -10\sqrt{3}$

29. $\sqrt{7}-5\sqrt{28}=\sqrt{7}-5\left(\sqrt{4}\cdot\sqrt{7}\right)=\sqrt{7}-5(2)\sqrt{7}=\sqrt{7}-10\sqrt{7}=(1-10)\sqrt{7}=-9\sqrt{7}$

31. $2\sqrt{54}-2\sqrt{24}=2\left(\sqrt{9}\cdot\sqrt{6}\right)-2\left(\sqrt{4}\cdot\sqrt{6}\right)=2(3)\sqrt{6}-2(2)\sqrt{6}=6\sqrt{6}-4\sqrt{6}=(6-4)\sqrt{6}=\ 2\sqrt{6}$

33. $9\sqrt{50}-3\sqrt{8}+\sqrt{2}=9\left(\sqrt{25}\cdot\sqrt{2}\right)-3\left(\sqrt{4}\cdot\sqrt{2}\right)+\sqrt{2}=9(5)\sqrt{2}-3(2)\sqrt{2}+\sqrt{2}=$

$45\sqrt{2}-6\sqrt{2}+\sqrt{2}=(45-6+1)\sqrt{2}=40\sqrt{2}$

35. $2\sqrt[3]{4}+5\sqrt[3]{4}=(2+5)\sqrt[3]{4}=7\sqrt[3]{4}$

37. $\sqrt[3]{20}-2\sqrt[3]{20}=(1-2)\sqrt[3]{20}=-\sqrt[3]{20}$

39. $3\sqrt{32}+\sqrt{18}+2\sqrt{75}=3\left(\sqrt{16}\cdot\sqrt{2}\right)+\left(\sqrt{9}\cdot\sqrt{2}\right)+2\left(\sqrt{25}\cdot\sqrt{3}\right)$

$=3(4)\sqrt{2}+(3)\sqrt{2}+2(5)\sqrt{3}=12\sqrt{2}+3\sqrt{2}+10\sqrt{3}$

$=(12+3)\sqrt{2}+10\sqrt{3}=15\sqrt{2}+10\sqrt{3}$

41. $3\sqrt{4x^3}-\sqrt{x}=3\left(\sqrt{4x^2}\cdot\sqrt{x}\right)-\sqrt{x}=3(2x)\sqrt{x}-\sqrt{x}$

$=6x\sqrt{x}-\sqrt{x}=(6x-1)\sqrt{x}$

43. $\sqrt{4x}+\sqrt{x^3}+\sqrt{2x^2}=\left(\sqrt{4}\cdot\sqrt{x}\right)+\left(\sqrt{x^2}\cdot\sqrt{x}\right)+\left(\sqrt{x^2}\cdot\sqrt{2}\right)$

$=2\sqrt{x}+x\sqrt{x}+x\sqrt{2}=(2+x)\sqrt{x}+x\sqrt{2}$

45. $7\sqrt{a^2b}-2\sqrt{a^2b}=7\left(\sqrt{a^2}\cdot\sqrt{b}\right)-2\left(\sqrt{a^2}\cdot\sqrt{b}\right)=7a\sqrt{b}-2a\sqrt{b}$

$=(7a-2a)\sqrt{b}=5a\sqrt{b}$

47. $3\sqrt{t}+2\sqrt{t}=(3+2)\sqrt{t}=5\sqrt{t}$

49. $\sqrt{x}-5\sqrt{x}=(1-5)\sqrt{x}=-4\sqrt{x}$

51. $\sqrt{9b}+\sqrt{4b}=\sqrt{9}\cdot\sqrt{b}+\sqrt{4}\cdot\sqrt{b}=3\sqrt{b}+2\sqrt{b}=(3+2)\sqrt{b}=5\sqrt{b}$

53. $\sqrt{4t^3}-8\sqrt{t}-\sqrt{4t}=\left(\sqrt{4t^2}\cdot\sqrt{t}\right)-8\sqrt{t}-\left(\sqrt{4}\cdot\sqrt{t}\right)=2t\sqrt{t}-8\sqrt{t}-2\sqrt{t}=(2t-8-2)\sqrt{t}=\ (2t-10)\sqrt{t}$

55. $4\sqrt[3]{x}-2\sqrt[3]{x}=(4-2)\sqrt[3]{x}=2\sqrt[3]{x}$

57. $9\sqrt[3]{x}+2\sqrt[3]{x}=(9+2)\sqrt[3]{x}=11\sqrt[3]{x}$

Geometry

59. Using Pythagorean Theorem: $a^2+b^2=c^2$, for $a=\sqrt{22}, b=\sqrt{33}$, we get $\left(\sqrt{22}\right)^2+\left(\sqrt{33}\right)^2=c^2 \Rightarrow$

$22+33=c^2 \Rightarrow 55=c^2 \Rightarrow \sqrt{55}=c \Rightarrow c=\sqrt{55}$ so the perimeter is $\sqrt{22}+\sqrt{33}+\sqrt{55}$ inches.

Applications

61. $m_1 = 0.09, m_2 = 0.04, R_1 + R_2 = 1500\sqrt{m_1} + 2000\sqrt{m_2}$

$\Rightarrow R_1 + R_2 = 1500\sqrt{0.09} + 2000\sqrt{0.04} \Rightarrow R_1 + R_2 = 1500(0.3)$

$+ 2000(0.2) \Rightarrow R_1 + R_2 = 450 + 400 \Rightarrow R_1 + R_2 = 850 \text{ ft}^3/\text{sec}$

8.4: Simplifying Radical Expressions

Concepts

1. $a^2 - b^2$

2. b

3. $(1-\sqrt{2})(1+\sqrt{2}) = 1^2 - (\sqrt{2})^2 = 1 - 2 = -1$

4. $(\sqrt{a}-\sqrt{b})(\sqrt{a}+\sqrt{b}) = (\sqrt{a})^2 - (\sqrt{b})^2 = a - b$

5. $4+\sqrt{11}$

6. $2-\sqrt{3}$

7. $\dfrac{\sqrt{11}}{\sqrt{11}}$

8. $\dfrac{4+\sqrt{11}}{4+\sqrt{11}}$

Multiplying Radical Expressions

9. $(4-\sqrt{2})(4+\sqrt{2}) = 4^2 - (\sqrt{2})^2 = 16 - 2 = 14$

11. $(8+\sqrt{11})(8-\sqrt{11}) = 8^2 - (\sqrt{11})^2 = 64 - 11 = 53$

13. $(\sqrt{5}-7)(\sqrt{5}+7) = (\sqrt{5})^2 - 7^2 = 5 - 49 = -44$

15. $(\sqrt{15}+2)(\sqrt{15}-2) = (\sqrt{15})^2 - 2^2 = 15 - 4 = 11$

17. $(\sqrt{x}-5)(\sqrt{x}+5) = (\sqrt{x})^2 - 5^2 = x - 25$

19. $(7-\sqrt{t})(7+\sqrt{t}) = 7^2 - (\sqrt{t})^2 = 49 - t$

21. $(\sqrt{y}-1)(\sqrt{y}-2) = \sqrt{y}\cdot\sqrt{y} - 1\cdot\sqrt{y} - 2\cdot\sqrt{y} + 1\cdot 2 = (\sqrt{y})^2 - 3\sqrt{y} + 2 = y - 3\sqrt{y} + 2$

23. $(3-\sqrt{x})(1+\sqrt{x}) = 3\cdot 1 - \sqrt{x}\cdot 1 + \sqrt{x}\cdot 3 - \sqrt{x}\cdot\sqrt{x} = 3 + 2\sqrt{x} - (\sqrt{x})^2 = 3 + 2\sqrt{x} - x$

25. $(\sqrt{x}-\sqrt{y})(\sqrt{x}+\sqrt{y}) = (\sqrt{x})^2 - (\sqrt{y})^2 = x - y$

27. $\left(\sqrt{r}-3\sqrt{t}\right)\left(\sqrt{r}+3\sqrt{t}\right)=\left(\sqrt{r}\right)^2-\left(3\sqrt{t}\right)^2=r-9t$

Rationalizing the Denominator

29. $\sqrt{17}+3$

31. $20-\sqrt{5}$

33. $\sqrt{x}+\sqrt{z}$

35. $\frac{2}{\sqrt{5}}\cdot\frac{\sqrt{5}}{\sqrt{5}}=\frac{2\sqrt{5}}{\sqrt{5}\cdot\sqrt{5}}=\frac{2\sqrt{5}}{5}$

37. $\frac{14}{\sqrt{14}}\cdot\frac{\sqrt{14}}{\sqrt{14}}=\frac{14\sqrt{14}}{\sqrt{14}\cdot\sqrt{14}}=\frac{14\sqrt{14}}{14}=\frac{14}{14}\cdot\sqrt{14}=1\cdot\sqrt{14}=\sqrt{14}$

39. $-\frac{3}{\sqrt{3}}\cdot\frac{\sqrt{3}}{\sqrt{3}}=-\frac{3\sqrt{3}}{\sqrt{3}\cdot\sqrt{3}}=-\frac{3\sqrt{3}}{3}=-\frac{3}{3}\cdot\sqrt{3}=-1\cdot\sqrt{3}=-\sqrt{3}$

41. $\frac{5}{\sqrt{10}}\cdot\frac{\sqrt{10}}{\sqrt{10}}=\frac{5\sqrt{10}}{\sqrt{10}\cdot\sqrt{10}}=\frac{5\sqrt{10}}{10}=\frac{1\sqrt{10}}{2}=\frac{\sqrt{10}}{2}$

43. $-\frac{4}{3\sqrt{7}}\cdot\frac{\sqrt{7}}{\sqrt{7}}=-\frac{4\sqrt{7}}{3\sqrt{7}\cdot\sqrt{7}}=-\frac{4\sqrt{7}}{3\cdot 7}=-\frac{4\sqrt{7}}{21}$

45. $\frac{3}{2\sqrt{3}}\cdot\frac{\sqrt{3}}{\sqrt{3}}=\frac{3\sqrt{3}}{2\sqrt{3}\cdot\sqrt{3}}=\frac{3\sqrt{3}}{2\cdot 3}=\frac{3}{3}\cdot\frac{\sqrt{3}}{2}=1\cdot\frac{\sqrt{3}}{2}=\frac{\sqrt{3}}{2}$

47. $\frac{2}{\sqrt{b}}\cdot\frac{\sqrt{b}}{\sqrt{b}}=\frac{2\sqrt{b}}{\sqrt{b}\cdot\sqrt{b}}=\frac{2\sqrt{b}}{b}$

49. $-\sqrt{\frac{3}{b}}=-\frac{\sqrt{3}}{\sqrt{b}}\cdot\frac{\sqrt{b}}{\sqrt{b}}=-\frac{\sqrt{3}\cdot\sqrt{b}}{\sqrt{b}\cdot\sqrt{b}}=-\frac{\sqrt{3b}}{b}$

51. $\frac{\sqrt{2x}}{\sqrt{3}}\cdot\frac{\sqrt{3}}{\sqrt{3}}=\frac{\sqrt{2x}\cdot\sqrt{3}}{\sqrt{3}\cdot\sqrt{3}}=\frac{\sqrt{6x}}{3}$

53. $\frac{\sqrt{4x^3}}{\sqrt{36x}}=\frac{\sqrt{4x^2}\cdot\sqrt{x}}{\sqrt{36}\cdot\sqrt{x}}=\frac{2x}{6}\cdot 1=\frac{x}{3}$

55. $\frac{1}{\sqrt{3}-1}\cdot\frac{\sqrt{3}+1}{\sqrt{3}+1}=\frac{\sqrt{3}+1}{\left(\sqrt{3}\right)^2-1^2}=\frac{\sqrt{3}+1}{3-1}=\frac{\sqrt{3}+1}{2}$

57. $\frac{\sqrt{5}}{\sqrt{5}+2}\cdot\frac{\sqrt{5}-2}{\sqrt{5}-2}=\frac{\left(\sqrt{5}\right)^2-2\sqrt{5}}{\left(\sqrt{5}\right)^2-2^2}=\frac{5-2\sqrt{5}}{5-4}=\frac{5-2\sqrt{5}}{1}=5-2\sqrt{5}$

59. $\frac{\sqrt{2}+1}{\sqrt{2}-1}\cdot\frac{\sqrt{2}+1}{\sqrt{2}+1}=\frac{\sqrt{2}\cdot\sqrt{2}+1\cdot\sqrt{2}+1\cdot\sqrt{2}+1\cdot 1}{\left(\sqrt{2}\right)^2-1^2}=\frac{\left(\sqrt{2}\right)^2+2\sqrt{2}+1}{2-1}=\frac{2+2\sqrt{2}+1}{1}=\frac{2\sqrt{2}+3}{1}=2\sqrt{2}+3$

61. $\frac{5-\sqrt{13}}{5+\sqrt{13}}\cdot\frac{5-\sqrt{13}}{5-\sqrt{13}}=\frac{5\cdot5-5\cdot\sqrt{13}-5\cdot\sqrt{13}+\sqrt{13}\cdot\sqrt{13}}{5^2-\left(\sqrt{13}\right)^2}=\frac{25-10\sqrt{13}+\left(\sqrt{13}\right)^2}{25-13}=$

$\frac{25-10\sqrt{13}+13}{12}=\frac{38-10\sqrt{13}}{12}=\frac{19-5\sqrt{13}}{6}$

63. $\frac{1}{2\sqrt{x}}\cdot\frac{\sqrt{x}}{\sqrt{x}}=\frac{1\sqrt{x}}{2\sqrt{x}\cdot\sqrt{x}}=\frac{\sqrt{x}}{2x}$

65. $\frac{\sqrt{x}-2}{\sqrt{x}+2}\cdot\frac{\sqrt{x}-2}{\sqrt{x}-2}=\frac{\sqrt{x}\cdot\sqrt{x}-2\cdot\sqrt{x}-2\cdot\sqrt{x}+2\cdot2}{\left(\sqrt{x}\right)^2-2^2}=\frac{\left(\sqrt{x}\right)^2-4\sqrt{x}+4}{x-4}=\frac{x-4\sqrt{x}+4}{x-4}$

67. $\frac{\sqrt{a}-2b}{\sqrt{a}+2b}\cdot\frac{\sqrt{a}-2b}{\sqrt{a}-2b}=\frac{\sqrt{a}\cdot\sqrt{a}-2b\cdot\sqrt{a}-2b\cdot\sqrt{a}+2b\cdot2b}{\left(\sqrt{a}\right)^2-(2b)^2}=\frac{\left(\sqrt{a}\right)^2-4b\sqrt{a}+4b^2}{a-4b^2}=\frac{a-4b\sqrt{a}+4b^2}{a-4b^2}$

69. $\frac{\sqrt{a}+\sqrt{b}}{\sqrt{a}-\sqrt{b}}\cdot\frac{\sqrt{a}+\sqrt{b}}{\sqrt{a}+\sqrt{b}}=\frac{\sqrt{a}\cdot\sqrt{a}+\sqrt{a}\cdot\sqrt{b}+\sqrt{a}\cdot\sqrt{b}+\sqrt{b}\cdot\sqrt{b}}{\left(\sqrt{a}\right)^2-\left(\sqrt{b}\right)^2}=\frac{\sqrt{a^2}+2\sqrt{ab}+\sqrt{b^2}}{a-b}=\frac{a+2\sqrt{ab}+b}{a-b}$

71. $\frac{\sqrt{2x}-\sqrt{y}}{\sqrt{2x}+\sqrt{y}}\cdot\frac{\sqrt{2x}-\sqrt{y}}{\sqrt{2x}-\sqrt{y}}=\frac{\sqrt{2x}\cdot\sqrt{2x}-\sqrt{2x}\cdot\sqrt{y}-\sqrt{2x}\cdot\sqrt{y}+\sqrt{y}\cdot\sqrt{y}}{\left(\sqrt{2x}\right)^2-\left(\sqrt{y}\right)^2}=$

$\frac{\left(\sqrt{2x}\right)^2-2\sqrt{2xy}+\left(\sqrt{y}\right)^2}{2x-y}=\frac{2x-2\sqrt{2xy}+y}{2x-y}$

73. $\frac{1}{\sqrt{b+1}+\sqrt{b}}\cdot\frac{\sqrt{b+1}-\sqrt{b}}{\sqrt{b+1}-\sqrt{b}}=\frac{\sqrt{b+1}-\sqrt{b}}{\left(\sqrt{b+1}\right)^2-\left(\sqrt{b}\right)^2}=\frac{\sqrt{b+1}-\sqrt{b}}{b+1-b}=\frac{\sqrt{b+1}-\sqrt{b}}{1}=\sqrt{b+1}-\sqrt{b}$

Geometry

75. Perimeter: If $P=2l+2w$ then, $2\left(\sqrt{91}+2\right)+2\left(\sqrt{91}-2\right)=(2+2)\sqrt{91}+2(2-2)=4\sqrt{91}$

Area: If $A=l\cdot w$ then, $\left(\sqrt{91}-2\right)\left(\sqrt{91}+2\right)=\left(\sqrt{91}\right)^2-2^2=91-4=87$

Applications

77. $L=2, T=2\pi\sqrt{\frac{L}{32}}\Rightarrow T=2\pi\sqrt{\frac{2}{32}}\Rightarrow T=2\pi\sqrt{\frac{1}{16}}\Rightarrow T=2\pi\cdot\frac{1}{4}\Rightarrow T=\frac{\pi}{2}\approx1.57$ sec

Checking Basic Concepts for Sections 8.3 & 8.4

1. (a) $5\sqrt{6}+5\sqrt{6}=(5+5)\sqrt{6}=10\sqrt{6}$

(b) $3\sqrt{75}-2\sqrt{27}=3\left(\sqrt{25}\cdot\sqrt{3}\right)-2\left(\sqrt{9}\cdot\sqrt{3}\right)=3(5)\sqrt{3}-2(3)\sqrt{3}=15\sqrt{3}-6\sqrt{3}=(15-6)\sqrt{3}=9\sqrt{3}$

(c) $\sqrt{16k}+\sqrt{25k}=\left(\sqrt{16}\cdot\sqrt{k}\right)+\left(\sqrt{25}\cdot\sqrt{k}\right)=4\sqrt{k}+5\sqrt{k}=(4+5)\sqrt{k}=9\sqrt{k}$

2. Perimeter: If $P = 2l + 2w$ then, $2\sqrt{45} + 2\sqrt{125} = 2\left(\sqrt{9} \cdot \sqrt{5}\right) + 2\left(\sqrt{25} \cdot \sqrt{5}\right) = 2(3)\sqrt{5} + 2(5)\sqrt{5} =$

$6\sqrt{5} + 10\sqrt{5} = (6+10)\sqrt{5} = 16\sqrt{5}$

3. (a) $\left(5+\sqrt{2}\right)\left(5-\sqrt{2}\right) = 5^2 - \left(\sqrt{2}\right)^2 = 25 - 2 = 23$

(b) $\left(\sqrt{x}-3\right)\left(\sqrt{x}+5\right) = \sqrt{x} \cdot \sqrt{x} - 3 \cdot \sqrt{x} + 5 \cdot \sqrt{x} - 3 \cdot 5 = \left(\sqrt{x}\right)^2 + 2\sqrt{x} - 15 = x + 2\sqrt{x} - 15$

4. (a) $\dfrac{5}{3\sqrt{5}} \cdot \dfrac{\sqrt{5}}{\sqrt{5}} = \dfrac{5\sqrt{5}}{3\sqrt{5} \cdot \sqrt{5}} = \dfrac{5\sqrt{5}}{3 \cdot 5} = \dfrac{5\sqrt{5}}{15} = \dfrac{1\sqrt{5}}{3} = \dfrac{\sqrt{5}}{3}$

(b) $\dfrac{3+\sqrt{7}}{3-\sqrt{7}} \cdot \dfrac{3+\sqrt{7}}{3+\sqrt{7}} = \dfrac{3 \cdot 3 + 3 \cdot \sqrt{7} + 3 \cdot \sqrt{7} + \sqrt{7} \cdot \sqrt{7}}{3^2 - \left(\sqrt{7}\right)^2} = \dfrac{9 + 6\sqrt{7} + \left(\sqrt{7}\right)^2}{9-7} =$

$\dfrac{9 + 6\sqrt{7} + 7}{2} = \dfrac{6\sqrt{7} + 16}{2} = \dfrac{3\sqrt{7} + 8}{1} = 3\sqrt{7} + 8$

(c) $\dfrac{1}{\sqrt{x}+\sqrt{y}} \cdot \dfrac{\sqrt{x}-\sqrt{y}}{\sqrt{x}-\sqrt{y}} = \dfrac{\sqrt{x}-\sqrt{y}}{\left(\sqrt{x}\right)^2 - \left(\sqrt{y}\right)^2} = \dfrac{\sqrt{x}-\sqrt{y}}{x-y}$

8.5: Equations Involving Radical Expressions

Concepts

1. Square each side.
2. Check your answers.
3. $\sqrt{x} = 4 \Rightarrow \left(\sqrt{x}\right)^2 = 4^2 \Rightarrow x = 16$ Check: $x = 16 \Rightarrow \sqrt{16} = 4 \Rightarrow 4 = 4$
4. $x = 8 \Rightarrow \sqrt{2(8)} - 1 = 3 \Rightarrow \sqrt{16} - 1 = 3 \Rightarrow 4 - 1 = 3 \Rightarrow 3 = 3$, Yes
5. $x = 4$, solution is 4; $x^2 = 16$, solution is 4 and -4, No
6. $a = \sqrt{b} \Rightarrow a^2 = \left(\sqrt{b}\right)^2 \Rightarrow a^2 = b \Rightarrow b = a^2$

Solving Radical Equations

7. $\sqrt{x} = 6 \Rightarrow \left(\sqrt{x}\right)^2 = 6^2 \Rightarrow x = 36$ Check: $x = 36 \Rightarrow \sqrt{36} = 6 \Rightarrow 6 = 6$, so $x = 36$

9. No solution, no square roots can equal a negative number.

11. $\sqrt{x-7} = 3 \Rightarrow \left(\sqrt{x-7}\right)^2 = 3^2 \Rightarrow x - 7 = 9 \Rightarrow x = 16$ Check: $x = 16 \Rightarrow \sqrt{16-7} = 3 \Rightarrow$

$\sqrt{9} = 3 \Rightarrow 3 = 3$, so $x = 16$

13. $\sqrt{1-k} = 10 \Rightarrow \left(\sqrt{1-k}\right)^2 = 10^2 \Rightarrow 1 - k = 100 \Rightarrow -k = 99 \Rightarrow k = -99$ Check: $k = -99 \Rightarrow$

$\sqrt{1-(-99)} = 10 \Rightarrow \sqrt{100} = 10 \Rightarrow 10 = 10$, so $k = -99$

15. $\sqrt{2t-3}=4 \Rightarrow \left(\sqrt{2t-3}\right)^2=4^2 \Rightarrow 2t-3=16 \Rightarrow 2t=19 \Rightarrow t=\frac{19}{2}$ Check: $t=\frac{19}{2} \Rightarrow$

$\sqrt{2\left(\frac{19}{2}\right)-3}=4 \Rightarrow \sqrt{19-3}=4 \Rightarrow \sqrt{16}=4 \Rightarrow 4=4$, so $t=\frac{19}{2}$

17. $\sqrt{3x}+1=5 \Rightarrow \sqrt{3x}=4 \Rightarrow \left(\sqrt{3x}\right)^2=4^2 \Rightarrow 3x=16 \Rightarrow x=\frac{16}{3}$ Check: $x=\frac{16}{3} \Rightarrow$

$\sqrt{3\left(\frac{16}{3}\right)}+1=5 \Rightarrow \sqrt{16}+1=5 \Rightarrow 4+1=5 \Rightarrow 5=5$, so $x=\frac{16}{3}$

19. $9-\sqrt{x}=1 \Rightarrow -\sqrt{x}=-8 \Rightarrow \sqrt{x}=8 \Rightarrow \left(\sqrt{x}\right)^2=8^2 \Rightarrow x=64$ Check: $x=64 \Rightarrow 9-\sqrt{64}=1 \Rightarrow$

$9-8=1 \Rightarrow 1=1$, so $x=64$

21. $\sqrt{b+1}=2\sqrt{b} \Rightarrow \left(\sqrt{b+1}\right)^2=\left(2\sqrt{b}\right)^2 \Rightarrow b+1=4b \Rightarrow 1=3b \Rightarrow \frac{1}{3}=b$ Check: $b=\frac{1}{3} \Rightarrow$

$\sqrt{\frac{1}{3}+1}=2\sqrt{\frac{1}{3}} \Rightarrow \sqrt{\frac{4}{3}}=2\sqrt{\frac{1}{3}} \Rightarrow \sqrt{4\cdot\frac{1}{3}}=2\sqrt{\frac{1}{3}} \Rightarrow 2\sqrt{\frac{1}{3}}=2\sqrt{\frac{1}{3}}$, so $b=\frac{1}{3}$

23. $\sqrt{4z+2}=\sqrt{3z} \Rightarrow \left(\sqrt{4z+2}\right)^2=\left(\sqrt{3z}\right)^2 \Rightarrow 4z+2=3z \Rightarrow z=-2$

Check: $z=-2 \Rightarrow \sqrt{4(-2)-2}=\sqrt{3(-2)} \Rightarrow \sqrt{-10}=\sqrt{-6}$; There are no solutions.

25. $3\sqrt{x}=12 \Rightarrow \sqrt{x}=4 \Rightarrow \left(\sqrt{x}\right)^2=4^2 \Rightarrow x=16$

Check: $x=16 \Rightarrow 3\sqrt{16}=12 \Rightarrow 3\cdot 4=12 \Rightarrow 12=12$, so $x=16$

27. $-2\sqrt{x}=4 \Rightarrow \sqrt{x}=-2$, the principal square root is nonnegative $\Rightarrow$ no solutions

29. $\sqrt{x-2}+1=5 \Rightarrow \sqrt{x-2}=4 \Rightarrow \left(\sqrt{x-2}\right)^2=4^2 \Rightarrow x-2=16 \Rightarrow x=18$

Check: $x=18 \Rightarrow \sqrt{18-2}+1=5 \Rightarrow \sqrt{16}+1=5 \Rightarrow 4+1=5 \Rightarrow 5=5$, so $x=18$

31. $2\sqrt{x-1}-3=1 \Rightarrow 2\sqrt{x-1}=4 \Rightarrow \sqrt{x-1}=2 \Rightarrow \left(\sqrt{x-1}\right)^2=2^2 \Rightarrow x-1=4 \Rightarrow x=5$

Check: $x=5 \Rightarrow 2\sqrt{5-1}-3=1 \Rightarrow 2\sqrt{4}-3=1 \Rightarrow 2(2)-3=1 \Rightarrow 4-3=1 \Rightarrow 1=1$, so $x=5$

33. $3\sqrt{3x+1}+x=7x \Rightarrow 3\sqrt{3x+1}=6x \Rightarrow \sqrt{3x+1}=2x \Rightarrow \left(\sqrt{3x+1}\right)^2=(2x)^2 \Rightarrow 3x+1=4x^2 \Rightarrow 4x^2-3x-1=0$

$\Rightarrow (4x+1)(x-1)=0 \Rightarrow 4x+1=0$ or $x-1=0 \Rightarrow x=-\frac{1}{4}$ or $x=1$

35. $2\sqrt{x+1}+3=9 \Rightarrow 2\sqrt{x+1}=6 \Rightarrow \sqrt{x+1}=3 \Rightarrow \left(\sqrt{x+1}\right)^2=3^2 \Rightarrow x+1=9 \Rightarrow x=8$

Check: $2\sqrt{8+1}+3=9 \Rightarrow 2\sqrt{9}+3=9 \Rightarrow 2(3)+3=9 \Rightarrow 9=9$, so $x=8$

37. $\sqrt{x+20}=x \Rightarrow \left(\sqrt{x+20}\right)^2=x^2 \Rightarrow x+20=x^2 \Rightarrow 0=x^2-x-20 \Rightarrow 0=(x-5)(x+4) \Rightarrow$

$x=5,-4$ Check: $\sqrt{5+20}=5 \Rightarrow \sqrt{25}=5 \Rightarrow 5=5$, and $\sqrt{-4+20}=-4 \Rightarrow \sqrt{16}=-4 \Rightarrow 4 \neq -4$, so $x=5$

39. $5\sqrt{2t-1}=3t \Rightarrow \sqrt{2t-1}=\frac{3}{5}t \Rightarrow \left(\sqrt{2t-1}\right)^2=\left(\frac{3}{5}t\right)^2 \Rightarrow 2t-1=\frac{9}{25}t^2 \Rightarrow 25(2t-1)=9t^2 \Rightarrow$

$50t-25=9t^2 \Rightarrow 0=9t^2-50t+25 \Rightarrow 0=(9t-5)(t-5) \Rightarrow t=\frac{5}{9},5$ Check:

$5\sqrt{2\left(\frac{5}{9}\right)-1}=3\left(\frac{5}{9}\right) \Rightarrow \quad 5\sqrt{\frac{10}{9}-1}=\frac{5}{3} \Rightarrow \quad 5\sqrt{\frac{1}{9}}=\frac{5}{3} \Rightarrow 5\left(\frac{1}{3}\right)=\frac{5}{3} \Rightarrow \frac{5}{3}=\frac{5}{3}$, and

$5\sqrt{2(5)-1}=3(5) \Rightarrow 5\sqrt{9}=15 \Rightarrow 5(3)=15 \Rightarrow 15=15$, so $t=\frac{5}{9},5$

41. $\sqrt{2x+1}=x-1 \Rightarrow \left(\sqrt{2x+1}\right)^2=(x-1)^2 \Rightarrow 2x+1=x^2-2x+1 \Rightarrow 0=x^2-4x \Rightarrow$

$0=x(x-4) \Rightarrow x=0,4$ Check: $\sqrt{2(0)+1}=0-1 \Rightarrow \sqrt{1}=-1 \Rightarrow 1 \neq -1$, and

$\sqrt{2(4)+1}=4-1 \Rightarrow \sqrt{9}=3 \Rightarrow 3=3$, so $x=4$

43. $z+1=\sqrt{37+2z} \Rightarrow (z+1)^2=\left(\sqrt{37+2z}\right)^2 \Rightarrow z^2+2z+1=37+2z \Rightarrow z^2-36=0 \Rightarrow$

$(z+6)(z-6)=0 \Rightarrow z=-6,6$ Check: $-6+1=\sqrt{37+2(-6)} \Rightarrow -5=\sqrt{25} \Rightarrow -5 \neq 5$, and

$6+1=\sqrt{37+2(6)} \Rightarrow 7=\sqrt{49} \Rightarrow 7=7$, so $x=6$

45. $\sqrt{(x-1)(x+2)}=2 \Rightarrow \left(\sqrt{(x-1)(x+2)}\right)^2=2^2 \Rightarrow (x-1)(x+2)=4 \Rightarrow x^2+x-2=4 \Rightarrow$

$x^2+x-6=0 \Rightarrow (x+3)(x-2)=0 \Rightarrow x=-3,2$ Check: $\sqrt{(-3-1)(-3+2)}=2 \Rightarrow$

$\sqrt{(-4)(-1)}=2 \Rightarrow \sqrt{4}=2 \Rightarrow 2=2$, and $\sqrt{(2-1)(2+2)}=2 \Rightarrow \sqrt{(1)(4)}=2 \Rightarrow \sqrt{4}=2 \Rightarrow \; 2=2$, so $x=-3,2$

47. $T=\sqrt{L}$ for $L \Rightarrow T^2=\left(\sqrt{L}\right)^2 \Rightarrow T^2=L \Rightarrow L=T^2$

49. $b=\sqrt{2a}$ for $a \Rightarrow b^2=\left(\sqrt{2a}\right)^2 \Rightarrow b^2=2a \Rightarrow \frac{1}{2}b^2=a \Rightarrow a=\frac{1}{2}b^2$

51. $D=4\sqrt{t+1}$ for $t \Rightarrow \frac{1}{4}D=\sqrt{t+1} \Rightarrow \left(\frac{1}{4}D\right)^2=\left(\sqrt{t+1}\right)^2 \Rightarrow \frac{1}{16}D^2=t+1 \Rightarrow \frac{1}{16}D^2-1=t \Rightarrow \; t=\frac{1}{16}D^2-1$

53. $c=\sqrt{a^2+b^2}$ for $b \Rightarrow c^2=\left(\sqrt{a^2+b^2}\right)^2 \Rightarrow c^2=a^2+b^2 \Rightarrow c^2-a^2=b^2 \Rightarrow \sqrt{c^2-a^2}=\sqrt{b^2} \Rightarrow$

$\sqrt{c^2-a^2}=b \Rightarrow b=\sqrt{c^2-a^2}$

55. $k=4+\sqrt{d}$ for $d \Rightarrow k-4=\sqrt{d} \Rightarrow (k-4)^2=\left(\sqrt{d}\right)^2 \Rightarrow (k-4)^2=d \Rightarrow d=(k-4)^2$

Numerical and Graphical Solutions

57. (a) From the table we see that $\sqrt{x}=2$ when $x=4$. See Figure 57a.

(b) The solution is the x-coordinate of the intersection point, or 4. See Figure 57b.

x	0	1	4	9	16	25
$\sqrt{x}$	0	1	2	3	4	5

Figure 57a

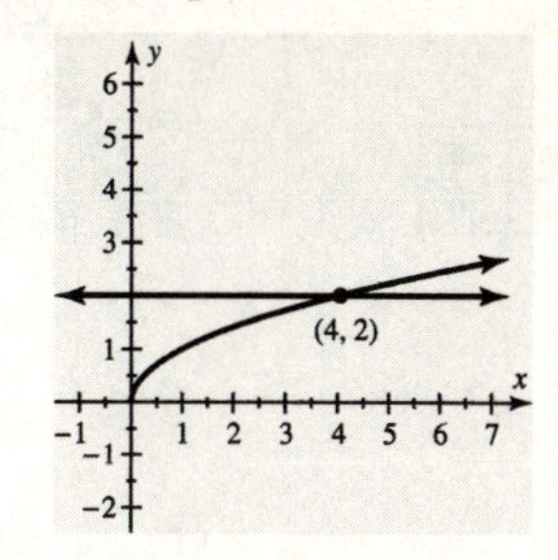

Figure 57b

59. (a) From the table we see that $\sqrt{x-2}=1$ when $x=3$. See Figure 59a.

(b) The solution is the x-coordinate of the intersection point, or 3. See Figure 59b.

x	2	3	6	11	18	27
$\sqrt{x-2}$	0	1	2	3	4	5

Figure 59a

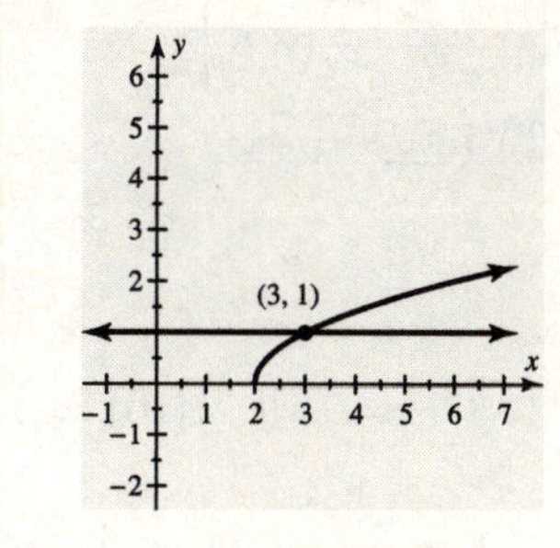

Figure 59b

61. (a) From the table we see that $\sqrt{x}+1=3$ when $x=4$. See Figure 61a.

(b) The solution is the x-coordinate of the intersection point, or 4. See Figure 61b.

x	0	1	4	9	16	25
$\sqrt{x}+1$	1	2	3	4	5	6

Figure 61a

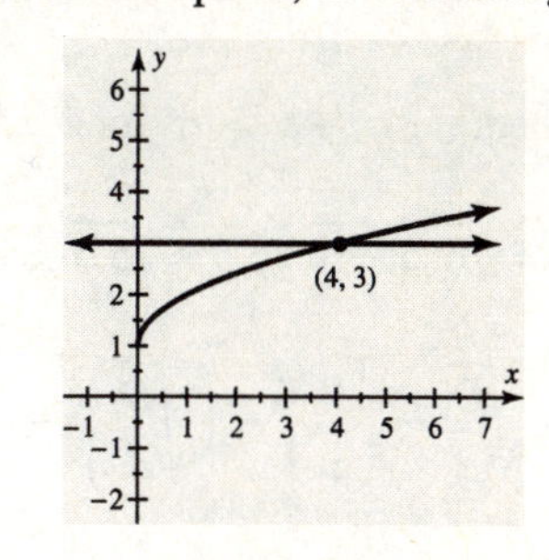

Figure 61b

63. (a) From the table we see that $2\sqrt{x}=2$ when $x=1$. See Figure 63a.

(b) The solution is the x-coordinate of the intersection point, or 1. See Figure 63b.

x	0	1	4	9	16	25
$2\sqrt{x}$	0	2	4	6	8	10

Figure 63a

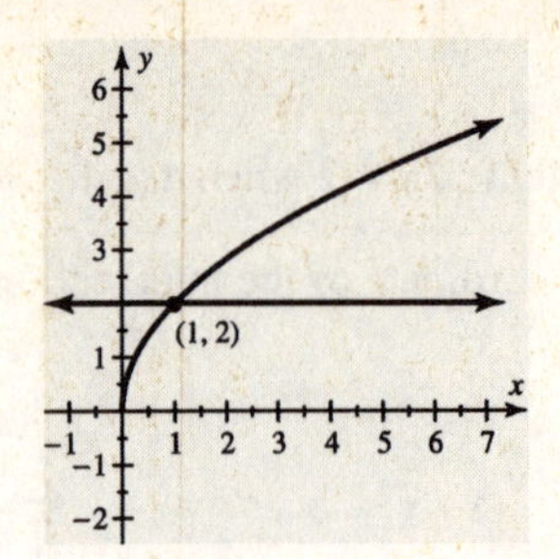

Figure 63b

65. 

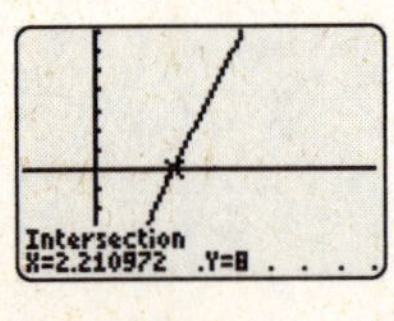

2.21

67.

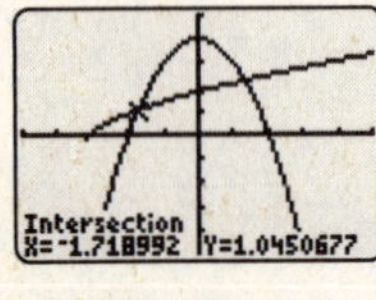

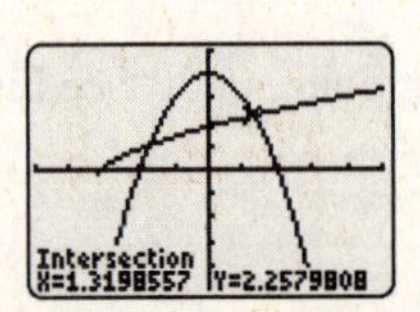

−1.72, 1.32

Applications

69. (a) If $h = 100$, then $D = 1.22\sqrt{100} \Rightarrow D = 1.22(10) \Rightarrow D = 12.2$ miles.

(b) If $D = 50$, then $50 = 1.22\sqrt{h} \Rightarrow \dfrac{50}{1.22} = \sqrt{h} \Rightarrow 40.984^2 = \left(\sqrt{h}\right)^2 \Rightarrow h =$ about 1680 feet.

(c) If $h = x$, then $D = 1.22\sqrt{x}$, if $4x$, then $D = 1.22\sqrt{4x} \Rightarrow D = 1.22(2)\sqrt{x}$, so the distance to the horizon doubles.

71. (a) Solving Pythagorean Theorem $a^2 + b^2 = c^2$ for b, $b^2 = c^2 - a^2 \Rightarrow \sqrt{b^2} = \sqrt{c^2 - a^2} \Rightarrow b = \sqrt{c^2 - a^2}$

(b) $b = \sqrt{(25)^2 - (7)^2} \Rightarrow b = \sqrt{625 - 49} \Rightarrow b = \sqrt{576} \Rightarrow b = 24$ inches

73. (a) $V = v\sqrt{\dfrac{D}{d}}$ for $D \Rightarrow \dfrac{V}{v} = \sqrt{\dfrac{D}{d}} \Rightarrow \left(\dfrac{V}{v}\right)^2 = \left(\sqrt{\dfrac{D}{d}}\right)^2 \Rightarrow \dfrac{V^2}{v^2} = \dfrac{D}{d} \Rightarrow \dfrac{V^2 d}{v^2} = D \Rightarrow D = \dfrac{V^2 d}{v^2}$

(b) $V = 50$mph, $v = 40$mph, $d = 200$ feet $\Rightarrow D = \dfrac{50^2(200)}{40^2} \Rightarrow D = \dfrac{2500(200)}{1600} \Rightarrow D = \dfrac{500{,}000}{1600} \Rightarrow$

$D = 312.5$ feet.

75. If $m = x$, then $R = k\sqrt{x}$ and if $m = 4x$ then $R = k\sqrt{4x} \Rightarrow R = k2\sqrt{x}$ which is two times bigger, so it increases by a factor of 2 or doubles.

77. If $k = 2000$, and $m = 0.09$ then $R = 2000\sqrt{0.09} \Rightarrow R = 2000(0.3) \Rightarrow R = 600$ cfs, it increases to 600 cfs.

8.6: Higher Roots and Rational Exponents

Concepts

1. index, radicand
2. odd
3. even
4. No, $\sqrt[3]{3} \approx 1.44$ because $1.44^3 \approx 3$
5. Negative number
6. real number
7. $\sqrt[n]{a} \cdot \sqrt[n]{b} = \sqrt[n]{ab}$
8. $\sqrt[n]{\dfrac{a}{b}} = \dfrac{\sqrt[n]{a}}{\sqrt[n]{b}}$
9. $a^{1/n} = \sqrt[n]{a}$
10. $a^{m/n} = \sqrt[n]{a^m}$ *or* $\left(\sqrt[n]{a}\right)^m$

Simplifying Expressions

11. $\sqrt[3]{-27} = -3$, because $(-3)^3 = -27$

13. $\sqrt[5]{243} = 3$, because $3^5 = 243$

15. $\sqrt[4]{16} = 2$, because $2^4 = 16$

17. $\sqrt[4]{10,000} = 10$, because $10^4 = 10,000$

19. $-\sqrt[5]{-32} = -(-2) = 2$, because $(-2)^5 = -32$

21. $-\sqrt[6]{1} = -\left(1^{1/6}\right) = -1$

23. $\sqrt[4]{5^4} = 5^{4/4} = 5^1 = 5$

25. $-\sqrt[4]{5^4} = -5^{4/4} = -5^1 = -5$

27. $\sqrt[3]{\dfrac{27}{8}} = \dfrac{\sqrt[3]{27}}{\sqrt[3]{8}} = \dfrac{3}{2}$, because $\dfrac{3^3 = 27}{2^3 = 8}$

29. $\sqrt[4]{\dfrac{20}{81}} = \dfrac{\sqrt[4]{20}}{\sqrt[4]{81}} = \dfrac{\sqrt[4]{20}}{3}$, because $3^4 = 81$

31. $\sqrt[4]{-16}$ is not a real number because $x^4 \geq 0$ for any real number x.

33. $\sqrt[3]{\dfrac{8}{27}} = \dfrac{\sqrt[3]{8}}{\sqrt[3]{27}} = \dfrac{2}{3}$ because $\dfrac{2^3 = 8}{3^3 = 27}$

35. $\sqrt[4]{\dfrac{7}{16}} = \dfrac{\sqrt[4]{7}}{\sqrt[4]{16}} = \dfrac{\sqrt[4]{7}}{2}$ because $2^4 = 16$

37. $\sqrt[4]{\frac{4}{9}} \cdot \sqrt[4]{\frac{4}{9}} = \sqrt[4]{\frac{4}{9} \cdot \frac{4}{9}} = \sqrt[4]{\frac{16}{81}} = \frac{\sqrt[4]{16}}{\sqrt[4]{81}} = \frac{2}{3}$ because $\frac{2^4 = 16}{3^4 = 81}$

39. $\sqrt[3]{16} \cdot \sqrt[3]{4} = \sqrt[3]{64} = 4$, because $4^3 = 64$

41. $\sqrt[6]{2} \cdot \sqrt[6]{32} = \sqrt[6]{64} = 2$, because $2^6 = 64$

43. $\sqrt[5]{16} \cdot \sqrt[5]{-2} = \sqrt[5]{-32} = -2$, because $(-2)^5 = -32$

45. $\frac{\sqrt[3]{3}}{\sqrt[3]{81}} = \sqrt[3]{\frac{3}{81}} = \sqrt[3]{\frac{1}{27}} = \frac{1}{3}$, because $\frac{1^3 = 1}{3^3 = 27}$

47. $\frac{\sqrt[4]{2}}{\sqrt[4]{32}} = \sqrt[4]{\frac{2}{32}} = \sqrt[4]{\frac{1}{16}} = \frac{1}{2}$, because $\frac{1^4 = 1}{2^4 = 16}$

49. $\frac{\sqrt[5]{-128}}{\sqrt[5]{4}} = \sqrt[5]{\frac{-128}{4}} = \sqrt[5]{-32} = -2$, because $(-2)^5 = -32$

Rational Exponents

51. $36^{1/2} = \sqrt{36} = 6$

53. $27^{1/3} = \sqrt[3]{27} = 3$

55. $(-1)^{1/3} = \sqrt[3]{-1} = -1$

57. $81^{1/4} = \sqrt[4]{81} = 3$

59. $32^{1/5} = \sqrt[5]{32} = 2$

61. $4^{3/2} = \sqrt{4^3} = \left(\sqrt{4}\right)^3 = 2^3 = 8$

63. $64^{2/3} = \sqrt[3]{64^2} = \left(\sqrt[3]{64}\right)^2 = 4^2 = 16$

65. $16^{5/4} = \sqrt[4]{16^5} = \left(\sqrt[4]{16}\right)^5 = 2^5 = 32$

67. $(-1)^{3/5} = \sqrt[5]{(-1)^3} = \left(\sqrt[5]{-1}\right)^3 = (-1)^3 = -1$

69. $49^{-1/2} = \frac{1}{\sqrt{49}} = \frac{1}{7}$

71. $(-27)^{-2/3} = \frac{1}{(-27)^{2/3}} = \frac{1}{\left(\sqrt[3]{-27}\right)^2} = \frac{1}{(-3)^2} = \frac{1}{9}$

73. $625^{-3/4} = \frac{1}{\sqrt[4]{625^3}} = \frac{1}{\left(\sqrt[4]{625}\right)^3} = \frac{1}{5^3} = \frac{1}{125}$

75. $16^{-5/4} = \frac{1}{\sqrt[4]{16^5}} = \frac{1}{\left(\sqrt[4]{16}\right)^5} = \frac{1}{2^5} = \frac{1}{32}$

Applications

77. If $W = 27$, then $A = 100\sqrt[3]{27^2} \Rightarrow A = 100\left(\sqrt[3]{27}\right)^2 \Rightarrow A = 100(3)^2 \Rightarrow A = 100(9) \Rightarrow A = 900 \text{ in}^2$

79. (a) $r = \sqrt{\dfrac{A}{\pi}} \Rightarrow r = \left(\dfrac{A}{\pi}\right)^{1/2}$

(b) If $A = 9\pi$, then $r = \sqrt{\dfrac{9\pi}{\pi}} \Rightarrow r^2 = \dfrac{9\pi}{\pi} \Rightarrow r^2 = 9 \cdot \dfrac{\pi}{\pi} \Rightarrow r^2 = 9 \cdot 1 \Rightarrow r^2 = 9 \Rightarrow \sqrt{r^2} = \sqrt{9} \Rightarrow r = 3$ feet.

Checking Basic Concepts for Sections 8.5 & 8.6

1. (a) $\sqrt{3x+4} - 2 = 2 \Rightarrow \sqrt{3x+4} = 4 \Rightarrow \left(\sqrt{3x+4}\right)^2 = 4^2 \Rightarrow 3x + 4 = 16 \Rightarrow 3x = 12 \Rightarrow x = 4$

Check: $x = 4 \Rightarrow \sqrt{3(4)+4} - 2 = 2 \Rightarrow \sqrt{12+4} - 2 = 2 \Rightarrow \sqrt{16} - 2 = 2 \Rightarrow 4 - 2 = 2 \Rightarrow 2 = 2$, so $x = 4$

(b) $\sqrt{24-2x} = -x \Rightarrow \left(\sqrt{24-2x}\right)^2 = (-x)^2 \Rightarrow 24 - 2x = x^2 \Rightarrow 0 = x^2 + 2x - 24 \Rightarrow$

$0 = (x+6)(x-4)$, so $x = -6, 4$ Check: $x = -6 \Rightarrow \sqrt{24 - 2(-6)} = -(-6) \Rightarrow \sqrt{24 - (-12)} = 6 \Rightarrow$

$\sqrt{36} = 6 \Rightarrow 6 = 6$, and $\sqrt{24 - 2(4)} = -4 \Rightarrow \sqrt{16} = -4 \Rightarrow 4 \neq -4$, so $x = -6$

2. Solve: $S = 4\sqrt{t} - 2$ for t, $S + 2 = 4\sqrt{t} \Rightarrow \dfrac{S+2}{4} = \sqrt{t} \Rightarrow \left(\dfrac{S+2}{4}\right)^2 = \left(\sqrt{t}\right)^2 \Rightarrow \dfrac{(S+2)^2}{16} = t \Rightarrow t = \dfrac{(S+2)^2}{16}$

3. (a) $\sqrt[3]{-27} = -3$, because $(-3)^3 = -27$

(b) $\sqrt[4]{16} = 2$, because $2^4 = 16$

4. (a) $\sqrt[4]{9} \cdot \sqrt[4]{9} = \sqrt[4]{81} = 3$, because $3^4 = 81$

(b) $\sqrt[3]{\dfrac{9}{64}} = \dfrac{\sqrt[3]{9}}{\sqrt[3]{64}} = \dfrac{\sqrt[3]{9}}{4}$, because $4^3 = 64$

(c) $\dfrac{\sqrt[3]{40}}{\sqrt[3]{5}} = \sqrt[3]{\dfrac{40}{5}} = \sqrt[3]{8} = 2$, because $2^3 = 8$

5. (a) $9^{1/2} = \sqrt{9} = 3$

(b) $16^{3/4} = \sqrt[4]{16^3} = \left(\sqrt[4]{16}\right)^3 = 8$

(c) $1000^{-1/3} = \dfrac{1}{\sqrt[3]{1000}} = \dfrac{1}{10}$

Chapter 8 Review Exercises

Section 8.1

1. $\pm\sqrt{36} = \pm 6$

2. $\pm\sqrt{400} = \pm 20$

3. $\pm\sqrt{7} = \pm 2.646$

4. $\pm\sqrt{12} = \pm 3.464$

5. real, rational

6. real, rational

7. real, irrational

8. None

9. $\sqrt{121} = 11$

10. $\sqrt{400} = 20$

11. $-\sqrt{49} = -7$

12. $-\sqrt{17} = -4.12$

13. $\sqrt[3]{6} = 1.82$ because $1.82^3 = 6$

14. $\sqrt[3]{-125} = -5$ because $(-5)^3 = -125$

15. Using Pythagorean Theorem: $a^2 + b^2 = c^2$, for $a = 4$ inches, $b = 5$ inches, we get $4^2 + 5^2 = c^2 \Rightarrow$

$16 + 25 = c^2 \Rightarrow 41 = c^2 \Rightarrow \sqrt{41} = \sqrt{c^2} \Rightarrow c = \sqrt{41}$ inches.

16. Using Pythagorean Theorem: $a^2 + b^2 = c^2$, for $a = 7$ feet, $c = 25$ feet, we get $7^2 + b^2 = 25^2 \Rightarrow$

$49 + b^2 = 625 \Rightarrow b^2 = 576 \Rightarrow \sqrt{b^2} = \sqrt{576} \Rightarrow b = 24$ feet.

17. Using the distance formula: $d = \sqrt{(x_2 - x_1)^2 + (y_2 - y_1)^2}$, for $(-3, 2), (3, -3)$, we get

$d = \sqrt{(3-(-3))^2 + (-3-2)^2} \Rightarrow \quad d = \sqrt{(6)^2 + (-5)^2} \Rightarrow d = \sqrt{36+25} \Rightarrow d = \sqrt{61} \Rightarrow \quad d = \sqrt{61} \approx 7.81.$

18. Using Pythagorean Theorem: $a^2 + b^2 = c^2$, for $a = \sqrt{20}, b = \sqrt{30}$, we get $\left(\sqrt{20}\right)^2 + \left(\sqrt{30}\right)^2 = c^2 \Rightarrow$

$20 + 30 = c^2 \Rightarrow 50 = c^2 \Rightarrow \sqrt{50} = \sqrt{c^2} \Rightarrow c = \sqrt{50} \Rightarrow c = \sqrt{25} \cdot \sqrt{2} \Rightarrow c = 5\sqrt{2}$. Therefore the

perimeter is $2\sqrt{20} + 2\sqrt{30} \Rightarrow 2\left(\sqrt{4} \cdot \sqrt{5}\right) + 2\sqrt{30} \Rightarrow 2(2)\sqrt{5} + 2\sqrt{30} \Rightarrow 4\sqrt{5} + 2\sqrt{30}$.

19. Using the distance formula: $d = \sqrt{(x_2 - x_1)^2 + (y_2 - y_1)^2}$, for $(2, -5), (7, 7)$, we get

$d = \sqrt{(7-2)^2 + (7-(-5))^2} \Rightarrow \quad d = \sqrt{(5)^2 + (12)^2} \Rightarrow d = \sqrt{25+144} \Rightarrow d = \sqrt{169} \Rightarrow \quad d = 13.$

20. Using the distance formula: $d = \sqrt{(x_2 - x_1)^2 + (y_2 - y_1)^2}$, for $(0, -3), (-1, 2)$, we get

$d = \sqrt{(-1-0)^2 + (2-(-3))^2} \Rightarrow \quad d = \sqrt{(-1)^2 + (5)^2} \Rightarrow d = \sqrt{1+25} \Rightarrow d = \sqrt{26} \Rightarrow \quad d = \sqrt{26} \approx 5.099.$

21. Using the distance formula: $d = \sqrt{(x_2 - x_1)^2 + (y_2 - y_1)^2}$, for $(6, -4), (-5, -4)$, we get

$$d = \sqrt{(-5-6)^2 + (-4-(-4))^2} \Rightarrow d = \sqrt{(-11)^2 + (0)^2} \Rightarrow d = \sqrt{121+0} \Rightarrow d = \sqrt{121} \Rightarrow d = 11.$$

22. Using the distance formula: $d = \sqrt{(x_2 - x_1)^2 + (y_2 - y_1)^2}$, for $(20, -30), (5, -10)$, we get

$$d = \sqrt{(5-20)^2 + (-10-(-30))^2} \Rightarrow d = \sqrt{(-15)^2 + (20)^2} \Rightarrow d = \sqrt{225+400} \Rightarrow d = \sqrt{625} \Rightarrow d = 25.$$

Section 8.2

23. $\sqrt{6} \cdot \sqrt{5} = \sqrt{6 \cdot 5} = \sqrt{30}$

24. $\sqrt{\frac{1}{3}} \cdot \sqrt{\frac{2}{3}} = \sqrt{\frac{1}{3} \cdot \frac{2}{3}} = \sqrt{\frac{2}{9}} = \frac{\sqrt{2}}{\sqrt{9}} = \frac{\sqrt{2}}{3}$

25. $\sqrt{4} \cdot \sqrt{64} = 2 \cdot 8 = 16$

26. $\sqrt{2x} \cdot \sqrt{2x} = \sqrt{2x \cdot 2x} = \sqrt{4x^2} = 2x$

27. $\sqrt{3t} \cdot \sqrt{27t} = \sqrt{3t \cdot 27t} = \sqrt{81t^2} = 9t$

28. $\sqrt{7a} \cdot \sqrt{2b} = \sqrt{7a \cdot 2b} = \sqrt{14ab}$

29. $\sqrt{45} = \sqrt{9} \cdot \sqrt{5} = 3\sqrt{5}$

30. $\sqrt{200} = \sqrt{100} \cdot \sqrt{2} = 10\sqrt{2}$

31. $\sqrt{t^5} = \sqrt{t^4} \cdot \sqrt{t} = t^2\sqrt{t}$

32. $\sqrt{5b^2} = \sqrt{b^2} \cdot \sqrt{5} = b\sqrt{5}$

33. $\sqrt{9x^2y} = \sqrt{9x^2} \cdot \sqrt{y} = 3x\sqrt{y}$

34. $\sqrt{49a^3b} = \sqrt{49a^2} \cdot \sqrt{ab} = 7a\sqrt{ab}$

35. $\sqrt{\frac{25}{36}} = \frac{\sqrt{25}}{\sqrt{36}} = \frac{5}{6}$

36. $\sqrt{5} \cdot \sqrt{8} = \sqrt{5 \cdot 8} = \sqrt{40} = \sqrt{4} \cdot \sqrt{10} = 2\sqrt{10}$

37. $\sqrt{\frac{3}{x}} \cdot \sqrt{\frac{27}{x}} = \sqrt{\frac{3}{x} \cdot \frac{27}{x}} = \sqrt{\frac{81}{x^2}} = \frac{\sqrt{81}}{\sqrt{x^2}} = \frac{9}{x}$

38. $\frac{\sqrt{64}}{\sqrt{4}} = \frac{8}{2} = 4$

39. $\frac{\sqrt{50}}{\sqrt{2}} = \sqrt{\frac{50}{2}} = \sqrt{25} = 5$

40. $\dfrac{\sqrt{3r^3}}{\sqrt{r}}=\dfrac{\sqrt{r^2}\cdot\sqrt{3}\cdot\sqrt{r}}{\sqrt{r}}=r\sqrt{3}\cdot\dfrac{\sqrt{r}}{\sqrt{r}}=r\sqrt{3}\cdot 1=r\sqrt{3}$

41. $\sqrt{\dfrac{x}{16}}\cdot\sqrt{\dfrac{x^2}{4}}=\sqrt{\dfrac{x^3}{64}}=\dfrac{\sqrt{x^2\cdot x}}{\sqrt{64}}=\dfrac{\sqrt{x^2}\cdot\sqrt{x}}{8}=\dfrac{x\sqrt{x}}{8}$

42. $\sqrt{3ab^3}\cdot\sqrt{27ab}=\sqrt{3ab^3\cdot 27ab}=\sqrt{81a^2b^4}=9ab^2$

43. $\dfrac{\sqrt{75ab}}{\sqrt{3ab}}=\sqrt{\dfrac{75ab}{3ab}}=\sqrt{25}\cdot\dfrac{\sqrt{ab}}{\sqrt{ab}}=5\cdot 1=5$

44. $\dfrac{\sqrt{8xy^3}}{\sqrt{2xy}}=\sqrt{\dfrac{8xy^3}{2xy}}=\sqrt{4y^2}\cdot\dfrac{\sqrt{xy}}{\sqrt{xy}}=2y\cdot 1=2y$

Section 8.3

45. $8\sqrt{7}-5\sqrt{7}=(8-5)\sqrt{7}=3\sqrt{7}$

46. $\sqrt{15}+2\sqrt{15}=(1+2)\sqrt{15}=3\sqrt{15}$

47. $2\sqrt{2}+\sqrt{8}=2\sqrt{2}+\left(\sqrt{4}\cdot\sqrt{2}\right)=2\sqrt{2}+2\sqrt{2}=(2+2)\sqrt{2}=4\sqrt{2}$

48. $3\sqrt{45}-2\sqrt{20}=3\left(\sqrt{9}\cdot\sqrt{5}\right)-2\left(\sqrt{4}\cdot\sqrt{5}\right)=3(3)\sqrt{5}-2(2)\sqrt{5}=9\sqrt{5}-4\sqrt{5}=(9-4)\sqrt{5}=5\sqrt{5}$

49. $5\sqrt[3]{9}-2\sqrt[3]{9}=(5-2)\sqrt[3]{9}=3\sqrt[3]{9}$

50. $9\sqrt[3]{10}+\sqrt[3]{10}=(9+1)\sqrt[3]{10}=10\sqrt[3]{10}$

51. $\sqrt{4t}+\sqrt{9t}=\left(\sqrt{4}\cdot\sqrt{t}\right)+\left(\sqrt{9}\cdot\sqrt{t}\right)=2\sqrt{t}+3\sqrt{t}=(2+3)\sqrt{t}=5\sqrt{t}$

52. $2\sqrt{ab}-\sqrt{4ab}=2\sqrt{ab}-\left(\sqrt{4}\cdot\sqrt{ab}\right)=2\sqrt{ab}-2\sqrt{ab}=0$

Section 8.4

53. $\left(9-\sqrt{6}\right)\left(9+\sqrt{6}\right)=9^2-\left(\sqrt{6}\right)^2=81-6=75$

54. $\left(2+\sqrt{t}\right)\left(2-\sqrt{t}\right)=2^2-\left(\sqrt{t}\right)^2=4-t$

55. $\dfrac{5}{\sqrt{5}}=\dfrac{5}{\sqrt{5}}\cdot\dfrac{\sqrt{5}}{\sqrt{5}}=\dfrac{5\sqrt{5}}{\sqrt{5}\cdot\sqrt{5}}=\dfrac{5\sqrt{5}}{5}=\dfrac{5}{5}\cdot\sqrt{5}=1\cdot\sqrt{5}=\sqrt{5}$

56. $-\dfrac{9}{3\sqrt{11}}=-\dfrac{9}{3\sqrt{11}}\cdot\dfrac{\sqrt{11}}{\sqrt{11}}=-\dfrac{9\sqrt{11}}{3\sqrt{11}\cdot\sqrt{11}}=-\dfrac{3\sqrt{11}}{11}$

57. $-\dfrac{4}{3\sqrt{x}}\cdot\dfrac{\sqrt{x}}{\sqrt{x}}=-\dfrac{4\cdot\sqrt{x}}{3\sqrt{x}\cdot\sqrt{x}}=-\dfrac{4\sqrt{x}}{3x}$

58. $\dfrac{\sqrt{5x^3}}{\sqrt{25x}}=\dfrac{\sqrt{x^2}\cdot\sqrt{5x}}{\sqrt{25}\cdot\sqrt{x}}=\dfrac{x\sqrt{5x}}{5\sqrt{x}}\cdot\dfrac{\sqrt{x}}{\sqrt{x}}=\dfrac{x\sqrt{5x}\cdot\sqrt{x}}{5\sqrt{x}\cdot\sqrt{x}}=\dfrac{x\sqrt{5x^2}}{5x}\Rightarrow\dfrac{x}{x}\cdot\dfrac{\sqrt{5}\cdot\sqrt{x^2}}{5}=1\cdot\dfrac{\sqrt{5}x}{5}=\dfrac{\sqrt{5}x}{5}$

59. $\frac{1}{\sqrt{5}-2}=\frac{1}{\sqrt{5}-2}\cdot\frac{\sqrt{5}+2}{\sqrt{5}+2}=\frac{\sqrt{5}+2}{\sqrt{5}^2-2^2}=\frac{\sqrt{5}+2}{5-4}=\frac{\sqrt{5}+2}{1}=\sqrt{5}+2$

60. $\frac{4-\sqrt{b}}{4+\sqrt{b}}=\frac{4-\sqrt{b}}{4+\sqrt{b}}\cdot\frac{4-\sqrt{b}}{4-\sqrt{b}}=\frac{4^2-8\sqrt{b}+\sqrt{b}^2}{4^2-\sqrt{b}^2}=\frac{16-8\sqrt{b}+b}{16-b}$

Section 8.5

61. $\sqrt{x}=5\Rightarrow\sqrt{x}^2=5^2\Rightarrow x=25$ Check: $x=25\Rightarrow\sqrt{25}=5\Rightarrow 5=5$

62. No solution, a radical cannot equal a negative number.

63. $\sqrt{2-t}=5\Rightarrow\left(\sqrt{2-t}\right)^2=5^2\Rightarrow 2-t=25\Rightarrow -t=23\Rightarrow t=-23$ Check: $t=-23\Rightarrow$

$\sqrt{2-(-23)}=5\Rightarrow\sqrt{25}=5\Rightarrow 5=5$

64. $\sqrt{3b+1}=5\Rightarrow\left(\sqrt{3b+1}\right)^2=5^2\Rightarrow 3b+1=25\Rightarrow 3b=24\Rightarrow b=8$ Check: $b=8\Rightarrow$

$\sqrt{3(8)+1}=5\Rightarrow\sqrt{25}=5\Rightarrow 5=5$

65. $\sqrt{x+4}-1=5\Rightarrow\sqrt{x+4}=6\Rightarrow\left(\sqrt{x+4}\right)^2=6^2\Rightarrow x+4=36\Rightarrow x=32$ Check: $x=32\Rightarrow$

$\sqrt{32+4}-1=5\Rightarrow\sqrt{36}-1=5\Rightarrow 6-1=5\Rightarrow 5=5$

66. $\sqrt{8x}=4\sqrt{x-4}\Rightarrow\frac{\sqrt{8x}}{4}=\sqrt{x-4}\Rightarrow\left(\frac{\sqrt{8x}}{4}\right)^2=\left(\sqrt{x-4}\right)^2\Rightarrow\frac{8x}{16}=x-4\Rightarrow\frac{x}{2}=x-4\Rightarrow$

$x=2x-8\Rightarrow 0=x-8\Rightarrow x=8$ Check: $x=8\Rightarrow\sqrt{8(8)}=4\sqrt{8-4}\Rightarrow\sqrt{64}=4\sqrt{4}\Rightarrow 8=4(2)\Rightarrow 8=8$

67. $\frac{1}{2}\sqrt{2b+4}+4=b\Rightarrow\sqrt{2b+4}=2(b-4)\Rightarrow\left(\sqrt{2b+4}\right)^2=(2b-8)^2\Rightarrow$

$2b+4=4b^2-32b+64\Rightarrow 0=4b^2-34b+60\Rightarrow 2\left(2b^2-17b+30\right)=0\Rightarrow$

$2(2b-5)(b-6)=0\Rightarrow 2b-5=0\Rightarrow 2b=5\Rightarrow b=\frac{5}{2}$, and $b-6=0\Rightarrow b=6$ Check: $b=6\Rightarrow$

$\frac{1}{2}\sqrt{2(6)+4}+4=6\Rightarrow\frac{1}{2}\sqrt{16}+4=6\Rightarrow\frac{1}{2}(4)+4=6\Rightarrow 2+4=6\Rightarrow 6=6$, and $b=\frac{5}{2}\Rightarrow$

$\frac{1}{2}\sqrt{2\left(\frac{5}{2}\right)+4}+4=\frac{5}{2}\Rightarrow\frac{1}{2}\sqrt{5+4}+4=\frac{5}{2}\Rightarrow\frac{1}{2}\sqrt{9}+4=\frac{5}{2}\Rightarrow\frac{1}{2}(3)+4=\frac{5}{2}\Rightarrow\frac{3}{2}+\frac{8}{2}=\frac{5}{2}\Rightarrow\frac{11}{2}\neq\frac{5}{2}$, so $b=6$

68. $x-7=\sqrt{2x+1}\Rightarrow(x-7)^2=\left(\sqrt{2x+1}\right)^2\Rightarrow x^2-14x+49=2x+1\Rightarrow x^2-16x+48=0\Rightarrow$

$(x-12)(x-4)=0\Rightarrow x-12=0\Rightarrow x=12$, and $x-4=0\Rightarrow x=4$ Check: $x=12\Rightarrow$

$12-7=\sqrt{2(12)+1}\Rightarrow 5=\sqrt{25}\Rightarrow 5=5$, and $x=4\Rightarrow 4-7=\sqrt{2(4)+1}\Rightarrow -3=\sqrt{9}\Rightarrow -3\neq 3$, so $x=12$

69. $A=\sqrt{4P}$ for $P\Rightarrow A^2=4P\Rightarrow\frac{A^2}{4}=P\Rightarrow P=\frac{A^2}{4}$

70. $P = 2\pi\sqrt{\frac{L}{g}}$ for $g \Rightarrow \frac{P}{2\pi} = \sqrt{\frac{L}{g}} \Rightarrow \left(\frac{P}{2\pi}\right)^2 = \left(\sqrt{\frac{L}{g}}\right)^2 \Rightarrow \frac{P^2}{4\pi^2} = \frac{L}{g} \Rightarrow \frac{P^2}{4\pi^2 L} = \frac{1}{g} \Rightarrow g = \frac{4\pi^2 L}{P^2}$

71. (a) From the table we see that $\sqrt{x}+2=2$ when $x=0$. See Figure 71a.

(b) The solution is the x-coordinate of the intersection point, or 0. See Figure 71b.

x	0	1	4	9	16	25
$\sqrt{x}+2$	2	3	4	5	6	7

Figure 71a

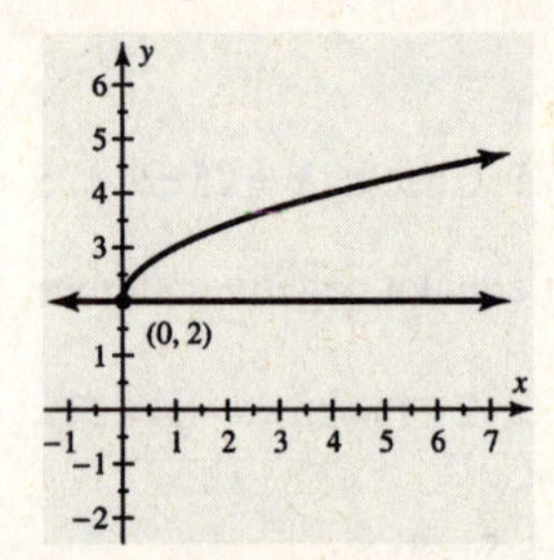

Figure 71b

72. (a) From the table we see that $\sqrt{x-1}=4$ when $x=17$. See Figure 72a.

(b) The solution is the x-coordinate of the intersection point, or 17. See Figure 72b.

x	1	2	5	10	17	26
$\sqrt{x-1}$	0	1	2	3	4	5

Figure 72a

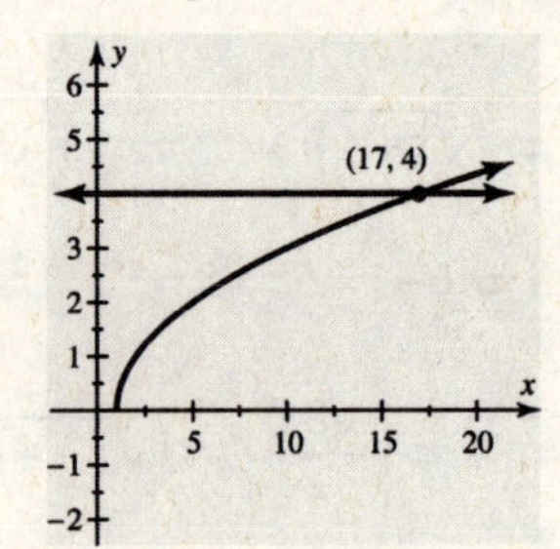

Figure 72b

Section 8.6

73. $\sqrt[3]{64} = 4$, because $4^3 = 64$

74. $\sqrt[3]{-1000} = -10$, because $(-10)^3 = -1000$

75. $-\sqrt[4]{81} = -3$, because $-\left(3^4\right) = -(81)$

76. $\sqrt[5]{32} = 2$, because $2^5 = 32$

77. $\sqrt[4]{7^4} = 7^{4/4} = 7^1 = 7$

78. $-\sqrt[6]{\frac{1}{64}} = -\frac{\sqrt[6]{1}}{\sqrt[6]{64}} = -\frac{1}{2}$, because $-\frac{1^6}{2^6} = -\frac{1}{64}$

79. $\sqrt[3]{-4}\cdot\sqrt[3]{16} = \sqrt[3]{-4\cdot 16} = \sqrt[3]{-64} = -4$, because $(-4)^3 = -64$

80. $\sqrt[4]{3}\cdot\sqrt[4]{27} = \sqrt[4]{3\cdot 27} = \sqrt[4]{81} = 3$, because $3^4 = 81$

81. $\frac{\sqrt[3]{2}}{\sqrt[3]{16}} = \sqrt[3]{\frac{2}{16}} = \sqrt[3]{\frac{1}{8}} = \frac{\sqrt[3]{1}}{\sqrt[3]{8}} = \frac{1}{2}$, because $\frac{1^3}{2^3} = \frac{1}{8}$

82. $\dfrac{\sqrt[4]{2}}{\sqrt[4]{32}}=\sqrt[4]{\dfrac{2}{32}}=\sqrt[4]{\dfrac{1}{16}}=\dfrac{\sqrt[4]{1}}{\sqrt[4]{16}}=\dfrac{1}{2}$, because $\dfrac{1^4}{2^4}=\dfrac{1}{16}$

83. $\sqrt[4]{\dfrac{2}{3}}\cdot\sqrt[4]{\dfrac{4}{27}}=\sqrt[4]{\dfrac{2}{3}\cdot\dfrac{4}{27}}=\sqrt[4]{\dfrac{8}{81}}=\dfrac{\sqrt[4]{8}}{\sqrt[4]{81}}=\dfrac{\sqrt[4]{8}}{3}$ because $3^4=81$

84. $\sqrt[3]{\dfrac{4}{27}}=\dfrac{\sqrt[3]{4}}{\sqrt[3]{27}}=\dfrac{\sqrt[3]{4}}{3}$ because $3^3=27$

85. $25^{3/2}=\sqrt{25^3}=\left(\sqrt{25}\right)^3=5^3=125$

86. $1000^{1/3}=\sqrt[3]{1000}=10$

87. $16^{-5/4}=\dfrac{1}{\sqrt[4]{16^5}}=\dfrac{1}{\left(\sqrt[4]{16}\right)^5}=\dfrac{1}{2^5}=\dfrac{1}{32}$

88. $(-8)^{-2/3}=\dfrac{1}{\sqrt[3]{(-8)^2}}=\dfrac{1}{\left(\sqrt[3]{-8}\right)^2}=\dfrac{1}{-2^2}=\dfrac{1}{4}$

Applications

89. $a^2+b^2=c^2\Rightarrow a=30\cdot 2.5, b=16\cdot 2.5\Rightarrow a=75, b=40$, we get $75^2+40^2=c^2\Rightarrow$

$5625+1600=c^2\Rightarrow 7225=c^2\Rightarrow\sqrt{7225}=\sqrt{c^2}\Rightarrow 85=c\Rightarrow c=85$ miles.

90. $a^2+b^2=c^2\Rightarrow a=400, b=300$, we get $400^2+300^2=c^2\Rightarrow 160{,}000+90{,}000=c^2\Rightarrow$

$250{,}000=c^2\Rightarrow\sqrt{250{,}000}=\sqrt{c^2}\Rightarrow 500=c\Rightarrow c=500$ feet.

91. $R=150\text{ bpm}\Rightarrow 150=\dfrac{885}{\sqrt{W}}\Rightarrow\dfrac{150}{885}=\dfrac{1}{\sqrt{W}}\Rightarrow\sqrt{W}=\dfrac{885}{150}\Rightarrow\left(\sqrt{W}\right)^2=5.9^2\Rightarrow W=$ about 35 lb

92. $T=4\text{ seconds}\Rightarrow 4=\sqrt{\dfrac{h}{4}}\Rightarrow 4^2=\left(\sqrt{\dfrac{h}{4}}\right)^2\Rightarrow 16=\dfrac{h}{4}\Rightarrow 64=h\Rightarrow h=64$ feet

93. (a) $T=D^{3/2}\Rightarrow T=\sqrt{D^3}\Rightarrow T=\left(\sqrt{D}\right)^3$

(b) $D=1.52\Rightarrow T=\left(\sqrt{1.52}\right)^3\Rightarrow T=1.233^3\Rightarrow T=1.87$ years

94. $V=v\sqrt{\dfrac{D}{d}}$ for $d\Rightarrow\dfrac{V}{v}=\sqrt{\dfrac{D}{d}}\Rightarrow\left(\dfrac{V}{v}\right)^2=\left(\sqrt{\dfrac{D}{d}}\right)^2\Rightarrow\dfrac{V^2}{v^2}=\dfrac{D}{d}\Rightarrow\dfrac{V^2}{v^2D}=\dfrac{1}{d}\Rightarrow d=\dfrac{v^2D}{V^2}$

95. $P=2\left(100-\sqrt{98}\right)+2\left(100+\sqrt{98}\right)=\left(200-2\sqrt{98}\right)+\left(200+2\sqrt{98}\right)=400+0=400$

$A=\left(100-\sqrt{98}\right)\left(100+\sqrt{98}\right)=100^2-\left(\sqrt{98}\right)^2=10{,}000-98=9902$

96. $D=10\Rightarrow 10=1.22\sqrt{h}\Rightarrow\dfrac{10}{1.22}=\sqrt{h}\Rightarrow 8.19=\sqrt{h}\Rightarrow 8.19^2=\sqrt{h}^2\Rightarrow h=$ about 67 feet

97. (a) $L=8 \Rightarrow P=2\pi\sqrt{\frac{8}{32}} \Rightarrow P=2\pi\sqrt{\frac{1}{4}} \Rightarrow P=2\pi\left(\frac{1}{2}\right) \Rightarrow P=\frac{2}{2}\cdot\pi \Rightarrow P=\ 1\cdot\pi \Rightarrow\ P=\pi$ seconds

(b) $P=2 \Rightarrow 2=2\pi\sqrt{\frac{L}{32}} \Rightarrow \frac{2}{2\pi}=\sqrt{\frac{L}{32}} \Rightarrow \left(\frac{2}{2\pi}\right)^2=\left(\sqrt{\frac{L}{32}}\right)^2 \Rightarrow \frac{4}{4\pi^2}=\frac{L}{32} \Rightarrow \frac{128}{4\pi^2}=L \Rightarrow$

$L=$ about 3.24 feet

98. (a) $L=27.5\sqrt[3]{W} \Rightarrow L=27.5W^{1/3}$

(b) $W=8 \Rightarrow L=27.5\sqrt[3]{8} \Rightarrow L=27.5(2) \Rightarrow L=55$ inches

Chapter 8 Test

1. $\pm\sqrt{64}=\pm 8$

2. real, irrational

3. $\sqrt{81}=9$, because $9^2=81$

4. $\sqrt[3]{-64}=-4$, because $(-4)^3=-64$

5. $\sqrt[5]{32}=2$, because $2^5=32$

6. $\sqrt{11}\approx 3.32$

7. $25^{3/2}=\sqrt{25^3}=\left(\sqrt{25}\ \right)^3=5^3=125$

8. $16^{-3/4}=\frac{1}{16^{3/4}}=\frac{1}{\sqrt[4]{16^3}}=\frac{1}{\left(\sqrt[4]{16}\right)^3}=\frac{1}{2^3}=\frac{1}{8}$

9. Using Pythagorean Theorem: $a^2+b^2=c^2$, for $a=5$ inches, $b=7$ inches, we get $5^2+7^2=c^2 \Rightarrow$

$25+49=c^2 \Rightarrow 74=c^2 \Rightarrow \sqrt{74}\ =\sqrt{c^2} \Rightarrow \sqrt{74}=c \Rightarrow c=\sqrt{74}$ inches.

10. Using the distance formula: $d=\sqrt{(x_2-x_1)^2+(y_2-y_1)^2}$, for $(-3,-2),(3,1)$, we get

$d=\sqrt{(3-(-3))^2+(1-(-2))^2} \Rightarrow\ d=\sqrt{(6)^2+(3)^2} \Rightarrow d=\sqrt{36+9} \Rightarrow d=\sqrt{45}$ or $\sqrt[3]{5}$.

11. $\mathrm{D}=a^2+b^2=c^2$, for $a=\sqrt{8}, b=\sqrt{32}$, we get $\left(\sqrt{8}\right)^2+\left(\sqrt{32}\right)^2=c^2 \Rightarrow 8+32=c^2 \Rightarrow 40=c^2 \Rightarrow$

$\sqrt{40}\ =\sqrt{c^2} \Rightarrow \sqrt{40}=c \Rightarrow c=\sqrt{4}\cdot\sqrt{10} \Rightarrow c=2\sqrt{10}$

$P=2\sqrt{8}+2\sqrt{32}=2\left(\sqrt{4}\cdot\sqrt{2}\ \right)+2\left(\sqrt{16}\cdot\sqrt{2}\ \right)=2(2)\sqrt{2}+2(4)\sqrt{2}=4\sqrt{2}+8\sqrt{2}=(4+8)\sqrt{2}=\ 12\sqrt{2}$

12. Using $d=\sqrt{(x_2-x_1)^2+(y_2-y_1)^2}$, for $(5,-4),(-2,3)$, we get $d=\sqrt{(-2-5)^2+(3-(-4))^2} \Rightarrow$

$d=\sqrt{(-7)^2+(7)^2} \Rightarrow d=\sqrt{49+49} \Rightarrow d=\sqrt{98} \Rightarrow d=\sqrt{49}\cdot\sqrt{2} \Rightarrow d=7\sqrt{2}\approx 9.899$.

13. $\sqrt{54}=\sqrt{9}\cdot\sqrt{6}=3\sqrt{6}$

14. $\sqrt{9t^3} = \sqrt{9t^2} \cdot \sqrt{t} = 3t\sqrt{t}$

15. $\sqrt{6} \cdot \sqrt{24} = \sqrt{6 \cdot 24} = \sqrt{144} = 12$

16. $\dfrac{\sqrt{90}}{\sqrt{10}} = \sqrt{\dfrac{90}{10}} = \sqrt{9} = 3$

17. $\sqrt{3a} \cdot \sqrt{27a} = \sqrt{3a \cdot 27a} = \sqrt{81a^2} = 9a$

18. $\dfrac{\sqrt{4r^5}}{\sqrt{r^3}} = \dfrac{\sqrt{4r^2} \cdot \sqrt{r^3}}{\sqrt{r^3}} = 2r\dfrac{\sqrt{r^3}}{\sqrt{r^3}} = 2r \cdot 1 = 2r$

19. $\sqrt{\dfrac{x}{3}} \cdot \sqrt{\dfrac{x}{3}} = \sqrt{\dfrac{x}{3} \cdot \dfrac{x}{3}} = \sqrt{\dfrac{x^2}{9}} = \dfrac{x}{3}$

20. $\sqrt{8ab^2} \cdot \sqrt{2a^2b} = \sqrt{8ab^2 \cdot 2a^2b} = \sqrt{16a^3b^3} = \sqrt{16a^2b^2} \cdot \sqrt{ab} = 4ab\sqrt{ab}$

21. $7\sqrt{5} + 2\sqrt{5} = (7+2)\sqrt{5} = 9\sqrt{5}$

22. $3\sqrt{8} - \sqrt{18} = 3(\sqrt{4} \cdot \sqrt{2}) - (\sqrt{9} \cdot \sqrt{2}) = 3(2)\sqrt{2} - 3\sqrt{2} = 6\sqrt{2} - 3\sqrt{2} = (6-3)\sqrt{2} = 3\sqrt{2}$

23. $2\sqrt[3]{11} - 3\sqrt[3]{11} = (2-3)\sqrt[3]{11} = -\sqrt[3]{11}$

24. $\sqrt{25ab} + \sqrt{9ab} = (\sqrt{25} \cdot \sqrt{ab}) + (\sqrt{9} \cdot \sqrt{ab}) = 5\sqrt{ab} + 3\sqrt{ab} = (5+3)\sqrt{ab} = 8\sqrt{ab}$

25. $\sqrt{8} + \sqrt{72} - \sqrt{27} = 2\sqrt{2} + 6\sqrt{2} - 3\sqrt{3} = (2+6)\sqrt{2} - 3\sqrt{3} = 8\sqrt{2} - 3\sqrt{3}$

26. $\sqrt{96} + 2\sqrt{24} = 4\sqrt{6} + 2(2)\sqrt{6} = 4\sqrt{6} + 4\sqrt{6} = (4+4)\sqrt{6} = 8\sqrt{6}$

27. $\sqrt{4b} - \sqrt{9b} + \sqrt{16b} = 2\sqrt{b} - 3\sqrt{b} + 4\sqrt{b} = (2-3+4)\sqrt{b} = 3\sqrt{b}$

28. $\sqrt{16x} + \sqrt{4x^3} = 4\sqrt{x} + 2x\sqrt{x} = (4+2x)\sqrt{x}$

29. $\dfrac{4}{2\sqrt{3}} = \dfrac{4}{2\sqrt{3}} \cdot \dfrac{\sqrt{3}}{\sqrt{3}} = \dfrac{4\sqrt{3}}{2\sqrt{3} \cdot \sqrt{3}} = \dfrac{4\sqrt{3}}{2(3)} = \dfrac{4\sqrt{3}}{6} = \dfrac{2\sqrt{3}}{3}$

30. $\dfrac{2-\sqrt{x}}{2+\sqrt{x}} = \dfrac{2-\sqrt{x}}{2+\sqrt{x}} \cdot \dfrac{2-\sqrt{x}}{2-\sqrt{x}} = \dfrac{2^2 - 4\sqrt{x} + \sqrt{x^2}}{2^2 - \sqrt{x^2}} = \dfrac{4 - 4\sqrt{x} + x}{4 - x}$

31. $\sqrt{2x-5} = 5 \Rightarrow (\sqrt{2x-5})^2 = 5^2 \Rightarrow 2x - 5 = 25 \Rightarrow 2x = 30 \Rightarrow x = 15$ Check: $x = 15 \Rightarrow$

$\sqrt{2(15)-5} = 5 \Rightarrow \sqrt{30-5} = 5 \Rightarrow \sqrt{25} = 5 \Rightarrow 5 = 5$, so $x = 15$

32. $\sqrt{x+30} = x \Rightarrow (\sqrt{x+30})^2 = x^2 \Rightarrow x + 30 = x^2 \Rightarrow x^2 - x - 30 = 0 \Rightarrow 0 = (x-6)(x+5) \Rightarrow x = 6$ or $x = -5$

Check: $x = 6 \Rightarrow \sqrt{6+30} = 6 \Rightarrow \sqrt{36} = 6 \Rightarrow 6 = 6$, so $x = 6$; $x = -5 \Rightarrow \sqrt{-5+30} = -5 \Rightarrow \sqrt{25} = -5 \Rightarrow 5 = -5 \Rightarrow x \neq -5$

33. $b = \sqrt{c^2 - a^2}$ for $c \Rightarrow b^2 = (\sqrt{c^2 - a^2})^2 \Rightarrow b^2 = c^2 - a^2 \Rightarrow a^2 + b^2 = c^2 \Rightarrow \sqrt{a^2 + b^2} = \sqrt{c^2} \Rightarrow c = \sqrt{a^2 + b^2}$

34. $a^2+b^2=c^2$ for $a=60\cdot\frac{3}{4}, b=40\cdot\frac{3}{4}\Rightarrow a=45, b=30$, we get $45^2+30^2=c^2\Rightarrow$

$2025+900=c^2\Rightarrow 2925=c^2\Rightarrow\sqrt{2925}=\sqrt{c^2}\Rightarrow 54\approx c\Rightarrow c\approx 54$ miles.

35. (a) $W=64\Rightarrow R=\frac{885}{\sqrt{64}}\Rightarrow R=\frac{885}{8}\Rightarrow R\approx 111$ bpm

(b) $R=177\Rightarrow 177=\frac{885}{\sqrt{W}}\Rightarrow\frac{177}{885}=\frac{1}{\sqrt{W}}\Rightarrow\sqrt{W}=\frac{885}{177}\Rightarrow\left(\sqrt{W}\right)^2=5^2\Rightarrow W=25$ pounds

(c) $R=\frac{885}{\sqrt{W}}\Rightarrow R=885W^{-1/2}$

(d) $R=\frac{885}{\sqrt{W}}$ for $W\Rightarrow\frac{R}{885}=\frac{1}{\sqrt{W}}\Rightarrow\sqrt{W}=\frac{885}{R}\Rightarrow\left(\sqrt{W}\right)^2=\left(\frac{885}{R}\right)^2\Rightarrow W=\left(\frac{885}{R}\right)^2$

Chapter 8 Extended and Discovery Exercises

1. $\sqrt{2x}-1=\sqrt{x+1}\Rightarrow\left(\sqrt{2x}-1\right)^2=\left(\sqrt{x+1}\right)^2\Rightarrow 2x-2\sqrt{2x}+1=x+1\Rightarrow -2\sqrt{2x}=-x\Rightarrow$

$\sqrt{2x}=\frac{x}{2}\Rightarrow\left(\sqrt{2x}\right)^2=\left(\frac{x}{2}\right)^2\Rightarrow 2x=\frac{x^2}{4}\Rightarrow 8x=x^2\Rightarrow 0=x^2-8x\Rightarrow x(x-8)=0\Rightarrow$

$x=0$, or $x-8=0\Rightarrow x=8$ Check: $x=0\Rightarrow\sqrt{2(0)}-1=\sqrt{0+1}\Rightarrow -1\neq 1$, and $x=8\Rightarrow$

$\sqrt{2(8)}-1=\sqrt{8+1}\Rightarrow\sqrt{16}-1=\sqrt{9}\Rightarrow 4-1=3\Rightarrow 3=3$, so $x=8$

2. $\sqrt{x}=\sqrt{x-7}+1\Rightarrow\sqrt{x}-1=\sqrt{x-7}\Rightarrow\left(\sqrt{x}-1\right)^2=\left(\sqrt{x-7}\right)^2\Rightarrow x-2\sqrt{x}+1=x-7\Rightarrow$

$-2\sqrt{x}=-8\Rightarrow\sqrt{x}=4\Rightarrow\left(\sqrt{x}\right)^2=4^2\Rightarrow x=16$ Check: $x=16\Rightarrow\sqrt{16}=\sqrt{16-7}+1\Rightarrow$

$4=\sqrt{9}+1\Rightarrow 4=3+1\Rightarrow 4=4$, so $x=16$

3. $\sqrt{t-2}=\sqrt{t+3}-1\Rightarrow\left(\sqrt{t-2}\right)^2=\left(\sqrt{t+3}-1\right)^2\Rightarrow t-2=t+3-2\sqrt{t+3}+1\Rightarrow$

$-6=-2\sqrt{t+3}\Rightarrow 3=\sqrt{t+3}\Rightarrow 3^2=\left(\sqrt{t+3}\right)^2\Rightarrow 9=t+3\Rightarrow t=6$ Check:

$\sqrt{6-2}=\sqrt{6+3}-1\Rightarrow\sqrt{4}=\sqrt{9}-1\Rightarrow 2=3-1\Rightarrow 2=2$, so $t=6$

4. $\sqrt{2t-2}+\sqrt{t}=7\Rightarrow\sqrt{2t-2}=-\sqrt{t}+7\Rightarrow\left(\sqrt{2t-2}\right)^2=\left(-\sqrt{t}+7\right)^2\Rightarrow$

$2t-2=t-14\sqrt{t}+49\Rightarrow t-51=-14\sqrt{t}\Rightarrow(t-51)^2=\left(-14\sqrt{t}\right)^2\Rightarrow t^2-102t+2601=196t\Rightarrow$

$t^2-298t+2601=0\Rightarrow(t-9)(t-289)=0\Rightarrow t=9$ or 289 Check: $t=9\Rightarrow$

$\sqrt{2(9)-2}+\sqrt{9}=7\Rightarrow\sqrt{16}+\sqrt{9}=7\Rightarrow 4+3=7\Rightarrow 7=7$, and $t=289\Rightarrow$

$\sqrt{2(289)-2}+\sqrt{289}=7\Rightarrow\sqrt{578-2}+\sqrt{289}=7\Rightarrow\sqrt{576}+\sqrt{289}=7\Rightarrow 24+17=7\Rightarrow 41\neq 7$, so $t=9$

5. $\sqrt{b+1}-\sqrt{b-6}=1 \Rightarrow \sqrt{b+1}=\sqrt{b-6}+1 \Rightarrow \left(\sqrt{b+1}\right)^2=\left(\sqrt{b-6}+1\right)^2 \Rightarrow$

$b+1=b-6+2\sqrt{b-6}+1 \Rightarrow 6=2\sqrt{b-6} \Rightarrow 3=\sqrt{b-6} \Rightarrow 3^2=\left(\sqrt{b-6}\right)^2 \Rightarrow$

$9=b-6 \Rightarrow b=15$ Check: $b=15 \Rightarrow \sqrt{15+1}-\sqrt{15-6}=1 \Rightarrow \sqrt{16}-\sqrt{9}=1 \Rightarrow 4-3=1 \Rightarrow 1=1$, so $b=15$

6. $\sqrt{b-7}=\sqrt{b}-1 \Rightarrow \left(\sqrt{b-7}\right)^2=\left(\sqrt{b}-1\right)^2 \Rightarrow b-7=b-2\sqrt{b}+1 \Rightarrow -8=-2\sqrt{b} \Rightarrow$

$4=\sqrt{b} \Rightarrow 4^2=\left(\sqrt{b}\right)^2 \Rightarrow b=16$ Check: $b=16 \Rightarrow \sqrt{16-7}=\sqrt{16}-1 \Rightarrow \sqrt{9}=\sqrt{16}-1 \Rightarrow$

$3=4-1 \Rightarrow 3=3$, so $b=16$

7. $\sqrt[3]{x}=3 \Rightarrow \left(\sqrt[3]{x}\right)^3=3^3 \Rightarrow x=27$ Check: $x=27 \Rightarrow \sqrt[3]{27}=3 \Rightarrow 3=3$, so $x=27$

8. $\sqrt[3]{2x}+1=3 \Rightarrow \sqrt[3]{2x}=2 \Rightarrow \left(\sqrt[3]{2x}\right)^3=2^3 \Rightarrow 2x=8 \Rightarrow x=4$ Check: $x=4 \Rightarrow \sqrt[3]{2(4)}+1=3 \Rightarrow$

$\sqrt[3]{8}+1=3 \Rightarrow 2+1=3 \Rightarrow 3=3$, so $x=4$

9. $\sqrt[3]{2x-2}+1=4 \Rightarrow \sqrt[3]{2x-2}=3 \Rightarrow \left(\sqrt[3]{2x-2}\right)^3=3^3 \Rightarrow 2x-2=27 \Rightarrow 2x=29 \Rightarrow x=\dfrac{29}{2}$

Check: $x=\dfrac{29}{2} \Rightarrow \sqrt[3]{2\left(\dfrac{29}{2}\right)-2}+1=4 \Rightarrow \sqrt[3]{29-2}+1=4 \Rightarrow \sqrt[3]{27}+1=4 \Rightarrow 3+1=4 \Rightarrow 4=4$, so $x=\dfrac{29}{2}$

10. $\sqrt[4]{t}=2 \Rightarrow \left(\sqrt[4]{t}\right)^4=2^4 \Rightarrow t=16$ Check: $t=16 \Rightarrow \sqrt[4]{16}=2 \Rightarrow 2=2$, so $t=16$

11. $\sqrt[4]{4x}+2=4 \Rightarrow \sqrt[4]{4x}=2 \Rightarrow \left(\sqrt[4]{4x}\right)^4=2^4 \Rightarrow 4x=16 \Rightarrow x=4$ Check: $x=4 \Rightarrow$

$\sqrt[4]{4(4)}+2=4 \Rightarrow \sqrt[4]{16}+2=4 \Rightarrow 2+2=4 \Rightarrow 4=4$, so $x=4$

12. $\sqrt[5]{5x-3}=2 \Rightarrow \left(\sqrt[5]{5x-3}\right)^5=2^5 \Rightarrow 5x-3=32 \Rightarrow 5x=35 \Rightarrow x=7$ Check: $x=7 \Rightarrow$

$\sqrt[5]{5(7)-3}=2 \Rightarrow \sqrt[5]{35-3}=2 \Rightarrow \sqrt[5]{32}=2 \Rightarrow 2=2$, so $x=7$

13. Raise each side to the sixth power: $\sqrt[3]{3x}=\sqrt{x} \Rightarrow \left(\sqrt[3]{3x}\right)^6=\left(\sqrt{x}\right)^6 \Rightarrow (3x)^2=x^3 \Rightarrow 9x^2=x^3 \Rightarrow$

$x^3-9x^2=0 \Rightarrow x^2(x-9)=0 \Rightarrow x=0, 9$

14. Perfect squares between 10 and 99 are: 16, 25, 36, 49, 64 and 81. The perfect cubes are: 27 and $64 \Rightarrow 64$ is the only perfect square and cube between 10 and 99.

15. Perfect cubes between 100 and 999 are: 125, 216, 343, 512 and 729. The only perfect square of those numbers is $729 \Rightarrow 729$ is the only perfect square and cube between 100 and 999.

Chapters 1-8 Cumulative Review

1. $60 = 2 \cdot 2 \cdot 3 \cdot 5$

2. $45 - 3^2 \cdot 8 \div 2 = 45 - 9 \cdot 8 \div 2 = 45 - 72 \div 2 = 45 - 36 = 9$

3. $\dfrac{2x^2}{5y} \div \dfrac{x}{15y^2} = \dfrac{2x^2}{5y} \cdot \dfrac{15y^2}{x} = \dfrac{30x^2y^2}{5xy} = \dfrac{30}{5} \cdot \dfrac{x^2}{x} \cdot \dfrac{y^2}{y} = 6 \cdot x \cdot y = 6xy$

4. $4t - 3 = 2 \Rightarrow 4t = 5 \Rightarrow t = \dfrac{5}{4}$

   Check: $4\left(\dfrac{5}{4}\right) - 3 = 2 \Rightarrow 5 - 3 = 2 \Rightarrow 2 = 2$, so $t = \dfrac{5}{4}$

5. $45\% = \dfrac{45}{100} = \dfrac{9}{20}$; $45\% = 45 \times 0.01 = 0.45$

6. $-2x + 1 \geq 4 \Rightarrow -2x \geq 3 \Rightarrow x \leq -\dfrac{3}{2}$

7. See Figure 7.

8. See Figure 8.

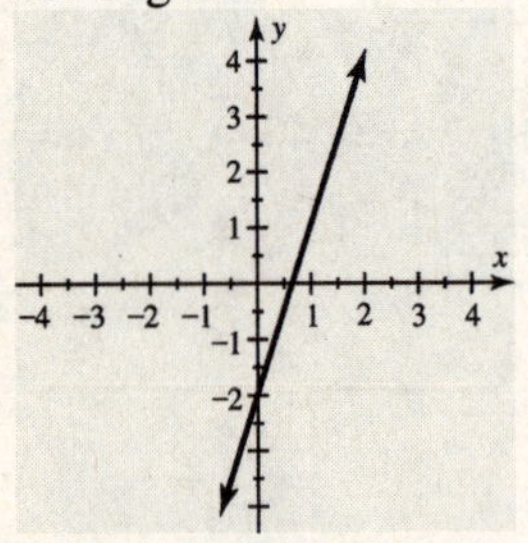

Figure 7

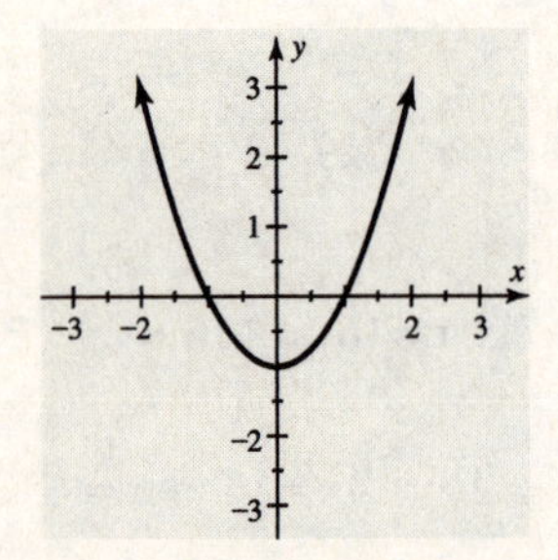

Figure 8

9. $3x - 2y = 6 \Rightarrow -2y = -3x + 6 \Rightarrow y = \dfrac{3}{2}x - 3$; the line perpendicular to this line has slope $m = -\dfrac{2}{3}$;

   $y - y_1 = m(x - x_1) \Rightarrow y - (-1) = -\dfrac{2}{3}(x - 1) \Rightarrow y + 1 = -\dfrac{2}{3}x + \dfrac{2}{3} \Rightarrow y = -\dfrac{2}{3}x - \dfrac{1}{3}$

10. Multiply the second equation by -1 and add the equations to eliminate the variable y.

$$\begin{array}{r} 2x + y = 4 \\ x - y = -1 \\ \hline 3x = 3, \end{array}$$

   $x = 1 \Rightarrow y = 2$, so (1, 2) is the solution.

11. $\left(4x^2 - 2\right) - \left(-2x^2 + 1\right) = 6x^2 - 3$

12. $\left(a^3b^5\right)\left(a^2b\right) = a^{3+2}b^{5+1} = a^5b^6$

13. $6x\left(-4x^2 + 2x\right) = -24x^3 + 12x^2$

14. $(x - 2)(x + 5) = x^2 + 5x - 2x - 10 = x^2 + 3x - 10$

15. $\frac{1}{4^{-3}} = \frac{4^3}{1} = 4^3$ or 64

16. $\left(3t^2\right)^{-3} = 3^{-3}t^{-6} = \frac{1}{3^3} \cdot \frac{1}{t^6} = \frac{1}{27t^6}$

17. $6.5 \times 10^4 = 65{,}000$

18. $\frac{9x^3 - 6x^2}{3x} = \frac{9x^3}{3x} - \frac{6x^2}{3x} = \frac{9}{3} \cdot \frac{x^3}{x} - \frac{6}{3} \cdot \frac{x^2}{x} = 3x^2 - 2x$

19. $x^2 + 2x - 15 = (x+5)(x-3)$ (FOIL)

20. $16z^2 - 9 = (4z-3)(4z+3)$ (difference of squares)

21. $x^3 - y^3 = (x-y)\left(x^2 + xy + y^2\right)$ (difference of cubes)

22. $x^4 + 4x^2 + 3 = \left(x^2 + 3\right)\left(x^2 + 1\right)$ (FOIL)

23. $x^2 - 5x - 14 = 0 \Rightarrow (x-7)(x+2) = 0 \Rightarrow x - 7 = 0,\ x = 7$ or $x + 2 = 0,\ x = -2$, so $x = -2, 7$

24. $x^4 + 16 = 8x^2 \Rightarrow x^4 - 8x^2 + 16 = 0 \Rightarrow \left(x^2 - 4\right)\left(x^2 - 4\right) = 0$

$\Rightarrow \left(x^2 - 4\right)^2 = 0 \Rightarrow x^2 - 4 = 0 \Rightarrow (x-2)(x+2) = 0 \Rightarrow x - 2 = 0,\ x = 2$ or $x + 2 = 0,\ x = -2$, so $x = -2, 2$

25. $\frac{1}{x-2} + \frac{1}{x^2-4} = \frac{1}{x-2} \cdot \frac{x+2}{x+2} + \frac{1}{(x-2)(x+2)} = \frac{x+2+1}{(x-2)(x+2)} = \frac{x+3}{x^2-4}$

26. $\frac{xy}{x+y} \div \frac{x}{x^2-y^2} = \frac{xy}{x+y} \cdot \frac{x^2-y^2}{x} = \frac{xy(x-y)(x+y)}{(x+y)x} = \frac{x}{x} \cdot \frac{x+y}{x+y} \cdot \frac{y(x-y)}{1} = 1 \cdot 1 \cdot \frac{y(x-y)}{1} = y(x-y)$

27. $\frac{\frac{1}{y} + \frac{1}{x}}{\frac{1}{y} - \frac{1}{x}} \cdot \frac{xy}{xy} = \frac{x+y}{x-y}$

28. $\frac{1}{x-1} + \frac{1}{x+1} = \frac{1}{x^2-1} \Rightarrow \frac{1}{x-1} \cdot \frac{x+1}{x+1} + \frac{1}{x+1} \cdot \frac{x-1}{x-1} = \frac{1}{x^2-1}$

$\Rightarrow \frac{x+1+x-1}{(x-1)(x+1)} = \frac{1}{x^2-1} \Rightarrow \frac{2x}{x^2-1} = \frac{1}{x^2-1}$

$\Rightarrow 2x\left(x^2-1\right) = x^2 - 1 \Rightarrow \frac{2x\left(x^2-1\right)}{x^2-1} = \frac{x^2-1}{x^2-1}$

$\Rightarrow 2x = 1 \Rightarrow x = \frac{1}{2}$

29. $\sqrt{121} = 11$

30. Using $d=\sqrt{(x_2-x_1)^2+(y_2-y_1)^2}$, for $(3,-4)$ and $(1,1)$, we get

$$d=\sqrt{(1-3)^2+(1-(-4))^2} \Rightarrow d=\sqrt{(-2)^2+(5)^2} \Rightarrow d=\sqrt{4+25} \Rightarrow d=\sqrt{29}$$

31. (a) $\sqrt{2t}\cdot\sqrt{8t}=\sqrt{2t\cdot 8t}=\sqrt{16t^2}=\sqrt{16\cdot t^2}=\sqrt{16}\cdot\sqrt{t^2}=4t$

(b) $\sqrt{200}+\sqrt{8}=\sqrt{100\cdot 2}+\sqrt{4\cdot 2}=\sqrt{100}\cdot\sqrt{2}+\sqrt{4}\cdot\sqrt{2}=10\sqrt{2}+2\sqrt{2}=(10+2)\sqrt{2}=12\sqrt{2}$

(c) $\dfrac{\sqrt[3]{16x}}{\sqrt[3]{2}}=\sqrt[3]{\dfrac{16x}{2}}=\sqrt[3]{8x}=\sqrt[3]{8}\cdot\sqrt[3]{x}=2\sqrt[3]{x}$

(d) $(1-\sqrt{2})(1+\sqrt{2})=(1)^2-(\sqrt{2})^2=1-2=-1$

(e) $81^{3/4}=(\sqrt[4]{81})^3=3^3=27$

(f) $32^{-1/5}=\dfrac{1}{32^{1/5}}=\dfrac{1}{\sqrt[5]{32}}=\dfrac{1}{2}$

32. $5\sqrt{2-x}+1=11 \Rightarrow 5\sqrt{2-x}=10 \Rightarrow \sqrt{2-x}=2 \Rightarrow (\sqrt{2-x})^2=2^2 \Rightarrow 2-x=4 \Rightarrow -x=2 \Rightarrow x=-2$

33. $x^2+y^2=4 \Rightarrow y^2=4-x^2 \Rightarrow \sqrt{y^2}=\sqrt{4-x^2} \Rightarrow y=\sqrt{4-x^2}$

34. $\sqrt[3]{-27}=-3$

35. $\dfrac{500}{8}=\dfrac{x}{13} \Rightarrow 6500=8x \Rightarrow x=812.5 \Rightarrow \$812.50$

36. Let x represent the minutes on the stair climber and $50-x$ represent the minutes rowing.

$10x+9(50-x)=470 \Rightarrow 10x+450-9x=470 \Rightarrow x=20$, so 20 minutes on the stair climber and 30 minutes rowing.

37. Let x represent the measure of the smallest angle. Then $x+x+2x=180 \Rightarrow 4x=180 \Rightarrow x=45$, so $45°, 45°, 90°$

38. Using Pythagorean Theorem: $a^2+b^2=c^2$, for $a=60(1.5)$ miles, $b=40(1.5)$ miles $\Rightarrow$

$a=90$ miles, $b=60$ miles $\Rightarrow 90^2+60^2=c^2 \Rightarrow 8100+3600=c^2 \Rightarrow 11{,}700=c^2 \Rightarrow$

$\sqrt{c^2}=\sqrt{11{,}700} \Rightarrow c\approx 108$ miles

Critical Thinking Solutions for Chapter 8

Section 8.2

- $\sqrt{x^{16}}=\left(\sqrt{x^8}\right)^2=x^8$, so $\sqrt{x^{16}}\neq x^4$ but $\sqrt{x^{16}}=x^8$

Section 8.4

- $$\frac{\sqrt{x}-\sqrt{y}}{\sqrt{x}+\sqrt{y}}=\frac{\sqrt{x}-\sqrt{y}}{\sqrt{x}+\sqrt{y}}\cdot\frac{\sqrt{x}+\sqrt{y}}{\sqrt{x}+\sqrt{y}}=\frac{\left(\sqrt{x}\right)^2-\left(\sqrt{y}\right)^2}{\left(\sqrt{x}\right)^2+2\sqrt{x\cdot y}+\left(\sqrt{y}\right)^2}=\frac{x-y}{x+2\sqrt{xy}+y}$$

Section 8.5

- First divide each side by 2 and then cube each side: $2\sqrt[3]{x}=6\Rightarrow\sqrt[3]{x}=3\Rightarrow\left(\sqrt[3]{x}\right)^3=3^3\Rightarrow x=27$

- Using $P=2\pi\sqrt{\frac{L}{32}}$ and quadrupling L gives us $P=2\pi\sqrt{\frac{4L}{32}}\Rightarrow P=2\pi\sqrt{4\cdot\frac{L}{32}}\Rightarrow P=2\pi(2)\sqrt{\frac{L}{32}}$ which is 2 times the original, so it doubles.

Chapter 9: Quadratic Equations

9.1: Parabolas

Concepts

1. parabola
2. The vertex
3. axis of symmetry
4. (0, 0)
5. See Figure 5. *Answers may vary.*

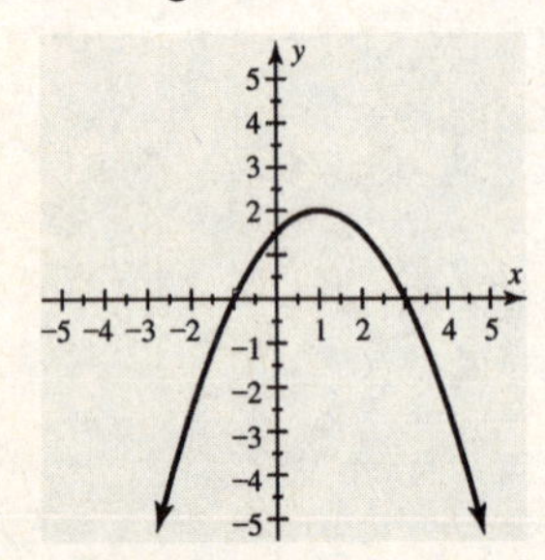

Figure 5

6. $-\dfrac{b}{2a}$
7. narrower
8. downward
9. ax^2+bx+c with $a \neq 0$
10. vertex
11. The equation is in the form $y=ax^2+bx+c$ with $a=5$. Since $a>0$, the parabola opens upward.
12. The equation is in the form $y=ax^2+bx+c$ with $a=-2$. Since $a<0$, the parabola opens downward.
13. Rewrite the equation as $y=-4x^2+3x-7$ with $a=-4$. Since $a<0$, the parabola opens downward.
14. Rewrite the equation as $y=2x^2-3x+8$ with $a=2$. Since $a>0$, the parabola opens upward.
15. Rewrite the equation as $y=-x^2-12x-36$ with $a=-1$. Since $a<0$, the parabola opens downward.
16. Rewrite the equation as $y=-x^2-2x+2$ with $a=-1$. Since $a<0$, the parabola opens downward.
17. Rewrite the equation as $y=x^2-4x+4$ with $a=1$. Since $a>0$, the parabola opens upward.
18. Rewrite the equation as $y=-x^2+6x-5$ with $a=-1$. Since $a<0$, the parabola opens downward.

Graphs of Quadratic Equations

19. The vertex is $(1,-2)$. The axis of symmetry is $x=1$. The parabola opens upward.

    The graph is increasing when $x \geq 1$ and decreasing when $x \leq 1$.

21. The vertex is $(-2, 3)$. The axis of symmetry is $x = -2$. The parabola opens downward.

    The graph is increasing when $x \leq -2$ and decreasing when $x \geq -2$.

23. (a) See Figure 23.

    (b) The vertex is $(0, -2)$. The axis of symmetry is $x = 0$.

25. (a) See Figure 25.

    (b) The vertex is $(0, 1)$. The axis of symmetry is $x = 0$.

27. (a) See Figure 27.

    (b) The vertex is $(1, 0)$. The axis of symmetry is $x = 1$.

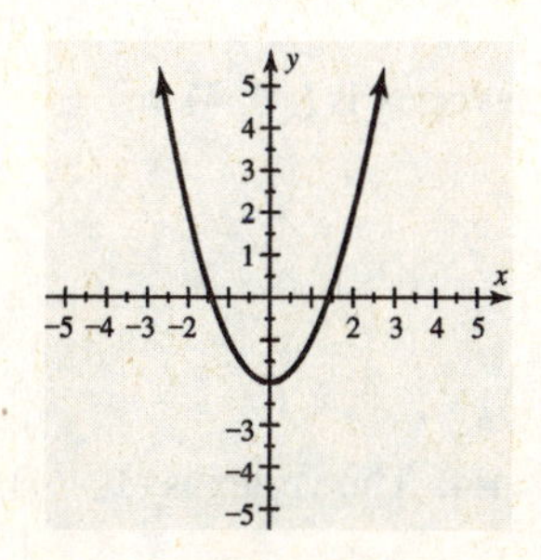

Figure 23

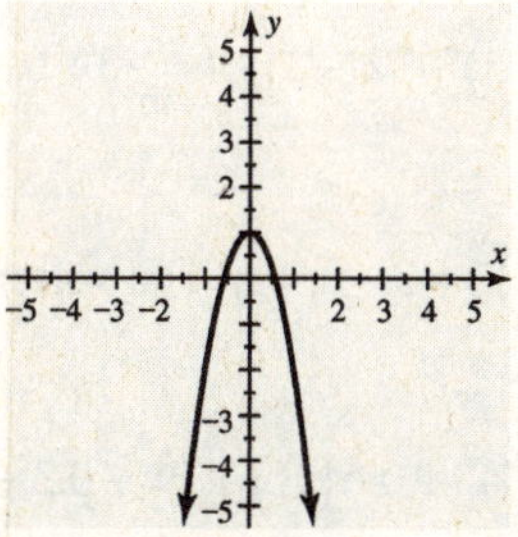

Figure 25

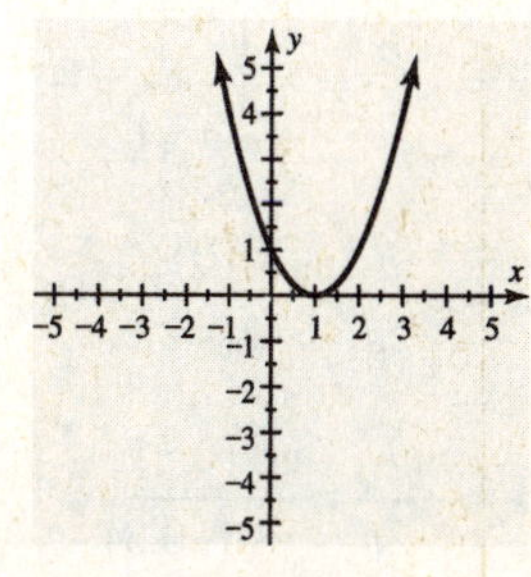

Figure 27

29. (a) See Figure 29.

    (b) The vertex is $(-0.5, -2.25)$. The axis of symmetry is $x = -0.5$.

31. (a) See Figure 31.

    (b) The vertex is $(0, -3)$. The axis of symmetry is $x = 0$.

33. (a) See Figure 33.

    (b) The vertex is $(1, 1)$. The axis of symmetry is $x = 1$.

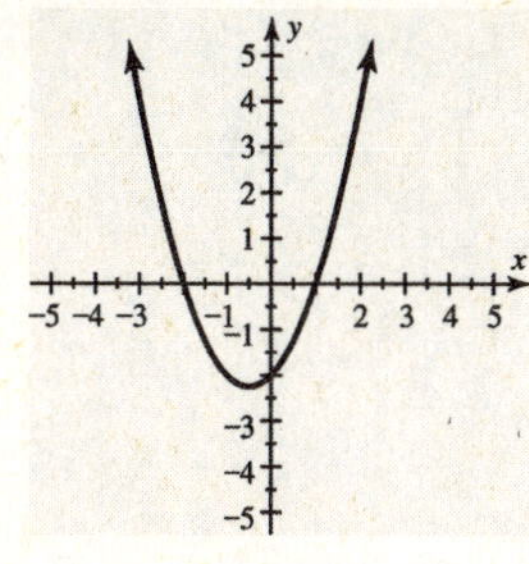

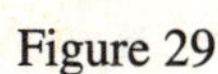

Figure 29

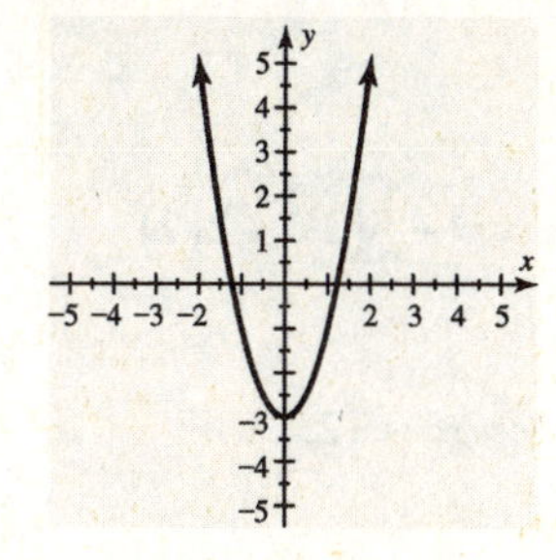

Figure 31

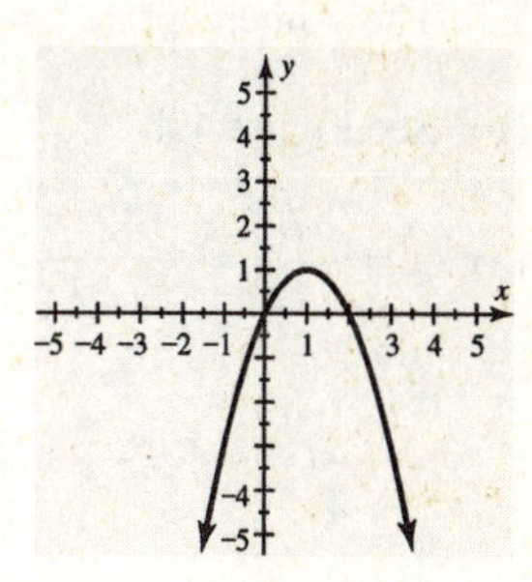

Figure 33

35. (a) See Figure 35.

    (b) The vertex is $(1, 1)$. The axis of symmetry is $x = 1$.

37. (a) See Figure 37.

    (b) The vertex is $(2, 4)$. The axis of symmetry is $x = 2$.

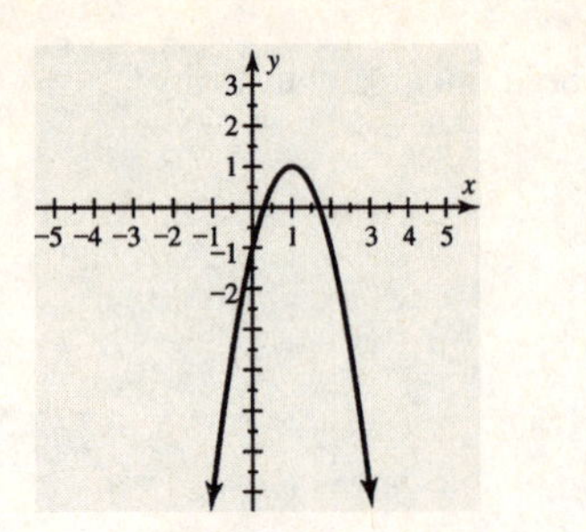

Figure 35

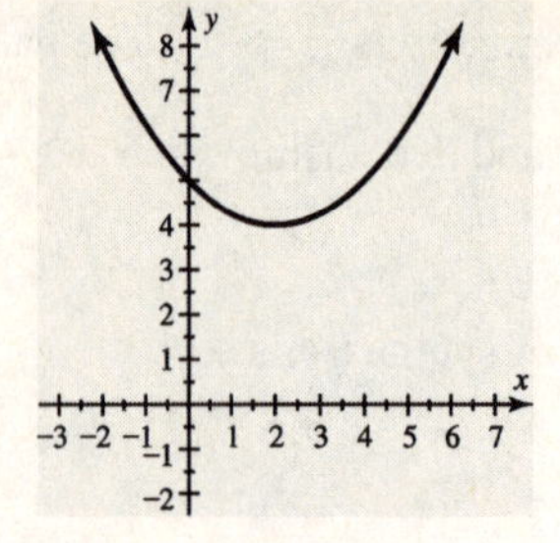

Figure 37

39. $x=-\dfrac{b}{2a}=-\dfrac{(-4)}{2(1)}=2,\ y=(2)^2-4(2)-2=4-8-2=-6$; The vertex is (2, –6).

41. $x=-\dfrac{b}{2a}=-\dfrac{(-2)}{2\left(-\frac{1}{3}\right)}=-3,\ y=-\dfrac{1}{3}(-3)^2-2(-3)+1=-3+6+1=4$; The vertex is (–3, 4).

43. $x=-\dfrac{b}{2a}=-\dfrac{(0)}{2(-2)}=0,\ y=3-2(0)^2=3-0=3$; The vertex is (0, 3).

45. $x=-\dfrac{b}{2a}=-\dfrac{(0.6)}{2(-0.3)}=1,\ y=-0.3(1)^2+0.6(1)+1.1=-0.3+0.6+1.1=1.4$; The vertex is (1, 1.4).

47. (a) $x=-\dfrac{b}{2a}=-\dfrac{0}{2(1)}=0,\ y=(0)^2-4=0-4=-4$; the vertex is $(0,-4)$.

The y-intercept is $y=(0)^2-4=0-4=-4$.

(b) See Figure 47b.

49. (a) $x=-\dfrac{b}{2a}=-\dfrac{2}{2(1)}=-\dfrac{2}{2}=-1,\ y=(-1)^2+2(-1)-1=1-2-1=-2$; the vertex is $(-1,-2)$.

The y-intercept is $y=(0)^2+2(0)-1=0+0-1=-1$.

(b) See Figure 49b.

51. (a) $x=-\dfrac{b}{2a}=-\dfrac{2}{2\left(-\frac{1}{2}\right)}=-\dfrac{2}{-1}=2,\ y=-\dfrac{1}{2}(2)^2+2(2)$

$=-\dfrac{1}{2}(4)+4=-2+4=2$; the vertex is $(2,2)$.

The y-intercept is $y=-\dfrac{1}{2}(0)^2+2(0)=0+0=0$.

(b) See Figure 51b.

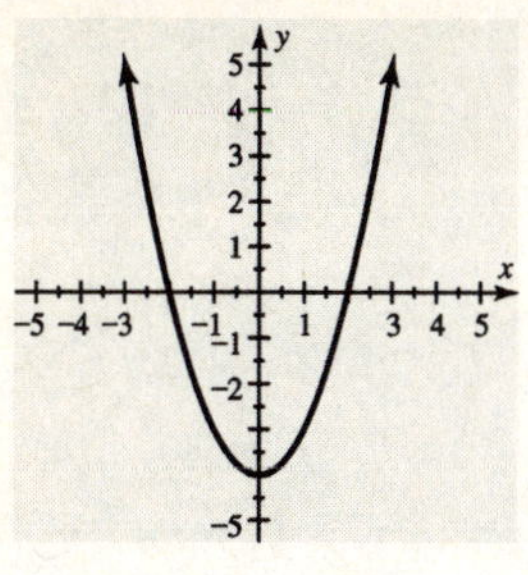

Figure 47b

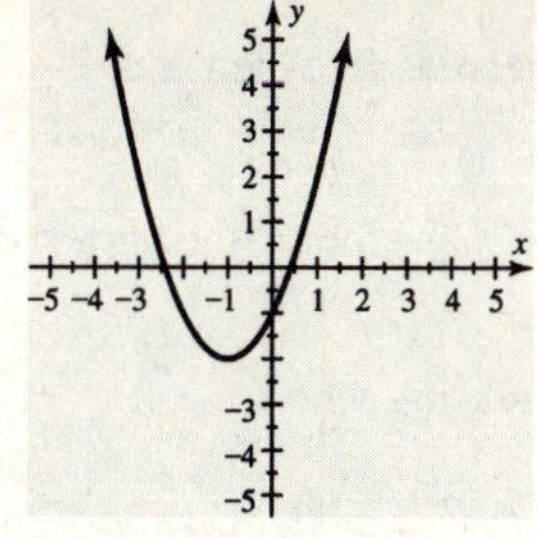

Figure 49b

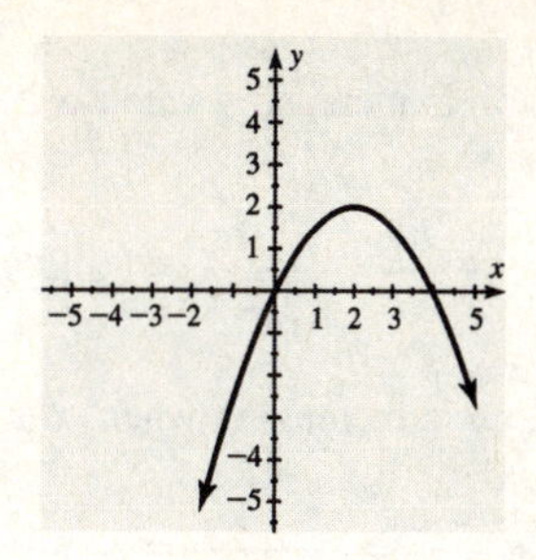

Figure 51b

53. See Figure 53. Compared to $y = x^2$, the graph is reflected across the x-axis.

55. See Figure 55. Compared to $y = x^2$, the graph is narrower.

57. See Figure 57. Compared to $y = x^2$, the graph is wider.

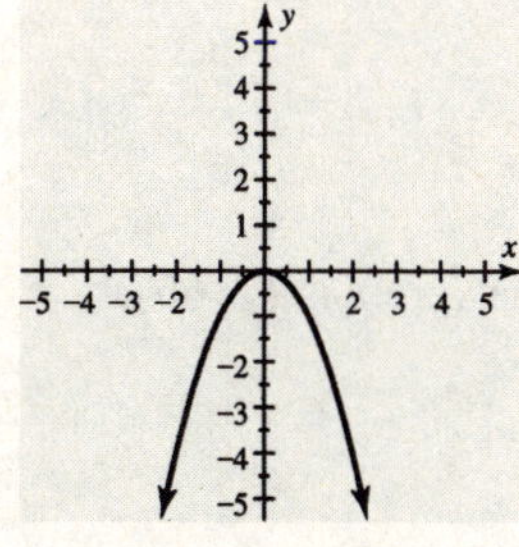

Figure 53

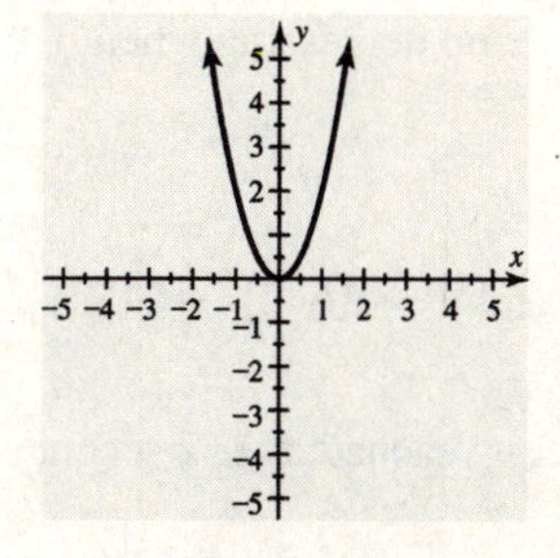

Figure 55

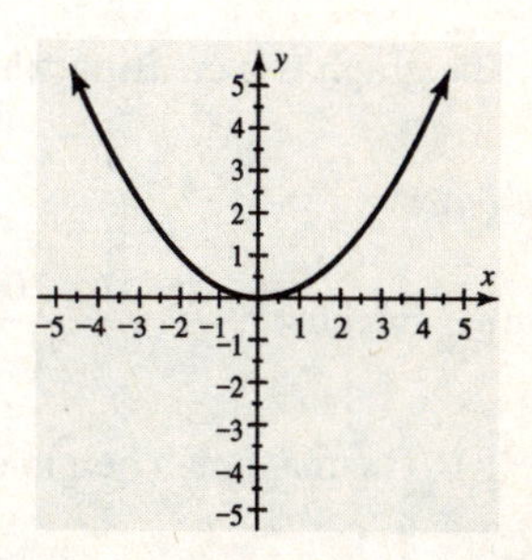

Figure 57

59. See Figure 59. Compared to $y = x^2$, the graph is reflected across the x-axis and is wider.

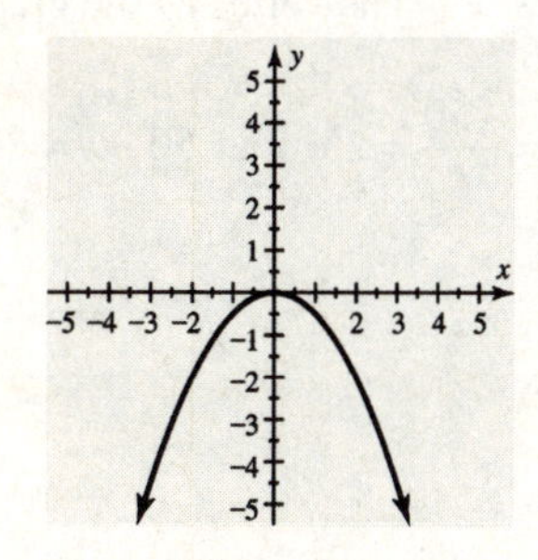

Figure 59

61. Because $-\frac{b}{2a} = -\frac{2}{2(1)} = -1$ and $y = (-1)^2 + 2(-1) - 1 = -2$, the vertex is $(-1, -2)$.

The graph is increasing when $x \ge -1$ and decreasing when $x \le -1$.

63. Because $-\frac{b}{2a} = -\frac{-5}{2(1)} = \frac{5}{2}$ and $y = \left(\frac{5}{2}\right)^2 - 5\left(\frac{5}{2}\right) = -\frac{25}{4}$, the vertex is $\left(\frac{5}{2}, -\frac{25}{4}\right)$.

The graph is increasing when $x \ge \frac{5}{2}$ and decreasing when $x \le \frac{5}{2}$.

65. Because $-\frac{b}{2a} = -\frac{2}{2(2)} = -\frac{1}{2}$ and $y = 2\left(-\frac{1}{2}\right)^2 + 2\left(-\frac{1}{2}\right) - 3 = -\frac{7}{2}$, the vertex is $\left(-\frac{1}{2}, -\frac{7}{2}\right)$.

The graph is increasing when $x \ge -\frac{1}{2}$ and decreasing when $x \le -\frac{1}{2}$.

67. Because $-\frac{b}{2a} = -\frac{2}{2(-1)} = 1$ and $y = -(1)^2 + 2(1) + 5 = 6$, the vertex is $(1, 6)$.

The graph is increasing when $x \le 1$ and decreasing when $x \ge 1$.

69. Because $-\frac{b}{2a} = -\frac{4}{2(-1)} = 2$ and $y = 4(2) - (2)^2 = 4$, the vertex is $(2, 4)$.

The graph is increasing when $x \le 2$ and decreasing when $x \ge 2$.

71. Because $-\frac{b}{2a} = -\frac{1}{2(-2)} = \frac{1}{4}$ and $y = -2\left(\frac{1}{4}\right)^2 + \left(\frac{1}{4}\right) - 5 = -\frac{39}{8}$, the vertex is $\left(\frac{1}{4}, -\frac{39}{8}\right)$.

The graph is increasing when $x \le \frac{1}{4}$ and decreasing when $x \ge \frac{1}{4}$.

Applications

73. (a) Because $-\frac{b}{2a} = -\frac{64}{2(-16)} = 2$ and $y = -16(2)^2 + 64(2) + 2 = 66$, the vertex is $(2, 66)$.

(b) The maximum height of 66 feet is reached after 2 seconds.

75. Let x represent one of the dimensions of the pen. Then $\frac{1}{2}(100 - 2x) = 50 - x$ represents the other dimension.

The area of the pen is the product of the dimensions, $x \cdot (50 - x) = 50x - x^2$. The vertex of the graph of

$y = 50x - x^2$ is the maximum value of the graph. $x = -\frac{b}{2a} = -\frac{50}{2(-1)} = -\frac{50}{-2} = 25$, and $50 - x = 25$.

The pen should be 25 ft by 25 ft.

9.2: Introduction to Quadratic Equations

Concepts

1. $x^2 + 3x - 2 = 0$; *Answers may vary.* A quadratic equation can have 0, 1 or 2 solutions.

2. ± 3

3. Factoring and square root property.

4. See Figure 4. *Answers may vary.*

5. See Figure 5. *Answers may vary.*

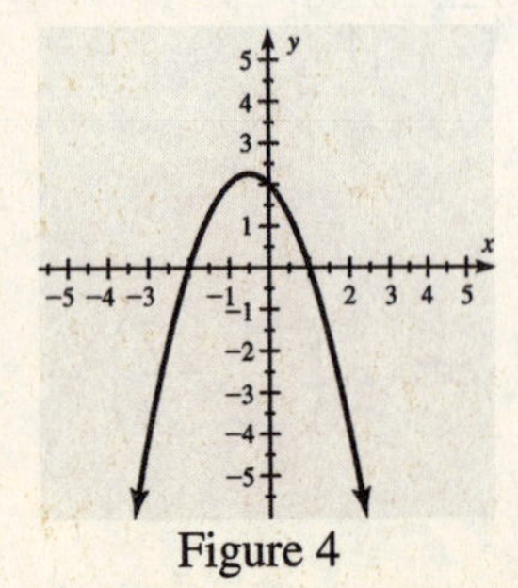
Figure 4

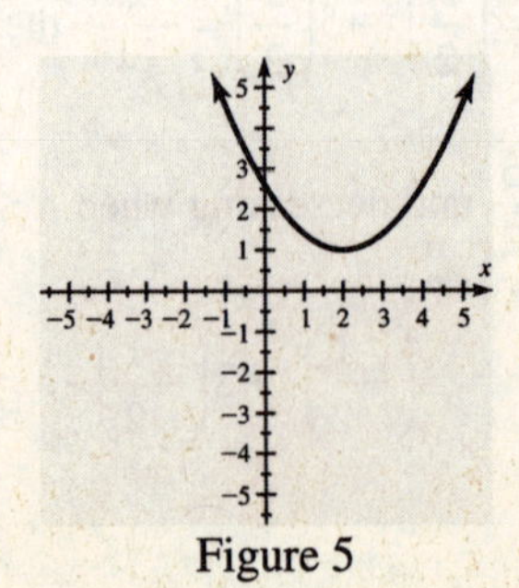
Figure 5

6. The equation has two solutions. The solutions are the x-intercepts.

7. Yes

8. Yes

9. No, there is no x^2 term.

10. No, the degree of this polynomial is 3.

11. Yes

12. Yes

13. No, the term $\sqrt{x}$ is not allowed in a quadratic equation.

14. No, the variable x is located in the denominator.

Solving Quadratic Equations

15. The solutions are the x-intercepts, –2, 1.

17. Since there are no x-intercepts, there are no real solutions.

19. The solutions are the x-values where $Y_1 = 0$: –2, 3.

21. The solutions are the x-values where $Y_1 = 0$: –0.5

23. $x^2 - 4x - 5 = 0 \Rightarrow (x+1)(x-5) = 0 \Rightarrow x+1 = 0$ or $x-5 = 0 \Rightarrow x = -1$ or 5

    A graph of $y = x^2 - 4x - 5$ (not shown) intersects the x-axis at -1 and 5.

25. $x^2 + 2x = 3 \Rightarrow x^2 + 2x - 3 = 0 \Rightarrow (x+3)(x-1) = 0 \Rightarrow x+3 = 0$ or $x-1 = 0 \Rightarrow x = -3$ or 1

    A graph of $y = x^2 + 2x - 3$ (not shown) intersects the x-axis at -3 and 1.

27. $x^2 = 9 \Rightarrow x^2 - 9 = 0 \Rightarrow (x+3)(x-3) = 0 \Rightarrow x+3 = 0$ or $x-3 = 0 \Rightarrow x = -3$ or 3

    A graph of $y = x^2 - 9$ (not shown) intersects the x-axis at -3 and 3.

29. $4x^2 + 1 = 0$ has no solutions. The graph (not shown) is a parabola that does not intersect the x-axis.

31. $x^2 + 2x + 1 = 0 \Rightarrow (x+1)(x+1) = 0 \Rightarrow x+1 = 0 \Rightarrow x = -1$.

    A graph of $y = x^2 + 2x + 1$ (not shown) intersects the x-axis at -1.

33. $x^2 + 2x - 35 = 0 \Rightarrow (x+7)(x-5) = 0 \Rightarrow$ Either $x+7 = 0 \Rightarrow x = -7$ or $x-5 = 0 \Rightarrow x = 5$

35. $6x^2 - x - 1 = 0 \Rightarrow (3x+1)(2x-1) = 0 \Rightarrow$ Either $3x+1 = 0 \Rightarrow x = -\frac{1}{3}$ or $2x-1 = 0 \Rightarrow x = \frac{1}{2}$

37. $2x^2 - 5x + 3 = 0 \Rightarrow (x-1)(2x-3) = 0 \Rightarrow$ Either $x-1 = 0 \Rightarrow x = 1$ or $2x-3 = 0 \Rightarrow x = \frac{3}{2}$

39. $x^2 = 144 \Rightarrow x = \pm\sqrt{144} \Rightarrow x = \pm 12$

41. $4x^2 - 64 = 0 \Rightarrow 4x^2 = 64 \Rightarrow x^2 = \frac{64}{4} \Rightarrow x^2 = 16 \Rightarrow x = \pm\sqrt{16} \Rightarrow x = \pm 4$

43. $(x+1)^2 = 25 \Rightarrow x+1 = \pm\sqrt{25} \Rightarrow x+1 = \pm 5 \Rightarrow x = -1 \pm 5 \Rightarrow x = -6$ or 4

45. $(x-1)^2 = 64 \Rightarrow x-1 = \pm\sqrt{64} \Rightarrow x-1 = \pm 8 \Rightarrow x = 1 \pm 8 \Rightarrow x = -7$ or 9

Solving Equations by More Than One Method

47. (a) $x^2 - 3x - 18 = 0 \Rightarrow (x+3)(x-6) = 0 \Rightarrow$ Either $x+3=0 \Rightarrow x=-3$ or $x-6=0 \Rightarrow x=6$

(b) Graph $Y_1 = X^2 - 3X - 18$ in [–5, 8, 1] by [–25, 5, 5]. See Figures 47a & 47b.

The solutions are the x-intercepts, $x = -3$ or $x = 6$.

(c) Table $Y_1 = X^2 - 3X - 18$ with TblStart = –6 and ΔTbl = 3. See Figure 47c.

Since $Y_1 = 0$ when $x = -3$ or when $x = 6$, the solutions are $x = -3$ or $x = 6$.

[–5, 8, 1] by [–25, 5, 5]

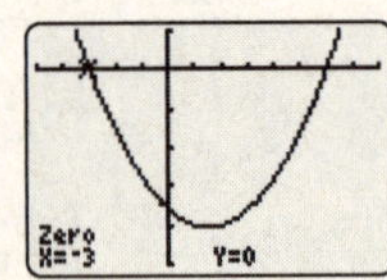

Figure 47a

[–5, 8, 1] by [–25, 5, 5]

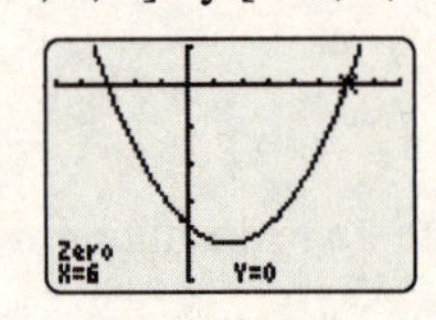

Figure 47b

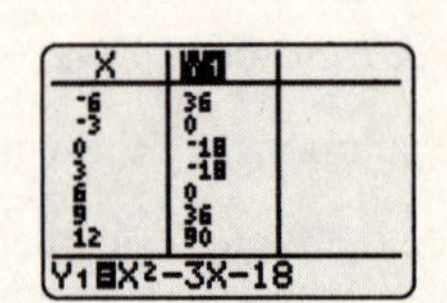

Figure 47c

49. (a) $x^2 - 8x + 15 = 0 \Rightarrow (x-3)(x-5) = 0 \Rightarrow$ Either $x-3=0 \Rightarrow x=3$ or $x-5=0 \Rightarrow x=5$

(b) Graph $Y_1 = X^2 - 8X + 15$ in [–10, 0, 1] by [–5, 10, 1]. See Figures 49a & 49b.

The solutions are the x-intercepts, $x = 3$ or $x = 5$.

(c) Table $Y_1 = X^2 - 8X + 15$ with TblStart = 1 and ΔTbl = 1. See Figure 49c.

Since $Y_1 = 0$ when $x = 3$ or when $x = 5$, the solutions are $x = 3$ or $x = 5$.

[–10, 0, 1] by [–5, 10, 1]

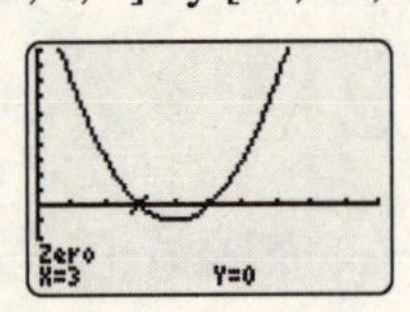

Figure 49a

[–10, 0, 1] by [–5, 10, 1]

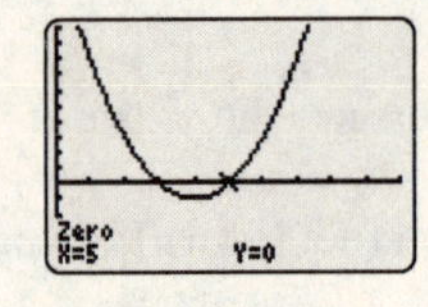

Figure 49b

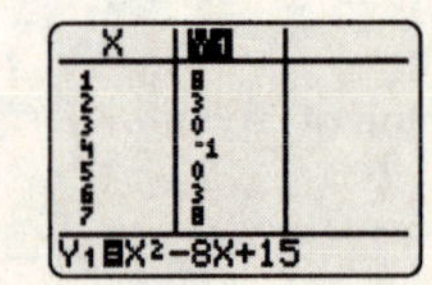

Figure 49c

51. (a) $4(x^2+35) = 48x \Rightarrow x^2 + 35 = 12x \Rightarrow x^2 - 12x + 35 = 0 \Rightarrow (x-5)(x-7) = 0 \Rightarrow$

Either $x-5=0 \Rightarrow x=5$ or $x-7=0 \Rightarrow x=7$

(b) Graph $Y_1 = 4(X^2+35) - 48X$ in [–2, 10, 1] by [–6, 6, 1]. See Figures 51a & 51b.

The solutions are the x-intercepts, $x = 5$ or $x = 7$.

(c) Table $Y_1 = 4(X^2+35) - 48X$ with TblStart = 3 and ΔTbl = 1. See Figure 51c.

Since $Y_1 = 0$ when $x = 5$ or when $x = 7$, the solutions are $x = 5$ or $x = 7$.

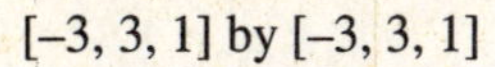

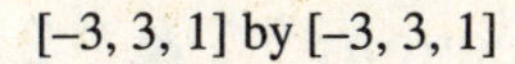

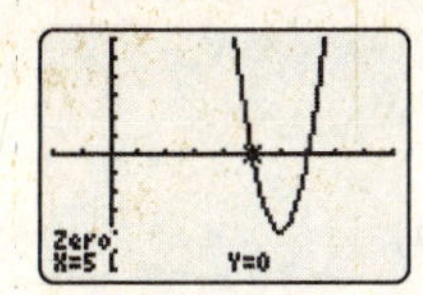

Figure 51a

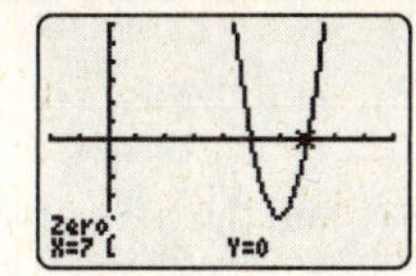

Figure 51b

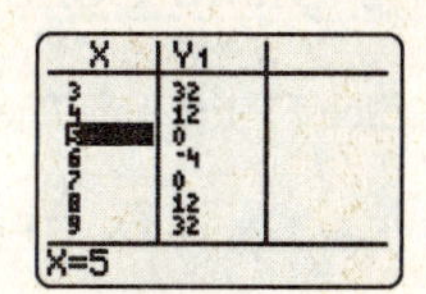

Figure 51c

53. (a) $1-2x-8x^2=0 \Rightarrow 8x^2+2x-1=0 \Rightarrow (4x-1)(2x+1)=0$

$\Rightarrow$ Either $4x-1=0 \Rightarrow x=\frac{1}{4}$ or $2x+1=0 \Rightarrow x=-\frac{1}{2}$

(b) Graph $Y_1=1-2X-8X^2$ in $[-3,3,1]$ by $[-3,3,1]$.

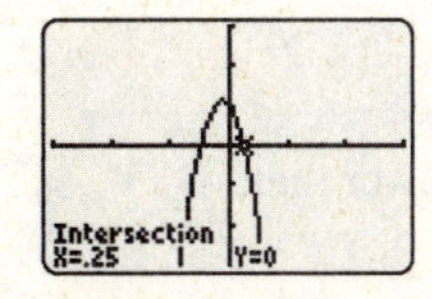

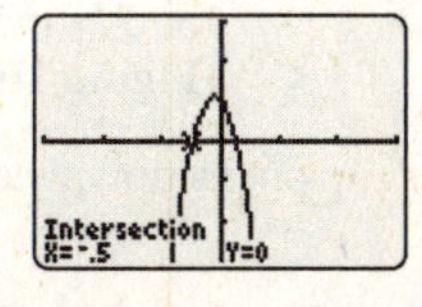

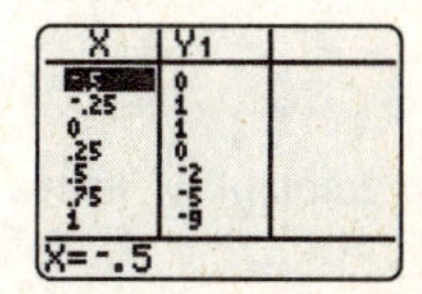

(c) Table $Y_1=1-2X-8X^2$ with TblStart $=-0.5$ and ΔTbl $=0.25$.

Since $Y_1=0$ when $x=-0.5$ or when $x=0.25$, the solutions are $x=-0.5=-\frac{1}{2}$ or $x=0.25=\frac{1}{4}$.

Applications

55. (a) $\frac{1}{2}x^2=450 \Rightarrow x^2=900 \Rightarrow x=\pm\sqrt{900} \Rightarrow x=30$ mph ($x=-30$ has no physical meaning)

(b) $\frac{1}{2}x^2=800 \Rightarrow x^2=1600 \Rightarrow x=\pm\sqrt{1600} \Rightarrow x=40$ mph ($x=-40$ has no physical meaning)

57. $60-16t^2=0 \Rightarrow -16t^2=-60 \Rightarrow t^2=\frac{60}{16} \Rightarrow t=\pm\sqrt{\frac{60}{16}} \Rightarrow t=\frac{\sqrt{60}}{4} \approx 1.9$ seconds. The value $t \approx -1.9$ seconds has no physical meaning. The toy takes about 1.9 seconds to hit he ground. This is not twice the time it takes to fall from a height of 30 feet.

59. Let x represent the height of the triangle. Then $x+3$ represents the base and $\frac{1}{2}x(x+3)$ represents the area.

$\frac{1}{2}x(x+3)=35 \Rightarrow x(x+3)=70 \Rightarrow x^2+3x-70=0 \Rightarrow (x-7)(x+10)=0 \Rightarrow x=7$ or -10

Since the value -10 has no physical meaning, the height is 7 inches and the base is $7+3=10$ inches.

61. $x=1, N=18.5(1)^2+30.1(1)+32.6=18.5+30.1+32.6$

$=81.2$; there were about 81,200 identity thefts in 2001.

Checking Basic Concepts for Sections 9.1 & 9.2

1. (a) See Figure 1a. The vertex is $(0,-2)$, and the axis of symmetry is $x=0$.

   (b) See Figure 1b. The vertex is $(1,-3)$, and the axis of symmetry is $x=1$.

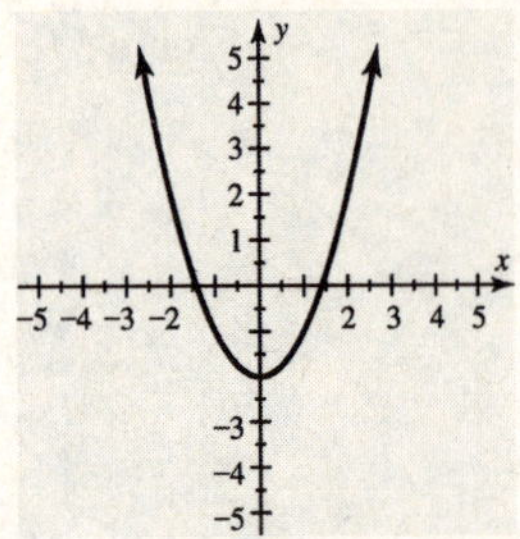

Figure 1a

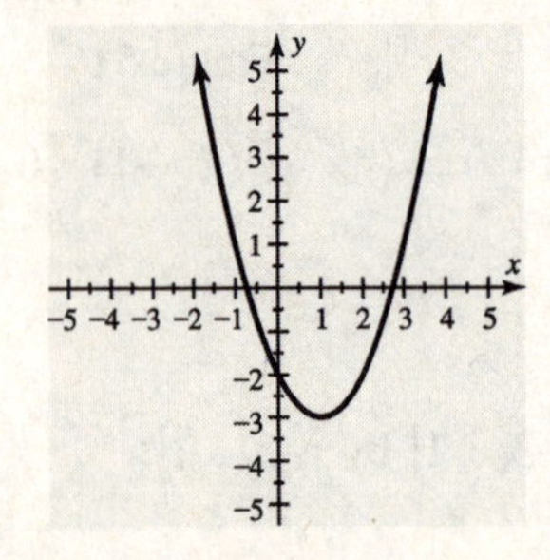

Figure 1b

2. The graph of y_1 opens upward whereas y_2 opens downward. Also, y_1 is narrower than y_2.

3. Because $-\frac{b}{2a}=-\frac{12}{2(-3)}=2$ and $y=-3(2)^2+12(2)-5=7$, the vertex is $(2,7)$.

   The graph is increasing when $x\le 2$ and decreasing when $x\ge 2$.

4. Symbolical:

   $2x^2-7x+3=0\Rightarrow(2x-1)(x-3)=0\Rightarrow$ Either $2x-1=0\Rightarrow x=\frac{1}{2}$ or $x-3=0\Rightarrow x=3$

   Graphical: Graph $Y_1 = 2X^2-7X+3$ in [–5, 5, 1] by [–5, 5, 1]. See Figures 4a & 4b. $x=\frac{1}{2}$ or $x=3$

[–5, 5, 1] by [–5, 5, 1]

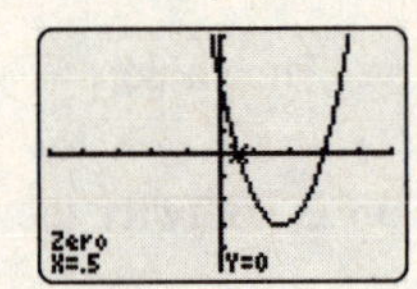

Figure 4a

[–5, 5, 1] by [–5, 5, 1]

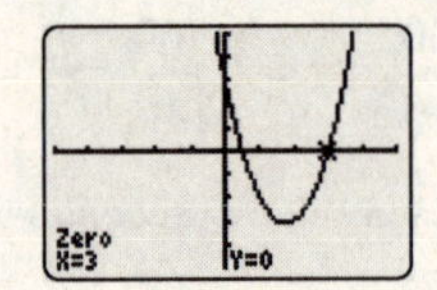

Figure 4b

5. $x^2=5\Rightarrow x=\pm\sqrt{5}$

9.3: Solving by Completing the Square

Concepts

1. $x+3$
2. $x-3$
3. $x+\frac{b}{2}$
4. $x-\frac{b}{2}$
5. The constant term equals the square of half the coefficient of the x-term.
6. completing the square

7. The square of half the coefficient of the x-term is 9.

8. $\left(\frac{b}{2}\right)^2$

Perfect Square Trinomials

9. No, because $20 \neq \left(\frac{-10}{2}\right)^2$.

11. Yes, because $4 = \left(\frac{4}{2}\right)^2$.

13. Yes, because $64 = \left(\frac{-16}{2}\right)^2$.

15. No, because $\frac{81}{2} \neq \left(\frac{-9}{2}\right)^2$.

17. The term needed to complete the square is $\left(\frac{-8}{2}\right)^2 = (-4)^2 = 16$. The resulting perfect square is $(x-4)^2$.

19. The term needed to complete the square is $\left(\frac{9}{2}\right)^2 = \frac{81}{4}$. The resulting perfect square is $\left(x+\frac{9}{2}\right)^2$.

Completing the Square

21. $\left(\frac{6}{2}\right)^2 = 3^2 = 9$

23. $\left(\frac{-7}{2}\right)^2 = \frac{49}{4}$

25. $x^2 - 2x = 15 \Rightarrow x^2 - 2x + 1 = 15 + 1 \Rightarrow (x-1)^2 = 16 \Rightarrow x - 1 = \pm\sqrt{16} \Rightarrow x - 1 = \pm 4 \Rightarrow x = 1 \pm 4 \Rightarrow x = -3$ or 5

27. $x^2 + 6x - 1 = 0 \Rightarrow x^2 + 6x + 9 = 1 + 9 \Rightarrow (x+3)^2 = 10 \Rightarrow x + 3 = \pm\sqrt{10} \Rightarrow x = -3 \pm \sqrt{10}$

29. $x^2 - 3x = 5 \Rightarrow x^2 - 3x + \frac{9}{4} = 5 + \frac{9}{4} \Rightarrow \left(x - \frac{3}{2}\right)^2 = \frac{29}{4} \Rightarrow x - \frac{3}{2} = \pm\sqrt{\frac{29}{4}} \Rightarrow x = \frac{3}{2} \pm \frac{\sqrt{29}}{2} = \frac{3 \pm \sqrt{29}}{2}$

31. $x^2 - 5x + 2 = 0 \Rightarrow x^2 - 5x = -2 \Rightarrow x^2 - 5x + \frac{25}{4} = -2 + \frac{25}{4} \Rightarrow \left(x - \frac{5}{2}\right)^2 = \frac{17}{4} \Rightarrow$

$x - \frac{5}{2} = \pm\sqrt{\frac{17}{4}} \Rightarrow x = \frac{5}{2} \pm \frac{\sqrt{17}}{2} = \frac{5 \pm \sqrt{17}}{2}$

33. $x^2 - 4 = 2x \Rightarrow x^2 - 2x = 4 \Rightarrow x^2 - 2x + 1 = 4 + 1 \Rightarrow (x-1)^2 = 5 \Rightarrow x - 1 = \pm\sqrt{5} \Rightarrow x = 1 \pm \sqrt{5}$

35. $2x^2 - 3x = 4 \Rightarrow x^2 - \frac{3}{2}x = 2 \Rightarrow x^2 - \frac{3}{2}x + \frac{9}{16} = 2 + \frac{9}{16} \Rightarrow \left(x - \frac{3}{4}\right)^2 = \frac{41}{16} \Rightarrow x - \frac{3}{4} = \pm\sqrt{\frac{41}{16}} \Rightarrow$

$x = \frac{3}{4} \pm \frac{\sqrt{41}}{4} = \frac{3 \pm \sqrt{41}}{4}$

37. $4x^2-8x=7 \Rightarrow 4x^2-8x-7=0 \Rightarrow x^2-2x-\frac{7}{4}=0 \Rightarrow x^2-2x+1=\frac{7}{4}+1 \Rightarrow (x-1)^2=\frac{11}{4} \Rightarrow$

$x-1=\pm\sqrt{\frac{11}{4}} \Rightarrow x=1\pm\frac{\sqrt{11}}{2} \Rightarrow x=\frac{2\pm\sqrt{11}}{2}$

39. $36x^2+18x=-1 \Rightarrow 36x^2+18x+1=0 \Rightarrow 36\left(x^2+\frac{1}{2}x+\frac{1}{16}\right)=-1+\frac{36}{16} \Rightarrow 36\left(x+\frac{1}{4}\right)^2=\frac{20}{16} \Rightarrow$

$\left(x+\frac{1}{4}\right)^2=\frac{20}{576} \Rightarrow x+\frac{1}{4}=\pm\sqrt{\frac{20}{576}} \Rightarrow x=-\frac{1}{4}\pm\frac{\sqrt{20}}{24}=\frac{-6\pm2\sqrt{5}}{24}=\frac{-3\pm\sqrt{5}}{12}$

Applications

41. Let x represent the width of the rectangle. Then the length is represented by $x+4$. Since the area is 16,

$x(x+4)=16 \Rightarrow x^2+4x=16 \Rightarrow x^2+4x+4=16+4 \Rightarrow (x+2)^2=20 \Rightarrow x+2=\pm\sqrt{20} \Rightarrow$

$x=-2+\sqrt{20}\approx 2.47$ inches (The value $-2-\sqrt{20}\approx -6.47$ has no physical meaning in this problem.)

9.4: The Quadratic Formula

Concepts

1. We use the quadratic formula to solve equations of the form $ax^2+bx+c=0$.
2. No solutions, one solution, or two solutions.
3. b^2-4ac
4. The quadratic equation has only one solution.
5. Factoring, square root property, completing the square and the quadratic formula.
6. No. When $b^2-4ac<0$, there are no real solutions.

The Quadratic Formula

7. $5x^2-4x+6=0 \Rightarrow a=5, b=-4, c=6$

9. $3x^2=2x-5 \Rightarrow 3x^2-2x+5=0 \Rightarrow a=3, b=-2, c=5$

11. $x^2=x \Rightarrow x^2-x=0 \Rightarrow a=1, b=-1, c=0$

13. $(x-3)(x+4)=0 \Rightarrow x^2+x-12=0 \Rightarrow a=1, b=1, c=-12$

15. Factoring: $x^2-2x+1=0 \Rightarrow (x-1)(x-1)=0 \Rightarrow x-1=0, x=1$

Quadratic formula: Let $a=1, b=-2, c=1$.

$$x=\frac{-b\pm\sqrt{b^2-4ac}}{2a}=\frac{-(-2)\pm\sqrt{(-2)^2-4(1)(1)}}{2(1)}=\frac{2\pm\sqrt{0}}{2}=\frac{2}{2}=1$$

17. Factoring: $x^2-2x-3=0 \Rightarrow (x-3)(x+1)=0 \Rightarrow x-3=0, x=3$ or $x+1=0, x=-1$, so $x=3,-1$

Quadratic formula: Let $a=1, b=-2, c=-3$.

$$x=\frac{-b\pm\sqrt{b^2-4ac}}{2a}=\frac{-(-2)\pm\sqrt{(-2)^2-4(1)(-3)}}{2(1)}=\frac{2\pm\sqrt{16}}{2}=\frac{2\pm4}{2}$$

$$=\frac{2+4}{2}\text{ or }\frac{2-4}{2}=3\text{ or }-1$$

19. Factoring: $2x^2-5x-3=0\Rightarrow(2x+1)(x-3)=0\Rightarrow 2x+1=0,\ x=-\frac{1}{2}$ or $x-3=0,\ x=3$, so $x=-\frac{1}{2},3$

Quadratic formula: Let $a=2, b=-5, c=-3$.

$$x=\frac{-b\pm\sqrt{b^2-4ac}}{2a}=\frac{-(-5)\pm\sqrt{(-5)^2-4(2)(-3)}}{2(2)}=\frac{5\pm\sqrt{49}}{4}$$

$$=\frac{5\pm7}{4}=\frac{5+7}{4}\text{ or }\frac{5-7}{4}=3\text{ or }-\frac{1}{2}$$

21. $$x=\frac{-11\pm\sqrt{(11)^2-4(2)(-6)}}{2(2)}=\frac{-11\pm\sqrt{169}}{4}=\frac{-11\pm13}{4}\Rightarrow x=-6\text{ or }x=\frac{1}{2}$$

Graph $Y_1=2X^2+11X-6$ in [–10, 5, 1] by [–25, 10, 5]. See Figures 21a & 21b. $x=-6$ or $x=\frac{1}{2}$

[–10, 5, 1] by [–25, 10, 5]

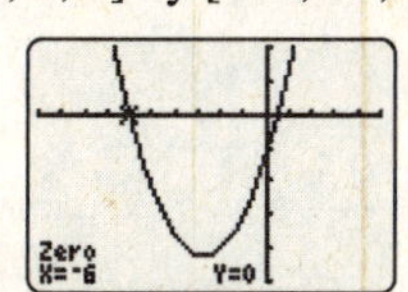

Figure 21a

[–10, 5, 1] by [–25, 10, 5]

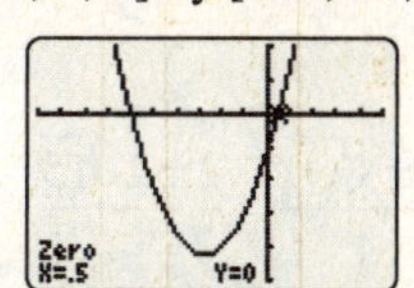

Figure 21b

23. $$x=\frac{-2\pm\sqrt{(2)^2-4(-1)(-1)}}{2(-1)}=\frac{-2\pm\sqrt{0}}{-2}=\frac{-2}{-2}\Rightarrow x=1$$

Graph $Y_1=X^2+2X-1$ in [–5, 5, 1] by [–5, 5, 1]. See Figure 23a. $x=1$

[–5, 5, 1] by [–5, 5, 1]

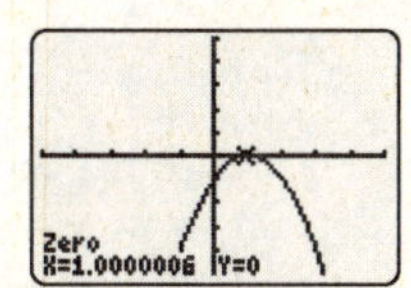

Figure 23a

25. $$x=\frac{-1\pm\sqrt{(1)^2-4(2)(1)}}{2(2)}=\frac{-1\pm\sqrt{-7}}{4}\Rightarrow\text{No real solutions.}$$

Graph $Y_1=2X^2+X+1$ in $[-5,5,1]$ by $[-5,5,1]$. No real solutions.

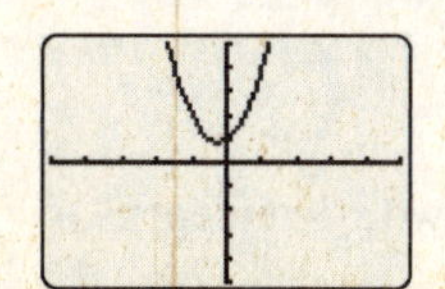

27. $x=\dfrac{-(-6)\pm\sqrt{(-6)^2-4(1)(-16)}}{2(1)}=\dfrac{6\pm\sqrt{100}}{2}=\dfrac{6\pm10}{2}\Rightarrow x=-2 \text{ or } x=8$

29. $x=\dfrac{-(-1)\pm\sqrt{(-1)^2-4(4)(-1)}}{2(4)}=\dfrac{1\pm\sqrt{17}}{8}$

31. $x=\dfrac{-2\pm\sqrt{(2)^2-4(-3)(-1)}}{2(-3)}=\dfrac{-2\pm\sqrt{-8}}{-6}\Rightarrow$ No real solutions.

33. $36x^2+9=36x\Rightarrow 36x^2-36x+9=0\ x=\dfrac{-(-36)\pm\sqrt{(-36)^2-4(36)(9)}}{2(36)}=\dfrac{36\pm\sqrt{0}}{72}=\dfrac{36}{72}=\dfrac{1}{2}$

35. $x=\dfrac{-(-6)\pm\sqrt{(-6)^2-4(2)(-2)}}{2(2)}=\dfrac{6\pm\sqrt{52}}{4}=\dfrac{6\pm2\sqrt{13}}{4}=\dfrac{2\left(3\pm\sqrt{13}\right)}{2(2)}=\dfrac{3\pm\sqrt{13}}{2}$

37. $x=\dfrac{-(-4)\pm\sqrt{(-4)^2-4(1)(1)}}{2(1)}=\dfrac{4\pm\sqrt{12}}{2}=\dfrac{4\pm2\sqrt{3}}{2}=\dfrac{2\left(2\pm\sqrt{3}\right)}{2}=2\pm\sqrt{3}$

39. $x=\dfrac{-\left(-\frac{1}{2}\right)\pm\sqrt{\left(-\frac{1}{2}\right)^2-4\left(\frac{3}{2}\right)\left(-\frac{3}{2}\right)}}{2\left(\frac{3}{2}\right)}=\dfrac{\frac{1}{2}\pm\sqrt{\frac{37}{4}}}{3}=\dfrac{\frac{1}{2}\pm\frac{\sqrt{37}}{2}}{3}=\dfrac{1\pm\sqrt{37}}{6}$

41. $x=\dfrac{-(-2)\pm\sqrt{(-2)^2-4(2)(-7)}}{2(2)}=\dfrac{2\pm\sqrt{60}}{4}=\dfrac{2\pm2\sqrt{15}}{4}=\dfrac{2\left(1\pm\sqrt{15}\right)}{4}=\dfrac{1\pm\sqrt{15}}{2}$

43. $x=\dfrac{-10\pm\sqrt{(10)^2-4(-3)(-5)}}{2(-3)}=\dfrac{-10\pm\sqrt{40}}{-6}=\dfrac{-10\pm2\sqrt{10}}{-6}=\dfrac{-2\left(5\pm\sqrt{10}\right)}{-6}=\dfrac{5\pm\sqrt{10}}{3}$

The Discriminant

45. (a) Since the parabola opens upward, $a>0$.

 (b) The solutions are the x-intercepts, $x=-1$ or $x=2$.

 (c) Since there are two real solutions, the discriminant is positive.

47. (a) Since the parabola opens upward, $a>0$.

 (b) Since there are no x-intercepts, there are no real solutions.

 (c) Since there are no real solutions, the discriminant is negative.

49. (a) Since the parabola opens downward, $a<0$.

 (b) The solution is the x-intercept, $x=2$.

 (c) Since there is one real solution, the discriminant is zero.

51. (a) $(1)^2-4(3)(-2)=25$

 (b) Since the discriminant is positive, there are two real solutions.

 (c) Graph $Y_1=3X^2+X-2$ in [−3, 3, 1] by [−3, 3, 1]. See Figure 51. There are two x-intercepts.

53. (a) $(-4)^2 - 4(1)(4) = 0$

(b) Since the discriminant is zero, there is only one real solution.

(c) Graph $Y_1 = X^2 - 4X + 4$ in [0, 4, 1] by [–3, 3, 1]. See Figure 53. There is one x-intercept.

[–3, 3, 1] by [–3, 3, 1]

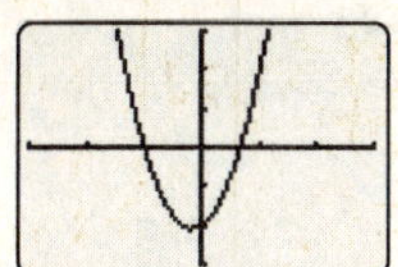

Figure 51

[0, 4, 1] by [–3, 3, 1]

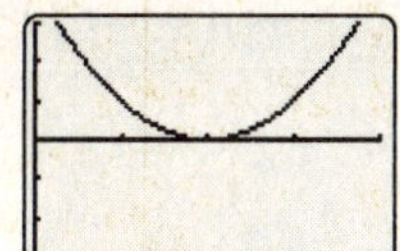

Figure 53

55. (a) $\left(\frac{3}{2}\right)^2 - 4\left(\frac{1}{2}\right)(2) = -\frac{7}{4}$

(b) Since the discriminant is negative, there are no real solutions.

(c) Graph $Y_1 = (1/2)X^2 + (3/2)X + 2$ in [–10, 10, 1] by [0, 10, 1]. See Figure 55. There are no x-intercepts.

57. (a) $(3)^2 - 4(1)(-3) = 21$

(b) Since the discriminant is positive, there are two real solutions.

(c) Graph $Y_1 = X^2 + 3X - 3$ in [–5, 5, 1] by [–8, 5, 1]. See Figure 57. There are two x-intercepts.

[–10, 10, 1] by [0, 10, 1]

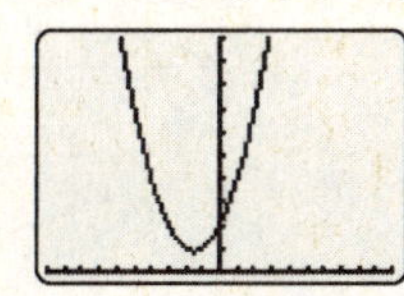

Figure 55

[–5, 5, 1] by [–8, 5, 1]

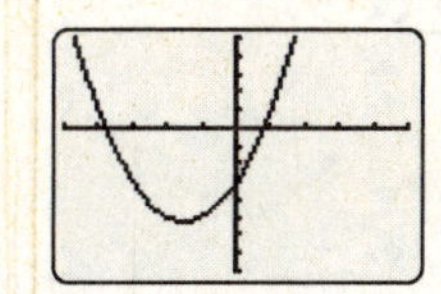

Figure 57

59. $x = \dfrac{-(-2) \pm \sqrt{(-2)^2 - 4(1)(-1)}}{2(1)} = \dfrac{2 \pm \sqrt{8}}{2} = \dfrac{2 \pm 2\sqrt{2}}{2} = \dfrac{2(1 \pm \sqrt{2})}{2} = 1 \pm \sqrt{2}$

61. $x = \dfrac{-(-1) \pm \sqrt{(-1)^2 - 4(-2)(3)}}{2(-2)} = \dfrac{1 \pm \sqrt{25}}{-4} = \dfrac{1 \pm 5}{-4} = -\dfrac{3}{2}$ or 1

63. $x = \dfrac{-1 \pm \sqrt{(1)^2 - 4(1)(5)}}{2(1)} = \dfrac{-1 \pm \sqrt{-19}}{2} \Rightarrow$ No real solutions. No x-intercepts.

65. $x = \dfrac{-(0) \pm \sqrt{(0)^2 - 4(1)(9)}}{2(1)} = \dfrac{0 \pm \sqrt{-36}}{2} \Rightarrow$ No real solutions. No x-intercepts.

67. $x = \dfrac{-4 \pm \sqrt{(4)^2 - 4(3)(-2)}}{2(3)} = \dfrac{-4 \pm \sqrt{40}}{6} = \dfrac{-4 \pm 2\sqrt{10}}{6} = \dfrac{2(-2 \pm \sqrt{10})}{6} = \dfrac{-2 \pm \sqrt{10}}{3}$

Applications

69. $\frac{1}{9}x^2+\frac{11}{3}x-80=0 \Rightarrow x^2+33x-720=0 \Rightarrow x=(x-15)(x+48)=0 \Rightarrow x=15$ mph

The value $x=-48$ has no physical meaning.

71. $\frac{1}{9}x^2+\frac{11}{3}x-390=0 \Rightarrow x^2+33x-3510=0 \Rightarrow x=(x-45)(x+78)=0 \Rightarrow x=45$ mph

The value $x=-78$ has no physical meaning.

73. $2.39x^2+5.04x+5.1=200 \Rightarrow 2.39x^2+5.04x-194.9=0$

$$x=\frac{-5.04\pm\sqrt{(5.04)^2-4(2.39)(-194.9)}}{2(2.39)}=\frac{-5.04\pm\sqrt{1888.6456}}{4.78} \Rightarrow x\approx 8.04;\text{ about 1992}$$

The value $x\approx -10.15$ has no meaning in this problem. Our answer agrees with the graph.

75. Let x represent the length of the garden. Then $\frac{1}{2}(200-2x)=100-x$ represents the width. The equation is

$x(100-x)=2475.$

$-x^2+100x-2475=0 \Rightarrow x^2-100x+2475=0 \Rightarrow (x-55)(x-45)=0$

$\Rightarrow x=45$ or $55 \Rightarrow 100-x=55$ or 45. The garden is 45 feet by 55 feet.

Checking Basic Concepts for Sections 9.3 & 9.4

1. (a) No, because $12\neq\left(\frac{6}{2}\right)^2$.

   (b) Yes, because $1=\left(\frac{2}{2}\right)^2$.

2. (a) $x^2-4x=-1 \Rightarrow x^2-4x+4=-1+4 \Rightarrow (x-2)^2=3 \Rightarrow x-2=\pm\sqrt{3} \Rightarrow x=2\pm\sqrt{3}$

   (b) $x^2-6x=-4 \Rightarrow x^2-6x+9=-4+9 \Rightarrow (x-3)^2=5 \Rightarrow x-3=\pm\sqrt{5} \Rightarrow x=3\pm\sqrt{5}$

3. (a) $2x^2=3x+1 \Rightarrow 2x^2-3x-1=0 \Rightarrow x=\frac{-(-3)\pm\sqrt{(-3)^2-4(2)(-1)}}{2(2)}=\frac{3\pm\sqrt{17}}{4}$

   (b) $9x^2-24x+16=0 \Rightarrow x=\frac{-(-24)\pm\sqrt{(-24)^2-4(9)(16)}}{2(9)}=\frac{24\pm\sqrt{0}}{18}=\frac{24}{18}=\frac{4}{3}$

4. (a) $(-5)^2-4(1)(5)=5$; Since the discriminant is positive, there are two real solutions.

   (b) $(-5)^2-4(2)(4)=-7$; Since the discriminant is negative, there are no real solutions.

   (c) $(-56)^2-4(49)(16)=0$; Since the discriminant is zero, there is one real solution.

9.5: Complex Solutions

Concepts

1. $2+3i$; *Answers may vary.*

2. No, any real number a can be written as $a+0i$.

3. i

4. -1

5. $i\sqrt{a}$

6. $a+bi$

7. 4

8. -5

Complex Numbers

9. $\sqrt{-3}=i\sqrt{3}$

11. $\sqrt{-36}=i\sqrt{36}=i\cdot 6=6i$

13. $\sqrt{-144}=i\sqrt{144}=i\cdot 12=12i$

15. $\sqrt{-12}=i\sqrt{12}=i\cdot\sqrt{4\cdot 3}=i\cdot\sqrt{4}\cdot\sqrt{3}=i\cdot 2\cdot\sqrt{3}=2i\sqrt{3}$

17. $\sqrt{-18}=i\sqrt{18}=i\cdot\sqrt{9\cdot 2}=i\cdot\sqrt{9}\cdot\sqrt{2}=i\cdot 3\cdot\sqrt{2}=3i\sqrt{2}$

19. $(4+3i)+(-2-3i)=(4+(-2))+(3+(-3))i=2+0i=2$

21. $2i+5i=(2+5)i=7i$

23. $(2-7i)-(1+2i)=(2-1)+(-7-2)i=1-9i$

25. $5i-(10-2i)=(0-10)+(5-(-2))i=-10+7i$

27. $4(5-3i)=20-12i$

29. $(-3-4i)(5-4i)=-15+12i-20i+16i^2=-15-8i+16(-1)=-15-8i-16=-31-8i$

31. $(-4i)(5i)=-4\cdot 5\cdot i^2=-20(-1)=20$

33. $3i+(2-3i)-(1-5i)=3i+2-3i-1+5i=1+5i$

35. $(2+i)^2=4+4i+i^2=4+4i-1=3+4i$

37. $2i(-3+i)=-6i+2i^2=-6i+2(-1)=-2-6i$

39. $(a+bi)(a-bi)=a^2-abi+abi-(bi)^2=a^2-b^2i^2=a^2-b^2(-1)=a^2+b^2$

Complex Solutions

41. $x^2+9=0\Rightarrow x^2=-9\Rightarrow x=\pm\sqrt{-9}\Rightarrow x=\pm 3i$

43. $x^2+80=0\Rightarrow x^2=-80\Rightarrow x=\pm\sqrt{-80}\Rightarrow x=\pm\sqrt{-16\cdot 5}\Rightarrow x=\pm 4i\sqrt{5}$

45. $x^2+\frac{1}{4}=0 \Rightarrow x^2=-\frac{1}{4} \Rightarrow x=\pm\sqrt{-\frac{1}{4}} \Rightarrow x=\pm\frac{1}{2}i$

47. $3x^2+2=x \Rightarrow 3x^2-x+2=0 \Rightarrow x=\frac{-(-1)\pm\sqrt{(-1)^2-4(3)(2)}}{2(3)}$

$\Rightarrow x=\frac{1\pm\sqrt{-23}}{6} \Rightarrow x=\frac{1\pm i\sqrt{23}}{6} \Rightarrow x=\frac{1}{6}\pm i\frac{\sqrt{23}}{6}$

49. $x^2=-6 \Rightarrow x=\pm\sqrt{-6} \Rightarrow x=\pm i\sqrt{6}$

51. $x^2-3=0 \Rightarrow x^2=3 \Rightarrow x=\pm\sqrt{3}$

53. $x^2+2=0 \Rightarrow x^2=-2 \Rightarrow x=\pm\sqrt{-2} \Rightarrow x=\pm i\sqrt{2}$

55. $x=\frac{-(-1)\pm\sqrt{(-1)^2-4(1)(2)}}{2(1)}=\frac{1\pm\sqrt{-7}}{2}=\frac{1\pm i\sqrt{7}}{2}=\frac{1}{2}\pm i\frac{\sqrt{7}}{2}$

57. $x=\frac{-3\pm\sqrt{3^2-4(2)(4)}}{2(2)}=\frac{-3\pm\sqrt{-23}}{4}=\frac{-3\pm i\sqrt{23}}{4}=-\frac{3}{4}\pm i\frac{\sqrt{23}}{4}$

59. $x=\frac{-(-4)\pm\sqrt{(-4)^2-4(1)(1)}}{2(1)}=\frac{4\pm\sqrt{12}}{2}=\frac{4\pm 2\sqrt{3}}{2}=2\pm\sqrt{3}$

9.6: Introduction to Functions

Concepts

1. function

2. y equals f of x

3. $f(2)=(2)^2+1=4+1=5$

4. domain

5. range

6. one

7. We use the vertical line test to identify graphs of functions.

8. $(3, 4)$; 3; 6

9. (a, b)

10. d

Representing and Evaluating Functions

11. $f(-1)=4(-1)-2=-6$; $f(0)=4(0)-2=-2$

13. $f(0)=\sqrt{0}=0$; $f\left(\frac{9}{4}\right)=\sqrt{\frac{9}{4}}=\frac{3}{2}$

15. $f(-5)=(-5)^2=25;\ f\left(\dfrac{3}{2}\right)=\left(\dfrac{3}{2}\right)^2=\dfrac{9}{4}$

17. $f(-8)=3;\ f\left(\dfrac{7}{3}\right)=3$

19. $f(-2)=5-(-2)^3=13;\ f(3)=5-(3)^3=-22$

21. $f(-5)=\dfrac{2}{-5+1}=\dfrac{2}{-4}=-\dfrac{1}{2};\ f(4)=\dfrac{2}{4+1}=\dfrac{2}{5}$

23. (a) Because there are 36 inches in 1 yard, the formula is $I(x)=36x$.

 (b) $I(10)=36(10)=360$. There are 360 inches in 10 yards.

25. (a) The area formula for a circle is $A(r)=\pi r^2$.

 (b) $A(10)=\pi(10)^2=100\pi\approx 314.2$. The area of a circle with radius 10 is about 314.2.

27. (a) Because there are 43,560 square feet in 1 acre, the formula is $F(x)=43{,}560x$.

 (b) $F(10)=43{,}560(10)=435{,}600$. There are 435,600 square feet in 10 acres.

29. $f=\{(1,3)(2,-4)(3,0)\};\ D=\{1,2,3\},\ R=\{-4,0,3\}$

31. $f=\{(a,b)(c,d)(e,a)(d,b)\};\ D=\{a,c,d,e\},\ R=\{a,b,d\}$

33. 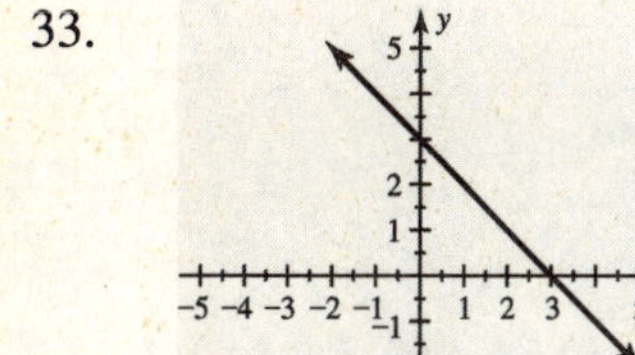
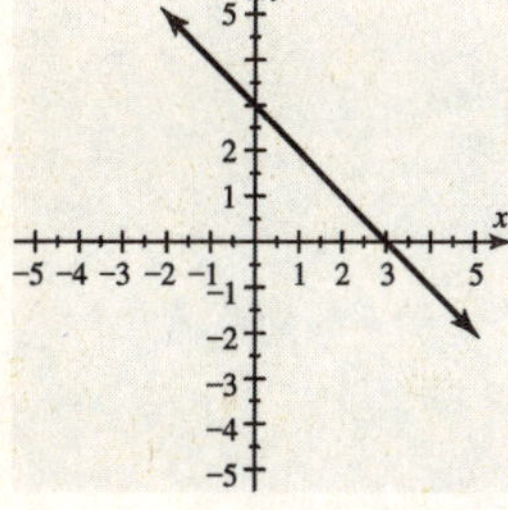

35. 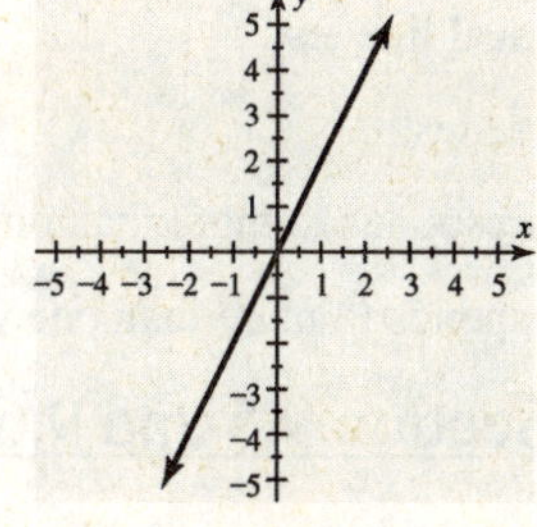

37. 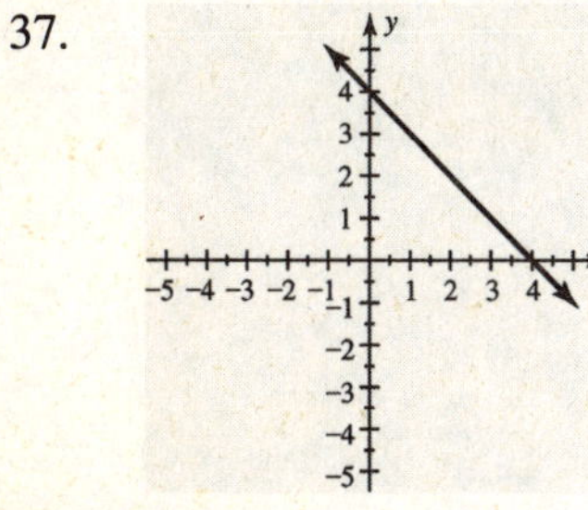

39. 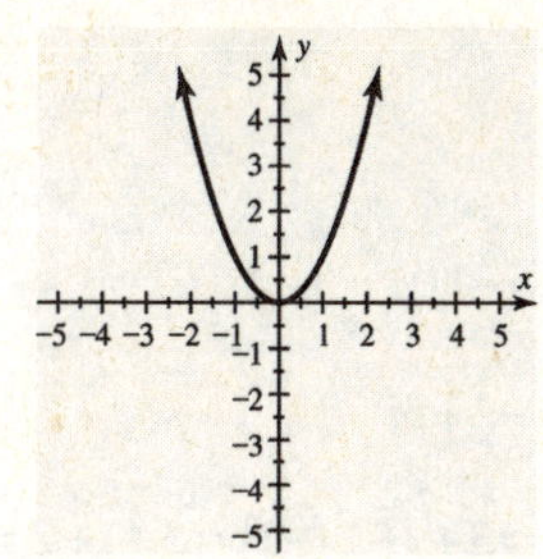

41. 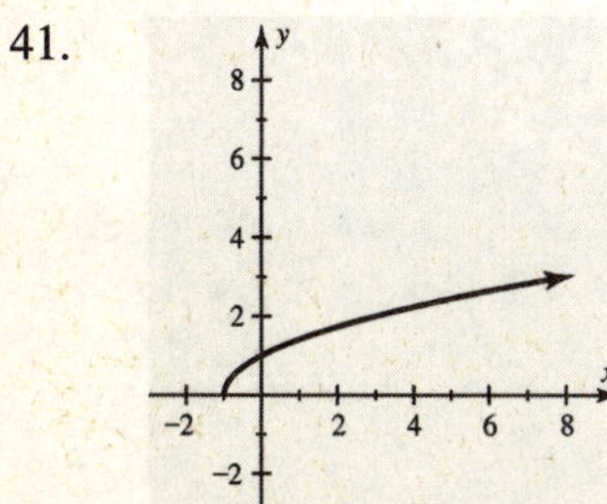

43. $f(0)=3;\ f(2)=-1$

45. $f(-2)=0;\ f(1)=2$

47. $f(1)=-4;\ f(2)=-3$

49. Subtract $\frac{1}{2}$ from the input x to obtain the output y.

51. Divide the input x by 3 to obtain the output y.

53. $P(1995)=5.7(1995-1990)+150=5.7(5)+150=28.5+150=178.5$; In 1995 the average price was $178,500.

55. Any real number is a valid input for this function. D: all real numbers.

57. Any real number is a valid input for this function. D: all real numbers.

59. The denominator of this function cannot equal zero. $D: x \neq 5$.

61. The denominator of this function will never equal zero because the variable is squared. D: all real numbers.

63. The radicand must be greater than or equal to zero. $D: x \geq 1$.

Identifying a Function

65. No. The value 1 in the domain corresponds to more than one value in the range.

67. Yes. Each value in the domain corresponds to exactly one value in the range.

69. Yes. The graph passes the vertical line test.

71. No. The graph does not pass the vertical line test.

73. Yes. The graph passes the vertical line test.

75. No. The value -2 in the domain corresponds to more than one value in the range.

77. No. The value 5 in the domain corresponds to more than one value in the range.

Checking Basic Concepts for Section 9.5 and 9.6

1. (a) $\sqrt{-64}=i\sqrt{64}=i(8)=8i$

   (b) $\sqrt{-17}=i\sqrt{17}$

2. (a) $(2-3i)+(1-i)=(2+1)+(-3+(-1))i=3-4i$

   (b) $4i-(2+i)=(0-2)+(4-1)i=-2+3i$

   (c) $(3-2i)(1+i)=3+3i-2i-2i^2=3+i-2(-1)=3+i+2=5+i$

3. (a) $x^2+25=0 \Rightarrow x^2=-25 \Rightarrow x=\pm\sqrt{-25} \Rightarrow x=\pm 5i$

   (b) $x=\dfrac{6\pm\sqrt{(-6)^2-4(1)(13)}}{2(1)}=\dfrac{6\pm\sqrt{-16}}{2}=\dfrac{6\pm 4i}{2}=\dfrac{6}{2}\pm\dfrac{4}{2}i=3\pm 2i$

   (c) $x=\dfrac{-1\pm\sqrt{1^2-4(1)(2)}}{2(1)}=\dfrac{-1\pm\sqrt{-7}}{2}=-\dfrac{1}{2}\pm i\dfrac{\sqrt{7}}{2}$

4. $f(x) = x^2 - 1$

5. $f(0) = 0; f(2) = 4$

Chapter 9 Review Exercises

Section 9.1

1. Vertex: $(-3, 4)$; Axis of symmetry: $x = -3$; Opens downward; Increasing: $x \le -3$; Decreasing: $x \ge -3$

2. Vertex: $(1, 0)$; Axis of symmetry: $x = 1$; Opens upward; Increasing: $x \ge 1$; Decreasing: $x \le 1$

3. (a) See Figure 3.

   (b) The vertex is (0, –2). The axis of symmetry is $x = 0$.

4. (a) See Figure 4.

   (b) The vertex is (2, 1). The axis of symmetry is $x = 2$.

5. (a) See Figure 5.

   (b) The vertex is (1, 2). The axis of symmetry is $x = 1$.

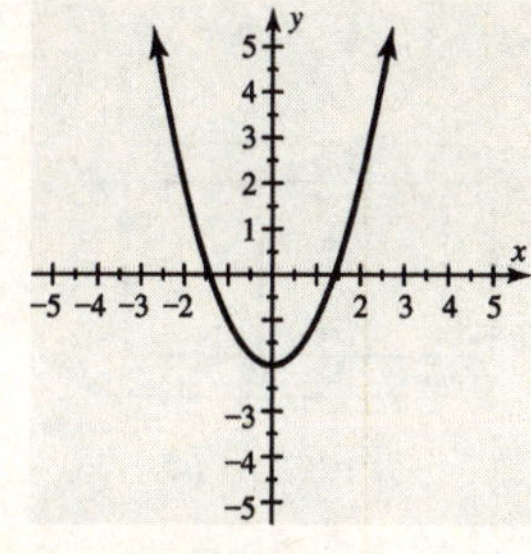

Figure 3

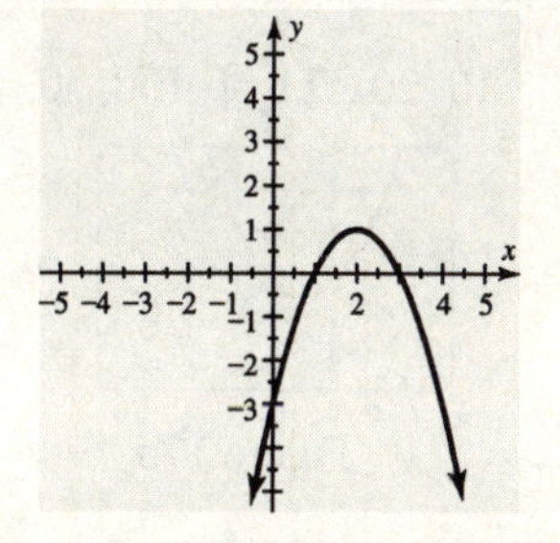

Figure 4

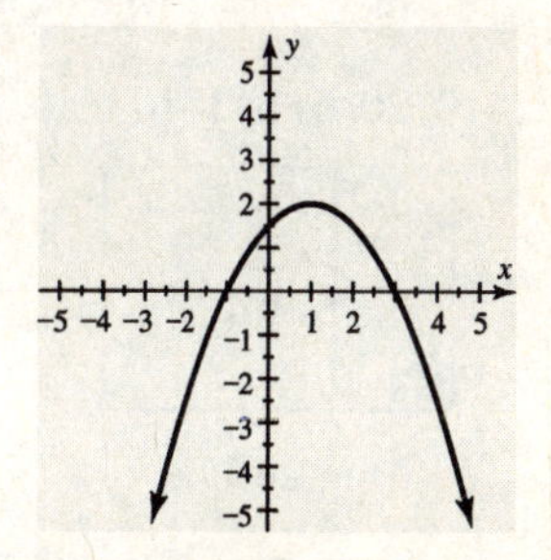

Figure 5

6. (a) See Figure 6.

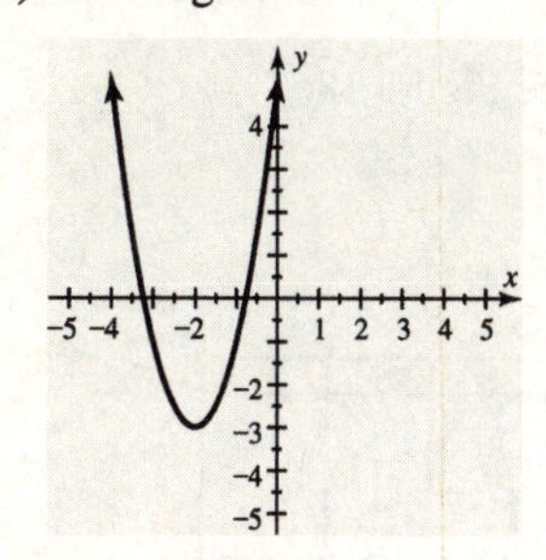

Figure 6

   (b) The vertex is (–2, –3). The axis of symmetry is $x = -2$.

7. $x = -\frac{b}{2a} = -\frac{(-4)}{2(1)} = 2,\ y = (2)^2 - 4(2) - 2 = -6$; The vertex is (2, –6).

8. $x = -\frac{b}{2a} = -\frac{(0)}{2(-1)} = 0,\ y = 5 - (0)^2 = 5$; The vertex is (0, 5).

9. $x = -\frac{b}{2a} = -\frac{(1)}{2(-\frac{1}{4})} = 2,\ y = -\frac{1}{4}(2)^2 + (2) = 1$; The vertex is (2, 1).

10. $x=-\frac{b}{2a}=-\frac{(2)}{2(1)}=-1,\ y=2+2(-1)+(-1)^2=1$; The vertex is (–1, 1).

Section 9.2

11. The solutions are the x-intercepts: –2, 3.

12. The solution is the x-intercept: –1.

13. There are no x-intercepts: No real solutions.

14. The solutions are the x-intercepts: –4, 6.

15. The solutions are the x-values where $Y_1=0$: –10, 5

16. The solutions are the x-values where $Y_1=0$: –0.5, 0.25

17. (a) Graph $Y_1=X^2-5X-50$ in [–10, 20, 5] by [–100, 20, 10]. See Figures 17a & 17b.

The solutions are the x-intercepts, $x=-5$ or $x=10$.

(b) Table $Y_1=X^2-5X-50$ with TblStart = –10 and ΔTbl = 5. See Figure 17c.

Since $Y_1=0$ when $x=-5$ or when $x=10$, the solutions are $x=-5$ or $x=10$.

[–10, 20, 5] by [–100, 20, 10]

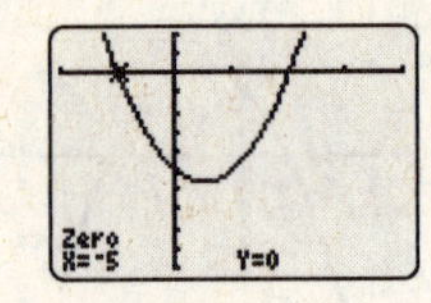

Figure 17a

[–10, 20, 5] by [–100, 20, 10]

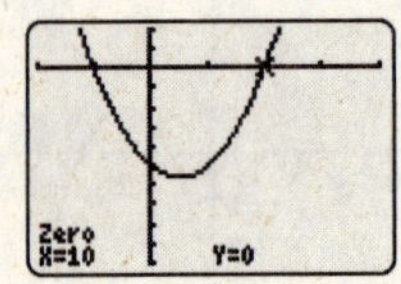

Figure 17b

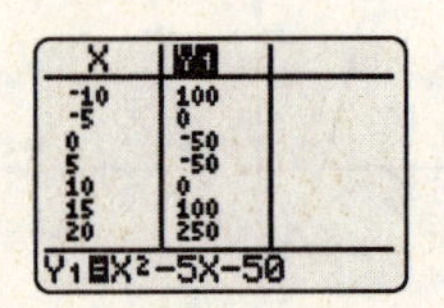

Figure 17c

18. (a) Graph $Y_1=(1/2)X^2+X-(3/2)$ in [–5, 5, 1] by [–5, 5, 1]. See Figures 18a & 18b.

The solutions are the x-intercepts, $x=-3$ or $x=1$.

(b) Table $Y_1=(1/2)X^2+X-(3/2)$ with TblStart = –4 and ΔTbl = 1. See Figure 18c.

Since $Y_1=0$ when $x=-3$ or when $x=1$, the solutions are $x=-3$ or $x=1$.

[–5, 5, 1] by [–5, 5, 1]

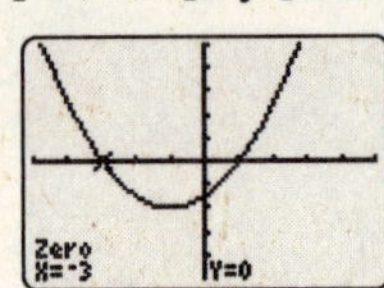

Figure 18a

[–5, 5, 1] by [–5, 5, 1]

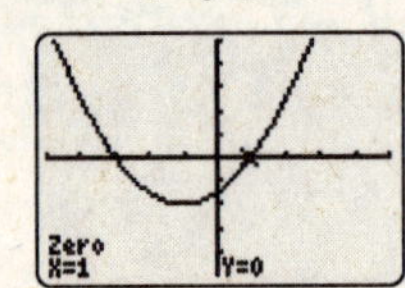

Figure 18b

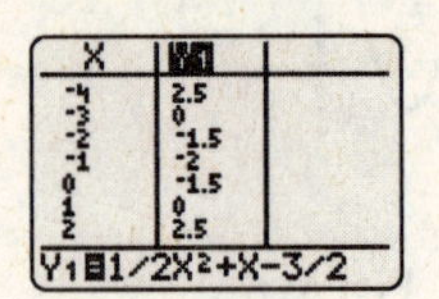

Figure 18c

19. (a) Graph $Y_1=(1/4)X^2+(1/2)X-2$ in [–5, 5, 1] by [–3, 3, 1]. See Figures 19a & 19b.

The solutions are the x-intercepts, $x=-4$ or $x=2$.

(b) Table $Y_1=(1/4)X^2+(1/2)X-2$ with TblStart = –8 and ΔTbl = 2. See Figure 19c.

Since $Y_1=0$ when $x=-4$ or when $x=2$, the solutions are $x=-4$ or $x=2$.

[−5, 5, 1] by [−3, 3, 1]

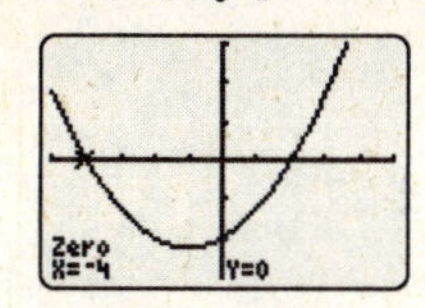

Figure 19a

[−5, 5, 1] by [−3, 3, 1]

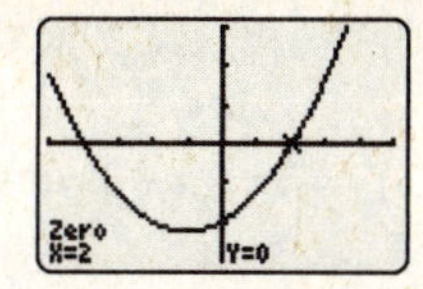

Figure 19b

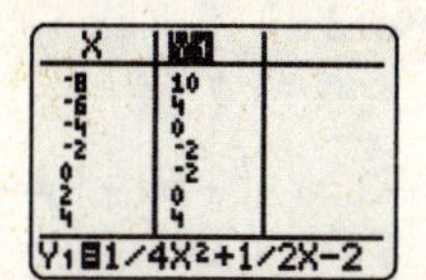

Figure 19c

20. (a) Graph $Y_1 = (1/4)X^2 - (1/2)X - (3/4)$ in [−5, 5, 1] by [−3, 3, 1]. See Figures 20a & 20b.

The solutions are the x-intercepts, $x = -1$ or $x = 3$.

(b) Table $Y_1 = (1/4)X^2 - (1/2)X - (3/4)$ with TblStart = −2 and ΔTbl = 1. See Figure 20c.

Since $Y_1 = 0$ when $x = -1$ or when $x = 3$, the solutions are $x = -1$ or $x = 3$.

[−5, 5, 1] by [−3, 3, 1]

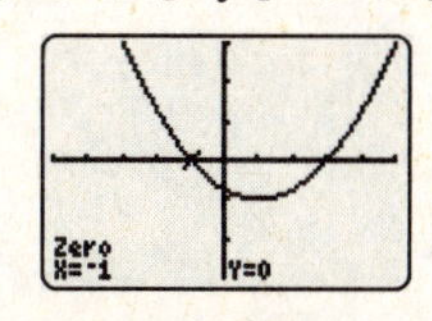

Figure 20a

[−5, 5, 1] by [−3, 3, 1]

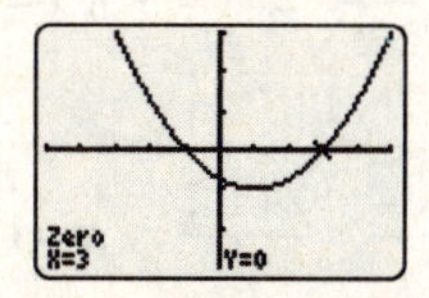

Figure 20b

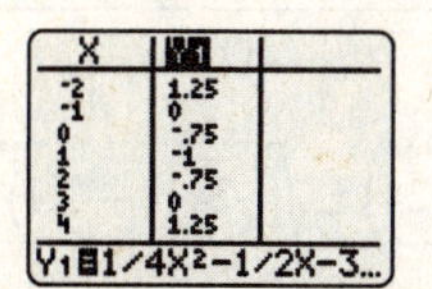

Figure 20c

21. $x^2 + x - 20 = 0 \Rightarrow (x+5)(x-4) = 0 \Rightarrow x = -5$ or $x = 4$

22. $x^2 + 11x + 24 = 0 \Rightarrow (x+8)(x+3) = 0 \Rightarrow x = -8$ or $x = -3$

23. $15x^2 - 4x - 4 = 0 \Rightarrow (5x+2)(3x-2) = 0 \Rightarrow x = -\frac{2}{5}$ or $x = \frac{2}{3}$

24. $7x^2 - 25x + 12 = 0 \Rightarrow (7x-4)(x-3) = 0 \Rightarrow x = \frac{4}{7}$ or $x = 3$

25. $x^2 = 100 \Rightarrow x = \pm\sqrt{100} \Rightarrow x = \pm 10$

26. $3x^2 = \frac{1}{3} \Rightarrow x^2 = \frac{1}{9} \Rightarrow x = \pm\sqrt{\frac{1}{9}} \Rightarrow x = \pm\frac{1}{3}$

27. $4x^2 - 6 = 0 \Rightarrow x^2 = \frac{6}{4} \Rightarrow x = \pm\sqrt{\frac{6}{4}} \Rightarrow x = \pm\frac{\sqrt{6}}{2}$

28. $5x^2 = x^2 - 4 \Rightarrow 4x^2 = -4 \Rightarrow x^2 = -1 \Rightarrow x = \pm\sqrt{-1} \Rightarrow$ No real solutions.

Section 9.3

29. No, because $8 \neq \left(\frac{-8}{2}\right)^2$.

30. Yes, because $4 = \left(\frac{-4}{2}\right)^2$.

31. Yes, because $100 = \left(\frac{20}{2}\right)^2$.

32. No, because $2 \neq \left(\frac{-2}{2}\right)^2$.

33. $x^2+6x=-2 \Rightarrow x^2+6x+9=-2+9 \Rightarrow (x+3)^2=7 \Rightarrow x+3=\pm\sqrt{7} \Rightarrow x=-3\pm\sqrt{7}$

34. $x^2-4x=6 \Rightarrow x^2-4x+4=6+4 \Rightarrow (x-2)^2=10 \Rightarrow x-2=\pm\sqrt{10} \Rightarrow x=2\pm\sqrt{10}$

35. $x^2-2x-5=0 \Rightarrow x^2-2x+1=5+1 \Rightarrow (x-1)^2=6 \Rightarrow x-1=\pm\sqrt{6} \Rightarrow x=1\pm\sqrt{6}$

36. $2x^2+6x-1=0 \Rightarrow x^2+3x+\frac{9}{4}=\frac{1}{2}+\frac{9}{4} \Rightarrow \left(x+\frac{3}{2}\right)^2=\frac{11}{4} \Rightarrow x+\frac{3}{2}=\pm\sqrt{\frac{11}{4}} \Rightarrow x=\frac{-3\pm\sqrt{11}}{2}$

Section 9.4

37. $x=\frac{-(-9)\pm\sqrt{(-9)^2-4(1)(18)}}{2(1)}=\frac{9\pm\sqrt{9}}{2}=\frac{9\pm3}{2} \Rightarrow x=3 \text{ or } x=6$

38. $x=\frac{-(-24)\pm\sqrt{(-24)^2-4(1)(143)}}{2(1)}=\frac{24\pm\sqrt{4}}{2}=\frac{24\pm2}{2} \Rightarrow x=11 \text{ or } x=13$

39. $x=\frac{-1\pm\sqrt{(1)^2-4(6)(-1)}}{2(6)}=\frac{-1\pm\sqrt{25}}{12}=\frac{-1\pm5}{12} \Rightarrow x=-\frac{1}{2} \text{ or } x=\frac{1}{3}$

40. $x=\frac{-(-5)\pm\sqrt{(-5)^2-4(5)(1)}}{2(5)}=\frac{5\pm\sqrt{5}}{10}$

41. $x=\frac{-(-8)\pm\sqrt{(-8)^2-4(1)(-5)}}{2(1)}=\frac{8\pm\sqrt{84}}{2}=\frac{8\pm2\sqrt{21}}{2}=\frac{2\left(4\pm\sqrt{21}\right)}{2}=4\pm\sqrt{21}$

42. $x=\frac{-(-6)\pm\sqrt{(-6)^2-4(2)(3)}}{2(2)}=\frac{6\pm\sqrt{12}}{4}=\frac{6\pm2\sqrt{3}}{4}=\frac{2\left(3\pm\sqrt{3}\right)}{2(2)}=\frac{3\pm\sqrt{3}}{2}$

43. (a) Since the parabola opens upward, $a>0$.

(b) The solutions are the x-intercepts, $x=-2$ or $x=3$.

(c) Since there are two unique solutions, the discriminant is positive.

44. (a) Since the parabola opens upward, $a>0$.

(b) The solution is the x-intercept, $x=2$.

(c) Since there is one solution, the discriminant is zero.

45. (a) Since the parabola opens downward, $a<0$.

(b) There are no x-intercepts. No real solutions.

(c) Since there are no real solutions, the discriminant is negative.

46. (a) Since the parabola opens downward, $a<0$.

(b) The solutions are the x-intercepts, $x=-4$ or $x=2$.

(c) Since there are two unique solutions, the discriminant is positive.

47. (a) $(-3)^2-4(2)(1)=1$

(b) Since the discriminant is positive, there are two real solutions.

(c) Graph $Y_1=2X^2-3X+1$ in [0, 2, 1] by [–1, 1, 1]. See Figure 47. There are two x-intercepts.

48. (a) $(2)^2-4(7)(-5)=144$

(b) Since the discriminant is positive, there are two real solutions.

(c) Graph $Y_1=7X^2+2X-5$ in [–2, 2, 1] by [–10, 5, 1]. See Figure 48. There are two x-intercepts.

49. (a) $3x^2+x=-2 \Rightarrow 3x^2+x+2=0;\ (1)^2-4(3)(2)=-23$

(b) Since the discriminant is negative, there are no real solutions.

(c) Graph $Y_1=3X^2+X+2$ in [–3, 3, 1] by [–5, 10, 1]. See Figure 49. There are no x-intercepts.

50. (a) $4.41x^2+9=12.6x \Rightarrow 4.41x^2-12.6x+9=0;\ (-12.6)^2-4(4.41)(9)=0$

(b) Since the discriminant is zero, there is one real solution.

(c) Graph $4.41X^2-12.6X+9$ in [0, 3, 1] by [–5, 10, 1]. See Figure 50. There is one x-intercept.

[0, 2, 1] by [–1, 1, 1] [–2, 2, 1] by [–10, 5, 1] [–3, 3, 1] by [–5, 10, 1] [0, 3, 1] by [–5, 10, 1]

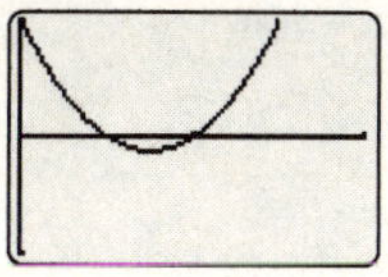

Figure 47

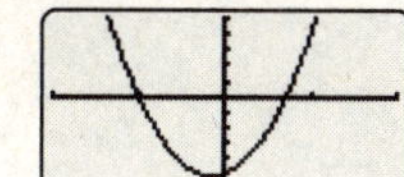

Figure 48

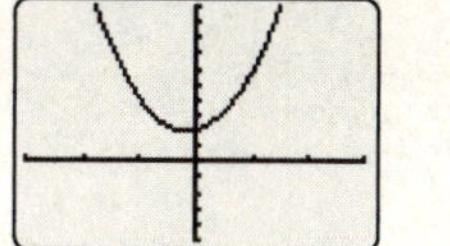

Figure 49

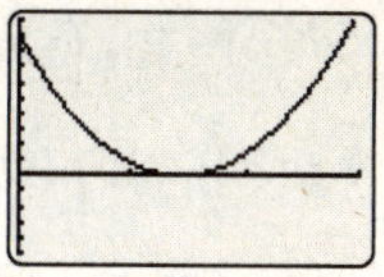

Figure 50

Section 9.5

51. $(1-2i)+2i=1+(-2+2)i=1+0i=1$

52. $(1+3i)-(3-i)=(1-3)+(3-(-1))i=-2+4i$

53. $(1-i)(2+3i)=2+3i-2i-3i^2=2+i-3(-1)=2+i+3=5+i$

54. $(1-i)^2(1+i)=(1-2i+i^2)(1+i)=(1-2i-1)(1+i)=-2i(1+i)=-2i-2i^2=2-2i$

55. $x=\dfrac{-1\pm\sqrt{(1)^2-4(1)(5)}}{2(1)}=\dfrac{-1\pm\sqrt{-19}}{2}=-\dfrac{1}{2}\pm i\dfrac{\sqrt{19}}{2}$

56. $2x^2+8=0 \Rightarrow 2x^2=-8 \Rightarrow x^2=-4 \Rightarrow x=\pm\sqrt{-4} \Rightarrow x=\pm 2i$

57. $x=\dfrac{-(-1)\pm\sqrt{(-1)^2-4(2)(1)}}{2(2)}=\dfrac{1\pm\sqrt{-7}}{4}=\dfrac{1}{4}\pm i\dfrac{\sqrt{7}}{4}$

58. $x=\dfrac{-(-2)\pm\sqrt{(-2)^2-4(7)(5)}}{2(7)}=\dfrac{2\pm\sqrt{-136}}{14}=\dfrac{2\pm 2\sqrt{-34}}{14}=\dfrac{2(1\pm\sqrt{-34})}{14}=\dfrac{1}{7}\pm i\dfrac{\sqrt{34}}{7}$

Section 9.6

59. $f(-2)=3(-2)-1=-7;\ f\left(\frac{1}{3}\right)=3\left(\frac{1}{3}\right)-1=0$

60. $f(-3)=5-3(-3)^2=5-3(9)=-22;\ f(1)=5-3(1)^2=2$

61. (a) Since there are 2 pints in a quart, $P(q)=2q$.

(b) $P(5)=2(5)=10$. There are 10 pints in 5 quarts.

62. (a) $f(x)=4x-3$

(b) $f(5)=4(5)-3=17$

63.

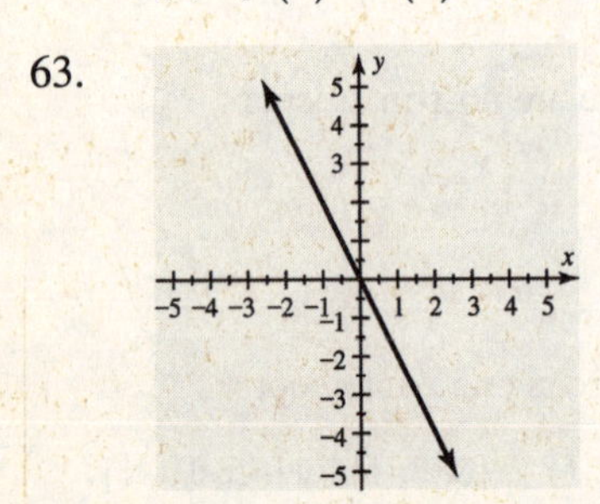

64.

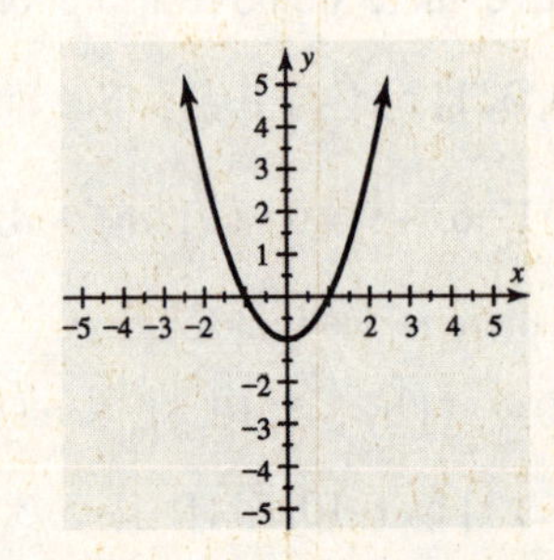

65. $f(0)=1;\ f(-3)=4$

66. $f(-2)=1;\ f(1)=-2$

67. Yes. The graph passes the vertical line test.

68. No. The graph does not pass the vertical line test.

69. Any real number is a valid input for this function. D: all real numbers.

70. The radicand must be greater than or equal to zero. $D: x \geq 0$.

71. The denominator of this function cannot equal zero. $D: x \neq 0$.

72. Any real number is a valid input for this function because the variable is squared. D: all real numbers.

Applications

73. (a) $\frac{x^2}{12}=144 \Rightarrow x^2=1728 \Rightarrow x=\sqrt{1728}\approx 41.6$ mph ($-\sqrt{1728}$ mph has no meaning).

(b) $\frac{x^2}{12}=300 \Rightarrow x^2=3600 \Rightarrow x=\sqrt{3600}=60$ mph (-60 mph has no meaning).

74. (a) $x(x+2)=143$ or $x^2+2x-143=0$

(b) $x^2+2x-143=0 \Rightarrow (x+13)(x-11)=0 \Rightarrow x=-13$ or 11. There are two possible number pairs.

Either $x=-13$, and the other number is -11, or $x=11$, and the other number is 13.

75. $\sqrt{123}\approx 11.1$; the screen is about 11.1 inches by 11.1 inches.

76. $x^2+(x+70)^2=130^2 \Rightarrow x^2+x^2+140x+4900=16,900 \Rightarrow 2x^2+140x-12,000=0 \Rightarrow$

$2(x-50)(x+120)=0 \Rightarrow x=50$ or $x=-120$. The solution is $x=50$ feet ($x=-120$ has no meaning).

Chapter 9 Test

1. $x=-\frac{b}{2a}=-\frac{1}{2(-\frac{1}{2})}=\frac{1}{1}=1$, $y=-\frac{1}{2}(1)^2+(1)+1=\frac{3}{2}$; Vertex: $\left(1,\frac{3}{2}\right)$; Axis of symmetry: $x=1$.

2. The solutions are the x-intercepts, $x=-1$ or $x=2$.

3. $3x^2+11x-4=0 \Rightarrow (x+4)(3x-1)=0 \Rightarrow x=-4$ or $x=\frac{1}{3}$

4. $2x^2=2-6x^2 \Rightarrow 8x^2=2 \Rightarrow x^2=\frac{1}{4} \Rightarrow x=\pm\sqrt{\frac{1}{4}} \Rightarrow x=-\frac{1}{2}$ or $x=\frac{1}{2}$

5. $x^2-8x=1 \Rightarrow x^2-8x+16=1+16 \Rightarrow (x-4)^2=17 \Rightarrow x-4=\pm\sqrt{17} \Rightarrow x=4\pm\sqrt{17}$

6. $x=\frac{-3\pm\sqrt{(3)^2-4(-2)(1)}}{2(-2)}=\frac{-3\pm\sqrt{17}}{-4}=\frac{3\pm\sqrt{17}}{4}$

7. (a) Since the parabola opens downward, $a<0$.

   (b) The solutions are the x-intercepts, $x=-3$ or $x=1$.

   (c) Since there are two real solutions, the discriminant is positive.

8. (a) $(4)^2-4(-3)(-5)=-44$

   (b) Since the discriminant is negative, there are no real solutions.

   (c) Graph $Y_1=-3X^2+4X-5$ in [–5, 5, 1] by [–20, 10, 5]. See Figure 8. It does not intersect the x-axis.

   [–5, 5, 1] by [–20, 10, 5]

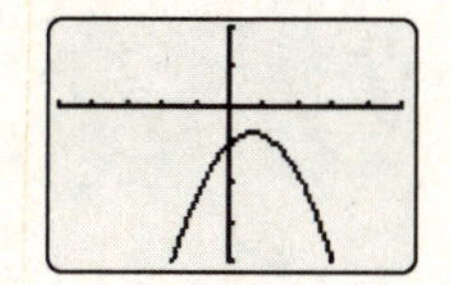

Figure 8

9. $(-5+i)+(7-20i)=(-5+7)+(1-20)i=2-19i$

10. $3i-(6-5i)=3i-6+5i=-6+(3+5)i=-6+8i$

11. $2i(2+3i)=(2i)(2)+(2i)(3i)=4i+6i^2=4i+6(-1)=4i-6=-6+4i$

12. $\left(\frac{1}{2}-i\right)\left(\frac{1}{2}+i\right)=\left(\frac{1}{2}\right)^2-i^2=\frac{1}{4}-(-1)=\frac{5}{4}$

13. $x^2+11=0 \Rightarrow x^2=-11 \Rightarrow x=\pm\sqrt{-11} \Rightarrow x=\pm i\sqrt{11}$

14. $x=\frac{8\pm\sqrt{(-8)^2-4(1)(25)}}{2(1)}=\frac{8\pm\sqrt{-36}}{2}=\frac{8\pm 6i}{2}=\frac{8}{2}\pm\frac{6}{2}i=4\pm 3i$

15. $f(4)=3(4)^2-\sqrt{4}=3(16)-2=48-2=46$;

The point $(4, 46)$ is on the graph of f.

16. $C(x)=4x;\ C(5)=4(5)=20$

17. (a) (b)

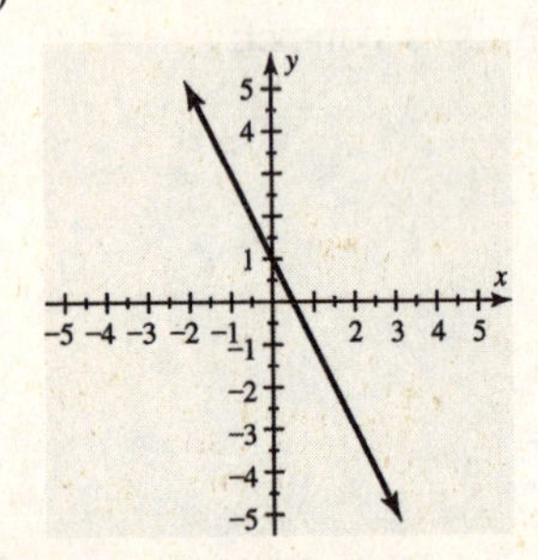

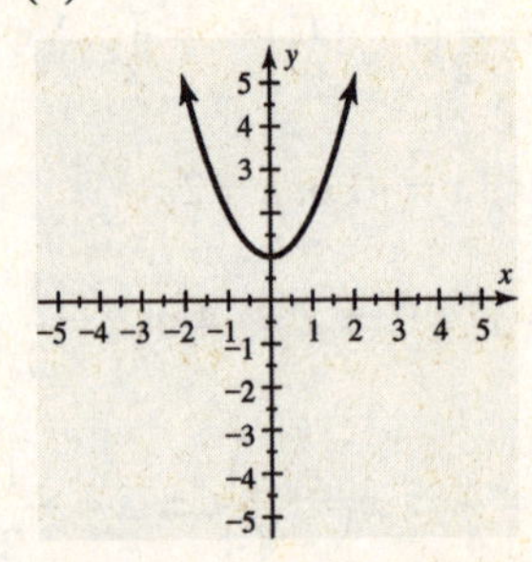

18. $f(-3)=0;\ f(0)=-3$

19. No, the graph does not represent a function because it fails the vertical line test.

20. (a) The domain of f is all real numbers.

    (b) The domain of f is all real numbers x where $x \neq 5$.

21. $\frac{x^2}{9}=250 \Rightarrow x^2=2250 \Rightarrow x=\sqrt{2250}\approx 47.4$ mph (the value $x\approx -47.4$ mph has no physical meaning).

Chapter 9 Extended and Discovery Exercises

1. (a) For the first 3 years of life, the likelihood of survival increases with age. After 3 years of life, it decreases with age.

   (b) Plot the data in [0, 10, 1] by [0, 75, 5]. See Figure 1b. A parabola could model this data.

   (c) Graph $Y_1=-3.57X+71.1$ in [0, 10, 1] by [0, 75, 5]. See Figure 1c.

   Graph $Y_2=-2.07X^2+17.1X+33$ in [0, 10, 1] by [0, 75, 5]. See Figure 1d.

   The equation y_2 models the data better.

   (d) Let $x=6.5$ to find the likelihood of a 5.5-year-old sparrowhawk living 1 more year.

   $y_2=-2.07(6.5)^2+17.1(6.5)+33\approx 56.7\,\%$

[0, 10, 1] by [0, 75, 5]

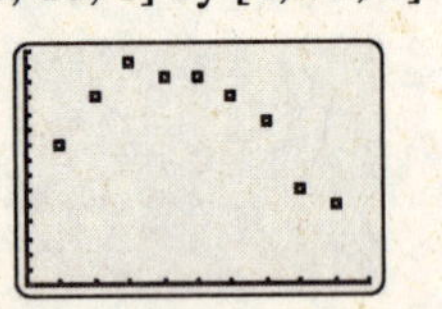

Figure 1b

[0, 10, 1] by [0, 75, 5]

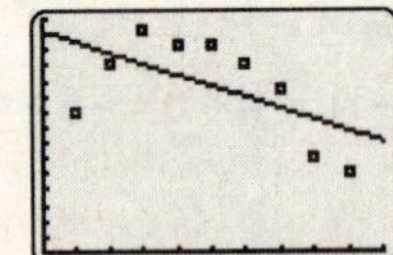

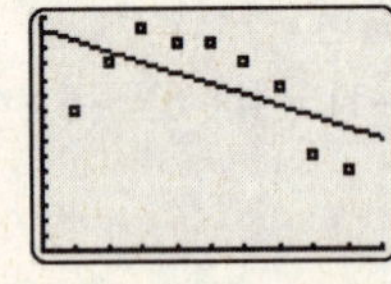

Figure 1c

[0, 10, 1] by [0, 75, 5]

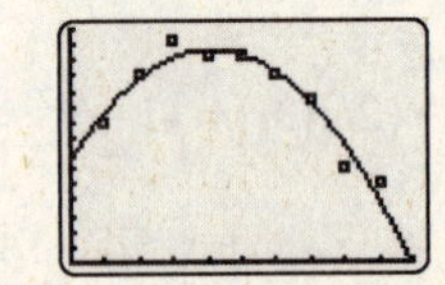

Figure 1d

2. The discriminant is $(-1)^2-4(10)(-3)=121=11^2$. The trinomial factors as $(2x+1)(5x-3)$.

3. The discriminant is $(-3)^2-4(4)(-6)=105$. The trinomial will not factor.

4. The discriminant is $(2)^2-4(3)(-2)=28$. The trinomial will not factor.

5. The discriminant is $(1)^2 - 4(2)(3) = -23$. The trinomial will not factor.

Chapters 1-9 Cumulative Review Exercises

1. Prime factors of $360 = 2 \cdot 2 \cdot 2 \cdot 3 \cdot 3 \cdot 5 \Rightarrow 2^3 \cdot 3^2 \cdot 5$.

2. $2n+7=n-2$, then $2n+7=n-2 \Rightarrow n+7=-2 \Rightarrow n=-9$

3. $\frac{2}{3}+\frac{4}{7}\cdot\frac{21}{28}=\frac{2}{3}+\frac{84}{196}=\frac{392}{588}+\frac{252}{588}=\frac{644}{588}=\frac{23}{21}$

4. $\frac{3}{5}\div\frac{6}{5}-\frac{2}{3}=\frac{3}{5}\times\frac{5}{6}-\frac{2}{3}=\frac{1}{2}-\frac{2}{3}=\frac{3}{6}-\frac{4}{6}=-\frac{1}{6}$

5. $30-4\div2\cdot6=30-(4\div2)\cdot6=30-(2)\cdot6=30-(2\cdot6)=30-12=18$

6. $\frac{3^2-2^3}{20-5\cdot2}=\frac{9-8}{20-10}\Rightarrow\frac{1}{10}$

7. $2(x+1)-6x=x-4 \Rightarrow 2x+2-6x=x-4 \Rightarrow -5x+2=-4 \Rightarrow -5x=-6 \Rightarrow x=\frac{6}{5}$

8. $x=2$. See Figure 8. Checking: $4-3(2)=-2 \Rightarrow 4-6=-2 \Rightarrow -2=-2$, yes.

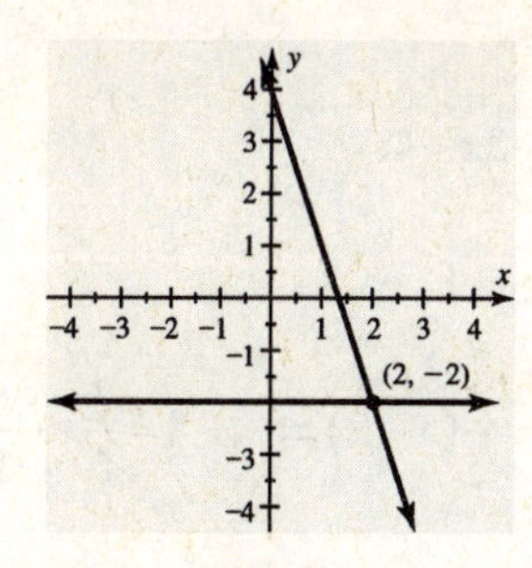

Figure 8

9. $124\%=\frac{124}{100}$, which reduced $=\frac{31}{25}$ and 1.24.

10. When $A=30$ miles^2 and $h=10$ miles then for $A=\frac{1}{2}bh$, $30=\frac{1}{2}b(10) \Rightarrow 30=5b \Rightarrow b=6$ miles.

11. $A=\frac{h}{2}(a+b)$ then $A=\frac{h}{2}a+\frac{h}{2}b \Rightarrow A-\frac{h}{2}a=\frac{h}{2}b \Rightarrow \frac{2}{h}\left(A-\frac{ha}{2}\right)=b \Rightarrow b=\frac{2a}{h}-A$

12. $6t-1<3-t, 7t<4 \Rightarrow t<\frac{4}{7} \Rightarrow \left\{t \middle| t<\frac{4}{7}\right\}$

13. See Figure 13.

14. See Figure 14. x-intercept: 4; y-intercept: 2.

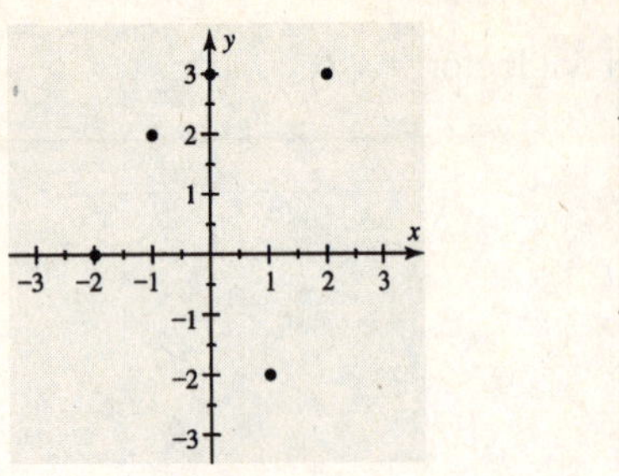

Figure 13

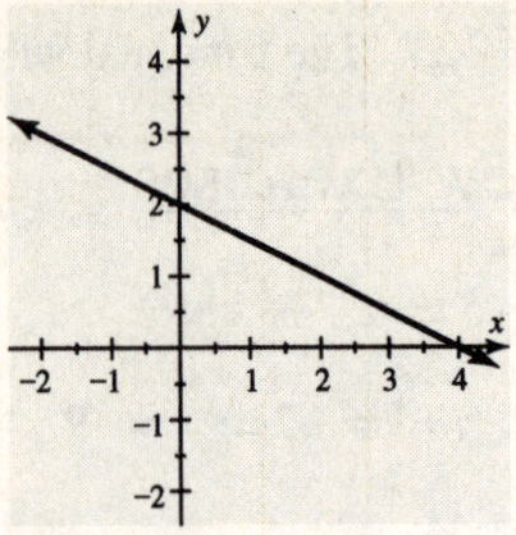

Figure 14

15. See Figure 15. x-intercept: -4; y-intercept: 3.

16. See Figure 16. x-intercept: -2; y-intercept: none.

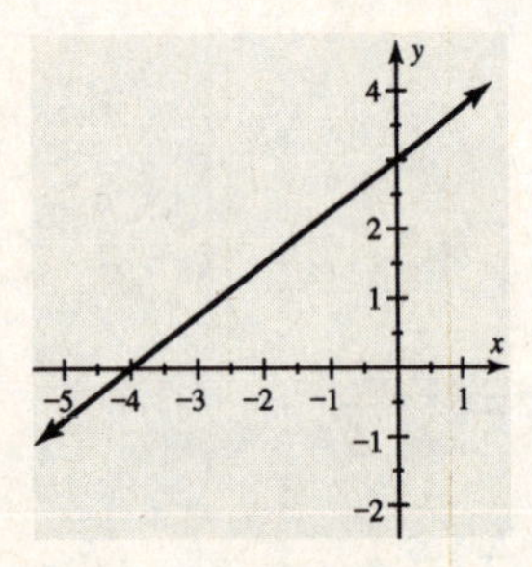

Figure 15

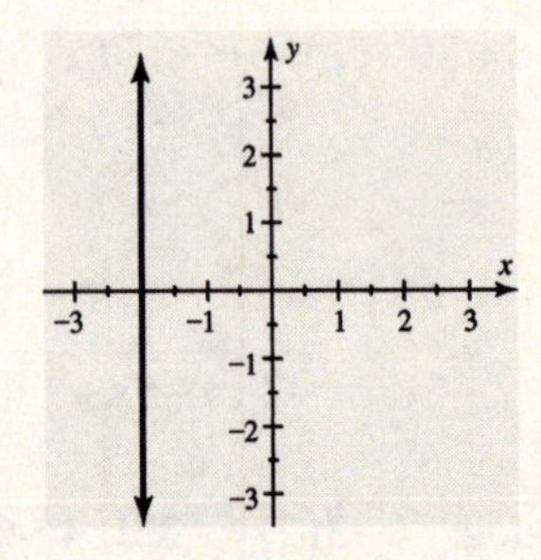

Figure 16

17. x-intercept: 2; y-intercept: 4; Therefore $m=\dfrac{4-0}{0-2}\Rightarrow m=-2\Rightarrow y=-2x+4$.

18. Using point-slope form: $y-3=-2(x+1)\Rightarrow y-3=-2x-2\Rightarrow y=-2x+1$

19. First find slope: $m=\dfrac{8-5}{2-(-3)}=\dfrac{3}{5}$. Now use point-slope form: $y-8=\dfrac{3}{5}(x-2)\Rightarrow y-8=\dfrac{3}{5}x-\dfrac{6}{5}\Rightarrow$

$y=\dfrac{3}{5}x+\dfrac{34}{5}$.

20. First put $x+2y=5$ into slope-intercept form: $x+2y=5\Rightarrow 2y=-x+5\Rightarrow y=-\dfrac{1}{2}x+\dfrac{5}{2}\Rightarrow\ m=-\dfrac{1}{2}$.

The slope of a line perpendicular to this would be $\dfrac{2}{1}$ or 2. Now using slope-intercept form

$y-1=2(x+1)\Rightarrow\ y-1=2x+2\Rightarrow y=2x+3$.

21. $P=500x+4000$

22. Using substitution; $-2a+b=-5\Rightarrow b=2a-5$ and so $4a-3(2a-5)=0$, $4a-6a+15=0$,

$-2a+15=0\Rightarrow a=7.5$, and $-2(7.5)+b=-5\Rightarrow b=10\Rightarrow(7.5,10)$.

23. See Figure 23.

24. See Figure 24.

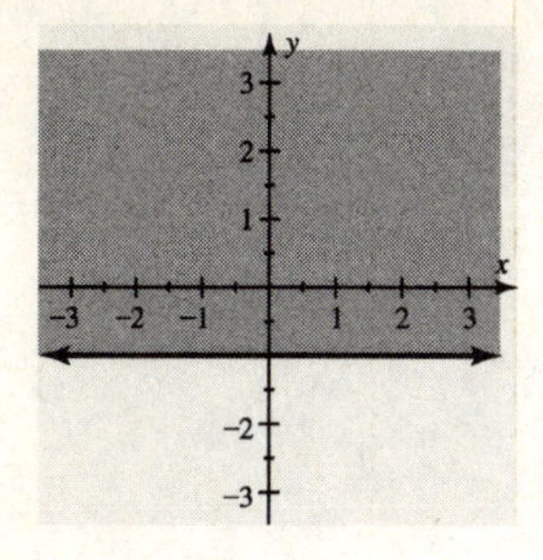

Figure 23

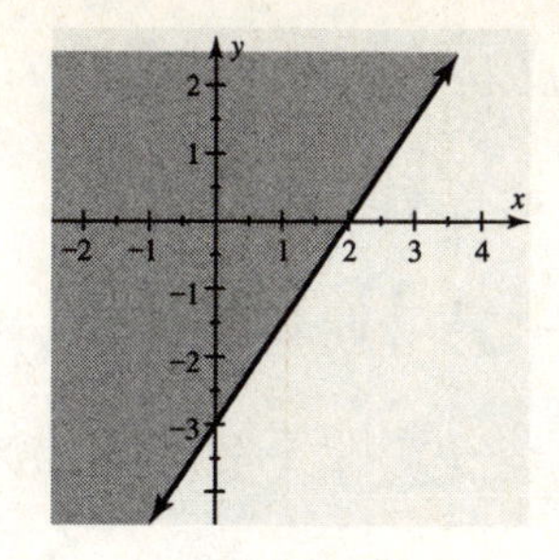

Figure 24

25. See Figure 25.

26. See Figure 26.

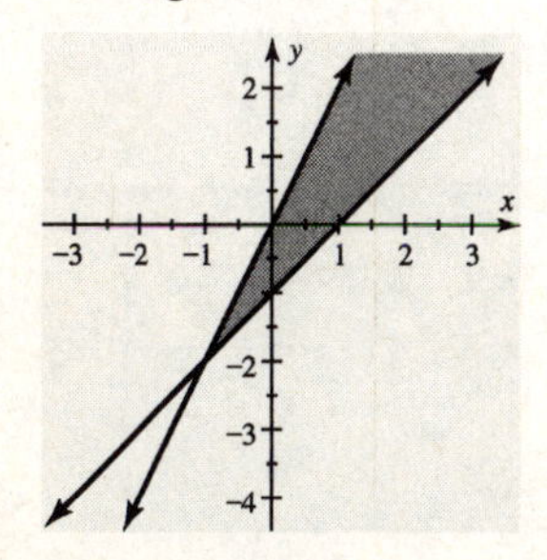

Figure 25

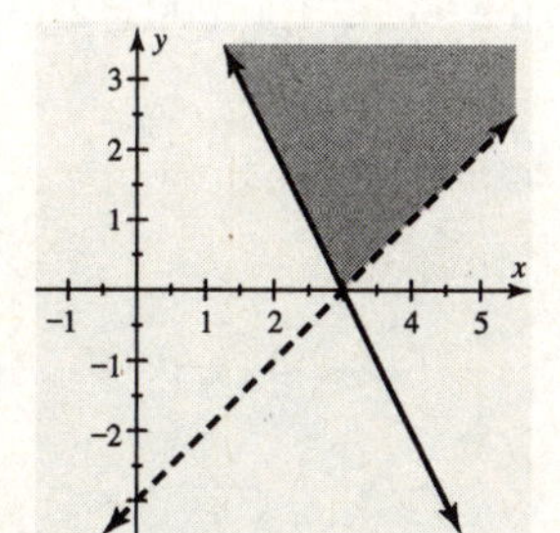

Figure 26

27. $\left(ab-6b^2\right)-\left(4b^2+8ab+2\right) \Rightarrow -7ab-10b^2-2$

28. $5-3^4=5-81 \Rightarrow -76$

29. $4\left(4^{-2}\right)\left(3^{-1}\right)\left(3^4\right)=\frac{4}{1}\cdot\frac{1}{4^2}\cdot\frac{1}{3}\cdot\frac{3^4}{1}=\frac{4}{1}\cdot\frac{1}{16}\cdot\frac{1}{3}\cdot\frac{81}{1}=\frac{27}{4}$

30. $\frac{2^{-4}}{4^{-2}}=\frac{4^2}{2^4}=\frac{16}{16}=1$

31. $\left(8t^{-3}\right)\left(3t^2\right)\left(t^5\right)=(8\cdot 3)\left(t^{-3+2+5}\right)=24t^4$

32. $2(rt)^4 \Rightarrow 2r^4t^4$

33. $\left(2t^3\right)^{-2}=\frac{1}{\left(2t^3\right)^2} \Rightarrow \frac{1}{2^2\cdot t^{3\cdot 2}}=\frac{1}{4t^6}$

34. $\left(4a^2b^3\right)^2(2ab)^{-3}=\left(4^2\cdot a^{2\cdot 2}\cdot b^{3\cdot 2}\right)\left(\frac{1}{2^3a^3b^3}\right)=\frac{16a^4b^6}{8a^3b^3}=2a^{4-3}b^{6-3}=2ab^3$

35. $\left(\frac{2rt^{-1}}{3r^{-2}t^3}\right)^4=\frac{2^4\cdot r^4\cdot t^{-1\cdot 4}}{3^4\cdot r^{-2\cdot 4}\cdot t^{3\cdot 4}}=\frac{16r^4t^{-4}}{81r^{-8}t^{12}}=\frac{16r^{4-(-8)}t^{-4-12}}{81}=\frac{16r^{12}t^{-16}}{81}=\frac{16r^{12}}{81t^{16}}$

36. $\left(\frac{2a^{-1}}{ab^{-2}}\right)^{-3}=\frac{2^{-3}\cdot a^{-1\cdot(-3)}}{a^{-3}b^{-2\cdot(-3)}}=\frac{2^{-3}a^3}{a^{-3}b^6}=\frac{a^{3-(-3)}}{2^3b^6}=\frac{a^6}{8b^6}$

37. $2a^2\left(a^2-2a+3\right)=2a^4-4a^3+6a^2$

38. $(a+b)\left(a^2-ab+b^2\right)=a^3-a^2b+ab^2+a^2b-ab^2+b^3=a^3+b^3$

39. $(5x+1)(x-7)=5x^2-35x+x-7=5x^2-34x-7$

40. $(y-3)(2y+3)=2y^2+3y-6y-9=2y^2-3y-9$

41. $(a+b)(a-b)=a^2-ab+ab-b^2=a^2-b^2$

42. $(2x+3y)^2=4x^2+6xy+6xy+9y^2=4x^2+12xy+9y^2$

43. Move the decimal point five places to the left, $1.5\times10^{-5}=0.000015$.

44. Move the decimal point six places to the left, $2{,}130{,}000=2.13\times10^{6}$.

45. $\dfrac{4x^3-8x^2+6x}{2x}=\dfrac{4x^3}{2x}-\dfrac{8x^2}{2x}+\dfrac{6x}{2x}=2x^2-4x+3$

46.

$$x^3-4x^2+3x-2+\frac{1}{x-5}$$

$$x-5\overline{)x^4-9x^3+23x^2-17x+11}$$

$$\underline{x^4-5x^3}$$

$$-4x^3+23x^2$$

$$\underline{-4x^3+20x^2}$$

$$3x^2-17x$$

$$\underline{3x^2-15x}$$

$$-2x+11$$

$$\underline{-2x+10}$$

$$1$$

47. $10ab^2-25a^3b^5=5ab^2\left(2-5a^2b^3\right)$

48. $y^3-3y^2+2y-6=\left(y^3-3y^2\right)+(2y-6)=y^2(y-3)+2(y-3)=(y-3)\left(y^2+2\right)$

49. $6z^2+7z-3=(2z+3)(3z-1)$

50. $4z^2-9=(2z)^2-3^2=(2z-3)(2z+3)$

51. $4y^2-20y+25=(2y-5)(2y-5)=(2y-5)^2$

52. $a^3-27=a^3-3^3=(a-3)\left(a^2+3a+9\right)$

53. $4z^4-17z+15=\left(z^2-3\right)\left(4z^2-5\right)$

54. $2a^3b+a^2b^2-ab^3=ab\left(2a^2+ab-b^2\right)=ab(2a-b)(a+b)$

55. $x-1=0\Rightarrow x=1$ and $x+2=0\Rightarrow x=-2$. Therefore $x=-2, 1$.

56. Factor $x^2-9x=0\Rightarrow x(x-9)=0$, so $x=0$ and $x-9=0\Rightarrow x=9$. Therefore $x=0, 9$.

57. Set $6y^2-7y=3$ equal to zero $\Rightarrow 6y^2-7y-3=0$. Now factor: $6y^2-7y-3=0\Rightarrow$

$(3y+1)(2y-3)=0$. So $3y+1=0\Rightarrow 3y=-1\Rightarrow y=-\dfrac{1}{3}$ and $2y-3=0\Rightarrow 2y=3\Rightarrow y=\dfrac{3}{2}$.

Therefore $y=-\dfrac{1}{3},\dfrac{3}{2}$.

58. Set $x^3=4x$ equal to zero $\Rightarrow x^3-4x=0$. Now factor: $x^3-4x=0\Rightarrow x\left(x^2-4\right)\Rightarrow$

$x(x+2)(x-2)=0$. So $x+2=0\Rightarrow x=-2$, and $x-2=0\Rightarrow x=2$. Therefore $x=-2,0,2$.

59. $\dfrac{x^2-16}{x+4}=\dfrac{(x+4)(x-4)}{x+4}=x-4$

60. $\dfrac{2x^2-11x-6}{6x^2-5x-4}=\dfrac{(2x+1)(x-6)}{(2x+1)(3x-4)}=\dfrac{x-6}{3x-4}$

61. $\dfrac{x-3}{16-x^2}$ will be undefined when $16-x^2=0$. $x^2-16=0\Rightarrow(x+4)(x-4)=0$. So $x+4=0\Rightarrow$

$x=-4$ and $x-4=0\Rightarrow x=4$. Therefore it will be undefined when $x=-4,4$.

62. $\dfrac{4(-2)+1}{(-2)-1}=\dfrac{-8+1}{-3}=\dfrac{-7}{-3}=\dfrac{7}{3}$

63. $\dfrac{x^2-3x+2}{x+7}\div\dfrac{x-2}{2x+14}=\dfrac{(x-2)(x-1)}{x+7}\div\dfrac{x-2}{2(x+7)}=\dfrac{(x-2)(x-1)}{x+7}\cdot\dfrac{2(x+7)}{x-2}=\dfrac{2(x-1)}{1}=2(x-1)$

64. $\dfrac{x}{2x+3}+\dfrac{x+3}{2x+3}=\dfrac{x+x+3}{2x+3}=\dfrac{2x+3}{2x+3}=1$

65. $\dfrac{5x}{x^2-1}-\dfrac{3}{x+1}=\dfrac{5x}{(x+1)(x-1)}-\dfrac{3}{x+1}=\dfrac{5x}{(x+1)(x-1)}-\dfrac{3(x-1)}{(x+1)(x-1)}=\dfrac{5x-3x+3}{(x+1)(x-1)}=\dfrac{2x+3}{(x-1)(x+1)}$

66. $\dfrac{\frac{2}{x}-\frac{2}{y}}{\frac{2}{x}+\frac{2}{y}}=\dfrac{\frac{2y}{xy}-\frac{2x}{xy}}{\frac{2y}{xy}+\frac{2x}{xy}}=\dfrac{\frac{2y-2x}{xy}}{\frac{2y+2x}{xy}}=\dfrac{2y-2x}{xy}\cdot\dfrac{xy}{2y+2x}=\dfrac{2y-2x}{2y+2x}=\dfrac{2(y-x)}{2(y+x)}=\dfrac{y-x}{y+x}$

67. $\dfrac{x+2}{5}=\dfrac{x}{4}\Rightarrow 4(x+2)=5x\Rightarrow 4x+8=5x\Rightarrow x=8$

68. $\dfrac{1}{3x}+\dfrac{5}{2x}=2\Rightarrow\dfrac{2}{6x}+\dfrac{15}{6x}=\dfrac{12x}{6x}$, so $2+15=12x\Rightarrow 17=12x\Rightarrow x=\dfrac{17}{12}$

69. $\dfrac{1}{x-2}+\dfrac{2}{x+2}=\dfrac{1}{x^2-4}\Rightarrow\dfrac{1}{x-2}+\dfrac{2}{x+2}=\dfrac{1}{(x+2)(x-2)}\Rightarrow\dfrac{x+2}{(x+2)(x-2)}+\dfrac{2(x-2)}{(x+2)(x-2)}=$

$\dfrac{1}{(x+2)(x-2)}$, so $x+2+2(x-2)=1\Rightarrow x+2+2x-4=1\Rightarrow 3x-2=1\Rightarrow 3x=3\Rightarrow x=1$

70. If y is inversely proportional to x then $y=\dfrac{k}{x}$. If $y=25$ when $x=4$ then $25=\dfrac{k}{4}\Rightarrow k=100$.

So if $x = 10$ then $y = \dfrac{100}{10} \Rightarrow y = 10.$

71. $8^2 + 3^2 = c^2 \Rightarrow 64 + 9 = c^2 \Rightarrow c^2 = 73 \Rightarrow c = \sqrt{73}$ feet

72. $a^2 + 24^2 = 25^2 \Rightarrow a^2 + 576 = 625 \Rightarrow a^2 = 49 \Rightarrow a = \sqrt{49} \Rightarrow a = 7$ miles

73. $d = \sqrt{(-2-4)^2 + (3-7)^2} = \sqrt{36+16} = \sqrt{52} = 2\sqrt{13}$

74. $d = \sqrt{(3-(-1))^2 + (-8-(-8))^2} = \sqrt{16+0} = \sqrt{16} = 4$

75. $\dfrac{\sqrt{80}}{\sqrt{5}} = \sqrt{\dfrac{80}{5}} = \sqrt{16} = 4$

76. $\sqrt{2x^3y} \cdot \sqrt{32xy} = \sqrt{2 \cdot 32x^3xyy} = \sqrt{64x^4y^2} = 8x^2y$

77. $8\sqrt[3]{7} - 3\sqrt[3]{7} = (8-3)\sqrt[3]{7} = 5\sqrt[3]{7}$

78. $\dfrac{\sqrt[4]{64}}{\sqrt[4]{4}} = \sqrt[4]{\dfrac{64}{4}} = \sqrt[4]{16} = 2$

79. $\sqrt{4-x} = 7 \Rightarrow \left(\sqrt{4-x}\,\right)^2 = 7^2 \Rightarrow 4 - x = 49 \Rightarrow x = -45$

Check: $\sqrt{4-(-45)} = \sqrt{49} = 7$

80. $x - 1 = \sqrt{2x+1} \Rightarrow (x-1)^2 = \left(\sqrt{2x+1}\right)^2 \Rightarrow x^2 - 2x + 1 = 2x + 1 \Rightarrow x^2 - 4x = 0 \Rightarrow x(x-4) = 0 \Rightarrow x = 0, 4$

Check: $0 - 1 \stackrel{?}{=} \sqrt{2(0)+1} \Rightarrow -1 \stackrel{?}{=} \sqrt{1} \Rightarrow -1 \neq 1$ (This answer does not check)

Check: $4 - 1 \stackrel{?}{=} \sqrt{2(4)+1} \Rightarrow 3 \stackrel{?}{=} \sqrt{9} \Rightarrow 3 = 3.$ The only solution is 4.

81. $x = -\dfrac{b}{2a} = -\dfrac{(-6)}{2(1)} = 3,\ y = (3)^2 - 6(3) + 14 = 5;$ The vertex is (3, 5).

82. $x = -\dfrac{b}{2a} = -\dfrac{(-4)}{2(2)} = 1,\ y = 2(1)^2 - 4(1) - 1 = -3;$ The vertex is $(1, -3)$.

83. $x^2 + 3x = 18 \Rightarrow x^2 + 3x - 18 = 0 \Rightarrow (x+6)(x-3) = 0 \Rightarrow x = -6, 3$

84. $x^2 - 2x = 2 \Rightarrow x^2 - 2x + 1 = 2 + 1 \Rightarrow (x-1)^2 = 3 \Rightarrow x - 1 = \pm\sqrt{3} \Rightarrow x = 1 \pm \sqrt{3}$

85. $3x^2 - 4 = 11 \Rightarrow 3x^2 = 15 \Rightarrow x^2 = 5 \Rightarrow x = \pm\sqrt{5}$

86. $x = \dfrac{12 \pm \sqrt{(-12)^2 - 4(4)(7)}}{2(4)} = \dfrac{12 \pm \sqrt{32}}{8} = \dfrac{12 \pm 4\sqrt{2}}{8} = \dfrac{4(3 \pm \sqrt{2})}{8} = \dfrac{3 \pm \sqrt{2}}{2}$

87. $x = \dfrac{4 \pm \sqrt{(-4)^2 - 4(1)(29)}}{2(1)} = \dfrac{4 \pm \sqrt{-100}}{2} = \dfrac{4 \pm 10i}{2} = \dfrac{4}{2} \pm \dfrac{10}{2}i = 2 \pm 5i$

88. $x=\dfrac{6\pm\sqrt{(-6)^2-4(9)(17)}}{2(9)}=\dfrac{6\pm\sqrt{-576}}{18}=\dfrac{6\pm 24i}{18}=\dfrac{6}{18}\pm\dfrac{24}{18}i=\dfrac{1}{3}\pm\dfrac{4}{3}i$

89. $f(4)=2(4)^2+\sqrt{4}=2(16)+2=34$

90. (a) $D=\{-3,0,2\}$

(b) The denominator cannot equal zero. $D: x\neq -6$.

(c) The radicand must be greater than or equal to zero. $D: x\geq -4$.

91.

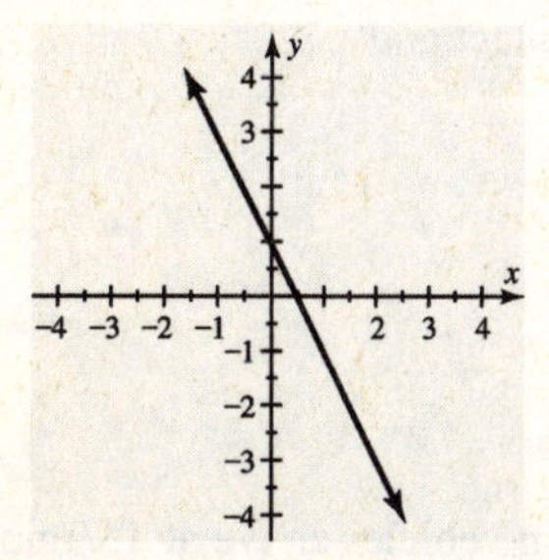

92.

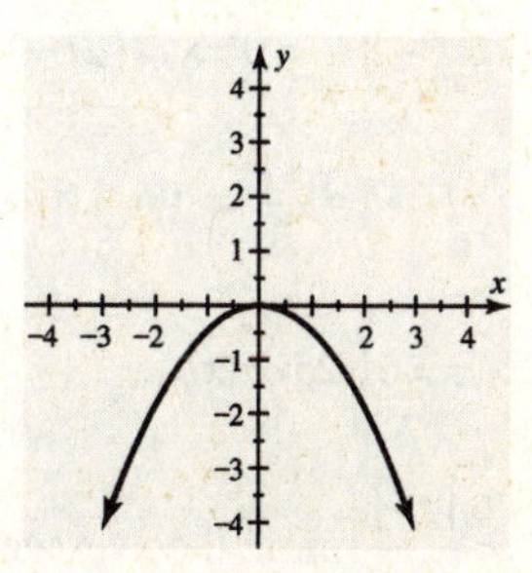

93. Since the constant change is 325 and since when $t=0$, $d=0$. Then $d=325t$.

94. (a) $I=\dfrac{1}{2}x+4$

(b) $\dfrac{1}{2}$

(c) Snow is falling at the rate of $\dfrac{1}{2}$ inch per hour.

(d) $I=\dfrac{1}{2}(4)+4\Rightarrow I(4)=2+4\Rightarrow I(4)=6$ inches of snow.

95. See Figure 95.

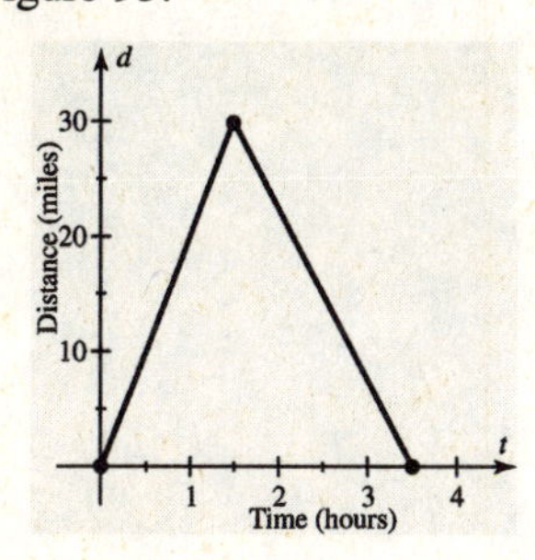

Figure 95

96. Let $x=$ number of adults and let $y=$ number of children. Then $x+y=8$ and $25x+15y=170$. Multiplying the first equation by (-15) and adding it to the second equation yields: $10x=50\Rightarrow x=5$, so $5+y=8\Rightarrow y=3$. Therefore there are 5 adults and 3 children.

97. If rate $\cdot$ time $= 1$ job, then the two rates are: $r \cdot 2 = 1 \Rightarrow r = \frac{1}{2}$ and $r \cdot (1.5) = 1 \Rightarrow r = \frac{2}{3}$. Now working together we add the rates so, $\frac{1}{2}t + \frac{2}{3}t = 1 \Rightarrow \frac{7}{6}t = 1 \Rightarrow t = \frac{6}{7}$. Therefore they can do the job together in $\frac{6}{7}$ hour.

98. (a) $h = 88(2) - 16(2)^2 \Rightarrow h = 176 - 16(4) \Rightarrow h = 176 - 64 \Rightarrow h = 112$ feet.

(b) $0 = 88t - 16t^2 \Rightarrow 16t^2 - 88t = 0 \Rightarrow 8t(2t - 11) = 0$. So $8t = 0 \Rightarrow t = 0$ at contact and $2t - 11 = 0 \Rightarrow 2t = 11 \Rightarrow t = \frac{11}{2} \Rightarrow t = 5.5$ seconds.

99. $\frac{7}{4} = \frac{x}{35} \Rightarrow 4x = 245 \Rightarrow x = 61.25$ feet.

100. (a) $T = \frac{1}{20 - 15} \Rightarrow T = \frac{1}{5}$. $\frac{1}{5}$ min.; the average wait is $\frac{1}{5}$ min. or 12 sec. when vehicles arrive at 15/min.

(b) See Figure 100.

x	5	10	15	19	19.9
T	$\frac{1}{15}$	$\frac{1}{10}$	$\frac{1}{5}$	1	10

Figure 100

(c) The wait time increases dramatically.

101. For $A = 74$, $-0.005 \le \frac{L - A}{A} \le 0.005 \Rightarrow -0.37 \le L - 74 \le 0.37 \Rightarrow 73.63 \le L \le 74.37$.

Critical Thinking Solutions for Chapter 9

Section 9.3

•

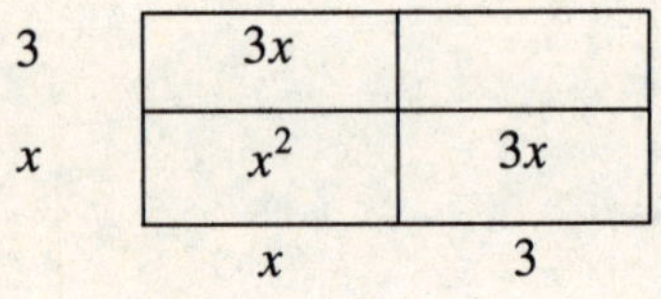

3	$3x$	
x	x^2	$3x$
	x	3

To complete the square for $x^2 + 6x$, add $3 \cdot 3 = 9$.